D0085030

Children's Exercise Physiology

SECOND EDITION

Thomas W. Rowland, MD
Baystate Medical Center

HUMAN KINETICS

Library of Congress Cataloging-in-Publication Data

Rowland, Thomas W.
Developmental exercise physiology / Thomas W. Rowland.--2nd ed.
p. ; cm.
Includes bibliographical references and index.
ISBN 0-7360-5144-9 (hard cover)
1. Exercise for children--Physiological aspects. 2. Motor ability in children--Physiological aspects.
[DNLM: 1. Exercise--physiology--Child. 2. Exercise--physiology--Infant. 3. Child Development--physiology. WE 103 R883d 2004] I. Title.
RJ133.R679 2004
612'.044'083--dc22

2004000740

ISBN: 0-7360-5144-9

This book is a revised edition of *Developmental Exercise Physiology,* published in 1996 by Human Kinetics.

Acquisitions Editor: Michael S. Bahrke, PhD; **Developmental Editor:** Renee Thomas Pyrtel; **Assistant Editors:** Ann M. Augspurger and Amanda M. Eastin; **Copyeditor:** Karen Bojda; **Proofreader:** Joanna Hatzopoulos Portman; **Indexer:** Betty Frizzéll; **Permission Manager:** Dalene Reeder; **Graphic Designer:** Robert Reuther; **Graphic Artist:** Kathleen Boudreau-Fuoss; **Photo Manager:** Kareema McLendon; **Cover Designer:** Keith Blomberg; **Photographer (cover):** Empics; **Photos (interior):** © Human Kinetics; **Art Manager:** Kelly Hendren; **Illustrators:** Craig Newsom and Jennifer Delmotte; **Printer:** Sheridan Books

Printed in the United States of America 10 9 8 7 6 5 4 3 2 1

Human Kinetics
Web site: www.HumanKinetics.com

United States: Human Kinetics
P.O. Box 5076, Champaign, IL 61825-5076
800-747-4457
e-mail: humank@hkusa.com

Canada: Human Kinetics
475 Devonshire Road Unit 100
Windsor, ON N8Y 2L5
800-465-7301 (in Canada only)
e-mail: orders@hkcanada.com

Europe: Human Kinetics
107 Bradford Road, Stanningley
Leeds LS28 6AT, United Kingdom
+44 (0) 113 255 5665
e-mail: hk@hkeurope.com

Australia: Human Kinetics
57A Price Avenue, Lower Mitcham
South Australia 5062
08 8277 1555
e-mail: liaw@hkaustralia.com

New Zealand: Human Kinetics
Division of Sports Distributors NZ Ltd.
P.O. Box 300 226 Albany
North Shore City, Aukland
0064 9 448 1207
e-mail: blairc@hknewz.com

CONTENTS

CHAPTER 5 Aerobic Fitness 89

CHAPTER 6 Cardiovascular Responses to Exercise 113

CHAPTER 7 Ventilation Responses 135

CHAPTER 8 Energy Demands of Weight-Bearing Locomotion 149

PREFACE

The past several decades have witnessed increasing interest in how children exercise. Health professionals want to know how to make children more active. Coaches seek means of training young athletes that are both safe and effective. Physicians and rehabilitation specialists wish for information on how exercise can be used to treat children with cardiopulmonary and musculoskeletal disorders.

This interest in exercise in youth has generated a growing volume of research data that indicate unique aspects of exercise physiology in the growing human. This book provides an overview of this state-of-the-art information. It endeavors, however, not simply to describe where we've been in pediatric exercise physiology but also where we might go. These new directions are presented in the context of issues that have arisen from research both in adults and animals but that have often not been addressed in the pediatric literature. These perspectives represent new opportunities for gaining insight into what makes children different from adults.

As might be expected from any fresh approach, it should be anticipated that this book will create more questions than answers. The limited amount of established dogma in these pages, in fact, may be disconcerting to the reader seeking definitive, evidence-based information. The author makes no apology for this scientific ambiguity (as it is, after all, not his fault). Our current understanding of many issues in pediatric exercise physiology is incomplete—which makes this field frustrating and an exciting challenge at the same time.

A great number of influences account for the physiologic responses to exercise as children grow. It is appropriate that this book begins in the first three chapters with a foundation of the most obvious: increases in body size and the hormonal effects of puberty. With a solid understanding of these determinants, the developmental aspects and training adaptations of aerobic fitness, anaerobic fitness, and strength can be more completely understood.

Some traditional caveats for reviews of pediatric exercise physiology need to be restated. A basic understanding of exercise physiology—indeed, of pediatric exercise physiology—by the reader is assumed. Details of testing methodology in children, for example, are largely ignored. Similarly, given the breadth of the content, no effort has been made to provide an inclusive overview of all subjects. In each chapter, the reader is directed to review articles that offer more in-depth discussions.

This book restricts itself almost exclusively to the physiology of the healthy child. How responses to exercise differ in youths with chronic disease or in child athletes are critical issues but are addressed in other publications. Here we focus on a base of scientific information from which these applied areas of clinical medicine and sport training can be derived. It should be recognized that this database is limited to children who are old enough to be tested effectively (i.e., usually over 9 or 10 years old) and who, in most cases, are willing to be recruited for exercise studies. While an increasing number of studies have involved girls, most of our information base describes the exercise responses of boys. Unfortunately, very little research in this field can comfortably be assumed to represent the total pediatric population. Instead, many "normative" data reflect a subset of motivated, older, usually male subjects.

The reader is cautioned, too, that drawing conclusions regarding cause-and-effect relationships in pediatric exercise physiology is particularly treacherous. In the course of normal maturation, a great number of variables change in concert. The extent to which one is caused by another, or to which they are both mutually related to a third factor, is often unclear. Indeed, devising means of defining the directions of causal "arrows" remains one of the major challenges of the field.

The author has attempted to maintain some consistency of definitions throughout but, admittedly, has not always compulsively done so. *Child* generally refers to a prepubertal subject, as opposed to *adolescent,* a teenager who has at least begun the pubertal process. The terms *pediatric exercise physiology* and *developmental exercise physiology* are considered to be synonymous. The different interpretations of the terms *peak* $\dot{V}O_2$ and $\dot{V}O_2max$ to describe aerobic fitness are explained in chapter 5. In describing individual studies, the term used by the respective authors has been honored. In other discussions, $\dot{V}O_2max$ is used, with apologies to those semantic purists who might find this term subversive.

This book is intended for a wide audience that includes students, health care providers, physical educators, public health professionals, exercise scientists, and sport administrators. It is designed to serve both as a useful reference source and as a textbook for courses involving pediatric exercise science. To this end, each chapter includes objectives and discussion questions, and a glossary has been added at the back of the book.

CREDITS

Figure I.2 Reprinted, by permission, from T.W. Rowland, 2001, "Pediatric exercise science: So What?" Pediatric Exercise Science 13: 207.

Introduction opening quote Copyright © 1969 Paul Simon. Used by permission of the publisher: Paul Simon Music.

Figure 1.2 Reprinted, by permission, from T.W. Rowland, 1998, "The case of the elusive denominator," Pediatric Exercise Science 10: 3.

Figures 1.3 and 4.8 Reprinted, by permission, from M.A. Holliday et al., 1967, "The relation of metabolic size to body weight," Pediatric Research 1: 188.

Figure 1.5 Reprinted, by permission, from J.R. Welsman and N. Armstrong, 2000, "Statistical techniques for interpreting body size: Related exercise performance during growth," Pediatric Exercise Science12: 116.

Figures 1.6, 4,1, and 4.2 Reprinted, by permission, from T.W. Rowland, 1996, Developmental exercise physiology (Champaign, IL: Human Kinetics), 10.

Figure 2.2 Reprinted from Journal of Pediatrics, Vol. 115, G. Costin et al., "Growth hormone secretory dynamics in subjects with normal stature," 537-544, Copyright (1989), with permission from Elsevier.

Figure 2.3 Reprinted from Pediatric Endocrinology, 2nd ed., Sperling, "Growth hormone/insulin-like growth factor secretion and action," 231, Copyright (2002), with permission from Elsevier.

Figure 2.6 Reprinted, by permission, from Eliakim et al., 1996, "Physical fitness, endurance training, and the growth hormone-insulin-like growth factor I System in adolescent females," The Journal of Clinical Endocrinology and Metabolism 81: 3986-3992. Copyright 1980, 1996, 2000, and 2002. The Endocrine Society.

Figure 2.7 Reproduced with permission from Nemet et.al., Pediatrics Vol. 110, Page 684, Copyright 2002.

Figure 2.8a and b From Mirwald, 1981, "Longitudinal comparison of aerobic power in active and inactive boys age 7-17 year," Annals of Human Biology 8: 405-414. Reprinted with permission from Taylor and Francis (http://www.tandf.co.uk/journals).

Figure 2.9a and b From Journal of Pediatrics, Vol. 122, Thientz, et al., 311, Copyright (1993), with permission from Elsevier.

Figures 3.1, 3.3, 3.4, 3.5, and 9.8 Reprinted, by permission, from R.M. Malina and C. Bouchard, 1991, Growth, maturation and physical activity (Champaign, IL: Human Kinetics), 341, 127, 143, 139, and 194.

Figure 3.2 Reprinted, by permission, from M.B. Horlick et al., 2000, "Effect of puberty on the relationship between circulating leptin and body composition" The Journal of Clinical Endocrinology and Metabolism 85: 2509-2518. Copyright 1980, 1996, 2000, and 2002. The Endocrine Society.

Figure 3.6 Reprinted, by permission, from G. Beunen and R.M. Malina, 1988, "Growth and physical performance relative to the timing of the adolescent spurt," Exercise and Sport Science Reviews 16: 503-540.

Figure 3.8a and b Reprinted, by permission, from M.P. Warren et al., 1980, "The effects of exercise on pubertal progression and reproductive function in girls," The Journal of Clinical Endocrinology and Metabolism 51: 1150-1157. Copyright 1980, 1996, 2000, and 2002. The Endocrine Society.

Figure 3.10 Reprinted, by permission, from E. Weinmann, 2002, "Gender-related differences in elite gymnasts," Journal of Applied Physiology 92: 2149.

Figure 4.3 Reprinted, by permission, from A. Berg et al., 1986, "Skeletal muscle enzyme activities in healthy young subjects," International Journal of Sports Medicine 7: 237.

Figures 4.4, 6.9, and 11.6 Reprinted, by permission, from S.F. Lewis and R.G. Haller, 1990, "Disorders of muscle glycongenolysis/glycolysis: The consequences of substrate-limited oxidative metabolism in humans." In Biochemistry of exercise VII, edited by A.W. Taylor et al. (Champaign, IL: Human Kinetics), 212, 151, and 98.

Figure 4.5 Reprinted, by permission, from Peterson et al., 1999, "Skeletal muscle metabolism during short-term high-intensity exercise in prepubertal and pubertal girls," Journal of Applied Physiology 87: 2153.

Figure 4.6 Reprinted, by permission, from O. Eriksson and B. Saltin, 1974, "Muscle metabolism during exercise in pubertal boys," Acta Paediatrica Belgica 28: 262.

Figure 4.9a and b Reprinted, by permission, from G.E. Duncan and E.T. Howley, 1999, "Substrate metabolism during exercise in children and the 'crossover concept,'" Pediatric Exercise Science 11: 12-21.

Figure 5.3 Reprinted, by permission, from T.W. Rowland, 1989, "Oxygen uptake and endurance fitness in children: A developmental perspective," Pediatric Exercise Science 1: 318.

Figure 5.5 Reprinted, by permission, from C.J.R. Blimkie, P. Roche, and O. Bar-Or, 1986, "The anaerobic-to-aerobic power ratio in adolescent boys and girls." In Children and Exercise XI, edited by J. Rutenfranz, R. Mocellin, and F. Klimt (Champaign, IL: Human Kinetics), 35.

Figure 5.6 Reprinted, by permission, from R. Benecke et al., 1996, "Maximal lactate steady state during the second decade of age," Medicine and Science in Sport and Exercise 28: 1474-1478.

Figure 5.7 Reprinted, by permission, from S.G. Fawkner and N. Armstrong, 2002, "Assessment of critical power with children," Pediatric Exercise Science 14: 259-268.

Figure 5.9 Reprinted, by permission, from Y. Armon et al., 1991, "Oxygen uptake dynamics during high-intensity exercise in children and adults," Journal of Applied physiology 70: 843.

Figures 6.1 and 6.4 From T.W. Rowland et al., 2003, "Circulatory responses to progressive exercise," International Journal of Sports Medicine 24: 3-4. Reprinted by permission of Georg Thieme Verlag.

Figure 6.2a and b Reprinted, by permission, from T.W. Rowland and J.W. Blum, "Cardiac dynamics during upright cycle exercise in boys," American Journal of Human Biology 12:749-757. Copyright © (2000, John Wiley & Sons, Inc.)

Figure 6.5 Reprinted, by permission, from T.W. Rowland et al., 2001, "Dynamics of left ventricular diastolic filling during exercise," Chest 120: 148.

Figure 6.6a and b Reprinted, by permission from C.C. Cheatham et al., 2000, "Cardiovascular responses during prolonged exercise at ventilatory threshold in boys and men," Medicine and Science in Sports and Exercise 32: 1080-1087.

Figure 6.7 Reprinted, by permission, from T. Rowland and R. Lisowski, 2003, "Determinants of diastolic cardiac filling during exercise," Journal of Sports Medicine and Physical Fitness 43:380-385.

Figure 6.8 Reprinted from American Journal of Cardiology, Vol. 44, D.A. Riopel, A.B. Taylor, and A.R. Hohn., "Blood pressure, heart rate, pressure–rate product and electrocardiographic changes in healthy children during treadmill exercise," 697-704, Copyright 1979, with permission from Excerpta Medica.

Figure 6.10 Reprinted, by permission, from K.L. Nau et al., 1990, "Acute intra-atrerial blood pressure response to bench press weight lifting in children," Pediatric Exercise Science 2:37-45.

Figure 6.11a and b Reprinted, by permission, from K.R. Turley et al., 2002, "Heart rate and blood pressure responses to static handgrip exercise of different intensities: Reliability and adult versus child differences," Pediatric Exercise Science 14: 45-55.

Figures 7.2 and 7.3 Reprinted, by permission, from T.W. Rowland, 1997, "Development of ventilatory responses to exercise in normal white children," Chest 11: 330.

Figure 7.5 Reprinted, by permission, from T. Ohkwwa et al., 1999, "Ventilatory response and arterial blood gases during exercise in children," Pediatric Research 45: 393.

Figure 7.6 Reprinted, by permission, from Y. Nagano et al., 1998, "Ventilatory control during exercise in normal children," Pediatric Research 43: 706.

Figures 7.7 and 7.8 Reprinted, by permission, from S. Ratel et al., 2002, "Acid-base balance during repeated cycling sprints in boys and men," Journal of Applied Physiology 92: 482.

Figure 7.9 From P.B. Laursen et al., "Incidence of exercise-induced arterial hypoxemia in prepubescent females," Pediatric Pulmonology 34:37-41 copyright © (2002 ,John Wiley and Sons, Inc.). Reprinted by permission of Wiley-Liss, Inc., a subsidiary of John Wiley & Sons, Inc.

Figure 7.10 Reprinted, by permission, from T.W. Rowland and T.A. Rimany, 1995, "Physiological responses to prolonged exercise in premenarcheal and adult females," Pediatric Exercise Science 7: 183-191.

Figures 8.1 and 8.2 Reprinted, by permission, from C.R. Taylor, N.C. Heglund, and G.M.O. Maloiy, 1982, "Energetics and mechanics of terrestrial locomotion: I. Metabolic energy consumption as a function of speed and body size in birds and mammals," J. Exp. Biol. 97:1-21and adapted from K. Schmidt-Nielsen, 1984, "Scaling: Why is animal size so important?" (Cambridge: Cambridge University Press), 165-181. Reprinted with the permission of Cambridge University Press.

Figures 8.3 and 8.4 Reprinted, by permission, from T.W. Rowland et al., 1987, "Green. Physiological responses to treadmill running in adult and prepubertal males," International Journal of Sports Medicine 8: 292-297.

Figures 8.6 and 8.7 Reprinted, by permission, from J.M. Workman and B.W. Armstrong, 1963, "Oxygen cost of treadmill walking," Journal of Applied Physiology 18:798-803.

Figures 8.8 and 8.9 Reprinted, by permission, from B. Schepens, P.A. Willems, and G.A. Cavagna, 1998, "The mechanics of running in children," Journal of Physiology 509:927-940.

Figure 8.10 Reproduced by permission, from T.W. Rowland, 1990, "Mechanical efficiency during cycling in prepubertal and adult males," International Journal of Sports Medicine 11:452-455.

Figure 8.11 Reprinted from Journal of Electromyogry and Kinesiology, Vol. 7, G. Frost et al., "Cocontraction in three age groups of children during treadmill locomotion," 179-186, Copyright (1997), with permission from Elsevier.

Figure 8.13 Reprinted, by permission, from L.M. Webber et al., 1989, "Serum creatine kinase activity and delayed onset muscle soreness in prepubescent children: A preliminary study," Pediatric Exercise Science 1: 351-359.

Figure 9.1 Reprinted, by permission, from E. Van Praagh, 2000, "Development of anaerobic function during childhood and adolescence," Pediatric Exercise Science 12:150-173.

Figure 9.2 Reprinted, by permission, from R.C. Bailey, 1995, "The level and tempo of children's physical activities: An observational study," Medicine and Science in Sports and Exercise 27:1033-1041.

Figures 9.4 and 9.5 Reprinted, by permission, from E. Doré, 2000, "Dimensional changes cannot account for all differences in short-term cycling power during growth," International Journal of Sports Medicine 21: 360-365.

Figure 9.6a and b Reprinted, by permission, from American Alliance for Health, Physical Education, Recreation, and Dance, 1976, AAHPERD youth fitness test manual Washington, DC, 19-20.

Figure 9.7 Adapted, by permission, from H. Hebestreit, B. Staschen, and A. Hebestreit, 2000, "Ventilatory threshold: A useful method to determine aerobic fitness in children?" Medicine and Science in Sports and Exercise 32:1964-1969.

Figures 10.1 and 10.2 Reprinted, by permission, from C.J.R. Blimkie and D.G. Sale, 1998, "Measuring maximal short-term power output during growth." In Pediatric anaerobic performance, edited by E. Van Praagh (Champaign, IL: Human Kinetics), 195, 205.

Figure 10.3a, b, and c Reprinted, by permission, from S. Jaric, D. Ugarkovic, and M. Kukolj, 2002, "Evaluation of methods for normalizing muscle strength in elite and young athletes," Journal of Sports Medicine and Physical Fitness 42:141-151.

Figure 10.4 Reprinted, by permission, from C.M. Neu, 2002, "Influence of puberty on muscle development at the forearm," American Journal of Physiology, Endocrinology, and Metabolism 283: E103-E107.

Figure 10.5 From C. Patten, G. Kamen, and D.M. Rowland, "Adaptations in maximal motor unit discharge rate to strength training in young and older adults," Muscle and Nerve 24:542-550, copyright © (2001, Jown Wiley and Sons, Inc.). Reprinted by permission of John Wiley & Sons, Inc.

Figure 10.6 Reprinted, by permission, from H. Kanehisa, 1995, "Strength and cross-sectional area of knee extensor muscles in children," European Journal of Applied Physiology 68:402-405. Copyright © Springer-Verlag.

Figure 10.7a and b Reprinted, by permission from H. Hebestreit et al., 1996, "Plasma metabolites, volume and electrolytes following 30-s high intensity exercise in boys and men," European Journal of Applied Physiology 72:566-569.

Figure 10.8 Reprinted, by permission, from C.J.R. Blimkie and D.G. Sale, 1998, "Strenght development and trainability during childhood." In Pediatric anaerobic performance, edited by E. Van Praagh (Champaign, IL: Human Kinetics), 209. Data from T. Fukunga and Y Kawakami.

Figure 10.9 Reprinted, by permission, from L.M. Webber et al., 1989, "Serum creatine kinase activity and delayed onset muscle soreness in prepubescent children: A preliminary study," Pediatric exercise science 1: 351-359.

Figure 11.2a and b Reprinted, by permission, from J.C. Ozmun, A.E. Mikesky, and P.R. Surburg, 1994, "Neuromuscular adaptations following prepubescent strength training," Medicine and Science in Sports and Exercise 26: 510-514.

Figure 11.4 Reprinted, by permission, from T.W. Rowland and A. Boyajian, 1995, "Aerobic response to endurance training in children," Pediatrics 96:654-658.

Figure 11.5a Reprinted, by permisison, from G. Koch and L. Rocker, 1977, "Plasma volume and intravascular protein masses in trained boys and fit young men," Journal of Applied Physiology 43:1085-1088.

Figure 11.7 Reprinted, by permission, from S. Nottin, 2002, "Central and peripheral cardiovascular adaptations to exercise in endurance-trained children," Acta Physiol. Scand. 175: 85-92.

Figure 11.8 Reprinted, by permission, from D.M. Cooper, 1994, "Effect of growth hormone suppression on exercise training and growth responses in young rats," Pediatric Research 35:223-227.

Figure 12.1 Reprinted, by permission, from B. Falk, 1992, "Sweat gland response to exercise in the heat among pre-, mid-, and late pubertal boys," Medicine and Science in Sports and Exercise 24:313-319

Figure 12.2 Reprinted, by permission, from F. Meyer, 1992, "Sweat electrolyte loss during exercise in the heat: Effects of gender and maturation," Medicine and Science in Sports and Exercise 24: 776-781.

Figure 12.3 From S.D.R. Galloway and R.J. Maughan, 1995, "Effects of ambient temperature on the capacity to perform prolonged exercise in men," Journal of Physiology 489: 35-36.

Figure 12.4 Reprinted, by permission, from B. Nielsen et al., 1990, "Muscle blood flow and muscle metabolism during exercise and heat stress," Journal of Applied Physiology 69: 1040-1046.

Figure 12.5a and b Reprinted, by permission, from J. Gonzalez-Alonso et al., 1997, "Dehydration markedly impairs cardiovascular function in hyperthermic endurance athletes during exercise," Journal of Applied Physiology 82: 1229-1236.

Figure 12.6 Reprinted from Journal of Thoracic and Cardiovascular Surgery, Vol. 44, T. Cooper, V.L. Williams, and C.R. Hanlon, "Cardiac and peripheral vascular responses to hyperthermia induced by blood stream heating," 667-673, copyright (1962), with permission from American Association for Thoracic Surgery.

Figure 12.7a and b Reprinted with permission from Jokinen et al., Pediatrics, Vol. 86, Page 285, Copyright 1990.

Figure 12.10 Reprinted, by permission, from A.M. Rivera Brown et al., 1999, "Drink composition, voluntary drinking, and fluid balance in exercising, trained, heat acclimatized boys," Journal of Applied Physiology 86: 78-84.

Figure 13.2 Adapted, by permission, from O. Bar-Or and D.S. Ward, 1989, "Rating of perceived exertion in children." In Advances in pediatric sport sciences. Volume 3 Biological Issues, edited by O. Bar-Or (Champaign, IL: Human Kinetics), 153.

Figure 13.3 Reprinted, by permission, from M. Weise et al., 2002, "Pubertal and gender-related changes in the sympathoadrenal system in healthy children," The Journal of Clinical Endocrinology and Metabolism 87: 5038-5043. Copyright 1980, 1996, 2000, and 2002. The Endocrine Society.

Figure 13.4 Reprinted, by permission, from T.W. Rowland et al., 1996, "Plasma norepinephrine responses to cycle exercise in boys and men," International Journal of Sports Medicine 17: 22-26.

Figure 13.5 Reprinted, by permission, from H. Ohuchi et al., 2000, "Heart rate recovery after exercise and cardiac autonomic nervous activity in children," Pediatric Research 47: 334.

Figure 13.6 Reprinted, by permission, from T.W. Rowland, 1990, Exercise and children's health (Champaign, IL: Human Kinetics), 35.

INTRODUCTION

The Challenges of Pediatric Exercise Physiology

> *"God only knows.*
> *God makes His plan.*
> *The information is unavailable*
> *to the mortal man."*
>
> Paul Simon, "Slip Slidin' Away"

The business of pediatric exercise physiology is highly complex. Its most essential feature, certainly, is recognizing the process of *change*. Just like cognitive, psychosocial, somatic, and almost all other biologic aspects one can think of, the physiologic factors that define locomotion during the course of childhood are in a continuous state of evolution.

Compare the physiologic features of a 12-year-old boy with those he had when he was 5. He is taller and heavier to be sure, and he now has greater maximal oxygen uptake, anaerobic capacity, muscular strength, running economy, aerobic scope, minute ventilation, cardiac output, stroke volume, ventilatory efficiency, and so on. Indeed, there are but a few measures of physiologic function that have not changed (maximal heart rate and muscular mechanical efficiency are two). Children's physiology is dynamic, and they cannot be regarded simply as miniature adults. It follows that all assessments of physiologic responses to exercise in children and young adolescents need to be considered in this context of change.

The pediatric exercise physiologist is challenged further by the observation that different children unfortunately do not follow the same *rates* of change as they progress through the growing years. We have early and late biologic maturers, such as child A and child B in figure I.1, whose individual differences in variables such as strength and speed at age 8 are not predictive of values at age 13.

Superimposed on these variations in rate of maturation are inherent (genetically-based) interindividual differences, which cause one child to be more fit than another, even when body size and level of biologic development are equivalent. Child C in figure I.1 is so blessed and will always be able to lift more weight for a longer time than child A.

We will see later in this book that the complexity created by these various physiologic patterns during growth can be compounded by the influence of extrinsic variables. Changes in body composition, level of habitual physical activity, and a period of exercise training all have the potential to modify the shape or shift the position of these developmental curves.

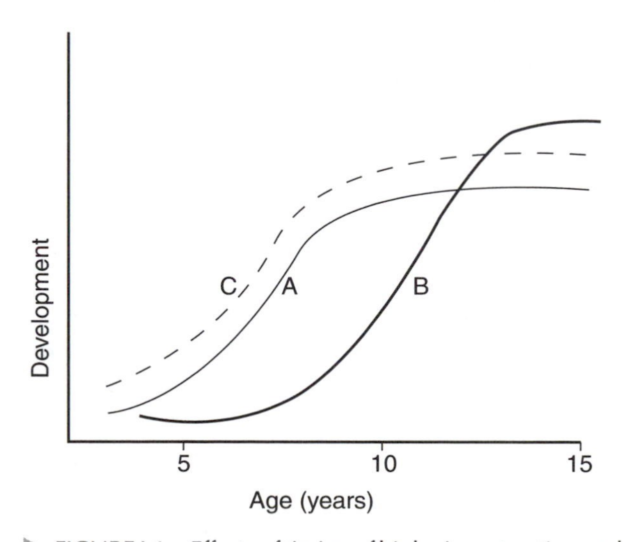

FIGURE I.1 Effects of timing of biologic maturation and inherent physical fitness on physiologic and performance development. Child A, early maturer; child B, late maturer; child C, inherently more highly fit.

Bending the Physiologic Development Curves

Pediatric exercise physiologists are challenged not only to describe the patterns of change in physiologic variables over time but also to explain their separate—or maybe mutual—determinant factors. We are also interested in the extent that these curves can be "bent" by extrinsic factors, perhaps most particularly by the role of physical activity and sport training.

Now the complexity deepens. Suppose that we want to know if a 12-week period of endurance training can alter $\dot{V}O_2$max in a group of 10-year-old children. We should structure such a study to provide us with valid results (i.e., appropriate exercise intensity, duration, and frequency), but how should we interpret the results? Suppose that $\dot{V}O_2$max goes up 10% in our training group compared with nontraining controls. Does this mean that training simply causes an increase in level in biological development without altering ultimate level of maturation (increase in progress along curve A in figure I.1)? Or, did training alter the rate of development of $\dot{V}O_2$max (shift from B to A) with no implication as to final level of aerobic fitness? Or, is the increase in $\dot{V}O_2$max from training independent of the rate of biological development (shift from A to C)?

The mechanisms by which aerobic fitness improved in response to training might (or might not) depend on which of the three preceding effects was operant. $\dot{V}O_2$max will increase with training in all three by increasing heart size and stroke volume, but perhaps the trigger is different depending on how biological development is affected. Do the actions of circulating growth hormone or local cardiac anabolic factors increase heart size? Or is the greater heart size secondary to a primary response of increased plasma volume, perhaps as a consequence of enhanced serum protein levels? Or could we be witnessing an increase in parasympathetic tone, with resting bradycardia and a resulting increase in left ventricular diastolic chamber size? That is, the stimulus → genetic response → phenotypic expression could relate to which way the developmental curve of $\dot{V}O_2$max is being bent.

The same kind of issue arises when we address a more real-world (but rarely considered) question: What is the objective of placing a child in an intensive sport training program? A child will normally improve in 50-yard-sprint times with age, even with no training whatsoever. If we place a child in a sprint training program, with associated plyometrics and resistance work, what are we trying to accomplish?

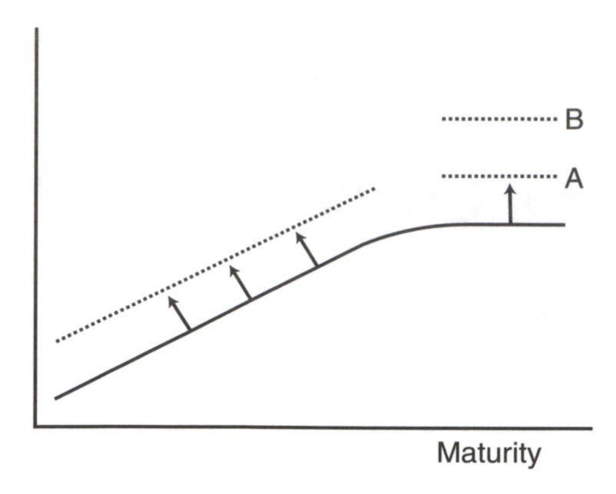

FIGURE I.2 Possible outcomes of exercise training in children (see text for explanation).
Reprinted by permission from T.W. Rowland 2001.

Suppose that the solid curve in figure I.2 represents normal improvement in sprint speed with age in a nonathletic child until maturity. Assume that sprint training by a mature individual will increase performance to level A. With sprint training of the child we are presumably trying to shift the developmental curve to the left (i.e., better performance at the same age). But to what end? Where will that curve advanced through training end up? At level A? Or will early sprint training in the child enhance the ultimate mature performance to level B?

This question has some important implications. If all this training does is shift a normal curve to the left, without any ultimate improvement in performance at maturity, does it provide any benefit to the child? If early sport training does nothing more than speed a child along to a predetermined genetic limit, it would make sense to concentrate early training on elementary skills, strategies, training education, and fun rather than subject the child to arduous workouts that might lead to injury or early burnout and withdrawal from sport.

Ontogeny and Phylogeny

The physiologic changes witnessed as a child grows represent *ontogenetic* changes: alterations that occur during the development of an individual from the immature to the mature biologic state. The differences we observe in a child from age 5 to age 12 are the effect of ontogeny. These changes should be expected to reflect the factors, both genetic and environmental, that influence growth and functional development of biologic systems.

Differences between two biologically mature individuals, on the other hand, are termed *phylogenetic*. Determinants of these differences are, for the most part, not those that influence rate of change but those that specify certain static characteristics of size and function within the limits set by that person's genetic endowment. (The genetic capacity of an adult to improve physical fitness with training is an example of an exception.) If phylogenetic differences in size, per se, are important in understanding exercise physiology (and we will see in chapter 1 that they certainly are), then growing children should be expected to be influenced by these factors as well. It may be difficult, in fact, to sort out the phylogenetic from the ontogenetic influences on physiologic responses to exercise as children grow.

For example, the relationship between oxygen uptake and body size in adult mammals of different species is such that absolute $\dot{V}O_2$ at rest relates to body mass by the exponent 0.75. This is a phylogenetically derived exponent that must be explained by some mechanism that influences metabolic rate and size in biologically mature animals, including humans. Since immature animals are of different sizes during growth, we might expect that this same phylogenetic mechanism applies. Still, the *change* in $\dot{V}O_2$ and its determinants as a child grows might be expected to (and in fact does) create a different ontogenetically derived scaling exponent. This is addressed in chapter 1, where we will see that differences between ontogenetic and phylogenetic scaling exponents for physiologic variables may provide insights into physiologic determinants in growing children.

Determinants of Physiologic Development

The multitude of factors that contribute to physiologic development during childhood can be divided into those that are related to increases in body size and others that have been traditionally considered size independent. The former are generally obvious and usually readily quantified: As the heart grows, so do maximal stroke volume, cardiac output, and oxygen uptake. With the enlargement of the cross-sectional area of muscles, the child becomes stronger.

Almost all factors that are affected by increases in body size follow the curves shown in figure I.3 for both sexes. During early childhood, a progressive, almost linear, increase is observed, with average values for boys slightly but consistently greater than for girls. At puberty, the added influence of sex hormones on somatic growth causes an acceleration in boys (from increased levels of circulating testosterone) and a plateau in girls as they reach sexual maturity. When physiologic variables are related to body size, of course, the developmental patterns in figure I.3 are altered. The appropriate means of relating values to body dimensions is addressed in chapter 1.

Size-independent factors, on the other hand, demonstrate a variety of patterns of change over the course of childhood. Maximal heart rate, myocardial contractility, and blood hemoglobin levels remain stable. Blood pressure and capacity for glycolytic metabolism increase. Some size-independent variables, such as breathing rate, are easily recognized and measured. Assessment of others, such as the neurologic input responsible for gains in muscle strength, is difficult, and their influence on phenotypic expression largely conjectural.

As discussed in chapter 1, the size-independent variables that relate to time (e.g., heart rate, breathing rate) may not be independent of body dimensions after all. There is evidence that physiologic time relates directly to body mass and that all time-related functions are similar in this respect.

Pediatric exercise physiologists seek to recognize the interplay and relative inputs of body size and

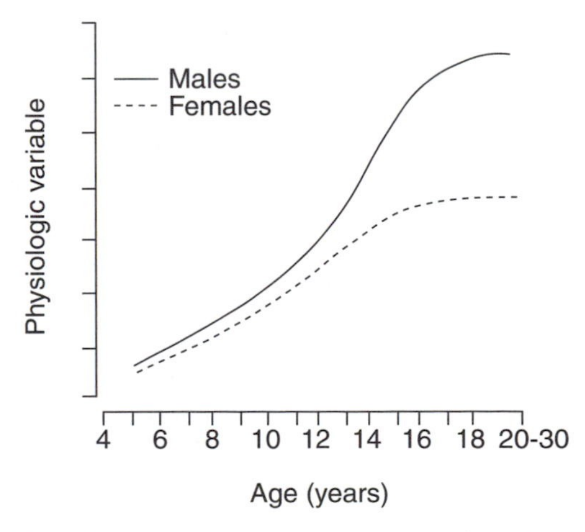

FIGURE I.3 The prototypical curves for physiologic development and performance fitness in boys and girls, typically observed for muscle size and strength, aerobic fitness ($\dot{V}O_2$max), and heart and lung size.

size-independent factors for any given physiologic outcome. For maximal cardiac output, the product of maximal heart rate and stroke volume, this is reasonably straightforward. The size-independent variable, maximal heart rate, does not change with age, while the size-dependent factor, maximal stroke volume, does and therefore accounts entirely for the rise in maximal cardiac output. (The secret of *why* Nature has chosen this particular pattern of relative inputs remains closely guarded.)

In other forms of fitness the contributions of and interplay between size and size-independent factors are less clear. Increases in strength with age are predominantly a reflection of increases in muscle size, but *allometric analysis* indicates that this explanation does not account entirely for the rise in strength during childhood. An understanding of the nature of the small but presumably important size-independent influence on the development of muscle strength in children remains a challenge.

Size-independent factors are more instrumental in the development of anaerobic fitness, as indicated by performance of *Wingate cycle testing.* Indeed, by comparing increases in absolute values of peak anaerobic power with those expressed per kilogram of body mass, one can reasonably conclude that increased muscle size and size-independent factors each contribute to about half of the development of anaerobic fitness in children.

The Principle of Developmental Symmorphosis

Taylor and Weibel introduced the idea that in any physiologic system the functional capacity of no single component should exceed that of any other part of the system (1). This is the principle of *symmorphosis,* which follows from the idea that it makes no sense for a part of a biologic system to be built to a greater functional capacity than the system as a whole. From a Darwinian viewpoint, there is no evolutionary impetus for this to occur.

As a corollary, it can be suggested that during ontogenetic growth, no component of a system should develop faster than the rest of the system as a whole. It makes no sense for the lungs to develop in size or function at a rate that exceeds that of the heart, or for the heart to grow to be a more effective pump than is needed to supply skeletal muscle of the current size and function. We in fact might expect that the components of a physiologic system, such as for oxygen delivery, should develop in concert. Indeed, this seems to be the case: Heart size in children is closely linked to lean body mass, and the ontogenetic scaling exponents for heart and lungs are similar.

This concept raises some interesting questions regarding the coordination of determinants of physiologic capacity *between* fitness types. Does developmental symmorphosis hold true between systems responsible for aerobic and anaerobic fitness? That is, do size-independent factors that contribute to anaerobic fitness develop at the same rate as muscle size, which contributes to $\dot{V}O_2$max, strength, and performance on Wingate cycle testing? The observation that absolute values for the different forms of fitness (aerobic, anaerobic, strength) are closely associated during growth, as discussed in the chapters that follow, suggests they do.

These observations support the validity of symmorphosis in the development of determinants of exercise physiology as children grow. They lead, moreover, to a new question: If these factors develop in concert, is this not an indicator that they are all ultimately under the direction of the same inherent controlling factor (or factors)? As discussed in chapter 1, such a factor that provides homogeneity of biologic development must be in some way connected with body size, and that influence controlling homogeneous development must affect not only the obvious factors related to dimension (e.g., heart stroke volume) but also those variables that are associated with time (e.g., heart rate). Thus the complexity—as well as the challenge and excitement—of pediatric exercise physiology continues to grow.

1

The Importance of Body Size

For he by geometric scale
Could take the size of pots of ale.

—Samuel Butler (1663)

In this chapter we discuss

- the effects of body size on physiologic function and
- the proper means of expressing physiologic variables relative to body size.

Adults are bigger than children. Older children are larger than younger children. And for the pediatric exercise scientist, size counts. In our age range of interest, the average 18-year-old boy weighs 70 kg, and the 5-year-old 20 kg. The height measurements are 175 and 115 cm, respectively, and body surface area 1.85 and 0.80 m^2. So, within this span of ages, the weights differ by a factor of 3.5, heights by 1.5, and surface areas by 2.3, which are equivalent to orders of magnitude of $10^{0.54}$, $10^{0.18}$, and $10^{0.36}$, respectively.

Within these size differences, the small child and young adult are not entirely geometrically similar. That is, their different body segments are not in the same proportion to total body size: Their shape is different, the importance of which becomes evident later in this chapter. The legs of young children, for instance, are short for their body height relative to older persons. This becomes evident when one compares the ratio of sitting to standing height, which is about 68% at birth and declines to 50% by maturity. The head is relatively large in small children, and the relationship between the breadth of shoulders and hips (bicristal–biacromial ratio) in males diminishes at puberty by 7% (17, pp. 39-65). These differences in proportionality are most exaggerated in very early childhood, and by age 10 children are geometrically similar to adults.

This chapter deals with how physiologic variables in children during rest and exercise are related to these variations in body size. The first section focuses on how the different associations between physiologic function and body dimensions can enhance our understanding of (a) the mechanisms by which body dimensions influence physiologic fitness and (b) the differences in physiologic features between individuals of different sizes. Given children's progression in body size as they grow, these considerations may help explain physiologic changes during biologic development.

The second section addresses the dilemma of how to best express physiologic variables in respect to body size. This is an often troublesome but critical aspect of pediatric exercise physiology, for "normalizing" variables to body size is necessary if we want to accurately compare physiologic factors between groups or longitudinally in the same individuals. Experience has proved the importance of selecting a proper means of accomplishing this, since inappropriate adjustments of physiologic variables for body size can produce misleading results and lead to erroneous conclusions.

Size and Function: Lessons From Allometry

In the spring of 1927, Sir Julian Huxley studied the size of birds' eggs. He was particularly intrigued by the observation that smaller birds tended to lay larger eggs relative to their body size than do larger birds (15). To be sure, the larger fowl produced bigger eggs. The typical egg of an ostrich weighs 1,700 g, in contrast to the hummingbird's egg, which tips the scales at 0.6 g (a factor of $10^{3.5}$). However, given the much greater mass of the adult female ostrich (113,380 g) than of the hummingbird (3.6 g), the ratio of egg mass to adult bird mass is 0.015 for the former and 0.167 for the latter. In proportion to adult body mass, then, the mass of the hummingbird egg is 11 times that of the ostrich. Subsequent investigators, considering the egg weights of birds of many sizes, were able to construct a mathematical equation for this phenomenon:

$$M_{egg} = 0.198 M_{adult}^{0.77}$$

where M designates mass in grams (8). This equation tells us that egg mass increases as adult female mass increases, but the rate of rise in egg size is greater than that of its producer.

Now, if we think about it, the egg and its size are rather important. The egg must include in its calcium shell all the nutrients, minerals, and water necessary for the growth of the fetal chick during the period of incubation (27). The shell itself must be large and porous enough to permit exchange of adequate oxygen and carbon dioxide yet prevent desiccation of its contents. Huxley and other researchers of egg size therefore suggested that the unhatched offspring of smaller birds have an exaggerated need for metabolic support relative to their body size compared with the offspring of larger birds.

This story is among the earliest examples of the use of *allometry* (meaning "of other and different measures") to establish the relationship between a variable (in this case, egg size) with body mass (of adult birds). It is also one of the first illustrations of how such an observed relationship can be used to understand physiologic differences (e.g., in metabolic rate) between animals. We will see in the course of this book how proper adjustment of physiologic variables to body size by allometric principles is important in understanding the changes in children's responses to exercise as they grow.

Understanding Allometry

Allometry is a method of mathematically expressing the extent to which a variable (be it physiologic, anatomic, or temporal) is related to a unit of body size, usually body mass, as size increases. Body mass has traditionally been used as the unit of size by biologists because of its ease of measurement and the difficulty in defining height in animals of such diverse morphology as trout, giraffes, and hummingbirds.

Allometric analysis, or scaling, is described by the equation

$$Y = aM^b$$

where *Y* is the variable to be related to mass M and *a* is the proportionality coefficient. The exponent *b* is of primary interest, since this scaling factor indicates the extent and direction of the relationship between changes in the variable *Y* and body mass. If *Y* increases in direct proportion to body mass, then $b = 1$. This type of relationship is observed, for instance, between heart size and increasing body mass among adult mammals. If $b = 0$, then body mass has no effect on *Y*, that is, the variable is independent of body mass. For example, maximal heart rate does not change with increases in body size during the childhood years, so $HR_{max} \sim M^{0.0}$.

If *Y* increases as body mass increases, but at a slower rate, then *b* will be greater than 0 but less than 1, as in the relationship of birds' egg size to maternal mass described earlier. A value of *b* that exceeds 1 tells us that the variable increases at a faster rate than mass (e.g., skeletal mass relates to total body mass of adult animals this way). If *Y* decreases as mass increases, then *b* will be negative. The decline in stride frequency as children grow in size is an example of this type of relationship.

The exponent *b* can provide us with information as to how *Y*/M changes as M increases. The allometric equation for resting $\dot{V}O_2$ in animals is

$$\dot{V}O_2\text{rest} = 0.19M^{0.75}$$

indicating that absolute values of resting oxygen uptake are greater in animals of larger size. Since the exponent is less than 1, however, smaller animals have a greater $\dot{V}O_2$ relative to their body mass than do larger ones. We can see this more clearly by dividing both sides of the equation by body mass (or $M^{1.00}$), which gives us

$$\dot{V}O_2\text{rest per kilogram} \sim M^{0.75}/M^{1.00} = M^{-0.25}$$

The exponent *b* is now negative, indicating that resting $\dot{V}O_2$ relative to body mass declines as mass increases. That is, the mouse has a $\dot{V}O_2$ per kilogram at rest that is much greater than that of the cow.

The proportionality coefficient *a* provides information regarding the actual value of the relationship between *Y* and M. For example, consider the equation derived in adult animals for absolute $\dot{V}O_2\text{max}$:

$$\dot{V}O_2\text{max} = 1.94M^{0.79}$$

Note that this equation is almost identical to that for resting $\dot{V}O_2$ except that coefficient *a* is now 1.94, whereas it was 0.19 in the resting equation. These two equations tell us that $\dot{V}O_2$ during maximal exercise increases in the same proportion to body mass as it does at rest. The difference in *a* indicates that $\dot{V}O_2$ during maximal exercise is 10 times that at rest.

The allometric equation is derived for a group of data by using logarithmic transformation of individual values of *Y* and mass to create the linear equation

$$\log Y = \log a + b \log M$$

In this form, which is a straight line, *b* is the slope, and *a* is the *Y* intercept. Figure 1.1 demonstrates how different values of *b* indicate the relationship of changes in *Y* to changes in mass when expressed in this linear form.

It is important to recognize that allometric exponents only characterize the relationship between changes in *Y* and in body size descriptively, without implying the mechanisms involved in this association. As Calder emphasized, "The allometric equations are empirical descriptions and no more . . . in that they relate some quantitative aspects of an animal's physiology, form, or natural history to its body mass, usually in the absence of theory or knowledge of causation" (8, p. 28).

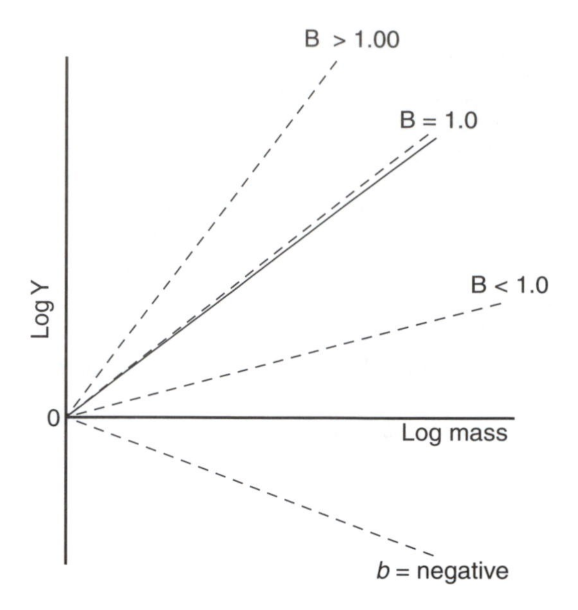

FIGURE 1.1 How different values of the mass scaling exponent b indicate the relationship of Y to M. The solid line is the line of identity.

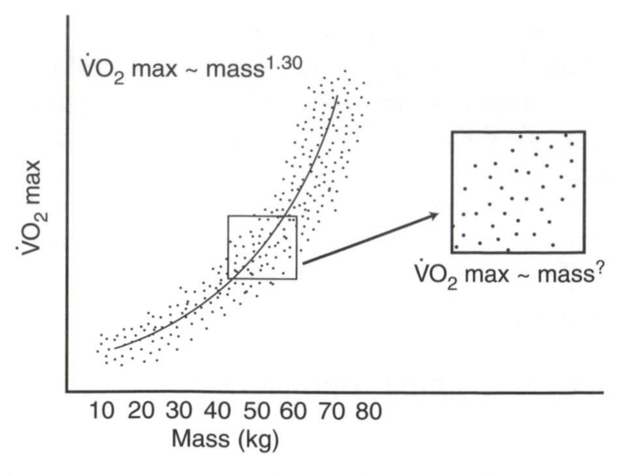

FIGURE 1.2 How limited differences in size in a population of subjects can obfuscate true allometric relationships.

Reprinted by permission from T.W. Rowland 1998.

In studies of large populations of animals of varying sizes, the scaling exponent b describes the average relationship of a variable Y to different body dimensions. But such a relationship cannot, de facto, be anticipated to be evident for all subsets of the data. For example, the allometric equation in mammals for resting metabolic rate (P_{met}) is

$$P_{met} = 70M^{0.75}$$

However, very small animals, such as the shrew, demonstrate an exponent b of 0.23.

Scaling exponents provide information on the relationships between variables and changes in body size for a population. While exponents may differ among subsets, the general relationship represented by these exponents presumably is dictated by some underlying physical, physiologic, or biochemical mechanism. Their ultimate value then lies in explaining the differences observed in the variable–mass relationship.

It is necessary to realize, too, that logarithmic transformation and allometric analysis of the data for a small range of body sizes, particularly from a limited number of subjects, may not accurately provide the big picture of the true relationship between a variable and body mass. This potential hazard needs to be recognized especially in allometric analysis of physiologic and anatomic variables in children, in whom the order of magnitude of size difference is about $10^{0.30}$, in contrast to that in large animal studies, typically $10^{4.0}$. Figure 1.2 illustrates this difficulty. Scaling exponent b for $\dot{V}O_2max$ in a small size–variable subset may not reflect the relationship observed in subjects with a larger order of size difference.

It has been argued that simple linear regression of untransformed data may have the same value as allometric analysis in representing variable–mass relationships when the range of body masses is small (8). The less size variability, the more likely this is to be true. As discussed later, arithmetic manipulation of allometric scaling exponents may also allow insights into relationships not obtained by regression analysis.

Use of Allometry

The allometric approach has proven to be highly useful in the hands of comparative biologists, who face the difficult task of comparing physiologic phenomena in animals of greatly different sizes. The past century has witnessed the publication of a vast number of allometric equations relating body mass to variables ranging from the diameter of bird tracheae ($M^{0.39}$) to the fertilization life span of spermatozoa ($M^{0.18}$). Some of these might be regarded as simply interesting. As discussed later, time can serve as a variable in allometric equations, and life span in animals in years (t_{life}) is expressed as

$$t_{life} = 11.6M^{0.20}$$

Calder has noted that, fortunately, humans fit outside this mathematical mean, for, if not, those of us over

27 years old—that is, $11.6(70\ kg)^{0.20}$—would have by now departed (8).

Other relationships derived empirically by allometric analysis have provided fuel for major biologic controversies. For example, we have seen from an earlier equationthat resting metabolic rate in animals varying in size from mice to elephants relates to body mass by the exponent 0.75. But this is not what it is "supposed to be." Core body temperature is similar in all mammals, and heat is lost from the body by skin surface area. Thus, the metabolic rate at rest should be related to body surface area (the so-called surface rule), which, by dimensionality theory, is equivalent to $M^{0.67}$. In the effort to reconcile this difference in theoretically and empirically derived exponents, a number of explanations have been put forth, including geometric dissimilarity, statistical artifact, elastic similarity, influence of gravity, improper measurement of body surface area, and so on. In general, these have been found, as Blaxter concluded, "not particularly convincing" (5, p. 146).

While simple equations relating a single variable to size have been valuable, the strength of the allometric approach lies in the ability to mathematically combine two allometric equations to gain insight on an additional variable. Before demonstrating this concept, it is useful to review the basic algebraic rules of manipulation of exponents:

$$(x^a)(x^b) = x^{(a+b)}$$

$$x^a/x^b = x^{(a-b)}$$

$$(x^a)^b = x^{ab}$$

How are these mathematical maneuvers useful? Suppose we want to know the effect of changes in body size on peripheral oxygen extraction, or the arterial-venous oxygen difference (AVO_2diff). Oxygen uptake is the product of cardiac output (Q) and AVO_2diff, so $AVO_2diff = \dot{V}O_2/Q$. We should therefore be able to answer this question by allometric analysis if we gather data on mass, $\dot{V}O_2$, and Q in our population of subjects.

Zoologists have already done this. Measurements in adult animals of varying sizes indicate the following:

$$\dot{V}O_2 \sim M^{0.75}$$

$$Q \sim M^{0.81}$$

Therefore

$$AVO_2diff \sim M^{0.75}/M^{0.81} = M^{-0.06}$$

indicating that AVO_2diff is essentially independent of body size in adult animals.

Or maybe we are interested in determining the maximal range of bird migration in relation to bird size. Maximal time (t) duration of bird migration is indicated by

$$t = 249M^{0.28}$$

while maximal velocity is

$$v_{max} = 15.7M^{0.17}$$

Maximal range, the product of the two, is thus expected to relate to body mass by the exponent 0.45.

Here is an example more applicable to exercise in children, and we'll return to it later, in the discussion in chapter 8 on running economy. To standardize studies of running mechanics in animals, comparisons are made at the speed of transition from trot to gallop. At this speed the stride length (L_{st}), defined as the distance from foot strike to next foot strike, relates to body mass as

$$L_{st} \sim M^{0.38}$$

The limb excursion distance (L_{exc}), or the length between the full forward position of the leg and rear extension in a single strike, is related to body mass as

$$L_{exc} = M^{0.15}$$

In walking, the stride length and the limb excursion distance are equal. In running, however, the stride length is greater, because of an airborne component. The relationship between the two measures during running is termed the *stride efficiency,* calculated as

$$L_{st}/L_{exc} \sim M^{0.38}/M^{0.15} = M^{0.23}$$

The results indicate that larger animals (in this case, antelope, wapiti, and bison) have a greater stride efficiency than smaller animals, and the equation provides the magnitude of that difference (8). Studies using this approach with children of different ages and sizes might provide some insights into the changes in running economy seen with age.

Phylogenetic and Ontogenetic Scaling

A primary rule in the allometric approach is that one must not combine phylogenetic and ontogenetic data (8). What this means is that in creating allometric equations, data obtained in different mature adult animals (phylogenetic scaling) should not be mixed, or compared with, measurements gathered longitudinally in growing organisms over time (ontogenetic scaling). As explained by Calder, scaling in the two may have entirely different meanings (8). Factors that influence or are affected by changes in growth during biologic development may not be identical to those that contribute to allometric relationships once maturity is reached. Ontogenetic analysis considers the phenotypic expression of the same genetic material over time (i.e., intraindividual developmental scaling). In a phylogenetic approach, time is not considered; the relationships of physiologic variables to body size are reflections of different genetic information among mature subjects (i.e., interindividual scaling). Comparing phylogenetic and ontogenetic allometric exponents for the same physiologic variable, however, might give some insights into the extent to which such genetic and developmental differences are biologically important.

In the animal research literature, significant discrepancies have been observed when mass scaling exponents for physiologic or anatomic variables during growth are compared with those between adult animals. This is most apparent in values of basal or resting metabolic rate, which is expected to relate to $M^{0.75}$ in mature animals (i.e., phylogenetic scaling). Brody described the allometric curves of basal rate versus mass for a large variety of growing animals to see if this relationship held up (7). For all these animals a biphasic curve of metabolic rate to mass was observed with growth: An early component related metabolic rate to body mass by $M^{1.00}$, then the curve reached a breakpoint at which it declined to approximately $M^{0.60}$. The breakpoint was not related temporally to any consistent life event (e.g., puberty, weaning), and its meaning remains conjectural.

The same phenomenon is observed in children. Summarizing several studies, Holliday et al. described an early basal metabolic rate–mass curve with an exponent of 1.02 until a child reaches a weight of 10 kg (14). Thereafter the scaling exponent declines to 0.58 (figure 1.3).

Scaling exponents for the size of specific organs during growth demonstrate a similar pattern. Lung weight scales to body weight in mature animals by 0.99, but in growing chickens, monkeys, and dogs the exponent is less, ranging from 0.58 to 0.92. Heart weight is related to body mass by an exponent of 0.85 during growth in children (28), but an exponent of 1.0 is observed among mature animals of different sizes. The reasons that these exponents are lower during growth than in maturity is obscure, but this phenomenon illustrates that the effect of body size on body metabolism is different during growth than in mature animals, even when the same size differences are being considered.

Allometric scaling studies in children can take one of three approaches. First, the only means of assessing true ontogenetic relationships is from data obtained from the same child longitudinally over a number of years. This, of course, involves the burdensome need for long-term studies, and, since only a limited number of data points are available (i.e., the number of years of measurements), the biologic accuracy of the derived scaling exponents might be questioned.

The second approach, creating allometric equations from a series of cross-sectional studies of children at different ages, is easier to do and simulates an ontogenetic scaling approach. This approach is not truly ontogenetic, however, since the subjects at different ages are not the same individuals.

An allometric equation constructed from cross-sectional data of children at a particular age is considered by some investigators to be a static, or phyloge-

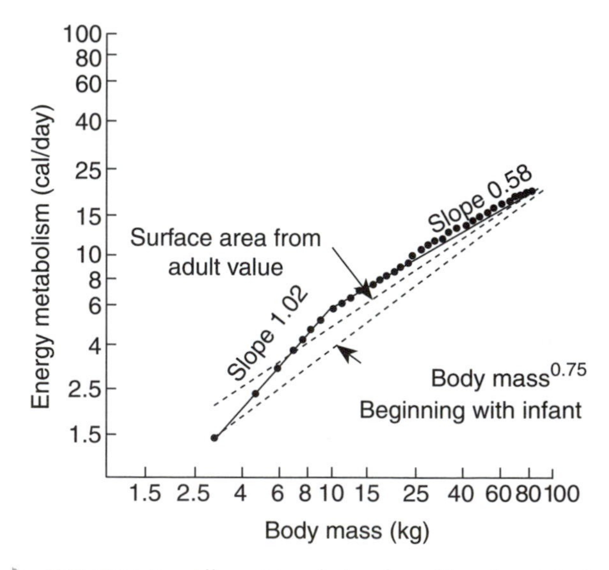

FIGURE 1.3 Allometric relationship of basal metabolic rate and body mass in growing children (from reference 14).

netic, approach. This does not seem to be altogether true, however, since one is considering children at different levels of biologic development.

Few studies have directly examined allometric scaling exponents in growing children, and these have all involved the measurement of maximal aerobic power. Beunen et al. determined $\dot{V}O_2max$ with treadmill testing annually for nine years in 73 boys who were initially seven years old (4). Static interindividual allometric exponents were obtained in a cross-sectional manner each year, and these were compared with intra-individual ontogenetic coefficients calculated from longitudinal measurements of $\dot{V}O_2max$ and body mass in the same child.

The interindividual scaling coefficient varied between 0.78 at 12 years and 1.22 at 16 years. Only at ages 11 and 12 was the expected value of 0.75 within the 95th percentile confidence limits of the measurements. The longitudinal coefficients ranged considerably, from 0.56 to 1.18 (mean 0.86).

In a similar study, Rowland et al. compared cross-sectional and longitudinal allometric scaling exponents for $\dot{V}O_2max$ in the same 20 children over the age span of 8 to 12 years (26). The cross-sectional exponents ranged from 0.40 to 0.56 (mean 0.52) in the boys and from 0.62 to 0.70 (mean 0.65) in the girls. The mean intra-individual longitudinal scaling exponent was 1.10 ± 0.30 for the boys and 0.78 ± 0.28 for the girls.

It is difficult to draw any conclusions regarding the relationships between ontogenetic and phylogenetic allometric scaling of $\dot{V}O_2max$ in humans from these limited data. What is clear, though, is the extensive variability of these scaling factors and their failure to strictly conform to theoretical expectations. The limited number of subjects and range in body size could account for these observations. As discussed later in this book, factors such as sex, body composition, and athleticism also may profoundly affect allometric analyses in the pediatric population.

Physiologic Time

When physiologic variables between small and large animals or between children and adults are compared, it is necessary (but often forgotten) to recognize that biologic events and functions operate in a different time context according to body size. The smaller the animal, the faster things go; that is, the speed of *physiologic time* is inversely related to body size. An analysis of how the time of physiologic events is linked to body size raises some intriguing questions regarding the means by which biologic functions are governed in the growth process.

Calder created a list of 40 allometric equations expressing the relationship of various physiologic times in mammals and birds with their body mass (8; table 1.1). Importantly, the times presented were for a wide range of seemingly unconnected biologic processes, including gestation period, respiratory and cardiac cycles, muscle twitch contraction cycle, and time of reproductive maturity. The fascinating finding was that *all* of these time variables were allometrically related to body mass by a similar exponent *b*, with a range of 0.25 to 0.39. The average exponent was 0.25 ± 0.05, which was within the 95% confidence limits of almost all the equations.

Remember that life span for animals relates to $M^{0.20}$, an exponent not dissimilar to the mass exponents for time functions described by Calder (8). That led Calder to suggest that "using maximum life span, rather than absolute time, it appears that each life comprises about the same number of physiologic events or actions; in other words, each animal lives its life faster or slower as governed by size, but accomplishes just much biologically whether large or small" (8, p. 141).

Schmidt-Nielsen (27) pointed out that a mouse that breathes 150 times per minute and an elephant that breathes 6 times per minute take about the same total number of breaths in a lifetime (3 years for the

TABLE 1.1 Allometric relationships between time functions and body mass in adult mammals

Function	Equation
Life span, in captivity (yr)	$11.6M^{0.20}$
Reproductive maturity (yr)	$0.75M^{0.29}$
Gestation period (d)	$65M^{0.25}$
Erythrocyte life span (d)	$23M^{0.18}$
Plasma albumin half-life (d)	$5M^{0.32}$
Glomerular filtration time (min)	$6.5M^{0.27}$
Blood circulation time (s)	$21M^{0.21}$
Respiratory cycle (s)	$1.1M^{0.26}$
Cardiac cycle (s)	$0.25M^{0.25}$
Twitch contraction cycle, soleus muscle (s)	$0.06M^{0.39}$

Derived from data compiled by Calder (8) from multiple sources.

mouse and 40 years for the elephant). This observation holds true for heart rate as well.

The other striking aspect of these data is the consistency of the relationship of physiologic time to body mass for biologic functions of marked diversity that have no obvious mechanistic common ground. Kidney glomerular filtration rate is hardly aware of gestation period, which in turn could care less about the twitch contraction cycle of the soleus muscle. How did all this uniformity in physiologic time evolve? What controls it? One cannot escape the conclusion that the physiologic time of these diverse biologic functions must have some common biologic basis.

These observations about physiologic time provide an alternative explanation for certain biologic phenomena. Take, for example, the controversy surrounding the discrepancy between the expected phylogenetic mass scaling exponent for basal metabolic rate (0.67) and that actually derived from empiric studies (0.75). It is assumed that 0.67 is "correct" according to dimensional theory but that 0.75 is actually observed because of some modifying factor. Debate about this factor has engendered a host of mechanistic explanations to explain the difference.

Suppose, however, that mammals operate by a strict inherent biologic clock that drives physiologic variables related to time (e.g., heart rate) and is controlled (somehow) by body mass expressed as $M^{0.25}$, as suggested by Calder (8). Differences in cardiac output, ventilation, and oxygen uptake between adult animals are all examples of volume rates, or volume per time. The volumes of the heart, lungs, and so on are related to mass as $M^{1.0}$. Therefore, they can be expressed as

$$\text{volume/time} = M^{1.0}/M^{0.25} = M^{0.75}$$

The result is a scaling exponent that corresponds to that observed in nature. The idea here is that the physiologic clock may serve as a primary rather than secondary determinant of physiologic processes, and this timing mechanism is geared to body mass. The fact that scaling of basal metabolic rate does not match expectations based on surface law may not represent a failure of the law but rather a controlling influence of physiologic time on biologic processes.

Does this idea have implications for understanding developmental exercise physiology? As a child grows, physiologic time slows. The ontogenetic scaling factor for this change has not been investigated. The change in resting respiratory frequency as a child's mass increases is related to $M^{-0.53}$ (which indicates that the time between breaths is related to $M^{0.53}$). This suggests that physiologic time in growing children might be different than in mature animals in the phylogenetic studies compiled by Calder (8).

Increases in physiologic variables such as cardiac output have traditionally been considered to result from the combined influences of growth in size (left ventricular dimension) and a size-independent factor (heart rate). But the allometric evidence suggests that biologic factors involving time, such as heart rate, are *not* size independent. They are in some way associated with body size, and changes in body size with growth should then be expected to control the pattern of change in these variables. According to this concept, increases in cardiac output, or any other volume-per-time process (ventilation, oxygen uptake), with growth are under the control of body size by two means: by an increase in volume, which grows in direct proportion to body mass ($M^{1.0}$), and by a time variable, which is also linked to body size but by another mass exponent ($M^{0.25}$ in adult animals). It may be that changes in children's body size with growth dictate changes in both volume and physiologic time.

The idea that body size controls physiologic variables during growth by influencing both volume and time is consistent with the concept of developmental symmorphosis (see the introduction). That is, by this mechanism, biologic responses to exercise would be expected to develop in concert. No single determinant of a physiologic system would be expected to show accelerated maturation and functional development compared with the other components of the system.

Some inconsistencies, however, do not fit this model, particularly with regard to exercise. Maximal heart rate, for instance, is independent of body mass with increasing age. It may be that the responses and adaptations to the stress of exercise can override basic regulatory effects of time that operate at rest.

Adventures in Allometry

In the course of this book we return to the use of allometric scaling in children to address issues of anaerobic, aerobic, and strength development, as well as exercise economy. What follows is a series of physiologic issues in animal studies for which an allometric approach has proven illuminating, and these scenarios may provide insights into human developmental exercise physiology. At present, however, experimental data in young people is lacking.

Uphill Running Is Easier for Smaller Animals

The use of allometry suggests that children should find uphill running less stressful than do adults. In a large group of mature mammals of different sizes, the energy cost of running on a flat surface per kilogram of body mass *(running economy)* is given by the equation

$$\dot{V}O_2\text{submax/kg} = 5.9M^{-0.33}$$

meaning that the larger the animal, the more economically (i.e., lower $\dot{V}O_2$ per kilogram) it runs. The energy cost of moving a kilogram of body mass vertically, however, is identical for all animals, regardless of size. Vertical work, expressed as $\dot{V}O_2$ per kilogram, is related to mass by $M^{0.0}$. The energy expended to move 1 kg of body weight up 1 m is 40 J, or 10 cal, or 2 ml O_2, regardless of whether it is a horse or a mouse. Since the smaller animal has a greater $\dot{V}O_2$ per kilogram on a flat surface compared with the larger one, running uphill against gravity adds a relatively smaller percentage of energy expenditure in the smaller animal.

Taylor et al. (30) tested out this premise in a comparative study of energy requirements during flat and uphill treadmill running in mice (mean mass 30 g) and chimpanzees (mean mass 18 kg). During level running the energy cost per kilogram of the mouse was eight times that of the chimpanzee. When the animals ran uphill on a treadmill at a 15° incline, the added energy requirement to raise 1 kg a vertical height of 1 m was similar in the two animals (approximately 5.2 J $\cdot$ kg^{-1} $\cdot$ m^{-1}). The authors calculated that if the net cost of lifting 1 kg vertically was the same as on the 15° incline, the cost at a speed of 2 km $\cdot$ h^{-1} would represent a 24% increase over flat running for the mouse but a 189% increase for the chimpanzee.

It is of interest to calculate the hypothetical findings that would be seen in children and adults participating in a similar study. Since the size difference between a child and adult is much smaller than between a mouse and a chimpanzee, a less dramatic difference in relative cost increase might be expected. Rowland et al. (24) found that $\dot{V}O_2$ per kilogram during horizontal running at 9.6 km $\cdot$ h^{-1} was 50 ml $\cdot$ kg^{-1} $\cdot$ min^{-1} for boys 9 to 13 years old and 40 ml $\cdot$ kg^{-1} $\cdot$ min^{-1} for young men ages 23 to 33 years. If the added cost was 2 ml $\cdot$ kg^{-1} for each 1-m rise, the energy cost of uphill running to this height would increase by 4% in the boys and 5% in the men. That this small difference has any biologic or performance significance seems questionable. Differences in metabolic stress between children and adults might be expected to be more dramatic (and meaningful) in activities that involve more vertical work (e.g., rock climbing).

Size Limits Cellular Oxygen Diffusion

The rate of oxygen entry into cells may limit aerobic metabolic rate as cell size increases (8). This may be important for children, whose growth of muscle tissue with age is largely a reflection of increases in cell size (hypertrophy).

There exists no pumping mechanism to drive oxygen into cells. Oxygen enters muscle cells by diffusion, the rate of which is defined by Fick's law of diffusion:

$$\text{Diffusion rate} = -DA[(PO_2)_a - (PO_2)_b]/x$$

where D is the diffusion coefficient (related to the physical nature of the gas and the protoplasm it is traversing), A is the area available for gas exchange, and PO_2 is the partial pressure of oxygen at points a and b, which are separated by distance x. Consider a to be at the surface of the cell and b its center, and assume that PO_2 at a and b remains unchanged. What will happen to the rate of diffusion of oxygen as the cell enlarges during growth?

According to simple geometry, if the distance x doubles in growth, A increases by a factor of $2^2 = 4$, while the mass or volume of the cell increases by a factor of $2^3 = 8$. Given that the metabolic rate is proportional to cell mass or volume, one would hope that diffusion rate would increase eight times as well. But from the diffusion equation, it will rise only by a factor of 2 (A/x). As a consequence, doubling cell diameter results in a decrease in the metabolic rate of protoplasm (relative to cell mass) to 25% of its pregrowth level. In allometric terms,

$$\text{Oxygen diffusion} \sim \text{cell area/cell diameter} \sim M^{0.67}/M^{0.33} = M^{0.33}$$

indicating that as cell size increases, metabolic rate per cell size declines. (Note that diffusion rate, a time function, is related to mass by an exponent expected for physiologic time.)

Might this limitation of oxygen diffusion by cell size contribute to the differences observed in whole-body resting metabolic rate relative to body mass as children grow? No information is available regarding ontogenetic changes in isolated cell metabolic rate. Porter and Brand, however, looked at this question from a phylogenetic perspective (22). They measured liver metabolic rate in nine species of adult animals of various body masses. The rate of cellular oxygen consumption ($\dot{V}O_2$ per unit mass of cells) fell as animal size increased ($\sim M^{-0.18}$). However, the decrease observed in cellular metabolic rate was found to be due not to increases in cell volume but rather to a decrease in the intrinsic metabolic activity of the cells.

It is important to note that in this study, as expected, the magnitude of difference in cell size between adult animals—although they varied markedly in body size—was small. The volume of the liver cells ranged from a mean of 1.0 µl in pigs to 2.2 µl in dogs. The findings indicate that there is no restriction of metabolic rate with increasing cell size, but the factor of difference in cell volume was only 2.0.

In children, the cross-sectional area of fibers in the vastus lateralis muscle increases by a factor of 5 between the ages of 1 year and 12 years (17), and muscle volume grows approximately 11-fold. Thus, the limits of oxygen diffusion rate as reflected by differences in cell metabolic rate related to cell size might become more apparent with the greater order of magnitude of difference in cell volume in children.

The Mass of Salmon (and Children) Influences Metabolic Scope

Early in this chapter we saw how the two allometric equations for resting and maximum $\dot{V}O_2$ reveal that, across the animal kingdom, oxygen uptake rises by a factor of 10 from rest to maximal exercise. This increase, the ratio of maximal to resting $\dot{V}O_2$, is termed the *metabolic scope,* and the equations indicate that this measure should be independent of differences in body mass.

Brett, however, demonstrated some interesting variations related to body size and metabolic scope in salmon (6). At rest, $\dot{V}O_2$ among salmon related to body mass thus:

$$\dot{V}O_2\text{rest} = 49M^{0.78}$$

As we have seen, this is similar to the relationship observed between resting metabolic rate and body mass throughout the adult animal kingdom. But when the salmon were swimming at a maximal sustained speed, the equation changed to

$$\dot{V}O_2\text{max} = 725M^{0.97}$$

indicating that maximal oxygen uptake increased in larger fish in direct proportion to body mass. These two equations also indicate that larger fish have a greater metabolic scope than smaller ones. The scope was calculated to be about 16 for the large fish and 4 for the small.

Interestingly, the same phenomenon is observed in children as they become larger with growth. Robinson found, for example, that metabolic scope increased in boys from 7 at age 6 years to 13 at age 17 years (23). Working backward, this indicates that the ontogenetic scaling exponents for $\dot{V}O_2$ during maximal exercise in children would be expected to be greater than those at rest (just as in the adult salmon).

While this makes for interesting descriptive reading, the biologic meaning remains uncertain. First of all, it is not clear if resting metabolic rate is an appropriate "currency" or multiplier by which to compare aerobic responses to exercise (see next section). It is intriguing to note that the magnitude of improvement in endurance performance in children with increasing age is very close to that observed in changes in metabolic scope. For instance, compare the 1.8-fold increase in metabolic scope that Robinson (23) reported in boys between the ages of 6 and 17 with the difference in their average 1-mi run times: 12:29 in boys 6 years old and 7:25 at age 17, a factor of 1.6 (2).

Capillary Flow Is Slower in Small Animals

When cardiac output is assessed as an indicator of heart function, values are expressed as a volume over time (i.e., liters per minute). The phylogenetic equation in animals for cardiac output at rest is

$$Q = 187M^{0.81}$$

From this equation we can derive some interesting information. For instance, is the time it takes for blood to circulate around the body related to body size? Circulation time (one round-trip) can be calculated by dividing blood volume (BV) by cardiac output (Q). Blood volume is allometrically relative to $M^{1.0}$, so circulation time (t) is

$$t = BV/Q \sim M^{1.0}/M^{0.81} = M^{0.19}$$

Once again we witness the marked similarity of mass scaling exponents for time-related variables, an expression of physiologic time. In this case, circulation time becomes greater as body size increases, and that increase in circulation time is similar to the decline in mass-specific metabolic rate ($\sim M^{-0.25}$). Schmidt-Nielsen calculated that circulation times were 4 s for a shrew, 50 s for a man, and over 2 min for an elephant (27).

What about the *velocity* of blood in the circulation (at rest) relative to body dimensions? Velocity in the aorta (v_{ao}) is equal to the volume of flow (Q) divided by its cross-sectional area, which relates to $M^{0.72}$. Therefore

$$v_{ao} \sim M^{0.81}/M^{0.72} = M^{0.09}$$

indicating that velocity of blood in the aorta is similar in all animals, large or small.

At any given distance in the circulatory system from the heart, the total flow rate (in liters per minute) in parallel vessels is identical. It would be desirable for the flow rate in the capillaries of highly metabolically active tissues to slow, giving more time for oxygen diffusion into the cells and carbon dioxide efflux out. Does this happen? And, if so, how can this be accomplished if the volume flow rate must remain constant?

Calder explained it this way (8): Volume rate of flow has the dimensions of length cubed per unit of time, while speed of flow is expressed as length per unit of time. The difference is length squared, which is the means of expressing the total cross-sectional area. At any given point in the circulatory system, the total cross-sectional area of the blood vessels is the product of individual vessel size cross-sectional area, the density of the vessels, and the cross-sectional size of the tissue. Capillary size scales to $M^{-0.11}$, capillary density to $M^{0.00}$, and limb cross-sectional area to $M^{0.72}$. This means that the cross-sectional area of the capillaries is

$$A_{cap} = (M^{-0.11})(M^{0.00})(M^{0.72}) = M^{0.61}$$

And the speed of blood in the capillaries is

$$v_{cap} = Q/A_{cap} \sim M^{0.81}/M^{0.61} = M^{0.20}$$

All this means that blood flow through the capillaries in smaller animals (which have a relatively greater metabolic rate) is slower than in large animals. And this occurs because the number of parallel capillaries (capillary density) is greater in small animals. The longer time for erythrocyte transit over a given distance in small animals provides greater opportunity for diffusional equilibrium of oxygen and carbon dioxide to be achieved with the cell. The greater capillary density in smaller animals also provides greater proximity between capillary and cell. These features are all consistent with the greater mass-specific metabolic rate in smaller animals.

These issues illustrate the substantial role that body size plays in defining physiologic variables. In principle, the same differences in physiologic variables related to size observed between adult animals should be expected to apply to size-related differences as children grow and between children and adults. What is uncertain, however, is the extent to which the relatively small size differences in growing humans (compared with the large differences in phylogenetic studies of animals) make the influences of size biologically important.

Most allometrically derived data in animals involves comparisons of anatomic or physiologic function in resting animals. Information as to how these findings are influenced by physical activity is limited, but, as the preceding examples show, that influence may be important to exercise fitness.

Another important observation is that ontogenetic changes in physiologic variables relative to size in humans might be expected to differ from the phylogenetic allometric relationships. Size changes in growing children can be presumed to affect physiologic variables through *both* ontogenetically and phylogenetically related mechanisms. As ambiguous as these issues seem, however, allometric analysis provides an opportunity for understanding basic mechanics that drive biologic factors that are relevant to children's exercise physiology.

Adjusting Physiologic Variables for Body Size

It is intrinsic to the nature of pediatric exercise physiologists to wish to perform *comparisons*. Sometimes it is important to compare a child's physiologic features with an expected norm. For instance, if a young patient with cystic fibrosis is found to have a $\dot{V}O_2$max of 1.78 L · min^{-1}, how does that compare to what it should be? This might be valuable for understanding the severity and prognosis of the child's disease or for recognizing response to medications. Or suppose you wanted to assess changes in muscle strength after a three-month resistance training program in a group of fifth-grade girls. Strength is found through testing to have improved by 6% after the training. But was that due to the training or only to the normal growth of mass and muscle strength over that period of time?

Comparisons can also be made between groups. Suppose a cohort of eight-year-old boys in San Francisco are tested for 50-yard sprint time while another age-matched group undergoes testing in Boston. How can we determine which group has the greater speed fitness? Speed fitness depends on a number of factors—leg length, muscle size, glycolytic capacity—that normally improve with age and body size, so we need some means of "leveling" the somatic maturation of the two groups.

The standard means of accomplishing these types of comparisons is to relate the physiologic variable in question to some measure of body size, usually body weight, height, or surface area. This makes good sense, since most determinants of physiologic fitness (ventilation, cardiac output, 1RM muscle strength) involve volumes (lung capacity, heart size, skeletal muscle mass), which are associated with total body dimensions. In some cases, too, body size represents the load on which physical work is being performed (e.g., running economy is expressed as the metabolic expenditure necessary to move body mass). In the end, most physiologic variables do, in fact, relate to body dimensions.

A physiologic variable is usually expressed relative to a marker of body size so that higher size-adjusted values imply greater capacity or fitness. More precisely, a higher size-adjusted value indicates that the specific tissues responsible for that physiologic variable are larger or functionally greater than those of another person of the same body size. If we were to take 20 boys who all weighed 40 kg, we would find variation in 1RM test scores of arm muscle strength, and we might conclude that some had superior strength fitness to that of others. This means that the cross-sectional area, pattern of motor unit activation, and twitch contractile properties of the biceps brachii muscle (i.e., the specific determinants of arm muscle strength) are greater relative to body mass in some children than in others. If we chose to change the denominator to lean body mass, the variation in arm strength fitness among the children would be reduced, and if 1RM scores were expressed relative to arm muscle cross-sectional area by magnetic resonance imaging, the variability would become much smaller.

A variety of body dimensions have been used as the denominator by which physiologic variables are adjusted for body size. Each measure has it own strengths and weaknesses in this role, and in some situations it is difficult to discern whether the use of one anthropometric measure for adjusting for body size over another has any true biologic advantage. Consequently, one single best way to adjust for body dimensions has yet to be identified. As Welsman and Armstrong concluded,

> It should be emphasized that there is no universally "correct" method of scaling, neither is any method necessarily "incorrect" in all instances. All of the mechanisms are constrained by underlying statistical assumptions that, if ignored, may confuse interpretations based upon them. Ultimately the choice of scaling technique depends upon the nature of the research question being addressed (31, p. 113).

Albrecht et al. described three criteria for determining if a dimensional factor X^b is an effective means of adjusting a physiologic variable Y for body dimensions, that is, whether it effectively eliminates the influence of body size on the variable in the population under study (1). First, the product-moment correlation coefficient between Y/X^b (the adjusted value for Y) and X is 0, or nearly so. Second, when graphed, the relationship between Y/X^b and X is a horizontal line (i.e., the slope of the linear regression line is 0). And third, the algebraic value of Y/X^b is equal to a constant.

The following sections briefly address the different approaches to adjusting physiologic variables to size, considering these criteria. Only general concepts are presented here. The chapters that follow address the use of methods for scaling to body size for specific anatomic and physiologic variables (cardiac, lung, muscle, etc.). Nevill (19) and Welsman and Armstrong (31) have contributed greatly to our understanding of these issues, and the reader is referred to their publications for more in-depth discussion.

Body Mass

Body mass has served as the traditional means of adjusting physiologic variables for body size: We measure $\dot{V}O_2$max per kilogram, peak anaerobic power per kilogram, hand grip strength per kilogram, and so on. In this regard, these variables are expressed relative to body mass as $M^{1.0}$, which is termed the *ratio standard.*

Body mass is readily measured, and at first glance its use to standardize physiologic measures is conceptually appealing. Body mass represents the volume of metabolically active tissue and should reflect differences in variables such as muscle volume, heart size, stride length, and body heat production. Also, as noted previously, energy requirements for performing weight-bearing physical work should vary according to the need to move the load created by body mass.

As is apparent from discussions earlier in this chapter, however, these concepts are not altogether accurate. Indeed, as a simple ratio standard, body mass often behaves poorly by failing to eliminate the effect of body size on the physiologic variable in question (1). The story of normalization of $\dot{V}O_2$max for body size is a good illustrative example. $\dot{V}O_2$max, while obviously affected by body size, does not increase in proportion to body mass as animals, including children, become larger. Consequently, $\dot{V}O_2$max varies with body mass in mature animals by approximately $M^{0.81}$. (Exercise studies in children have demonstrated a wide variety of mass scaling exponents for $\dot{V}O_2$max, but, on the average, the value is similar.)

This means that in any population of children, $\dot{V}O_2$max per kilogram of body mass is greater in smaller children than in larger ones. If we plot $\dot{V}O_2$ per kilogram against body mass, a negative relationship is observed (figure 1.4a), indicating that body mass as a denominator does not eliminate the effect of body size on $\dot{V}O_2$max. That makes it unfit to serve as a means of adjusting body size for inter- or intraindividual comparisons of $\dot{V}O_2$max. What is required instead is some marker of body size that causes the relationship of adjusted values for $\dot{V}O_2$max and body mass to be a flat line (figure 1.4b).

The amount of error engendered by use of body mass as the means of adjusting $\dot{V}O_2$max or any other physiologic variable *Y* for body size depends on the magnitude of the mass scaling exponent *b* in the true allometric relationship between *Y* and body mass. If the variable *Y* is expressed in the population as $M^{0.40}$, the use of mass to size-adjust *Y* results in a greater error than if *Y* relates to $M^{0.85}$.

The other factor that influences the degree of error is the size range of the subjects involved in the investigation. Figure 1.5 demonstrates the relationship between peak $\dot{V}O_2$ (in milliliters per kilogram per minute) and body mass in 12-year-old boys and girls (31). As expected, a negative relationship is observed: Smaller children have higher values than larger ones. The slope is sufficiently steep that differences are observed even within a relatively narrow range of body sizes, say, between 40 and 50 kg. It is obvious, too, that the error in using body mass as a denominator for $\dot{V}O_2$max in these children will be exaggerated if the study in question involves children whose body mass ranges from 30 to 60 kg.

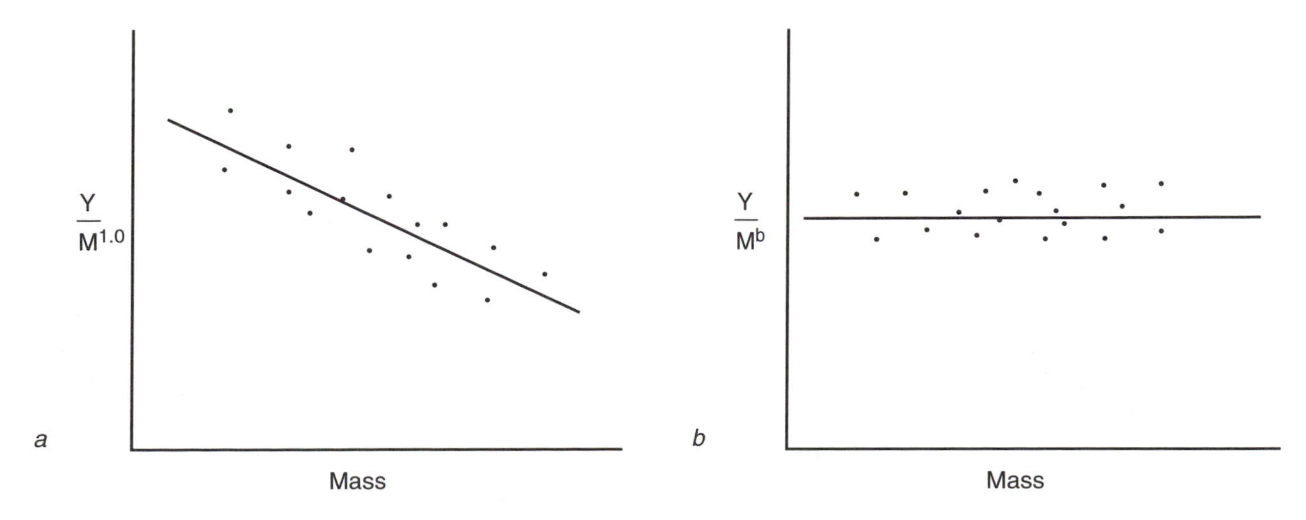

FIGURE 1.4 *(a)* Inappropriate ratio standard $M^{1.0}$ does not eliminate the effect of body mass on variable *Y* in subjects with different body masses. *(b)* When *Y* is adjusted by the proper mass scaling exponent *b*, no effect is observed on Y/M^b with increasing mass.

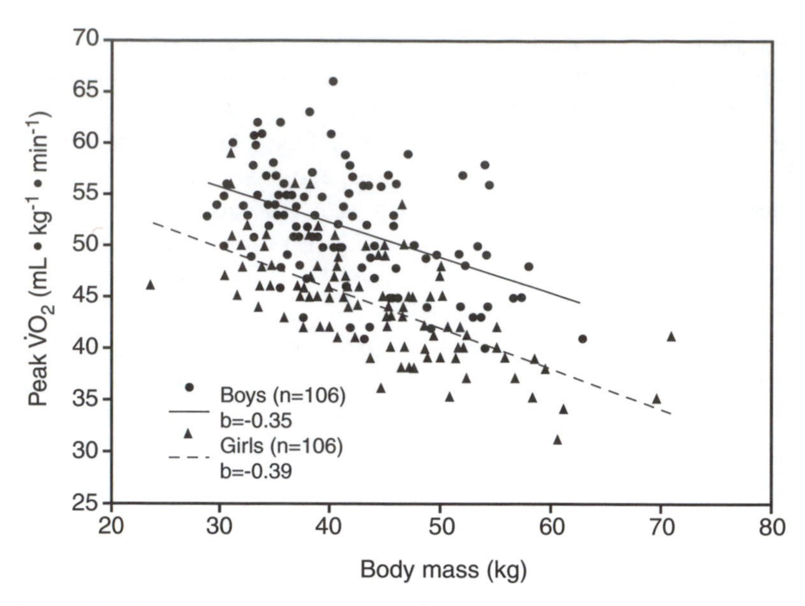

FIGURE 1.5 Comparison of peak $\dot{V}O_2$ to body mass in boys and girls (from reference 31).

Reprinted by permission from J.R. Welsman and N. Armstrong 2000.

There are some situations when the use of body mass as a ratio standard (i.e., expressed to the power 1.0) may be appropriate. When Nevill et al. (21) related 5-km run performance to both $\dot{V}O_2max$ and body mass, they found that

$$\text{Speed (in m} \cdot \text{s}^{-1}) = 84(\dot{V}O_2max)^{1.01}(M^{-1.03})$$

This indicates that the best predictor of speed on a 5-km run is $\dot{V}O_2max$ expressed relative to body mass in kilograms, the ratio standard. From this observation, the authors concluded that the use of $M^{1.00}$ may be an appropriate adjustment factor for weight-bearing athletic performance, but not for other physiologic comparisons, such as the relationship of $\dot{V}O_2max$ with risk factors for coronary artery disease.

In some situations, the use of body mass as a denominator for a particular physiologic variable is controversial. Running economy during submaximal treadmill testing is usually defined by $\dot{V}O_2$ per kilogram of body mass. It has been argued that the ratio standard ($M^{1.0}$) is appropriate in this situation, since body mass (in its entirety) is the load to be propelled, and the energy expenditure to move that load is being measured.

Others have noted that during submaximal treadmill running (in children as well as adults), the empirically observed relationship between $\dot{V}O_2$ and body mass is proportional to $M^{0.75}$. That exponent is, in fact, not greatly different from that observed at rest and at maximal exercise. The implication is that it should be interpreted the same way; that is, errors are introduced by using $M^{1.0}$ as the denominator when comparing running economy in children of different sizes. In this regard the study by Brett of salmon, whose swimming is a non-weight-bearing activity, is particularly interesting (6). $\dot{V}O_2$ at rest was found to be related to total fish mass by the scaling exponent 0.78. At 25% maximal speed the exponent rose to 0.85, at 50% it rose to 0.89, and at top speed ($\dot{V}O_2max$) it rose to 0.97 (the ratio standard). This suggests that mass—in conjunction with the force of gravity—should not be interpreted simply as a fixed load in studies of economy.

Interestingly, too, this study provides a clue that children running on a treadmill would show the same phenomenon of changes in scaling exponents with increasing speed as did the salmon. Since metabolic scope (the ratio of $\dot{V}O_2max$ to $\dot{V}O_2rest$) increases with age and size, the mass scaling exponent *b* for $\dot{V}O_2$ at rest must be different from that at maximal exercise (as seen in the fish). Although this has not been tested, the scaling exponent would be expected to increase with greater treadmill speeds. The interpretation of such a finding, however, would await clarification.

Other Ratio Standards

Any other anthropometric measure with a scaling exponent of 1.0 (i.e., a ratio standard) could also serve as a means of size adjustment for physiologic variables. Difficulties similar to those seen with use of body mass, however, might be anticipated.

Body Height

Body height is another simply measured ratio standard (i.e., $H^{1.0}$), which, in contrast to body mass, is not expected to be influenced by body composition. That is, an obese and a lean child might be equally tall but differ by a factor of 2 in body mass. The advantages of the use of body height as a size-adjusting factor, rather than mass, may depend on the research question being addressed.

Body height as a ratio standard shares the same weaknesses as body mass. Let's assume that, in ignorance, we expect $\dot{V}O_2max$ to be related to $M^{1.0}$. But according to dimensionality theory, it should be related to $M^{0.67}$ (volume per unit time) and in reality is more like $M^{0.81}$. Height is related to mass by $M^{0.33}$. So the respective differences in height exponents relative to $\dot{V}O_2max$ are expected to be 3.0 for the ratio standard, 2.0 according to dimensionality theory, and 2.4 in the real world. That is, the same error is incurred by using height as the ratio standard for $\dot{V}O_2max$ as by using weight: Smaller children will have higher values of size-adjusted maximal aerobic power than larger youngsters.

Body Surface Area

The use of body surface area (BSA) initially gained acceptance as a means of adjusting metabolic rate for body size. The rationale was straightforward: Animals of widely different sizes have nearly the same core body temperature. Heat is lost through the body's surface. Thus, the metabolic rate, which controls the production of heat, should be related to BSA, or $M^{0.67}$ (the surface rule).

We saw earlier that the actual phylogenetic scaling exponent for resting metabolic rate is $M^{0.75}$. This discrepancy notwithstanding, BSA has been commonly used in clinical medicine to determine drug dosages, calculate body fluid replacement volumes, and measure body functions such as urinary creatinine clearance.

BSA is a ratio standard, but it has been proven valid and useful in a number of physiologic contexts, including size adjustment for cardiac output and stroke volume. This is probably because $BSA^{1.0} \sim M^{0.67}$, which approximates empiric scaling exponents calculated for certain physiologic measures in exercise studies of children. Whether this relationship is coincidental or has some biologic meaning is uncertain.

It is important to recognize that the ratio of BSA to body mass decreases as children grow (figure 1.6). The magnitude of this decline is appreciable. The BSA-to-mass ratio is 4.0 at age 5 years and 2.7 at age 16 years. This bears particular importance for issues of thermoregulation in children, which are addressed in chapter 12.

Body surface area represents an interesting problem as a size denominator because it is not easily measured. In fact, the history of BSA as means of standardizing physiologic variables is chronicled by the multiple efforts toward its determination. These have included measurements of a polyethylene film placed over a frozen animal carcass (don't try this at home), plucked fresh bird skins, lacquered skin cut in sections, and measurement using an integrated roller (8). (In his book *Bioenergetics and Growth,* Brody (7) provides a photograph of such a device being rolled along lines painted on the flank of a clearly indifferent cow with a moderately distended udder.)

In humans, BSA is usually estimated by the formula of DuBois and DuBois, which was created in 1916 based on body molds of inflexible paper from nine subjects (11). In most subjects, areas of only one leg and arm were estimated, and nose, ears, and gluteal folds were not included.

Lean Body Mass

For certain physiologic measures, values expressed in respect to lean body mass (LBM) are particularly appropriate. LBM, which encompasses all body tissues except fat, includes the fitness-related organs (heart, lung) as well as the mass of skeletal muscle. LBM, then, should be more closely related than body mass to measures such as $\dot{V}O_2max$, cardiac output, and muscle strength.

In an early study, Miller and Blyth assessed different means of adjusting basal metabolic rate (BMR) for body size in 48 college students (18). The correlation between BMR and lean body mass (determined by underwater weighing) was $r = .92$, greater than that for BSA ($r = .84$) and body mass ($r = .85$). When body weight and BSA were adjusted for the effect of LBM, the correlations of BMR and BSA with body mass were $r = .45$ and $r = .14$, respectively.

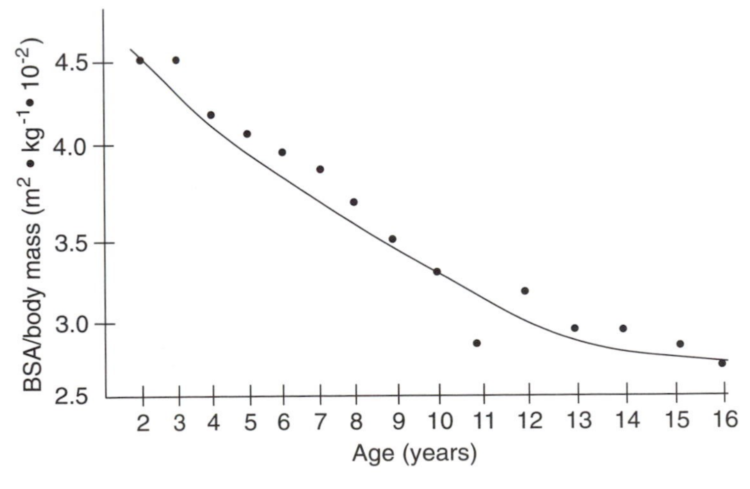

FIGURE 1.6 Changing ratio of body surface area to mass in growing children.

Reprinted by permission from T.W. Rowland 1996.

The authors concluded that it was "apparent that surface area and body weight may derive much of their validity as metabolic reference standards from their high degree of correlation with lean body mass . . . [which serves as] a constant fraction of the 'active body mass'" (p. 314).

Similar findings have been demonstrated for exercise measures. Janz et al., for example, reported that LBM was superior to body mass in eliminating the effect of body size on values of submaximal and peak $\dot{V}O_2$ in a five-year longitudinal study of children (16). Daniels et al. showed that LBM accounted for 75% of the variance in left ventricular mass in 201 children 6 to 17 years old (10).

That LBM cannot be readily measured is a drawback, and, as a ratio standard ($LBM^{1.0}$), it carries the same risks of spurious results as does body mass. In addition, this denominator might not be considered applicable when total body mass is the load to which the physiologic variable is applied (e.g., $\dot{V}O_2$ during submaximal running). Use of LBM in this case would be, in effect, a "mathematical adiposectomy," eliminating the influence of body fat on weight load (13).

Allometry

It is apparent from figure 1.5 that mass as a ratio standard ($M^{1.0}$) is not going to work as a means of normalizing Y for body size, because $Y/M^{1.0}$ changes (i.e., becomes smaller) as body mass increases. This results in spuriously high values for size-adjusted Y in smaller subjects. What is needed instead is a denominator M^b to which Y can be related that results in no difference in Y/M^b as body mass changes in the study population (as in figure 1.4b).

Allometric analysis provides an exponent b of body mass that causes mass and Y to rise proportionally with increasing body size. Consequently, values of Y expressed relative to M^b remain stable across the size range of the subjects. Adjusted Y, or Y/M^b, then serves as an accurate and appropriate means of comparing values of Y among individuals of different body sizes. Using Y/M^b "levels the playing field" for comparisons of the physiologic variable in question because this relationship is independent of body size.

The mass exponent for any group of subjects is determined by creating the allometric equation for values of Y and M in the same way as described at the beginning of this chapter. A linear regression equation is created from individual log-transformed values of M and Y. In this form $\log Y = \log a + b \log M$, where b is the slope and turns out to be the scaling exponent for M in the allometric equation $Y = aM^b$.

At this point, comparisons of Y between groups can be performed in two possible ways. One can apply analysis of covariance (ANCOVA) to the log-linear regression equations for the two groups of subjects. Welsman and Armstrong (31) used their data for $\dot{V}O_2max$ in boys and girls to demonstrate this approach, as illustrated in figure 1.5. The slope of the two sex-related regression equations is not significantly different. ANCOVA revealed that peak values of $\dot{V}O_2$ in relation to body mass were greater in the boys.

Alternatively, individual values of Y/M^b can be used in the same manner as the ratio standard is used to perform statistical comparisons between groups of subjects. In the previous example, for instance, the common slope b for the boys and girls was 0.66. Now all the absolute values of peak $\dot{V}O_2$ in both groups can be related to $M^{0.66}$ and compared by t test. When this is done, average values of peak $\dot{V}O_2/M^{0.66}$ per minute for the boys and girls are 182 ± 17 and 159 ± 14, respectively ($p < .05$).

When should the allometric approach be used instead of the ratio standard? From the preceding considerations, the best answer seems to be that allometry is necessary to avoid errors created by the ratio standard when the 95th percentile limits for the slope of the log-linear regression equation for any set of data do not include the value 1.0.

Some researchers have chosen to use theoretical exponents for scaling physiologic variables in studies of children, such as expressing values of $\dot{V}O_2max$ to $M^{0.75}$. However, empirically derived mass exponents for $\dot{V}O_2max$ in various studies of children vary widely (from 0.37 to 1.07). Use of "off-the-shelf" mass exponents such as 0.75 for $\dot{V}O_2max$ in any given data set might therefore create significant error. Although the empiric exponent calculated for a particular data set might not reflect "true" biologic relationships between mass and $\dot{V}O_2max$ (because of the small number of subjects, variations in body composition and athleticism, etc.), it does reflect the relationship between mass and maximal aerobic power in that population of subjects. For this reason, calculation and use of empirically derived scaling exponents rather than those derived by theoretical expectations seems prudent.

This allometric approach is equally useful for other anthropometric measures, such as height, body surface area, or lean body mass. Shephard et al. (29), for example, used the allometric equation $\dot{V}O_2 = a(H^b)$ to assess longitudinal changes in $\dot{V}O_2$max in children ages 6 to 12 years. Rowland et al. (25) reported allometric scaling factors for maximal cardiac output in relation to different size variables in 24 girls (mean age 12.2 ± 0.5 years). The scaling exponents for Q_{max} were $M^{0.59}$, $H^{1.76}$, and $BSA^{1.08}$. When Q was expressed as the ratio standard ($Q/M^{1.0}$, $Q/H^{1.0}$, and $Q/BSA^{1.0}$) the Pearson correlation coefficients with mass, height, and BSA were r = .54, .22, and .07, respectively. When values of Q were adjusted for the appropriate scaling factor ($Q/M^{0.59}$, $Q/H^{1.76}$, and $Q/BSA^{1.08}$), the correlation coefficients were r = –.01, –.01, and .01, respectively. These findings indicated that any of the allometrically adjusted denominators was useful in normalizing maximal cardiac output to body size, while the only ratio standard that was appropriate for this purpose was $BSA^{1.0}$.

Discrepancies in results obtained by different methods of size "normalization" illustrate the potential for error created by misuse of scaling factors. In some cases this involves our understanding of basic concepts of normal development in exercise physiology. For example, when $\dot{V}O_2$max is expressed relative to body mass in boys, values remain stable throughout childhood and early adolescence (see chapter 5). This implies that the mechanisms controlling maximal aerobic power during the prepubertal years are size independent. However, when these data are adjusted allometrically for size, $\dot{V}O_2$max in boys continuously rises with age. This suggests that factors dependent on size contribute to the development of physiologic aerobic fitness.

Regression Standards

Comparisons between two groups of subjects can also be examined by creating least-squares regression equations relating the physiologic variable Y to body size X for each set of subjects:

$$Y = a + bX$$

The slope b and intercept a terms can then be compared using ANCOVA (with mass as the covariate). This approach requires that the relationship between Y and X be linear and that the slopes of the regression equation be parallel. Differences in the intercept and computed adjusted means indicate differences between groups.

Error is introduced by this approach if the data are heteroscedastic, that is, if the degree of variability of values about the mean changes through the range of body size. As Welsman and Armstrong noted, allometric scaling incorporates multiplicative instead of additive error as seen with regression standards and therefore controls for heteroscedasticity (31).

Studies in children have demonstrated how relationships revealed by use of the ratio standard can be altered by the linear regression approach. Eston et al. compared treadmill running economy in 10 prepubertal male cross country runners (mean age 10.4 ± 0.5 years) and an equal number of trained young adult males (12). Values of $\dot{V}O_2$ per kilogram were significantly greater in the boys than the men, indicating inferior running economy. ANCOVA revealed no differences in the slope or elevation of the regression lines of mass versus submaximal $\dot{V}O_2$. The authors concluded that "there was no real difference in the dynamics of oxygen uptake for men and boys" and that "any apparent difference in oxygen uptake between groups may be attributable to inappropriate analysis rather than inherent physiological differences" (p. 239).

Cooper and Berman argued, however, that "both ratios and regression methods are valid but designed to answer different questions" (9). They considered ratio standards useful in understanding the relationships between size and function in a given individual, while regression analysis reveals mechanisms that account for differences between two study groups.

Multilevel Modeling

Multilevel modeling is a statistical approach that is particularly useful in analyzing longitudinal data sets (31). This method has therefore been useful in assessing the influence of various factors on the development over time of aerobic and anaerobic fitness as well as muscle strength in children.

In this technique, data are reviewed in a hierarchical structure, usually at two levels. For instance, the set of data from individual measurement sessions might be at level 1, and values for individual subjects over time in level 2. The multilevel modeling method permits a description of changes in a variable in the mean population, while at the same time indicating how individual responses deviate from the mean at both levels of the analysis. For example, for height data the terms in level 1 can describe how individual growth rates vary about that of the population mean, while at level 2 the measurements of each individual can be expressed relative to that subject's own curve of development. This approach is particularly effective for longitudinal assessment of exercise variables, for it allows an examination of the influences of extrinsic variables (sex, sexual maturation, age) on the development of physiologic factors as children grow.

Cross-sectional data sets can be analyzed by linear regression or log-linear regression (allometry analysis). The same holds true for multilevel modeling. The linear regression equation can be extended with additive error terms (e.g., age, height), or a log-linear allometric structure can be used. Nevill et al. (20) provided evidence that the latter approach yielded a statistical solution in longitudinal measures of strength and aerobic power in children that "was not only physiologically plausible but a much better statistical fit, required fewer fitted parameters than the original solution, and appropriately controlled for the heteroscedastic data" (20, p. 122).

Physiologic Scope

In adults, cardiovascular fitness is often estimated in the clinical setting by endurance time on standard exercise testing protocols. These times are equated to a certain number of METs, or the ratio between the $\dot{V}O_2$ at rest and that at the point of exhaustion. One MET is equivalent to an oxygen uptake of 3.5 $ml \cdot kg^{-1} \cdot min^{-1}$. Using this as the fitness "currency," then, an endurance time that reflects a $\dot{V}O_2$ of 35 $ml \cdot kg^{-1} \cdot min^{-1}$ would be interpreted as 10 METs. This can be considered a marker of the subject's level of aerobic fitness and can be used as a guideline for prescription of physical activities (3).

The ratio of $\dot{V}O_2max$ to resting $\dot{V}O_2$, or the *metabolic scope,* as a possible index of aerobic fitness has been examined in children as well. The concept is immediately attractive since it eliminates the influence of body size and all the complexity and variables associated with allometric scaling or multilevel modeling. Moreover, as indicated earlier, metabolic scope increases dramatically in children as they grow, doubling in boys between the ages of 6 and 17 years.

The true meaning, interpretation, and value of metabolic scaling, however, remains clouded. A primary issue is the validity of resting metabolic rate as a currency by which exercise values can be related. The determinants of resting metabolic rate are different from those during exercise, with a big shift in the contribution of contracting skeletal muscle. Resting metabolic rate increases with body size but declines with respect to body mass or BSA as children grow, and small differences in resting values would dramatically change metabolic scope. The studies of metabolic scope in children seem to indicate that nearly all the changes seen as children age are a result of the decline in relative basal metabolic rate (23).

There is sufficient ambiguity regarding the interpretation of changes in metabolic scope in children, then, to preclude its use at the present time as a size-independent marker of physiologic fitness. It is possible, however, that future research may shed light into its relevance for children's exercise physiology.

Conclusions

Broad experience with the allometric approach in studies of animals clearly confirms the biologic importance of body size. These data indicate, too, that the relationships of certain physiologic and anatomic variables with body dimensions may be unique to the variable in question. Such differences may shed light on the underlying mechanisms by which these variables affect physical fitness. Moreover, factors not previously considered to be size related, particularly those that operate in respect to time, may be influenced by differences in body dimensions.

The applicability of these principles to defining physiologic differences in children in response to exercise as they grow remains to be determined. Such insights may be limited by the relatively small range in the size of growing children, and ontogenetic relationships may or may not reflect those that have been established by phylogenetic studies. Still, the basic biologic concepts that define how size influences physiology must hold for children as well as animals.

Developmental exercise physiologists have become aware of the critical importance of the proper scaling of physiologic variables in respect to body size in young subjects. Identifying the optimal means for doing so remains a project in process. Clearly there exists no single proper means of normalization for body size, and the investigator is faced with the necessity of recognizing the most effective means of size adjustment for each individual study.

Discussion Questions and Research Directions

1. What can differences between ontogenetic and phylogenetic scaling exponents tell us about developmental changes in exercise physiology (as opposed to the effect of size alone)?
2. How fast does physiologic time "pass" during the childhood years?

   Is it linked to body mass in the same way as observed in adult animals (i.e., ~ $M^{0.25}$)? If so, what is the biologic mechanism that relates time and mass?
3. Does the concept of developmental symmorphosis hold true for physiologic variables in growing children? If so, how does body size "orchestrate" parallel developmental patterns in different systems involved in exercise?
4. What aspects of a particular research question dictate the appropriate means of normalizing physiologic variables for body size? What factors (sex, body composition, athleticism) alter the relationship of physiologic variables with body size in a given subject population?
5. What is the biologic basis for mass exponents? That is, *why* does $\dot{V}O_2max$ relate to $M^{0.81}$ instead of the ratio standard?

2

Growth and Exercise

Although growth is a familiar process, we have no idea why tissues grow or stop growing, how the growth of one tissue or organ is related to the others, or how the growth of a tissue is related to growth of its constituent cells. . . . So, if I met the good fairy godmother of science, I should ask her "What controls the growth of tissues in vivo*?", and I know the answer would be fascinating.*

—Jonathon Slack (1996)

In this chapter we discuss

- the role of somatic growth in the development of physical fitness and
- how physical activity and playing sports may influence body growth.

An understanding of the process of growth is essential for appreciating developmental changes in exercise physiology. Determinants of exercise physiology and performance are closely coupled with somatic growth, and, as we shall see later in this chapter, physical activity itself may in turn influence the growth process. Factors that influence this exercise–growth link, then, are of particular interest.

The "mysterious mechanisms of growth" (79) are far from understood. However, enormous progress was achieved during the 1990s in recognizing the multitude of factors responsible for human growth as well as the genetic basis for controlling the phenotypic expression of these agents. This newfound information has made it abundantly clear that constructs of how people grow that were accepted as recently as 20 years ago were grossly oversimplified. Indeed, it seems that each new insight, rather than clarifying the picture, indicates quite dramatically the profound complexity of the growth process.

The period of postnatal growth, from birth to roughly age 17 or 18 years, is the length of time available for biologic processes to mature to the adult state. Given this time constraint, we are interested in not only the extent of somatic development during the growing years but also its tempo, or rate of change. It has been long recognized that the relative time allotted for this biologic maturation in humans well exceeds that of other mammals (47), an observation that has served as the basis for philosophic and biologic discourse for the past 300 years.

In their comparison study of 21 species of anthropoid primates, Leigh and Park (47) found that the human growth period is longer relative to body size because of an extremely long early growth period (from birth to age at the beginning of the adolescent growth spurt). A number of explanations have been offered, the most popular being that the high intelligence of humans necessitates a longer period to acquire a complex information base.

The need for extended development of locomotor capacity in humans has also been suggested. The evolutionary shift to upright bipedal locomotion created challenges to both the neuromuscular and cardiovascular systems, the integrative solution to which may necessitate an extended period of development. As Leigh and Park noted, "temporal delays in ontogeny to enable 're-wiring' may have been key components of the response of hominid ancestors to selection favoring bipedalism" (47, p. 333).

Another temporal aspect of somatic growth that may influence long-term population changes in exercise capacity is the secular change that has been observed in growth maturation (35, 85). Patterns of growth in children in many populations over the past 200 years generally indicate a progressive increase in height and weight as well as a shortening of the growing period. The former can be attributed mainly to increases in leg length, while earlier age at menarche is the most clear-cut indicator of the accelerated tempo of biologic maturation.

In Europe and North America the increase in height between 1880 and 1980 was about 1.5 cm per decade in children and 2.5 cm per decade in adolescents (51). Average age at menarche became earlier by 4.5 months per decade between 1920 and 1960 in Belgian girls. In most developed countries, these trends, which are considered to reflect improved health and nutritional status, have slowed dramatically in the past several decades.

It is clear, too, that despite the smoothness of growth curves, human growth fluctuates considerably over time. H.G. Wells stated this nicely in his 1925 novel *The Food of the Gods,* noting that growth "went on with bursts and intermissions . . . as if every living thing had first to accumulate a force to grow, grew with vigor only for time and then had to wait for a space before it could go on growing again" (89). Wales cited this reference in an intriguing review of the variables that influence this inconsistency of the growth process, including season of the year (growth can occur above, below, or at the expected rate in three-month measurements), number of daylight hours, and spinal compression from standing (86). Differences are seen as well in the growth rates of the individual body segments. The lower legs of children, for example, grow linearly but can exhibit unexplained differential surges in length compared with other body segments.

Influence of Growth Factors on Physical Fitness

Physical growth is the most important factor in the development of physical responses to exercise during the childhood years. Moreover, differences in the *rate* of growth are largely responsible for interindividual differences in physical performance in the pediatric age group. A 16-year-old boy has three times greater maximal oxygen uptake ($\dot{V}O_2max$) than he had back at age 5. Grip strength in girls increases three-fold over the same age span. A teenager runs more economically (i.e., with a lower $\dot{V}O_2$ per kilogram) at a given speed than when he was a small child.

These are all expressions principally of increases in body size. Between the ages of 6 and 16 years, the lungs in males grow from a total capacity of 1,937 ml to 5,685 ml, and the weight of the heart grows from 95 g to 258 g. These increases are manifest in the development of maximal minute ventilation and stroke volume as children grow. Muscle strength improves as a consequence of increased volume of muscle tissue. Estimated total body muscle mass in girls grows from 7 kg at age six years to 23 kg by adolescence. As leg length increases with age, stride frequency at a given speed declines, resulting in a lower overall oxygen requirement for running.

The GH/IGF-I Axis

The past three decades have witnessed a virtual explosion of research information regarding the determinants of physical growth during childhood and adolescence. These data have greatly expanded our thinking about the processes that surround normal growth. Factors once believed to act independently have been found to be influenced by others. Actions of certain growth determinants are seen to be regulated by multiple factors, including other growth-promoting agents. A model of factors that promote overall body growth has been replaced by the identification of determinants of specific growth processes. By whatever means, however, the combined outcome of this complexity is a genetically based program of growth in physical size that serves as the principal determinant of the development of children's physiologic response to exercise as they grow. These growth factors themselves may play important roles in the anatomic and physiologic responses to physical training.

The central basis for physical growth during childhood is the *growth hormone (GH)/insulin-like growth factor I (IGF-I)* axis (figure 2.1). The primary agent within this axis is growth hormone (GH), a single-chain, 191-amino acid protein that is produced by the anterior lobe (adenohypophysis) of the pituitary gland. Growth hormone is secreted in a pulsatile pattern in response to the interplay of two regulatory peptides, which are produced in the nearby hypothalamus. *Growth hormone–releasing hormone (GHRH)* serves to stimulate growth hormone production, while *somatostatin* acts to inhibit GH release.

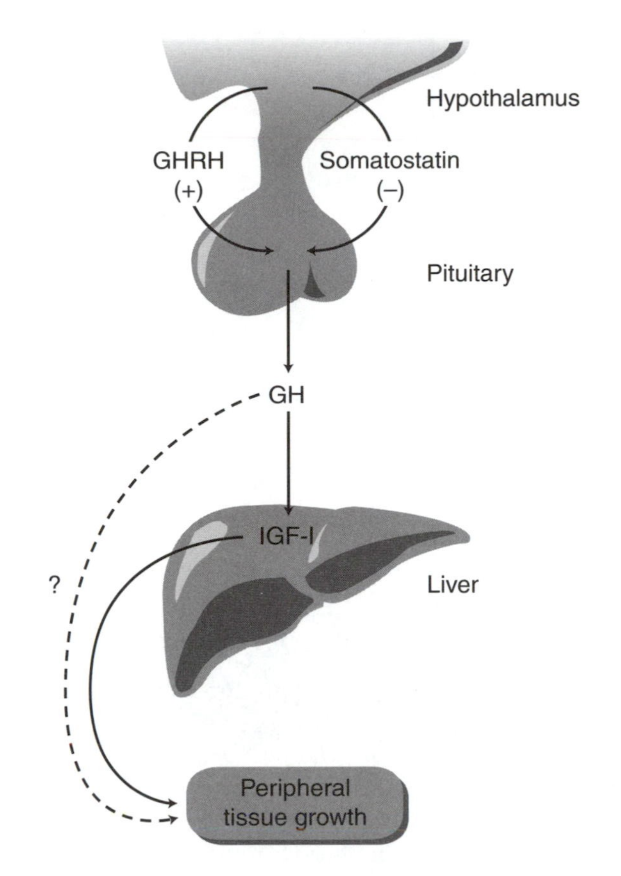

FIGURE 2.1 The GH/IGF-I axis.

Control of these two regulators of growth hormone production is poorly understood. It is obvious, however, that a host of factors can be involved, some biochemical (neurotransmitters, thyroid-releasing hormone, calcitonin, vasopressin, corticotrophin) and others physical or emotional (stress, sleep, fasting, exercise).

Growth hormone is produced by the pituitary in pulsatile bursts that occur about every two hours, most prominently at night during sleep (17). Peak serum GH concentrations typically reach approximately 2.0 ng $\cdot$ ml^{-1} but between secretory bursts may be almost undetectable (<0.1 ng $\cdot$ ml^{-1}). During waking hours, growth hormone pulses are lower and more irregular, particularly in prepubertal children.

Because of this marked variability, isolated blood samples have limited value for determination of GH

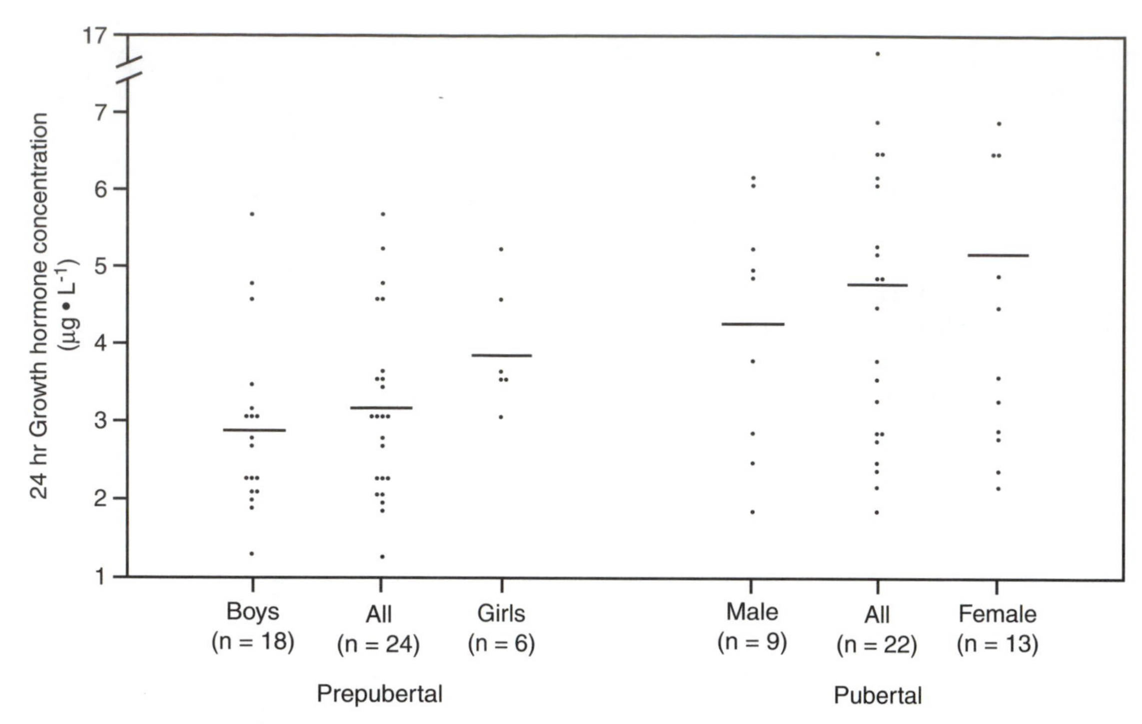

FIGURE 2.2 Average 24-hour growth hormone concentrations in 24 prepubertal and 22 pubertal youths (from reference 17).

concentration. Serial blood samples can provide an estimate of 24-hour GH production, but even those values are highly variable among groups of normal subjects. Consequently, 24-hour growth hormone levels have little value in identifying GH deficiency or in comparisons to determine whether levels are related to factors such as age, sexual maturation, and physical fitness. Compared to prepubertal children, however, postpubertal subjects demonstrate higher mean values for all indices of GH secretion (pulse amplitude, frequency of pulsation, 24-hour concentration; figure 2.2).

Given the difficulty of interpreting GH concentrations, the clinical assessment of GH "reserve," particularly in the diagnosis of GH deficiency, has relied on measuring GH response to a standardized pharmacologic or physiologic stimulus, or both (68). These have included physiologic stimulus such as fasting, sleep, and exercise as well as pharmacologic stimuli such as levodopa, clonidine, insulin, and glucagon. While certain "normal" GH responses to these provocative tests have been established, their value remains controversial. It has been emphasized that provocative tests for stimulation of GH do not mimic the usual physiologic state, and the accuracy and repeatability of these tests has been questioned. As with 24-hour GH levels, responses to provocative tests are generally greater in pubertal than prepubertal individuals.

The effects of GH are both (a) anabolic, including stimulation of epiphyseal and osteoblastic activity in bone (linear growth) and increased amino acid transport and nitrogen retention in muscle (development of lean tissue) and (b) metabolic, with lipolytic, insulin-resistant actions. It is not clear, however, whether these effects represent direct actions on GH via GH cell receptors or whether they are instead mediated by the actions of peptides called insulin-like growth factors (IGF-I and IGF-II; 13, 20). Of these, IGF-I (sometimes called somatomedin C) plays the most important role in the growing child.

Certain data support the hypothesis that GH released by the pituitary gland stimulates production of IGF-I in the liver and other tissues. IGF-I is then transported in the bloodstream both in the free form and bound to specific proteins (IGF-binding proteins, or IGFBP) to effect actions of increased protein synthesis and cell differentiation that lead to growth of bone, cartilage, and skeletal muscle.

Others believe that GH and IGF-I have independent actions on cellular growth and that IGF-I can be formed locally at the site of target tissues. By whichever mechanism, however, it appears that IGF-I is the major mediator of growth of skeletal and lean tissue in children (81). Serum concentrations of IGF-I rise during the course of childhood in both boys and girls (figure 2.3). Values then decline in the adult years. These levels of IGF-I do reflect GH status, as evidenced by the low concentrations seen in patients with GH deficiency that rise in response to GH treatment.

Information regarding the roles of GH and IGF-I in normal growth has been gained from humans with acquired or genetic abnormalities that affect their production. Patients with growth hormone deficiency or lack of GH receptors (Laron syndrome) demonstrate growth failure, with osseous delay and metabolic effects such as hypoglycemia. Rare patients with primary IGF-I deficiency demonstrate, in addition, mental retardation, microcephaly, and deafness (68).

On the other hand, children who have excessive secretion of growth hormone demonstrate accelerated linear growth, or gigantism. Giants can reach heights of over 250 cm (8 ft) and may also suffer from behavioral and visual problems (68). These cases are very rare and are usually caused by a pituitary adenoma. In older patients whose closed epiphyseal plates prevent extensive linear bone growth, excessive growth hormone production results in acromegaly, with enlargement of the skull and distal extremities, coarse facial features, cardiomyopathy, and visual and neurologic abnormalities.

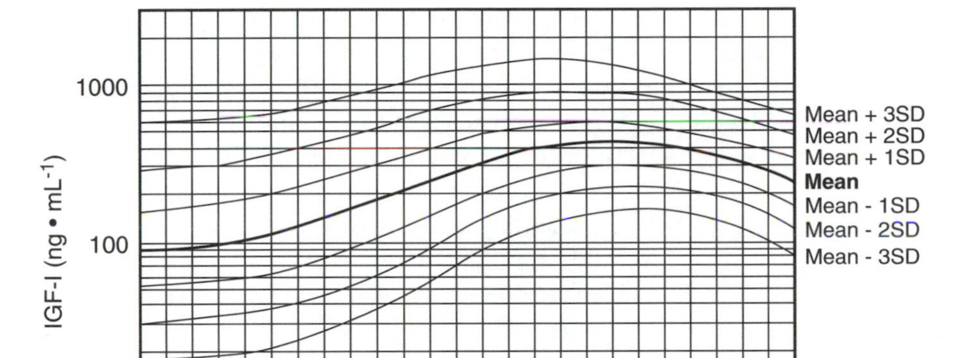

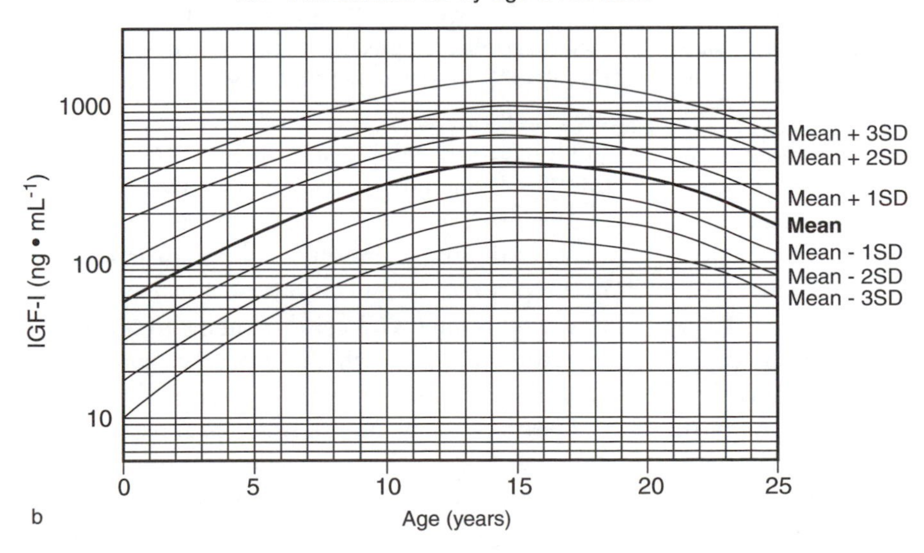

FIGURE 2.3 Serum IGF-I levels by age in boys and girls (from reference 68).
Reprinted by permission from Sperling 2002.

Factors Influencing GH and IGF-I

The multitude of factors that affect the actions of GH and IGF-I attest to the complexity of the growth process. As with most biologic processes, the final pathway of somatic growth reflects the interaction and complementary effects of both environmental and genetic influences.

Environmental Factors

Both acute physical activity and repeated bouts of exercise (leading to improved fitness) can affect components of the GH/IGF-I axis. At puberty, increases in GH and IGF-I production are linked to the onset of activity of the sex hormones at both hypothalamic and pituitary levels. These interactions are discussed in greater detail in the following chapter.

Appropriate nutrition is necessary for the anabolic actions of both GH and IGF-I. In states of overnutrition, such as obesity, GH secretion and release in response to provocative stimuli are depressed. This seems paradoxical, since obese youth are characterized by accelerated growth of not only body fat content but also height and lean body mass. At the same time, however, GH cell receptor activity increases and correlates well with body mass index. Obese children usually demonstrate normal or elevated IGF-I levels, but these are not associated with body mass index (BMI) or skinfold thicknesses (68).

Obesity is characterized by elevated serum insulin levels with associated insulin resistance. It has been suggested that increased IGF-I concentrations in response to this hyperinsulinemia may inhibit GH production by negative feedback on the pituitary gland. Attia et al. compared total and free IGF-I levels and IGFBP levels in obese versus lean adolescents and young adults (2). IGFBP in the obese individuals was markedly reduced and negatively related to basal insulin concentrations. Both circulating GH and total IGF-I levels were lower in the obese subjects, but free IGF-I was higher. The authors suggested that the higher ratio of free to total IGF-I in the obese subjects may have contributed to their anabolic characteristics.

Growth failure associated with undernutrition can reflect global caloric deprivation (marasmus) or inadequate protein intake (kwashiorkor). Serum GH concentrations have been reported to be normal in the former condition and increased in the latter. IGF-I levels are typically reduced in both states. With undernutrition there is a decline in the number of GH cell receptors, and target tissues become resistant to the anabolic actions of GH. This effect and the lowered IGF-I cause an overall inhibition of growth and protein synthesis (6). Rosenfeld and Cohen commented that the elevated GH concentration might indicate an adaptive response by which lipolysis spares protein degradation (68). The depressed IGF-I levels, on the other hand, "may represent a mechanism by which precious calories are shifted from use in growth to survival requirements of the organism" (p. 298).

Patients with anorexia nervosa typically demonstrate lower IGF-I but higher GH levels than normally nourished youth (32). These conditions can be associated with reductions in linear growth. It is not clear whether this is due to a primary pituitary/hypothalamic dysfunction or to cellular GH resistance with reduced negative feedback by IGF-I on GH production (33). GH and IGF-I values rapidly normalize with weight gain during treatment (60).

Genetic Influences

While environmental factors clearly modify the growth process, the timing and magnitude of somatic maturation during childhood and adolescence is largely under genetic control. Understanding the genetic basis of growth factor expression is of interest to pediatric exercise scientists because (a) the development of exercise physiology and fitness is largely related to changes in body size and (b) measures of physical fitness—aerobic, anaerobic, and strength—are governed to a significant extent by a person's genetic endowment.

The genetic influence on height is greater than that of weight (54). About 60% of height during later childhood and adolescence can be accounted for by heredity; for body weight the genetic contribution is approximately 40%. The tempo of biologic maturation is also heavily influenced by genetic factors, including timing of the adolescent growth spurt, age at peak height velocity, and progression of bone ossification. That is, early biologic maturers tend to beget children who mature early, and late maturers, children who mature late.

Recent research has revealed more details of the molecular genetic basis of actions of the GH/IGF-I axis. The GH gene contains 3,000 nucleotides and is located on chromosome 17. The active portion of the gene is preceded by a "promoter" segment of 300 nucleotides, which serve as a target for regulatory factors. This allows GHRH and other somatotrophic agents to recognize and act on the GH gene so that normal GH secretion can occur in the pituitary (63).

Human GH gene transcription is triggered by GHRH through the action of cyclic AMP. As a negative feedback mechanism, IGF-I acts as an inhibitory factor to this transcription, as do insulin and thyroid hormone (69). The end-organ and IGF-I-promoting effects of secreted GH are also regulated by the GH receptor population, which is under separate genetic control on chromosome 13.

Investigations of growth factor gene expression in growing children can be expected to provide insight into the determinants of the tempo and magnitude of somatic maturation. In turn, this information should help clarify the mechanisms behind developmental changes in exercise physiology and fitness. Later sections of this chapter discuss how physical activity and sport training might alter this gene expression and positively or negatively influence growth.

Other Growth Determinants

An increasing number of growth factors outside the GH/IGF-I axis that contribute to somatic growth during childhood are being identified. These determinants of growth demonstrate primary, independent anabolic actions but in many cases also serve to control—and are controlled by—components of the GH/IGF-I axis.

Insulin

Insulin secreted by the beta cells of the pancreas plays a prominent role in the regulation of glucose and lipid utilization but also has substantial anabolic effects. This hormone promotes amino acid uptake in muscle cells, probably contributes directly to protein synthesis, and limits protein breakdown. Besides these direct effects on growth, insulin contributes to somatic maturation by additional indirect mechanisms. Insulin acts in a synergistic fashion with growth hormone to increase protein synthesis in muscle. This effect may be explained by insulin's stimulation of the transcription of the growth hormone receptor gene (58).

In addition, IGF-I activity is facilitated by insulin (38). Insulin increases the production of receptors for IGF-I and in this way contributes to the effect of IGF-I on linear bone growth. Expression of IGF-I mRNA depends on insulin. This hormone also inhibits the production of IGFBP, which results in greater IGF-I activity.

A number of clinical situations illustrate the anabolic effects of insulin and its role in promoting growth. Subnormal growth can be seen in patients with insulin deficiency from type 1 diabetes mellitus, and growth retardation can be a sign of poor metabolic control of this disease. Newborn babies of insulin-dependent diabetic mothers demonstrate macrosomia, hypertrophic cardiomyopathy, and visceromegaly. These conditions are believed to reflect fetal intrinsic hyperinsulinemia in response to placental transfer of elevated glucose levels. Similar physical findings are observed in other neonatal disturbances that cause elevated serum insulin levels (nesidioblastosis, Beckwith-Wiedemann syndrome; 68).

Some patients who have undergone surgical removal of a tumor at the base of the brain (craniopharyngioma) have no GH secretion but still demonstrate normal IGF-I levels and appropriate somatic growth (30). In fact, many become obese and demonstrate elevated insulin levels. The obesity appears to result from deranged appetite control centers in the hypothalamus, and hyperinsulinemia occurs as a secondary response to obesity-induced insulin resistance. It has been hypothesized that both growth and IGF-I are maintained in these patients by their increased insulin levels.

A similar scenario is evident in young patients with exogenous obesity (38). Their condition is not one simply of excess body fat, since these individuals demonstrate greater lean body mass, taller stature, and accelerated biologic maturation compared with their lean peers. Increases in body fat content are associated with insulin resistance, and serum insulin levels rise as an adaptive response. The anabolic characteristics of obesity may at least in part represent the actions of elevated insulin levels, since insulin resistance to glucose uptake in this condition does not affect insulin actions on protein metabolism.

Thyroid Hormone

Depressed levels of thyroid hormone in patients with hypothyroidism are associated with short stature, growth retardation, and delayed bone maturation. On the other hand, excessive thyroid activity, as seen in patients with thyrotoxicosis, results in increased bone development and accelerated growth. As with insulin, these anabolic effects represent a combination of not only direct actions on growth tissues but also stimulation of GH secretion and promotion of action of IGF-I (88).

Other Growth-Stimulating Factors

An increasing number of factors that have growth-promoting actions on specific tissues have been identified (68). Vascular-epithelial growth factor (VEGF) controls angiogenesis and the development of vascular processes. Neurologic function is regulated by specific growth determinants such as nerve growth factor (NGF) and brain-derived neurotrophic factor (BDNF). Hepatocyte growth factor (HGF) targets the development of liver parenchymal cells.

These growth stimulants can be secreted by glands and transported in the blood to target tissues (endocrine activity), or they can be produced in adjacent cells (paracrine activity) or even within the target tissue itself (autocrine activity). IGF-I is included in the list of such agents, as it can be manufactured in tissues both locally and independently of GH, with resulting autocrine or paracrine action. It is likely that an increasing number of such tissue-specific growth determinants will be identified, including those that play an integral role in exercise responses.

Biologic Age

We have seen that a program of hormonal activity that stimulates somatic growth proceeds during the childhood and adolescent years. This evolution of body size increase is one example of biologic maturation, a series of developmental changes that culminates in the adult stage, or complete biologic maturation. At any point in time, a child's level of somatic growth can thus be considered in terms of biologic age, or degree of biologic maturation.

The extent of biologic maturation can be estimated by percentage of estimated adult height and weight (morphologic characteristics), wrist bone age as measured by radiography (skeletal biologic age), and—after the onset of puberty—by the progression of appearance of secondary sexual characteristics (level of sexual maturation). These are all indicators of different aspects of biologic growth, but since all are affected by similar anabolic hormonal influences, they typically cluster and track together as children grow. While these markers provide a general indication of biologic maturation, Beunen noted that "no single system provides a complete description of the maturation of a single child" (5).

In the prepubertal years, bone age and height (as a percentage of adult stature) are closely linked. The tempo of biologic maturation is altered at puberty by the influence of sex hormones, but a tight relationship is still observed among factors such as peak height velocity, menarcheal age, Tanner stage (a measure of sexual maturation; see chapter 3), and skeletal bone age (correlation coefficients typically range from .60 to .80) (54). There is evidence, however, that the mechanisms controlling the tempo of biologic maturation during the prepubertal years may be different from those after attainment of puberty (5).

Several aspects of biologic maturation are particularly important for understanding the developmental changes in exercise physiology in children and adolescents. These are discussed in the following sections.

Biologic Age May Not Parallel Chronologic Age

As any sixth-grade teacher can attest, body size, habitus, and physical fitness vary dramatically in any group of 12-year-old children. That is because each child is on a different curve of biologic growth: Some are early maturers, some late, and some in between. Boys and girls who mature early are taller and heavier than their peers and demonstrate greater lean body mass and heart size. They tend to perform better on motor tasks and are likely to be more successful in sport competition (at least at that age). Late maturers, on the other hand, may be inferior in all these aspects until they "catch up" in later adolescence. This variability in the timing of biologic development, which is most evident in the early- to mid-adolescent years, poses a number of practical dilemmas for those who deal with youth exercise and sports.

How should competitors in youth sport teams be matched? The potential folly of the traditional means of classifying teams by chronologic age is obvious as the 150-pound early-maturing football tackle bears down on the 60-pound late-maturing tailback in the 10- to 12-year-old category. Some better marker of biologic maturation is needed to preserve the spirit of fair competition and avoid injuries.

This issue is particularly relevant for contact sports such as hockey and football. Roy et al. illustrated this dramatically in two groups of ice hockey players, both of similar age (71). The "small" group of boys weighed an average of 37.1 ± 3.8 kg and had a mean height of 147 ± 6 cm, while the "large" group measured 74.3 ± 8.2 kg and 179 ± 4 cm. Grip strength was 27.7 ± 4.8 kg in the small players and 56.5 ± 6.5 kg in the large. Impact force was 1,010 ± 111 and 1,722 ± 326 N in the two groups, respectively.

In their review of this issue, Malina and Beunen outlined the strengths and weaknesses of alternative means of matching opponents (53). Classification by body size is appealing, since mismatch of height and weight seem to pose the greatest risk. Children matched for body size, however, might still exhibit marked differences in experience, body composition, emotional maturity, and skill. Tanner staging for matching by level of sexual maturation would not allow for variation in timing of stages and might raise privacy issues. Moreover, this approach would be of no value in prepubertal athletes. Skeletal age is the most accurate single means of estimating biologic maturation, but it is impractical because of the radiation exposure to athletes and the need for expertise in interpretation (5). The level of a child's social, emotional, and cognitive development needs to be considered in any matching scheme.

In short, this problem does not have an easy solution. Matching by skill, stature, and weight "within relatively narrow chronological ages" has been suggested as the best compromise approach for young (prepubertal) athletes (54).

The variability in tempo of biologic maturation has also created problems in interpreting performance on standardized fitness tests conducted in schools. Children's performance has been related to health-related norms, with the expectation that a subnormal performance signals a need for remedial increases in physical activity and fitness. But how does one compare a child who scores high on number of sit-ups or the mile run with another child who fails these tests? Is the first student simply an earlier maturer than the second? Could it be that the child who scored low is actually a late maturer who is performing well for her level of biologic maturation?

The issue of early versus late biologic maturation also arises in the context of predicting and channeling athletic talent at an early age. Coaches are eager to identify the child with potential abilities in a particular sport so that skill might be nurtured from a young age. The variable rate of biologic maturation makes this a difficult task. The college baseball coach picking out the 10-year-old who can throw a ball the fastest or hardest may simply be identifying the most biologically mature, with no assurance of his ultimate pitching skill.

Tracking studies to determine the accuracy with which a child's future skill performance can be predicted have produced conflicting results. When studies span more than five years, correlations for variables such as $\dot{V}O_2$max, sprint speed, and muscular strength are typically .30 to .45 (52). Matsudo emphasized that the influence of puberty on body dimensions and composition explained much of the success of physical performance prediction over time (57).

Clustering of Different Forms of Physical Fitness

The various types of physical fitness—aerobic, anaerobic, strength—are all influenced by changes in body size, presumably by the same hormonal mechanisms. The development of aerobic fitness involves growth of the heart, lungs, and circulatory system. Increases in strength with age are largely mediated through skeletal muscle hypertrophy. Improvements in spring times and performance on Wingate cycle testing are affected by increases in leg length and muscle size.

It is no surprise, then, that these very different fitness measures should share the same tempo of biologic maturation in a given child. That is, children's absolute values of physiologic and performance markers of these various forms of fitness are expected to demonstrate a close relationship.

Comparative studies have borne this out. Suei et al., for example, reported correlation coefficients of .80 between absolute values of total work on Wingate testing and vertical jump, .60 for Wingate work and isometric torque production, and .85 for jump and torque production in prepubertal subjects (82). Falk and Bar-Or described a close relationship between peak aerobic and anaerobic power (r = .50 to .78) in a longitudinal study of prepubertal boys (28). This close matching of performance and absolute physiologic measures—an indication of their mutual reliance on body size during growth—may have bearing on the question of "metabolic nonspecialization" in children (see chapter 4).

Divergences Between Physiologic Development and Somatic Growth

Increase in body size is the primary factor driving the changes in exercise physiology and performance in the childhood years. This fact notwithstanding, many physiologic variables are observed not to track over time with greater body dimensions. This is explained by the added influences of size-independent variables on the development of physiologic responses to exercise, the relative contributions of which depend on the type of physical fitness under consideration.

The development of anaerobic fitness, for example, is revealed by changes over time in peak and mean power measured during Wingate cycle testing. Increases in anaerobic power are expected as children grow as a result of greater muscle volume, since power production reflects both length and cross-sectional area of muscle sarcomeres (74). Other size-independent factors, however, might also contribute to the development of anaerobic power, including variations in glycolytic metabolism, muscle architecture, and neuromuscular recruitment and coordination patterns.

Absolute values of mean and peak anaerobic power increase with age in both boys and girls. When values are expressed relative to body size (i.e., mass in kilograms), measures of anaerobic fitness still rise during childhood, albeit at a slower rate, about half that seen in absolute values. This observation provides evidence that both somatic maturation and changes in size-independent factors contribute substantially to the development of anaerobic fitness as children grow. This issue is discussed further in chapter 9.

A similar, although less dramatic, picture is observed with muscular strength. Strength is clearly more closely related to muscle size, and dimensional theory indicates that if this were the only factor responsible, strength would increase during the childhood years in relation to height raised to the power 2.0. Studies examining this relationship have indicated, however, that strength generally increases faster during childhood than can be explained by dimensional theory (i.e., faster than height squared). It is likely that size-independent neural factors (improvements in motor unit activation and myelination, increased coordination of muscle synergists and antagonists) are primarily responsible.

Maximal oxygen uptake, on the other hand, appears to increase in prepubertal children specifically from somatic growth. According to the Fick equation, $\dot{V}O_2max$ reflects the combined influences of maximal stroke volume, heart rate, and arterial-venous oxygen difference. The latter two factors—both size-independent—do not change during the course of childhood. Increases in $\dot{V}O_2max$, then, can be attributed solely to the determinants of maximal stroke volume. Maximal stroke volume is related to resting stroke volume, and the factor that defines aerobic fitness in both children and adults is left ventricular diastolic size (filling volume, or preload).

Growth factors responsible for increases in left ventricular chamber size therefore appear to be the sole factor responsible for the increase in absolute $\dot{V}O_2max$ during the prepubertal years. At puberty, blood hemoglobin concentration and arterial oxygen content rise in males, adding a size-independent increase in arterial-venous oxygen difference to increased stroke volume to produce improvements in $\dot{V}O_2max$

These observations indicate that the contributions of not only somatic maturation but also a variety of size-independent factors must be considered to understand children's exercise physiology. The complexity is compounded by the interplay of different size-dependent and size-independent factors as determinants of fitness during growth. Katzmarzyk et al. (42) studied a large group of 7- to 12-year-old children to delineate the role of biologic maturation, independent of increases in body size, on the development of grip strength and motor fitness (35-yd dash, standing long jump, and baseball throw). Body mass was the major factor explaining variance in strength, while the motor tasks were most closely related to the residual effects of the regression of skeletal age on chronologic age. The authors considered skeletal maturation in the children to be a surrogate marker of neuromuscular maturation.

Growth of Fitness-Specific Tissues

The anabolic actions of growth factors are directed to specific tissues, whose increase in size is important in the development of physical fitness as children grow. In the development of aerobic fitness, increased dimensions of the respiratory and circulatory systems, particularly the heart, are instrumental. Improvements in strength and anaerobic fitness, on the other hand, are expected to be tied more closely to increases in muscle bulk.

Growth hormone can profoundly affect the functional capacity of the cardiac, respiratory, and muscular systems in children, as evidenced by observed improvements in physical fitness after its use in clinical settings. Hutler et al. reported the effect of growth hormone treatment on exercise tolerance in 10 prepubertal children with cystic fibrosis (39). The children were randomly assigned to either a control group or a group that received growth hormone treatment for the initial six months and then to the other group for the next six months. Compared with the control condition, growth hormone treatment produced a significant increase in absolute peak $\dot{V}O_2$ (+19%), peak ventilation (+14%), and peak oxygen pulse (+18%). In this study 71% of the improvement in peak $\dot{V}O_2$ was explained by changes in lean body mass and forced expiratory volume.

Cardiac Growth

Insights into the effect of growth factors on heart size and function have been derived from animal studies as well as investigations of humans—both children and adults—who have deficiencies of growth factors, particularly deficiencies involving the GH/IGF-I axis. This has been the subject of several reviews (15, 48, 73).

There is abundant evidence that both GH and IGF-I stimulate cardiac growth, with associated improvements in both myocardial contractility and energy economy. GH also causes peripheral vasodilatation, probably via the effect of IGF-I on nitric oxide release.

Cardiac size and contractility are depressed in patients with GH deficiency, conditions that resolve with GH treatment. Cittadini et al. found depressed ventricular ejection fraction, cardiac output, and exercise performance in 11 young adults with childhood-onset GH deficiency (14). In healthy control subjects, ejection fraction rose from 66% to 76% with exercise, while in the GH-deficient patients, little change was seen from a resting value of 54%. After six months of GH treatment, exercise tolerance rose from 7.2 to 9.4 minutes, and resting ejection fraction increased to 61%, while cardiac output responses to exercise became similar to those of the controls.

Administration of GH to children with GH deficiency does not stimulate cardiac size or function beyond that paralleling total body growth. Rowland et al. found no differences in left ventricular wall thickness or chamber size when related to body size in echocardiograms of 13 children with GH deficiency who had been treated with GH (mean dose 0.17 IU kg^{-1} three times per week) for 13 to 46 months (70). Crepaz et al. compared echocardiographic findings in 22 children ages 3 to 17 years with GH deficiency who were given GH for an average of 14 months (19). Compared with size-matched healthy children, no differences were observed in left ventricular chamber size, wall thickness, contractility, diastolic function, cardiac index, or systemic vascular resistance.

Studies of the cardiac effects of GH administered to children without GH deficiency probably serve as a better model for the role of GH in the normal course of cardiac development with somatic growth. Daubeney et al. reported echocardiographic measures of heart size in 15 children with short stature (but no GH deficiency) before and after four years of GH treatment, compared with an untreated group (22). No anthropometric or ventricular-dimension differences were observed between the two groups at baseline. After four years of GH treatment, however, the treated children exhibited greater height, weight, and lean body mass as well as larger left ventricular end-diastolic dimension (41 ± 5 mm vs. 36 ± 5 mm) and left ventricular mass (93 ± 33 g vs. 73 ± 26 g). Posttreatment left ventricular mass expressed relative to lean body mass was similar to that of the controls, however, suggesting that increases in heart size reflected changes in lean body mass with GH treatment. No alterations in ventricular shortening fraction (an indicator of systolic contractile function) were observed following hormone treatment.

Similar results were observed in short children by Barton et al., who found that increases in cardiac dimensions with GH treatment paralleled those of body size (4). These findings in short but otherwise normal children mimic those in rats with GH-secreting tumors, who show accelerated heart growth in proportion to increases in body dimensions (72). They also simulate empiric observations during the growth of normal children: increasing heart size in relation to body dimensions (particularly lean body mass) without changes in myocardial contractility.

In abnormal hearts, GH increases myocardial function and improves energy efficiency. In the presence of left ventricular hypertrophy and failure, GH administered to both humans and animals causes a shift in myocardial myosin composition (increased V3 isoform and enhanced troponin I and myosin light chain 2), all changes associated with greater efficiency of myocardial contraction. These salutary effects of GH suggest its potential as a therapeutic modality for patients with depressed myocardial function (41). Fazio et al. administered recombinant GH to seven adult patients with idiopathic dilated cardiomyopathy and mild to severe congestive heart failure (29). Serum IGF-I concentrations doubled in response. Ventricular wall thickness increased, and a reduction was observed in end-diastolic dimension. Cardiac output during exercise improved 31%, while myocardial oxygen uptake at rest fell by one third.

The cardiac effects of IGF-I are similar to those of GH. IGF-I enhances myocardial protein synthesis and contractile function, both in normal hearts and in those with myocardial dysfunction (23). For example, Donath et al. found that IGF-I administered to healthy adults resulted in an 18% rise in cardiac output in association with a 9% increase in ventricular ejection fraction (23). Maximal oxygen uptake and exercise endurance, however, were not altered.

A host of locally acting growth factors have been identified that can mediate cardiac myocyte hypertrophy (at least in vitro), including cytokines, catecholamines, angiotensin II, and IGF-I (36). The increase in heart muscle cell size (i.e., hypertrophy) from such factors in response to hemodynamic overload may, however, not be an appropriate model for assessment of increases in cell size that occur with normal growth. Stress-related responses involve specific alteration in gene expression, which alters the phenotypic expression of contractile proteins. The extent to which such alterations occur with normal growth in humans is uncertain. It is of interest, however, that stress-induced changes in forms of myosin are often characterized by a reversal to fetal gene expression.

Increases in contractility in response to IGF-I exposure are dose dependent and reach about 20% to 25% above baseline. Alterations of calcium flux appear to be responsible. This improvement in myocardial contractile function occurs despite absence of change in ATP content or high-energy phosphates, thus indicating greater energy efficiency.

Muscular Growth

Growth of skeletal muscle mass is one of the most firmly established activities of the GH/IGF-I axis (84). While animal studies indicate that GH has some primary effects, IGF-I appears to be predominantly responsible for the growth and development of muscle cells seen with exogenous or endogenous exposure to GH. This activity reflects both augmented protein synthesis through increased amino acid transport and inhibited cellular protein degradation.

Short stature is the most common feature of children with growth hormone deficiency, but these patients also demonstrate delayed motor development and reduced muscle strength. These conditions are typically reversed by GH treatment. Brat et al. measured isometric muscle force and endurance in a small group of children with GH deficiency before and after 10 and 24 months of GH treatment (8). Before treatment, the muscle force of children with GH deficiency was 56% to 62% of that of matched healthy children. Strength measurements rose to 75% to 78% after 10 months of treatment and to 87% to 93% at 24 months.

A deficiency of GH is not necessary for these treatment effects. Leger et al. (45) assessed the effects of three years of GH treatment ($0.2\ IU \cdot kg^{-1} \cdot d^{-1}$) on muscle mass in 14 prepubertal children who were born small for gestational age (SGA). Muscle mass in this study was determined by magnetic resonance imaging. Results were compared with longitudinal measurements of children of normal stature. The increase in muscle cross-sectional area was 72 ± 5 % in the GH-treated SGA children compared with 22 ± 5 % in the controls.

The pharmacologic dose of GH in this study was approximately twice that used for replacement therapy in children who are GH deficient. In a previous publication, Leger et al. described a more rapid change in muscle size in SGA children than in GH-deficient children (46). This caused the authors to suggest that it is possible that the effect of GH on muscle in children is dose dependent.

Paracrine and autocrine factors acting locally and independently of GH may also be important for muscle growth. For example, mechanogrowth factor (MGF), produced in response to muscle stretch, has been demonstrated to stimulate muscle protein synthesis (34).

Adams reviewed data regarding the role of IGF-I in mediating skeletal muscle adaptations to work stress (1). In rats, IGF-I stimulates muscle protein synthesis, increases amino acid uptake, and suppresses protein degradation. Furthermore, IGF-I has a separate action in promoting cell proliferation (mitogenesis) and differentiation in the embryonic development of skeletal muscle. Increasing evidence exists that these same two processes are involved in muscle adaptive responses to increased workload.

These investigations suggest that paracrine or autocrine actions rather than circulating IGF-I stimulate anabolic effects leading to muscle hypertrophy with work stress. In addition, IGF-I has been demonstrated to stimulate proliferation of satellite cells: small, mononucleated stem cells that have the capacity to proliferate and differentiate into myoblasts in response to work stress or injury.

Accounting for Interindividual Fitness Differences in Children

So far, the development of physiologic and physical fitness in children has been examined as an outcome of increases in body size. Somatic maturation is largely responsible for longitudinal changes in exercise functional capacity during the growing years. This analysis, however, does not address factors that are responsible for individual differences in physical fitness.

If the somatic expression of growth factors was the only determinant of development of physiologic and performance fitness, all children of the same size would exhibit identical levels of fitness. This is clearly not the case. Any physiologic factor varies significantly in respect to a given appropriate measure of body size. How can we explain such variations in fitness relative to the same body dimensions in any group of children of even identical chronologic or biologic age? Several possible answers can be considered.

1. Specific factors, which act independently of determinants of height and weight, may target their effects on the growth of fitness-related tissues. Such agents have been recognized, in fact. A number of peptide growth factors, particularly fibroblast growth factor 6 and epidermal growth factor, have been identified in cardiac myocytes (7). These factors have been implicated in hypertrophic responses of the heart to systolic overwork (systemic hypertension) as well as fetal cardiac organogenesis. In any event, they provide evidence that localized, tissue-specific factors exist that can stimulate heart growth.

2. GH receptors may be selectively more populous or active in the fitness-related tissues of children who exhibit superior levels of fitness. This would result in disproportionate growth of heart or muscle size, for example, relative to height and weight.

3. Size-independent factors may operate more prominently in children with greater fitness. As noted earlier, the development of size-independent factors may contribute significantly to the evolution of exercise fitness during childhood. The nature and magnitude of these size-independent influences vary among types of fitness. Interindividual variations in fitness could then relate to greater or lesser input from these factors, presumably on a genetic basis. The child with 50-yd dash performance superior to that of another child her own age and size might possess greater glycolytic capacity to provide anaerobically derived energy. A more developed level of neuromuscular input to muscles might explain the stronger child. This would not, however, explain differences in aerobic fitness, in which the influence of lung or heart size on $\dot{V}O_2max$ is predominant.

This issue is a critical one for developmental exercise physiologists, for it bears on the central question of identifying limiting factors to physical and physiologic performance by growing children. Because these interindividual differences in fitness are observed in nontrained youth in populations homogeneous in nutrition, socioeconomic status, and body composition, it is reasonable to conclude that, by whatever phenotypic mechanism, these differentiating factors are under genetic control. Much progress has been made in determining specific gene loci that influence physical fitness (65). It is possible that this research will identify specific gene markers that define individual levels of fitness via one or more of the preceding mechanisms.

Effects of Exercise on Growth

The previous section addressed the influence of somatic growth and its determinants on children's physiologic responses to exercise. We now turn the arrow around to look at cause and effect in the opposite direction: How might physical activity during childhood and adolescence, particularly athletic participation, affect physical growth? This question is a major concern of parents, coaches, and physical educators alike. Can the stresses of intense physical activity in the growing years impair linear and visceral growth? If so, should limits be placed on young people's participation in athletic training programs?

The concept that increased physical stresses might stunt growth arose initially from animal studies in which swim-trained rats demonstrated delayed bone growth and from an early report indicating delayed statural maturation in children undertaking hard labor in poor socioeconomic conditions (3). More recently, the concern has been reinforced by studies suggesting that intense training might delay linear growth in young female gymnasts (83).

At the same time, it is accepted that increased physical activity and musculoskeletal stress are important for *promoting* growth in children (3, 6). Moreover, children's involvement in sport training might provide particular long-term health benefits (e.g., stimulation of bone growth and density may ameliorate the risk of future osteoporosis). The following discussion focuses on the biologic means by which exercise by children might affect—positively or negatively—their normal patterns of growth. For the most part, however, analysis of how growth responses to repeated exercise (i.e., sport training) might improve physical fitness and accelerate specific forms of physiologic development is deferred until chapter 11.

Physical activity might influence growth in children by three possible mechanisms (6; figure 2.4): (a) Exercise draws on caloric stores and competes

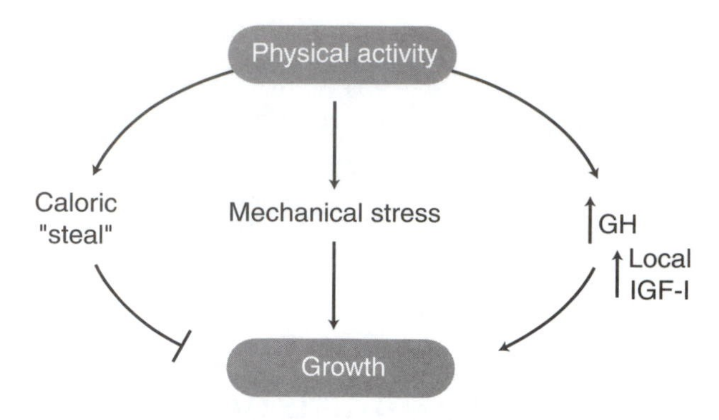

FIGURE 2.4 Potential mechanisms for the effects of physical activity on growth.

with the energy demands of normal growth for available nutrients. Through "caloric stealing," physical activity may thus potentially impair growth on a nutritional basis. (b) Physical activity serves as a potent stimulus for production of growth factors. However, the mechanisms behind this action—as well as its implications for positive growth—are not clear. (c) Muscular activity creates local mechanical stresses that trigger musculoskeletal growth. In some cases, intermediary apocrine and autocrine agents may mediate this process.

Competition for Nutrition

During childhood and early adolescence the energy needs of exercise are superimposed on the energy needs for not only homeostasis and tissue repair but also somatic growth. That the requirements for normal growth can lose in this conflict is evidenced by the poor growth seen in active children in nutritionally marginal conditions in underdeveloped countries. This phenomenon may also contribute to the delayed statural maturation associated with negative caloric balance in activities where slender body habitus is important, such as gymnastics and ballet dancing (31, 83).

Undernutrition with negative caloric balance is associated with depression of serum IGF-I levels in male (80) as well as female gymnasts (40). Smith et al. (80) demonstrated a significant decline in IGF-I levels after a seven-day exercise period that put young adult males in a negative caloric balance (causing weight loss of 0.5–2.0 kg). A similar decline in IGF-I was observed when an equivalent caloric imbalance was created by dietary restriction without exercise. Thus, energy deficiency caused by either exercise or diet caused a fall in IGF-I levels.

Roemmich and Sinning demonstrated increases in GH secretion but a decline in growth hormone–binding protein (GHBP) and IGF-I in undernourished adolescent wrestlers (67). All values quickly reverted to normal at the end of the competitive season. No significant effect was observed on linear bone growth or sexual maturation.

Response of Growth Factors

Acute and chronic exercise trigger alterations in the GH/IGF-I axis and in the production of other growth factors. The extent to which these alterations stimulate growth or reparative processes through anabolic activity or provide metabolic support (e.g., substrate utilization) in children is uncertain.

Growth Hormone

Acute bouts of exercise stimulate the release of GH. In fact, exercise serves as one of the most effective provocative agents for clinically evaluating GH production capacity. The rise in GH with acute exercise is not immediate. Levels typically begin to rise 10 to 15 minutes after the beginning of a 30-minute exercise bout and peak at the end of exercise. Level of physical fitness affects this acute GH response. Trained adults demonstrate blunted GH production compared with nontrained individuals, and a three- to six-week training period has been demonstrated to lower the GH response to acute exercise (90). Interestingly, however, 24-hour GH secretion and amplitude of GH pulses in adult athletes are greater than those of nonathletes.

The extent of GH response depends linearly on the intensity of exercise, but the intensity of exercise necessary to evoke a rise in GH varies considerably between individuals. In general, however, exercise above 50% $\dot{V}O_2$max is necessary to trigger a substantial rise. This suggests that some physiologic trigger related to intensity of muscle contraction serves as the stimulus for pituitary GH release.

The neuroendocrine mechanism by which acute exercise triggers and regulates GH is not known. Both alpha- and beta-adrenergic pathways appear to play a role, since pharmacologic beta-blockage enhances GH response to exercise, while phentolamine (an alpha-receptor antagonist) depresses GH release. Administration of pyridostigmine, a promoter of the cholinergic system, augments GH response to exercise, probably by inhibiting the action of somatostatin.

A rise in body temperature has been suggested as a mediator of GH release in response to acute exercise (9, 10, 87). Weeke and Gunderson could find no rise in GH secretion when exercise was performed in a cold environment (87). Brennen et al., however, reported no differences in GH response to acute exercise at 40°C and 23°C ambient temperature (9).

Dietary factors appear to strongly alter GH responses to acute exercise. Cappon et al. found that GH responses to exercise bouts following a high-fat drink were 50% lower than those following a placebo (12). Cooper suggested that this might be "a possible mechanism whereby not only the quality and quantity of caloric intake, but also the hormonal response to a particular diet, may play a role in attenuating the protein-anabolic and lipolytic effects of exercise" (16, p. 18).

Hopkins et al. provided evidence that IGF-I activity during prolonged exercise is not related to carbohydrate levels (37). Nineteen-year-old males cycling at 60% $\dot{V}O_2$max to fatigue demonstrated a fall in serum glucose concentration that was prevented by carbohydrate supplementation during a second trial. IGF-I levels did not change significantly in either trial. IGFBP-I levels rose with exercise, however, and this increase occurred whether or not glucose levels declined.

The biologic meaning of this GH response to exercise is not immediately obvious. Acute bouts of exercise—and repeated exercise, or training programs—do not result in measurable increases in linear growth (e.g., height). Since the peak of GH release occurs near the end of a brief exercise bout, its role as a metabolic support (such as by increasing the availability of free fatty acids via lipolysis) seems questionable. It may be that this GH release is anticipatory of the need for reparative anabolic actions in response to the musculoskeletal stresses of exercise.

The rise of GH seen with acute exercise does not, of course, necessarily imply increases in GH activity. Such changes could be associated with down-regulation of GH receptors (16) or with alterations in degradation, clearance, or affinity to circulating binding proteins.

Several reports have described "standardized" exercise protocols for GH provocation in the clinical testing of children (44, 75, 77). Sartorio et al. (75) found that plasma GH levels with acute exercise rose by 600% to 1200% in short children with normal resting GH, while in children with GH deficiency, levels increased by only 94%.

Seip et al. described their observations of 10 healthy children 9 to 15 years old who performed a 15-minute cycle bout at 70% of predicted maximal work rate on two occasions (77). Serum GH concentrations rose from 2.2 ± 2.8 to 27.1 ± 6.9 ng · ml^{-1} on the first test and from 6.5 ± 7.8 to 19.9 ± 12.3 ng · ml^{-1} on the second. While this report emphasized the feasibility of such testing, it also demonstrated that considerable variability in GH stimulation might be expected. Such variability has been reported by others (27). Seip et al. suggested that since the work rates in their two tests were equivalent, variation in secretion or pituitary response to GHRH or somatostatin might be responsible (77).

Both peak GH response to exercise and the magnitude of difference between exercise and resting levels are lower in prepubertal children than in those who have achieved sexual maturity. No qualitative or quantitative differences in these responses related to sex have been seen.

Wirth et al. measured blood levels of GH before and after submaximal exercise in 41 swimmers ages 8 to 18 years (91; figure 2.5). Subjects were divided into prepubertal, pubertal, and postpubertal stages of sexual maturation by Tanner staging. The exercise protocol involved pedaling for 15 minutes at 70% $\dot{V}O_2$max. As indicated in figure 2.5, both resting and incremental levels of GH in response to exercise were lowest in the prepubertal group, and this was observed in girls and boys equally. Sexually mature

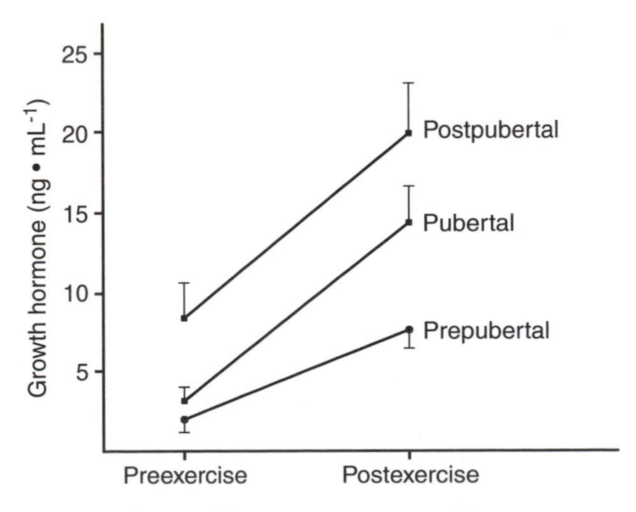

FIGURE 2.5 Growth hormone concentrations at rest and following submaximal exercise in prepubertal, pubertal, and postpubertal youth (from reference 91).

adolescents demonstrated a rise in growth hormone at least twice as great in magnitude as that of the prepubertal group; the authors suggested that the greater GH response may be a manifestation of the influence of sex steroids, particularly estrogen, at the time of puberty.

To test this idea, Marin et al. administered a standard treadmill exercise test to 84 healthy boys and girls at all stages of puberty (56). A randomized subset of 11 prepubertal children received ethinyl estradiol—an estrogen-like hormone—for two days prior to testing, while the rest received a placebo. As in the study by Wirth et al. (91), GH response to exercise increased with greater pubertal stage. The peak level of GH in Tanner stage V subjects (17.2 ± 14.7 ng · ml^{-1}) was three times greater than that of the prepubertal group (5.7 ± 4.1 ng · ml^{-1}).

GH levels in the 11 prepubertal children pretreated with ethinyl estradiol rose to 18.7 ± 9.2 ng · ml^{-1} with exercise, while mean concentration with exercise in the placebo-treated children was 6.9 ± 4.2 ng · ml^{-1}. These findings support the concept that sex steroids are responsible for the exaggerated response of GH to exercise at puberty.

Other Growth Factors

The responses of IGF-I to exercise are less predictable and sometimes seemingly paradoxical. In general, IGF levels do not change or increase mildly with bouts of acute exercise in adults (23, 37). In most studies of older individuals, concentrations are directly related to aerobic fitness (i.e., $\dot{V}O_2max$); that is, IGF levels are higher in more-fit subjects. In a study investigating the influences of anthropometric, dietary, and fitness factors on the age-related decline in IGF-I in adults, the sole factors independently related to IGF-I were $\dot{V}O_2max$ (r = .29) and leisure-time activity (r = .24; reference 64). IGF-I has been demonstrated to rise after a period of endurance training in adult men (66). As noted earlier, however, IGF levels decline when training is associated with negative caloric balance.

Eliakim and colleagues provided insights into maturational influences on growth factor responses to exercise in a series of studies in pre- and postpubertal youth (25). These researchers reported the effects of a five-week training program on growth factors in 15- to 17-year-old adolescent females compared with nontraining controls. Preintervention cross-sectional analysis indicated significant correlations of overnight GH secretion and GHBP with $\dot{V}O_2max$ (r = .36 and

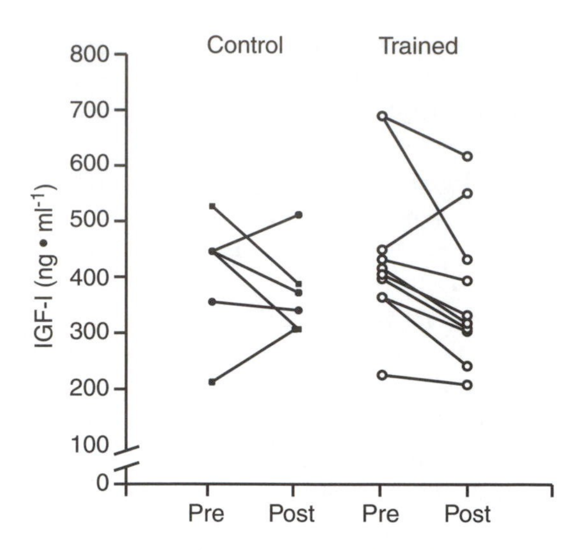

FIGURE 2.6 Effect of a five-week endurance training program on serum IGF-I level in adolescent females. A statistically significant decrease is seen in the trained subjects compared with controls (from reference 25).
Reprinted by permission from Eliakim et al. 1996.

.42, respectively). However, no relationship was seen between IGF-I and $\dot{V}O_{22}max$.

At pretraining testing there were no significant differences between the two groups in any of the hormone concentrations. With training, which improved both $\dot{V}O_2max$ and muscle size, no changes were seen in GH secretion, and levels of IGF-I (figure 2.6) and IGFBP-3 declined. This fall in IGF occurred in the absence of weight loss. These observations suggest that exercise training may lead to a catabolic state characterized by reductions in IGF-I.

Different findings were described in a similar study involving a group of prepubertal girls (26). A five-week training program of team sports and running games was conducted two sessions daily, 45 minutes each. In untrained girls before the training began, thigh muscle volume and peak $\dot{V}O_2$ correlated closely with IGF-I (r = .58 and .44, respectively). IGFBP-1 was negatively related to thigh muscle volume (r = –.71), while GHBP was associated with thigh muscle volume (r = .45) but not peak $\dot{V}O_2$. IGF-I did not change with training (but, unexplainably, increased in the controls), while significant reductions were seen in GHBP.

In 43 adolescent males (age 16 ± 0.7 years, 70% at Tanner stage V), peak $\dot{V}O_2$ per kilogram was positively correlated with overnight GH (r = .41), and both peak $\dot{V}O_2$ and thigh muscle volume relative to body mass were negatively correlated with GHBP (r = -0.33 and -0.45, respectively) (24). No correlation between aerobic fitness and IGF-I or IGFBP was seen. As in adolescent girls, a five-week training program triggered a significant fall in IGF-I (–12 ± 4 %) without evidence of negative caloric balance (i.e., no weight loss).

These findings fail to indicate a stimulating effect of exercise training on IGF-I levels in either prepubertal children or sexually mature adolescents. In fact, decreases in IGF concentrations were seen in both postpubertal groups, a finding expected with a negative caloric balance induced by training. In both cases, however, no weight loss was observed. An association with aerobic fitness ($\dot{V}O_2max$) was observed in the young girls but not in postpubertal male or female subjects.

The meaning of these discrepant results is unclear. In the prepubertal girls, IGF-I was related to aerobic fitness but not to the process of acquiring fitness. The implication is that the two issues (i.e., inherent fitness and the fitness effect of training) represent, at least from a hormonal standpoint, two separate attributes. Nemet et al. considered that the decline of IGF-I with exercise might be caused by the release of proinflammatory cytokines, which inhibit anabolic activity of the GH/IGF axis (61). [Cytokines are a family of proteins that mediate cellular interactions to immune responses. The muscle damage and soreness following endurance and resistance exercise are associated with a rise in cytokine activity, which regulates neutrophil and monocyte migration into impaired muscle tissue (62).]

Nemet et al. measured anabolic growth factors and proinflammatory cytokines interleukin-6, tumor necrosis factor-alpha, and interleukin-1 beta in 11 healthy boys ages 14 to 18 years before and after a 1.5-hour wrestling session (61). Following exercise, total and bound IGF-I both fell by an average of 11.2%, but there was no change in unbound IGF-I (figure 2.7). Levels of IGFBP-1, which inhibits action of IGF-I, rose eightfold. Dramatic increases (30% to 795%) were seen in all the cytokines. These findings suggest that acute bouts of exercise in male adolescents trigger predominantly a catabolic response.

The same picture has been observed in prepubertal children. A similar investigation of hormonal responses was conducted in 17 healthy 8- to 11-year-old children before and after a 1.5-hour soccer practice (76). Proinflammatory cytokines rose 18% to 125%. Circulating IGF-I fell by 6.4 ± 3.2 %, with a 156% mean increase in IGFBP-I.

Mechanical Stress

The influences of physical stress on the musculoskeletal system have long been appreciated. Immobilization or zero gravity in space results in well-recognized muscle atrophy and loss of bone mineral content, while loading of the musculoskeletal system produces the opposite effect, stimulating bone and muscle growth. Excessive stresses on muscle and bone are counterproductive, however, with ultimate tissue breakdown and injury.

This dose–response curve of a progression of positive biologic response up to a threshold limit was referred to in 1950 by Hans Selye as the *general adaptation syndrome* (78). Selye emphasized that both favorable (anabolic) and unfavorable (catabolic) adaptations occur in response to mechanical stress, and that rest and recovery are important in avoiding adverse outcomes.

As Borer noted, "Hypertrophic and cellular growth response to increased mechanical loading is linear in the lower ranges of stresses and declines or becomes negative when acting forces exceed the limit of adaptive biologic response" (6, p. 377). But the means by which mechanical stresses on bone and muscle stimulate growth is not fully understood. Particularly unclear is the role of local apocrine and paracrine hormonal responses in mediating growth adaptations to mechanical stress.

There is some evidence, in fact, that such mechanisms are operant. Goldspink and Yang at the University of London cloned the cDNA of two isoforms of IGF, which were created from the IGF gene by alternate splicing techniques (34). One of these was detected only when skeletal muscle was mechanically stimulated and was termed *mechanogrowth factor (MGF)*. They provided evidence that MGF induces

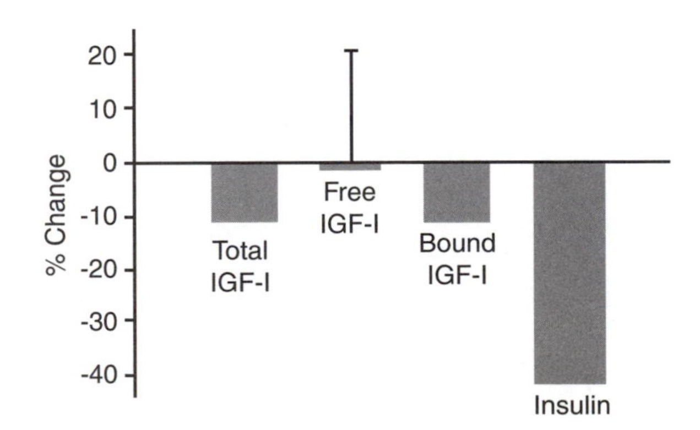

FIGURE 2.7 Changes in IGF levels in response to wrestling practice (from reference 61).
Reproduced with permission from Nemet et al. 2002.

protein synthesis locally in muscle and is triggered principally by muscle stretch. This research may thus describe a link between physical activity (i.e., muscle stretch) and gene expression for muscle growth.

Growth in Young Athletes: Empiric Observations

The preceding mechanical stress findings hold true for children as well as adults, but for growing youngsters extra dimensions must be recognized. First, the tissues undergoing mechanical stress are in the process of biologic growth and functional development. In fact, the outcomes of this biologic maturation mimic the adaptive responses to the stresses induced by physical activity. A period of resistance training by a 14-year-old boy will augment his 1RM lift performance, and so will a one-year period of biologic development during which he spends his free time on the sofa watching television.

In addition, a number of features of the growing child may serve to lower the threshold of negative stress responses to exercise (43). These include lack of flexibility due to disproportionate growth of the muscle-tendon-bone unit (particularly in adolescents), looser periosteum and tendon attachments in immature bones, and weakness and susceptibility to damage at the epiphyseal growth plate.

Two important questions then arise: How much mechanical stress is important for normal somatic growth in children? And at what point does excessive stress on muscle and bone in growing individuals become counterproductive? The latter question has particular bearing on whether limits should be placed on training volume and intensity for elite-level child athletes. The purpose of sport training, after all, is to impose high levels of stress on fitness tissues (heart and lungs with endurance training, skeletal muscle with resistance training). From the body's response to this microtrauma of training the benefits of improved physiologic and performance fitness are expected. But is there a limit beyond which this training-induced stress might interfere with the normal growth and development of children? While it is not difficult to raise hypothetical concerns, the best answers lie in critically observing growth patterns in growing athletes who engage in intense levels of training and competition.

Levels of habitual physical activity within the range of the normal nonathletic population of children do not affect statural growth. In a longitudinal study of 25 boys from ages 7 to 17 years, Mirwald et al. found no differences in height or height gain relative to activity or inactivity as determined by a questionnaire (59; figure 2.8).

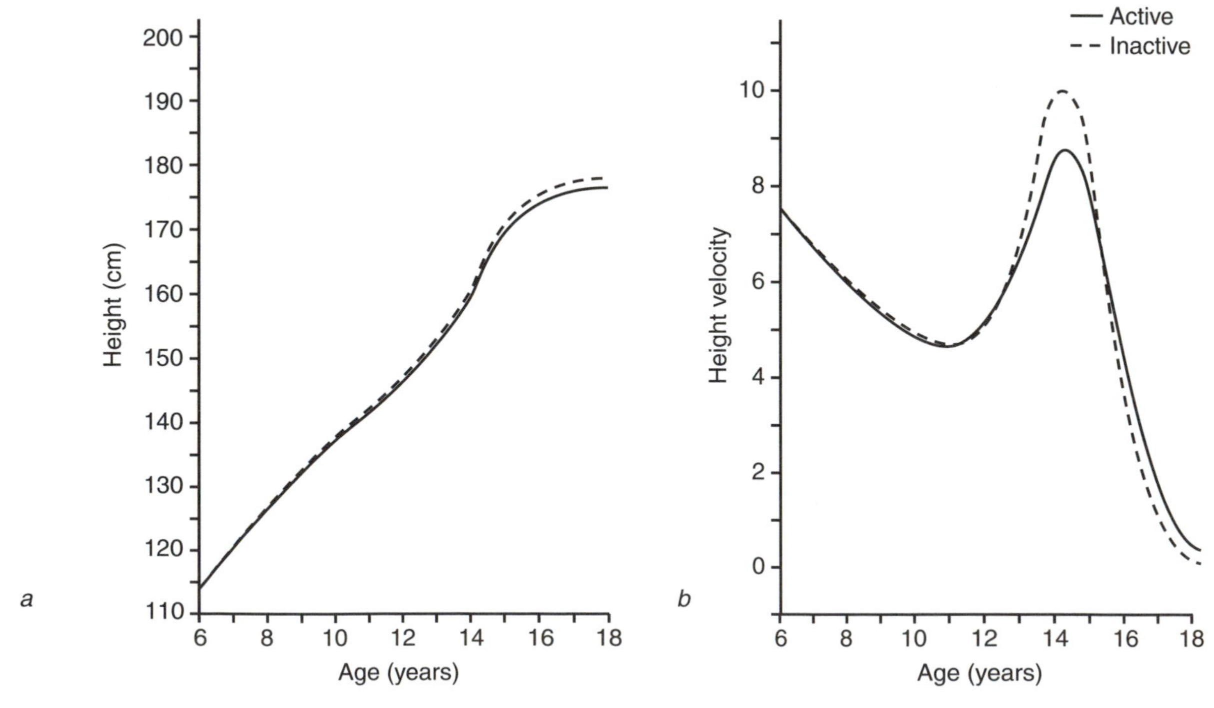

FIGURE 2.8 *(a)* Height and *(b)* height velocity curves by age for active and inactive boys (from reference 59).
Reprinted by permission from Mirwald 1981.

The impact of sport training on somatic growth in children has been extensively reviewed by Malina (49, 50). A considerable volume of cross-sectional and longitudinal data indicate that, with few exceptions (i.e., gymnasts and dancers), child and adolescent athletes experience normal statural growth compared with their nonathletic peers. That is, early, intensive sport training does not result in deviation—either positive or negative—from the normal pattern of linear growth. These investigations have indicated, too, that the rate of skeletal maturation and the timing of peak height velocity are for the most part unaffected by early sport training.

Such studies are complicated by the fact that athletes represent a genetically and phenotypically select group of children and adolescents (6, 49). Their body size and composition and level of skeletal maturation may preselect them for success in specific sports. Boys who are talented in sports such as football and hockey, for example, tend to be advanced in their biologic maturation. Female gymnasts, figure skaters, and ballet dancers, on the other hand, are favored in their performance by slow growth rates and delayed skeletal and sexual development.

Reports of diminished linear growth rates in gymnasts, who benefit from a short, light frame and slender body habitus, have in fact drawn considerable attention and concern. Investigations of young female gymnasts have consistently demonstrated delayed skeletal, somatic, and sexual maturation compared with nontraining girls. Theintz et al. compared measurements of 22 training female gymnasts (mean age 12.3 ± 0.2 years) with those of competitive swimmers every six months for two years (83). Growth velocities were less for the gymnasts (5.48 ± 0.32 cm · yr^{-1}) than for the swimmers (8.0 ± 0.50 cm · yr^{-1}), and the gymnasts also demonstrated a significant decrease in height standard deviation score and stunting of leg-length growth over time (figure 2.9).

Courteix et al. evaluated skeletal maturation and somatic growth longitudinally over three years in 10 elite female gymnasts (beginning at age 10.1 ± 1.3 years) and in 14 nonexercising girls (18). Each year the gymnasts demonstrated significantly lower bone age, height, fat, and lean body mass compared with the controls. However, when expressed as a percentage change, somatic development was similar over the three years in both groups.

Georgopoulos et al. studied 255 rhythmic gymnasts competing in the 13th European Championships (31). Skeletal maturation for the group was delayed by a mean of 1.3 years. Age at menarche

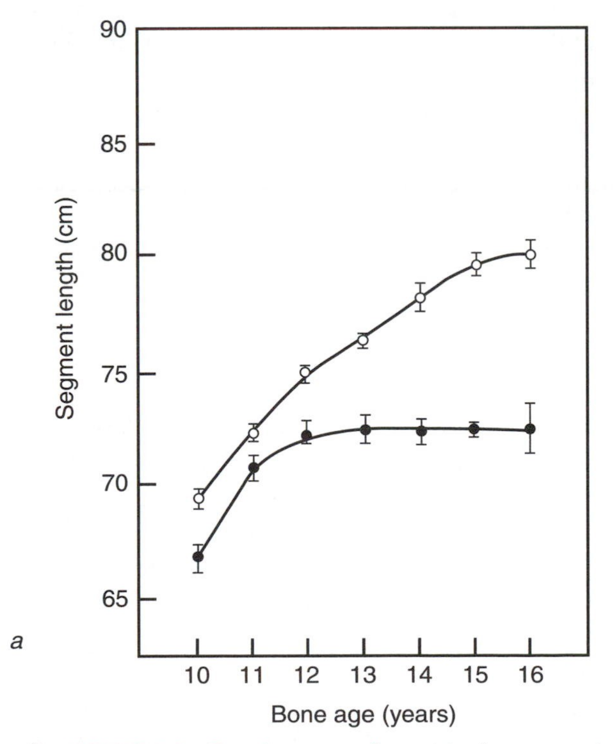

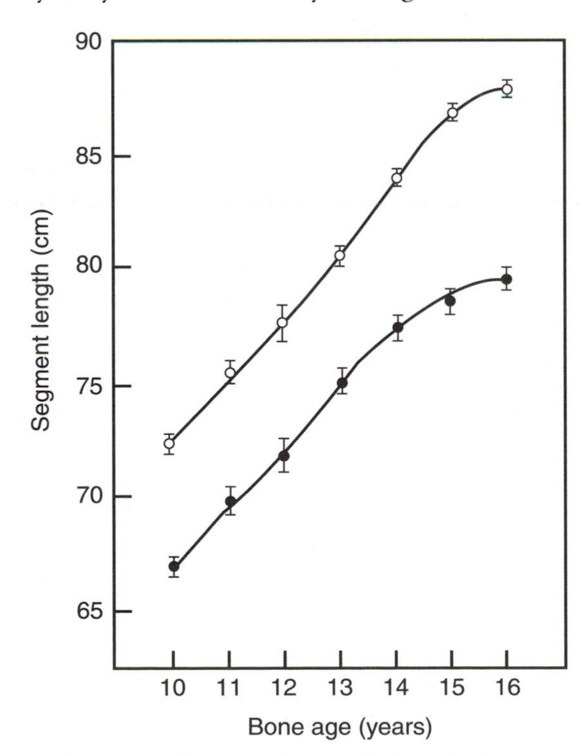

FIGURE 2.9 Development of upper body segment (open circles) and lower body segment (dark circles) with increasing bone age in *(a)* gymnasts and *(b)* swimmers (from reference 83).

was later than in nonathletic girls and was positively correlated with the intensity of training.

Jahreis et al. used IGF-I concentrations as a marker of pulsatile GH secretion in highly competitive 11-year-old female gymnasts (40). In a preliminary study of 43 gymnasts, IGF-I concentrations after three days of intensive training were all below mean levels expected for chronologic age, but most were within one standard deviation. In pre- and posttraining measurements in nine girls, IGF-I fell from 188 $\pm$ 56 ng $\cdot$ ml^{-1} to 146 $\pm$ 46 ng $\cdot$ ml^{-1}. In a second experiment involving 16 gymnasts, mean IGF-I levels fell by 25% over three days of training.

The authors considered the low baseline IGF-I levels to reflect genetic preselection of these athletes for small body size and delayed bone age maturation. They suggested that the fall in IGF-I with training, however, was related to energy deficiency during extended exercise, low thyroid hormone, the anti-insulin effect of GH, increased IGF-I binding protein, or some combination of these.

The relative contributions of genetic predisposition, undernutrition (negative caloric balance), and physical training per se to these findings in young female gymnasts remain unclear. Some researchers are convinced that the growth characteristics of these athletes reflect disturbances explained by intense training at very young ages. Jahreis et al., for instance, concluded that "the hormonal alterations found in gymnasts characterize an 'exercise-induced' delay in development . . . reflected in delayed growth, a retardation in bone age, markedly delayed sexual maturation, and, in part, . . . (a) danger to the skeletal and locomotor systems" (40, p. 98). Malina, however, considered preselection factors to be more likely responsible: "The adolescent growth pattern of female and male gymnasts is similar to that of short, normal, slow maturing children and/or late maturing children with short parents" (50, p. 143).

Caine et al. presented a literature review of 3 case reports, 18 cross-sectional studies, and 14 prospective and historical cohort studies addressing the growth of female gymnasts during training (11). Overall, gymnasts in these reports demonstrated diminished height and weight for age and low levels of serum growth factors. Attenuated growth during periods of training was followed by catch-up growth when training was reduced or discontinued. This suggests that growth patterns are not entirely related to genetic preselection.

An accompanying review of reports of energy intake in young gymnasts confirmed the impression that these athletes consume a suboptimal number of calories. Intake was often lower than the national recommendations by 275 to 1,200 kcal per day. The authors concluded that a cause-and-effect relationship between gymnastics training, independent of dietary and preselection factors, and delayed growth had not been demonstrated.

Less research attention has been directed to male gymnasts, who typically are short but demonstrate normal growth velocity. Daly et al. measured diet, hormonal responses, and stature over a 10-month period in a group of 16 intensely training boy gymnasts (21). Measurements were compared with those of 17 nontraining boys. No differences in rate of growth or levels of IGF-I were observed between the two groups, and no correlation was seen between IGF-I and diet.

Conclusions

Somatic development must be considered the driving force behind improvements in physiology and performance capacity as children grow. But no clear understanding exists of the factors that promote body growth over time nor of the determinants that differentiate the growth of fitness-related tissues between individual children. It is accepted that these are largely genetically based, and continued progress in the field of molecular genetics can be expected to provide insights into the cause of differences in phenotypic expression of both the magnitude and the tempo of growth during childhood. This understanding should, in turn, shed light on individual changes in exercise and physical fitness during childhood and adolescence.

Both acute and chronic exercise stimulate perturbation of growth factors as well as their binding proteins and receptor sites. The biologic significance of these responses is unclear at the present time. Variations in growth factor levels with exercise may indicate caloric imbalance or excessive catabolic state and may therefore serve as biochemical markers of excessive training (16).

Current data indicate that vigorous athletic training in children does not affect normal patterns of growth. The only exception may be the slow linear growth observed in highly trained gymnasts, which in some way reflects the contributions of genetic predisposition, training stress, and undernutrition. While the relative influence of each of these factors is controversial, these findings do signal the need to ensure that young athletes do not train in a state of negative caloric balance. As Manfield and Emans concluded, studies of gymnasts "support the wisdom of limiting the intensity of training and optimizing nutritional regimens during childhood and early adolescence, especially during the pubertal growth spurt" (55, p. 239).

Discussion Questions and Research Directions

1. What are the relative contributions of genetic preselection, caloric deficiency, and training stress on the slow biologic development of young gymnasts and dancers? What are the long-term health implications of this trend?
2. What mechanisms govern the growth of fitness-related tissues during childhood? What are the relative roles of the GH/IGF-I axis and local apocrine and paracrine hormonal actions?
3. How do genetic factors control the phenotypic expression of fitness-related tissues in youth?
4. What good markers of biologic maturation level could be used to appropriately match opponents in youth sports?
5. What factors account for the tempo of biologic maturation?
6. What are the determinants of interindividual differences in size or physiologic function that influence physical fitness?
7. What is the biologic significance of GH responses to exercise?

3

The Impact of Puberty

Brown-Sequard (1889), the renowned physiologist, prepared a dog testicular extract and administered it subcutaneously to himself. He was convinced that he had gained in vigor and capacity for work from the treatment, but it is now known that his aqueous extract was devoid of hormone.

—E.B. Astwood (1965)

In this chapter we discuss

- the normal process of puberty and its assessment,
- the impact of hormonal changes at puberty on physical fitness and how these changes are affected by sex, and
- the influence of exercise and sport participation on pubertal progression and reproductive function.

If one considers that the term *pediatric* encompasses persons up to the age of 20 years, the age span of interest to children's exercise physiologists can be divided into two roughly equal segments. First comes the period of growth up until about age 12 years, primarily directed by actions of the GH/IGF-I axis. Then, in the second period, these anabolic effects are supplemented by the influences of the sex hormones during the process of puberty, or reproductive maturation.

These hormonal changes at puberty exhibit a profound effect on exercise physiology and performance. These influences are largely sex specific, defining clear-cut differences between sexes that were marginally evident in the prepubertal years. Moreover, the anatomic and physiologic features that appear with puberty serve to characterize the differences between child and adult physiologic responses to exercise. Understanding the hormonal influences at puberty on exercise outcomes therefore requires careful consideration of levels of sexual maturation in physiologic studies involving older children and adolescents.

This chapter describes the endocrinologic basis for puberty and the impact of these hormonal changes on exercise physiology and performance. As in the previous chapter, it is important to consider the puberty–exercise relationship from the opposite direction too: How do acute and chronic exercise (i.e., sport training) influence the process of sexual maturation in boys and girls?

The Process of Puberty

Puberty is the succession of anatomic and physiologic changes in early adolescence that mark the transition period from the sexually immature to the fully fertile state. This process is characterized by not only development of reproductive function but also alterations in body size, composition, and function in response to actions of the sex hormones estrogen (in the female) and testosterone (in the male). The basic control of the actions of these two sex steroids is similar, but the outcomes are largely distinct, thus defining sex characteristics in adolescence. These characteristics include fat accumulation and bone maturation in females and linear growth and development of muscle bulk in males (102).

In the female, estradiol is the most active of the family of estrogen hormones, synthesized in the theca and granulosa cells of the ovaries. Production of estradiol is controlled by pulsatile release of two gonadotropins, follicle-stimulating hormone (FSH) and luteinizing hormone (LH), secreted by the anterior lobe of the pituitary gland, which in turn is stimulated by gonadotropin-releasing hormone (GnRH) in the hypothalamus.

Before puberty, a "gonadostat" in the central nervous system is exquisitely sensitive to the negative feedback influence of small amounts of circulating estrogen, and GnRH is suppressed. As a result, in young girls, LH secretion and FSH secretion are approximately 3% and 15%, respectively, of values normally seen in mature females. At the onset of puberty, the feedback sensitivity threshold of the gonadostat is markedly reduced, and, GnRH triggers production of increasing amounts of FSH and LH and ovarian release of estrogen.

FSH and LH also stimulate the ovary to produce progesterone, which acts to prepare the lining of the uterus for implantation of the fertilized ovum. The secretion of FSH and LH—and consequently of estrogen and progesterone—are periodic, explaining the timing of the menstrual cycle. In the early follicular phase of the cycle, circulating levels of both estrogen and progesterone are low, and both rise in the midluteal phase.

At puberty, then, the principal endocrinologic event is the augmented secretion of hypothalamic GnRH. In response to the magnified stimulation of LH and FSH, estradiol levels rise to from 15 to 35 $pg \cdot ml^{-1}$. Serum concentrations of estradiol begin to rise by approximately 10 years of age and continue to increase through each stage of puberty, as indicated by the development of secondary sexual characteristics (discussed later). Levels are particularly accelerated in the year before onset of menarche (age at first menses).

Differences in blood levels of estradiol between boys and girls are minimal but detectable in the prepubertal years (72). Using a very sensitive recombinant cell bioassay, Klein et al. reported that mean serum estradiol levels were 2.2 and 0.3 pmol · L^{-1} in prepubertal girls and boys, respectively (65). Estrogen levels do rise at the time of puberty in males as well, but values in females are about five times greater.

The clinical expression of puberty begins in girls between the ages of 10 and 11 years and typically lasts 4 to 5 years. Increased accumulation of adipose tissue is reflected by a higher percentage of body fat in girls by age 7 years. Breast development (thelarche) begins at 10.5 to 11.0 years, and menarche occurs approximately 2 years later. The timing and tempo of this schedule of pubertal development, however, varies widely between individuals.

Information concerning age of menarche was recently provided by the Third National Health and Nutrition Examination Survey of 2,510 girls who were 8 to 20 years of age (29). Fewer than 10% of girls start to menstruate before the age of 11 years, while 90% have reached menarche by age 13.75 years. The median age of menarche in the United States is currently 12.4 years, with menarche occurring significantly earlier in non-Hispanic black girls than in non-Hispanic white and Mexican-American girls.

Age of menarche, despite a trend toward younger ages in the past, is not significantly different from that reported in U.S. girls 30 years ago. This observation refutes the argument that the previous progression toward earlier age of menarche reflected the increasing incidence of childhood obesity (5). Since 1970 no further decrease in age of menarche has been evident despite an unabated epidemic of obesity.

In preparation for reproductive function, the rise in estrogen level stimulates ovulation, maturation of the female reproductive tract, and breast development. But the actions of estrogen are far more extensive, including effects on body fat content, mood, arteriolar vasodilatation, production of coagulation factors by the liver, skin characteristics, and risk factors for coronary artery disease (50). Estrogen is also responsible for a set of physical features termed *secondary sexual characteristics,* which include breast development and changes in the amount and pattern of pubic hair.

These secondary sexual characteristics form the basis of a classification system of pubertal development defined by and named for J.M. Tanner (see reference 80 for a review). Based on features of breast development and appearance of pubic hair and genitalia, girls are classified from Tanner stage I (preadolescent) to Tanner stage V (mature). As noted earlier, considerable variability is observed in the rate at which girls pass through these pubertal stages. One girl may complete a stage in six months, while another may take as long as two years.

The onset of puberty is about two years later in boys than in girls. During the prepubertal years, blood levels of testosterone, the primary male androgenic sex hormone, are very low (<20 pg · dl^{-1}). A rapid rise that coincides with timing of the adolescent growth spurt is observed at the time of puberty. Testosterone levels in the mature male are generally at least 20 times greater than those before puberty (figure 3.1).

In males, hypothalamic release of GnRH flows via hypophyseal circulation to the anterior pituitary, which produces FSH and LH. In the testes, LH acts on the Leydig cells to produce testosterone, while FSH stimulates the Sertoli cells to manufacture sperm. Spermatogenesis is usually evident by age 14 years. The action of testosterone on the hypothalamus creates a negative feedback loop similar to that described earlier for the gonadostat in females (52).

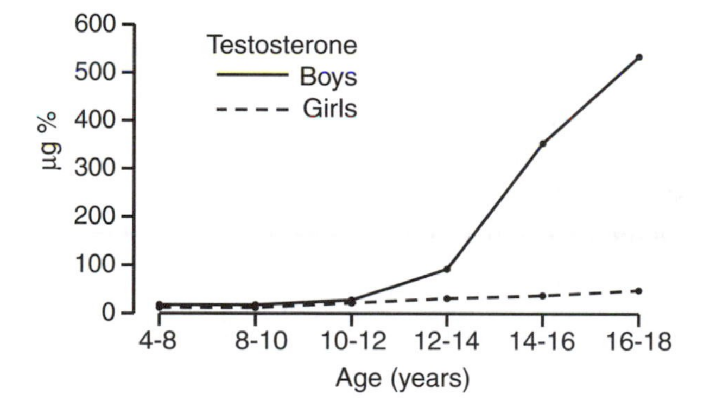

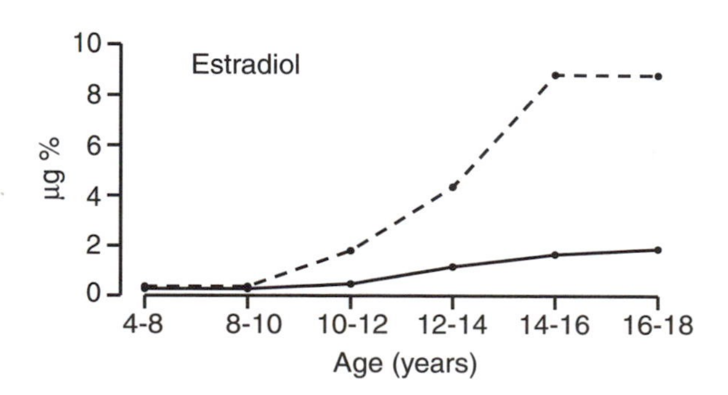

FIGURE 3.1 Changes with age in serum levels of testosterone and estradiol in children (from reference 80). Reprinted by permission from R.M. Malina and C. Bouchard 1991.

As estrogen does in the female, testosterone in the male acts to develop sexual function (penile and testicular development, sperm production) as well as somatic characteristics, particularly increases in skeletal muscle mass and linear growth. With puberty, the male develops facial, axillary, and pubic hair and experiences changes in sweat and sebaceous glands and deepening of the voice. Tanner staging in males is based on changes in genitalia and pubic hair. Pubertal development in males can also be estimated by testicular size. Small amounts of testosterone, resulting from enzymatic conversion of androgenic precursors in the adrenal gland, are also evident in mature females.

Leptin and the Initiation of Puberty

An explanation for the stimulus that "awakens" the hypothalamic-pituitary-gonadal axis at the onset of puberty has long intrigued but eluded investigators. The central theme in this story is the liberation of the GnRH generator from the inhibitory influence of feedback agents. At one time or another a wide variety of factors have been suggested as the inciting agent for this process: opioids, neuropeptide Y, galanin, corticotrophin-releasing factor, noradrenaline, dopamine, serotonin, melatonin, and gamma-aminobutyric acid (46). However, as Brook concluded, "numerous attempts have been made to identify modulators which either suppress or initiate human puberty, but there has really been no advance. The simple truth is that we have no idea how puberty is initiated" (20, p. 53).

This fact notwithstanding, considerable interest has surrounded recent evidence that leptin may act to influence the initiation of pubertal events. This is an intriguing notion, since, if it is true, leptin would serve as a link between reproductive function and nutritional status, which, as discussed later, might help explain the effects of exercise training on pubertal onset and menstrual function.

Leptin initially gained attention for its role as an appetite suppressant in obese mice. Secreted by fat cells, leptin was found to act on the hypothalamus of these animals to decrease food intake, inspiring hope that it might serve as a therapeutic agent for human obesity. Unfortunately, its action in humans is not so clear-cut, and the expectation that leptin might be an effective means of weight reduction in humans has so far ended in disappointment. In humans, serum leptin level is related exponentially to body fat content, suggesting that leptin insensitivity occurs as a person becomes more obese (31).

The leptin story has taken a new direction because of evidence that it might serve as a link between peripheral energy stores (i.e., body fat) and regulation of reproductive capacity (31). The idea that such a connection should occur at all is based on the evolutionary concept that fertile women who are about to conceive and deliver babies should possess sufficient caloric stores to satisfy the energy demands of pregnancy. This Darwinian standpoint, then, suggests that the activity of the hypothalamic-pituitary-gonadal axis becomes critically sensitive to caloric balance and amount of fat stores.

Most of the data that support a role for leptin in the initiation of puberty again come from studies in mice. Leptin levels are related directly to animal fat content, and concentrations fall during fasting. Animals who are deprived of food demonstrate, in addition, pituitary hypofunction and delayed sexual maturation (1). Leptin administered to normal prepubertal mice accelerates the onset of signs of puberty, but only if food intake is restricted to 70% of normal (26, 27). Kiess et al. thus described leptin as a "metabolic gate" to permit sexual maturation only under adequate nutritional conditions (64).

Evidence exists to suggest a similar function of leptin in human beings. The rare humans who have a congenital deficiency of leptin do not experience puberty until given exogenous leptin therapy (90). A surge of leptin concentration has been observed in boys just before pubertal onset (81). In a longitudinal study, Ahmed et al. found that leptin levels rose progressively with increasing age in both boys and girls as they approached puberty (2). This suggested to the authors that there might exist a leptin level threshold (different among individuals) that triggers the beginning of puberty.

Horlick et al. examined the relationship among level of sexual maturation, gonadal hormones, fat mass, and circulating leptin concentrations through the course of puberty in 6- to 19-year-old subjects (57). Plasma leptin levels were positively related to fat mass at all levels of sexual maturation in both males and females. Plasma leptin concentrations expressed relative to fat mass were not different between males and females except at Tanner stages IV and V, when values were greater in females (figure 3.2). This "sexual dimorphism" in leptin levels in late puberty thus cannot be explained by sex differences

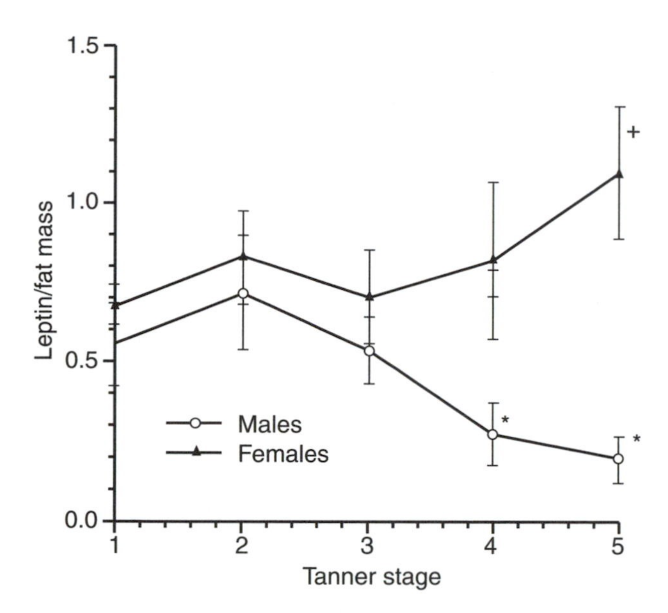

▶ FIGURE 3.2 Differences in serum leptin levels (expressed relative to lean body mass) at different pubertal stages in males and females (from reference 57). *P<0.05 compared to females at the same Tanner stage and compared to males at earlier Tanner stages; +P<0.05 compared to females at Tanner stage 1.

Reprinted by permission from Horlick et al. 2000.

in body composition (e.g., fat content) that occur with puberty.

Several clinical situations also support the possibility of a link between leptin, body composition, and reproductive function in humans. Leptin levels in girls with precocious puberty are elevated compared with those of normal prepubertal girls (91). Boys with constitutional delay of somatic and sexual development have hypoleptinemia (48). And low levels of leptin have been demonstrated in patients with anorexia nervosa, who are characterized by low body fat content and amenorrhea (69).

Serum testosterone appears to have a negative effect on leptin production. This is observed not only in vitro with cultured adipocytes but also clinically when adolescents with delayed puberty are treated with testosterone. In their review, Roemmich and Rogol commented that it was not possible to determine if this represented a direct inhibitory effect of androgens or a secondary effect via changes in body fat (98). Estrogens, on the other hand, stimulate increases in leptin levels. Roemmich et al. found that, even after correction for body fat content, serum leptin and estrogen concentrations in girls were directly related (97).

In summary, these data suggest, as Clayton and Trueman concluded, that leptin is important in human pubertal development (31). Its role, however, appears to be permissive rather than as a primary agent initiating puberty.

The role that leptin might play in triggering puberty is conveniently consistent with the early hypothesis that a critical amount of body fat is necessary to stimulate the onset of puberty. In the 1960s, Frisch (43) presented the hypothesis that the machinery of the hypothalamic-pituitary-gonadal axis does not commence until a girl achieves a certain amount of body fat (around 17% of body weight). At that time this idea was based on evidence for the metabolic effect of adipose tissue on estrogen activity. Although "enormously attractive," however, this concept never gained wide acceptance (114). The critical review by Scott and Johnston in 1982 concluded that it "does not appear that [menarche and the maintenance of menstrual cycles] have much to do with a critical fat percentage" and that "the critical weight (fat) hypothesis cannot be accepted" (114).

Now, 40 years later, the evolving picture of leptin as a chemical intermediary between adiposity and reproductive function lends credence to the basic critical-fat hypothesis. The idea is further supported by recent studies indicating that a negative caloric balance (rather than physical stress itself) is the primary factor responsible for secondary amenorrhea with sport training (77).

This issue has drawn the attention of exercise scientists because of (a) sport training's recognized effects of both reducing body fat and interrupting normal menstrual patterns, (b) the association of children's intensive early sport participation with later age of menarche, and (c) changes in serum leptin concentrations that occur with certain forms of exercise. It is tempting to consider that leptin might serve as a mediator by which exercise induces changes in body composition and reproductive function. We return to this issue later in the chapter, in the discussion of the effects of exercise training on pubertal development.

Assessment of Pubertal Status

Unquestionably, the hormonal changes at puberty can substantially affect physiologic responses to exercise and physical performance in youth. It follows that consideration of level of sexual maturation is often critical to the validity of comparison studies of young subjects. Consider an investigation of the effects of soccer play versus swim training on the bone density of young females. Estrogen level, reflecting sexual development, serves as a strong stimulant of bone development during puberty. If the subjects are classified by chronologic age rather than level of sexual maturation, comparisons between the two groups might be spurious.

Sometimes matching subjects by stage of sexual development is important because of what is *not* known about pubertal influences. For instance, suppose we wanted to know the effects of sprint training on anaerobic fitness as measured by Wingate cycle testing. The sexual-pubertal influences on such a response are unknown, but it is highly likely that differential effects related to hormonal influences by sex and level of sexual maturation would be important.

Proper assessment of pubertal status therefore is an important challenge for exercise scientists. A number of different means have been used to accomplish this in exercise research studies (32).

Secondary Sexual Characteristics

In 1962 Tanner and colleagues developed a classification scheme for defining pubertal development based on the appearance of certain secondary sexual characteristics in males and females (117). The scheme involves five stages, ranging from prepubertal (stage I) to full sexual maturity (stage V), based on photographs or line drawings of genitalia and pubic hair in boys and breast development and pubic hair in girls.

Although it is the most common means of assessing level of sexual maturation, this approach has its drawbacks. As pointed out by Malina and Bouchard, subjects just entering a stage and those leaving it are given the same rating, although they may differ significantly in level of sexual maturation (80).

Then there is the concern regarding invasion of personal privacy, as well as the presumed need for such an assessment to be performed by a health professional. While assessment by a health professional is considered the optimal means of Tanner staging, parental and self-reports have also been shown to be highly accurate. Brooks-Gunn et al. compared the validity of maternal ratings of the Tanner stage of their children to scores by trained physicians (21). The correlation between the mothers' assessments of their daughters' stage and the physicians' assessments was $r = .85$.

Studies comparing subjects' self-scoring (using photographs or line drawings) and scoring by physicians have also generally shown high correlations ($r = .77$ to .91; reference 38). Face-to-face descriptions of the stage characteristics and use of a mirror appear to improve the accuracy of self-reports of sexual maturation (32).

Menarche

The age when a girl first experiences menses has often been used as a well-defined landmark for establishing pubertal status. It does not, of course, provide any information regarding *level* of sexual development; a girl is either premenarcheal or postmenarcheal. And it is clear that premenarcheal girls cannot necessarily be considered prepubertal. Onset of menses occurs rather late in the pubertal progression, and menarche may occur as much as two years after thelarche (initial breast development).

Age at menarche is an important psychologic as well as physiologic event and is therefore usually accurately identified by adolescent girls. Koo and Rohan reported that 77% of girls were able to recall the date of menarche within one month of the date they reported three years earlier (68).

Spermarche

The use of spermarche as an indicator of pubertal development has been limited by social and ethical considerations. Ji described the relationship between age of first ejaculation emission with markers of somatic growth and physical fitness in a large number of Chinese boys ages 9 to 18 years (60). Boys who had experienced spermarche at any given age, compared with those who had not, were advanced in anthropometric factors and indicators of motor fitness (50-m dash, standing long jump, endurance run, and pull-ups). However, these differences disappeared by the age of 16 years. Boys who were prespermarcheal at any given age showed a greater potential for increasing leg length in late adolescence. This report thus indicates that spermarcheal status relates to growth and motor performance as other markers of biologic maturation do.

Serum Hormone Levels

Serum testosterone levels have been utilized as a marker of sexual development in males. Testosterone level variability is circadian (diurnal) as well as seasonal. Concentrations rise with physical activity, and considerable interindividual differences have been observed in serum levels. Nevertheless, with careful consideration of timing, measurement of serum testosterone levels to indicate level of sexual maturation may be reasonable.

Physiologic and Anatomic Expression of Sexual Maturation

The hormonal alterations during puberty become manifest in a remarkable multitude of anatomic and physiologic changes. This section examines how these expressions of sexual maturation might influence the physiologic responses to exercise, particularly as related to sex.

GH/IGF-I Axis, Sex Hormones, and Linear Growth

Puberty is marked by a dramatic acceleration of linear growth. Height velocity peaks during the midteen years, and the peak is later in males than females. Data from 22 studies compiled by Beunen and Malina indicate similar values for peak height velocity (PHV) in both North American and European youth (15). Average age at PHV was approximately 12 years in girls and 14 years in boys, with an overall standard deviation of about 1 year. In these reports, the mean PHV was 7.9 ± 0.6 cm · yr^{-1} in girls and 9.2 ± 0.6 cm · yr^{-1} in boys. (This contrasts with stable values of about 5.5 cm · yr^{-1} in the prepubertal period.) Children who had an earlier growth spurt were more likely to have had a higher one; that is, a negative relationship was observed between magnitude of PHV and age at PHV. However, no association was seen between PHV and adult stature.

While increases in testosterone and estrogen were once considered responsible for this increase in height at puberty, it is now recognized that these hormones work synergistically with increases in GH and IGF-I to stimulate growth (101). Part of this evidence is the observation that the adolescent growth spurt is impaired in patients who have growth hormone deficiency (83). Martha et al. demonstrated that mean 24-hour GH levels in late puberty were twice those before puberty and that this was a reflection of larger rather than more frequent pulses (83). Hindmarsh et al. demonstrated in turn that GH pulse amplitude and height velocity are closely related (56).

Parker et al. suggested that this increase in GH might be a direct effect of augmented secretion of sex hormones; these researchers found that testosterone administered to prepubertal boys stimulated production of IGF-I (92). And Ulloa-Aguirre and colleagues described dramatic increases in GH secretion after three months of testosterone treatment in early prepubertal boys with constitutional delay in growth (119). The action of testosterone on GH secretion appears to disappear during late puberty, however. GH and IGF-I levels decline at this time, while sex hormone concentrations remain stable.

Evidence exists that estrogen can have the same effect in girls. Mauras et al. (84) found that patients with Turner syndrome (who have gonadal dysgenesis) given low doses of estradiol for five weeks showed a significant rise in 24-hour GH production and pulse amplitude (but no changes in plasma IGF-I levels). Marin et al. compared GH responses to a standard treadmill exercise test in prepubertal children, who were randomized to receive estradiol or a placebo for two days prior to testing (82). The subjects treated with estrogen demonstrated a rise of GH to 18.7 ± 9.2 ng · ml^{-1}, compared with 6.9 ± 4.2 ng · ml^{-1} in those given placebo. It has been suggested that the faster rise in height velocity and earlier age of PHV in girls occurs because estradiol is a better modulator of GH secretion than testosterone (20).

The increased linear growth at puberty therefore is largely influenced by the GH/IGF-I axis under the stimulation of the sex hormones. Any possible direct role of androgens and estrogen on height development is uncertain. Rogol also noted, "Any straightforward relationship between growth velocity and the circulating GH concentrations, or attributes of GH neurosecretion, is diffused by the added components of GH binding proteins, circulating IGF-I and its binding proteins" (101).

Puberty and Insulin Resistance

Puberty is accompanied by significant changes in insulin production and sensitivity. Specifically, normal puberty is characterized by an increase in cellular resistance to insulin action that is not accompanied by alterations in other insulin actions, such as lipolysis and stimulation of protein metabolism. Insulin-stimulated glucose uptake (determined under hyperinsulinemic conditions) has been reported to be reduced by 20% to 45% during the adolescent years (4, 24).

The cause and physiologic significance of this phenomenon remain to be clarified. Any proposed etiologic agent or process must be reconciled with the time course of insulin resistance as puberty progresses. Insulin resistance peaks at Tanner stage II and then to returns to baseline prepubertal levels at sexual maturation (stage V; 87).

Current information fails to support a specific effect of sex hormones or body composition on the pubertal changes in insulin sensitivity. As the most obvious clue, body fat and levels of hormones both rise throughout puberty, while insulin sensitivity declines and then increases (49). The role of increases in growth hormone, however, may be important. GH has prominent anti-insulin actions, and the rise in GH levels during puberty could well explain the observed insulin resistance and enhanced insulin response to glucose during puberty.

Insulin resistance during puberty is compensated for by elevated insulin levels, such that glucose homeostasis is not expected to be disturbed. The insulin response to an intravenous glucose load in adolescents increases two to three times more than that of prepubertal and adult subjects (24). Caprio (24) postulated that this hyperinsulinemic response acts to suppress levels of IGF-binding protein (IGFBP-1). This would result in increased blood concentrations (and anabolic actions) of free IGF-I. In this scenario, then, changes in insulin receptors at puberty would serve as a means of modulating the GH/IGF-I axis and stimulating somatic growth.

Goran and Gower conducted a longitudinal study of insulin actions in 60 children at Tanner stage I (mean age 9.2 ± 1.4 years) and again two years later (49). Among the half of the subjects who had advanced to Tanner stage III or IV, insulin sensitivity (determined by a tolbutamide-modified intravenous glucose tolerance test) fell by 32%. This decline was not observed in those who were still prepubertal (Tanner stage I). Changes in insulin sensitivity were unrelated to alterations in fasting hormone levels (estradiol, testosterone, FSH, IGF-I, leptin, and cortisol) and to changes in body fat.

Arslanian and Suprasongsin administered testosterone for four months to adolescents with delayed puberty (9). This produced increases in lean body mass and levels of circulating testosterone and growth hormone, but no changes in insulin sensitivity were observed.

How, or if, the changes in insulin resistance that occur at puberty affect glucose utilization during acute or repetitive exercise remains to be clarified. In healthy humans an acute bout of exercise is accompanied by increased sensitivity to insulin, and increased glucose uptake by muscle cells occurs at the same time that insulin levels in the circulation actually fall (123). Findings in athletes or after a period of endurance training are similar. That is, the trend is opposite to that observed in puberty.

Differential Sex Hormone Effects on Fitness Variables

If testosterone is the "male" hormone and estrogen the "female," their actions should distinguish masculinity from femininity, muscle growth and virility from body fat accumulation and fecundity. It is apparent, however, that in some developmental processes during puberty, such as bone growth and development, similar changes are observed in both sexes. As Frank pointed out, "It makes little teleological sense to imagine two separate and distinct hormone-receptor pathways evolving to mediate the same physiologic effect that occurs in the pubertal skeleton in both males and females" (42, p. 627). This line of thinking—supported by recent experimental data—has led some to suggest that actions of the sex hormones are not, in fact, always sex specific. That is, testosterone and other androgenic hormones are active in females, and estrogens mediate anatomic and physiologic alterations in males. The following sections examine the effects of sex hormones on tissues and functions that relate to physiologic and physical fitness. In some cases, the traditional view that testosterone and estrogen effects are sex specific holds. In others, these hormones appear to cross the lines of sex.

Body Composition

Perhaps the most obvious differential effect of the sex hormones at puberty is the stimulation of the development of muscle mass by testosterone and body fat by estrogen. At puberty in males, increases in not only stature but also muscle mass accelerate. The typical muscle mass in an 11-year-old boy's body is 15 kg. With puberty-induced growth, this increases to 35 kg by the time he is 17 (15, 80).

Females are different. They experience smaller increases in stature and muscle mass but a significant accumulation of body fat. An 11-year-old girl has about the same muscle mass as her male counterpart (15 kg). But at age 17 she will possess only 22 kg of muscle. The ratio of fat-free mass to stature is similar in boys and girls at age 11 (0.21), but at age 17 it rises to 0.34 in males but only 0.26 in females. Muscle mass represents 53% of the weight of an average 17-year-old male but 42% of that of a 17-year-old female. In the prepubertal years the respective values are approximately 45% and 43% (figures 3.3 and 3.4).

These differences are an expression of the influence of sex on muscle fiber size (i.e., hypertrophy) rather than cell number. The average cross-sectional area of the vastus lateralis muscle is about 26% greater in 16-year-old boys than in 16-year-old girls.

Typical gains in fat mass during puberty in females (7.1 kg) are twice those in males (3.2 kg; 80). Percentage body fat is greater in females than in males even before puberty (e.g., 21% vs. 15% at age 11), but this gap is magnified during adolescence. At age 17, female body fat percentage is twice that of males (figure 3.4).

The greater body fat content in females is a manifestation of both greater number and size of fat cells. Adipocyte size is independent of sex until age 11, when a relatively greater increase is observed in males. At age 17, the average fat cell diameter is about 20% greater in females. A similar pattern is observed in cell number (figure 3.5).

The most obvious detrimental effect of this fat accumulation in females is to limit the performance of weight-bearing activities (e.g., running, chin-ups). That is, the added adipose tissue represents an additional inert load that must be transported. Up to one third of the variance in performance of the 1-mi run by a group of average sixth-graders can be accounted for by variations in body fat content (106).

The physiologic effect of the greater accumulation of adipose tissue in females at puberty is not so clear. Up to moderate levels of obesity at least, body fat does not have any detrimental effect on cardiac functional reserve during cycle testing (105).

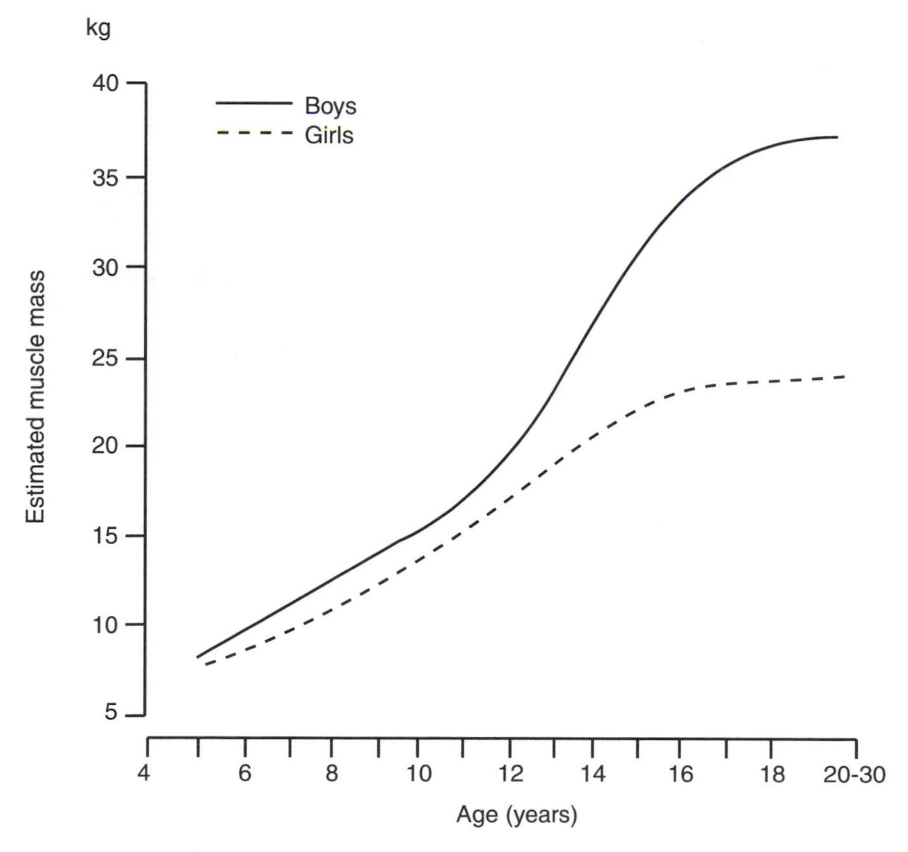

FIGURE 3.3 Increases in muscle mass with age from studies of creatinine excretion in boys and girls (from reference 80).
Reprinted by permission from R.M. Malina and C. Bouchard 1991.

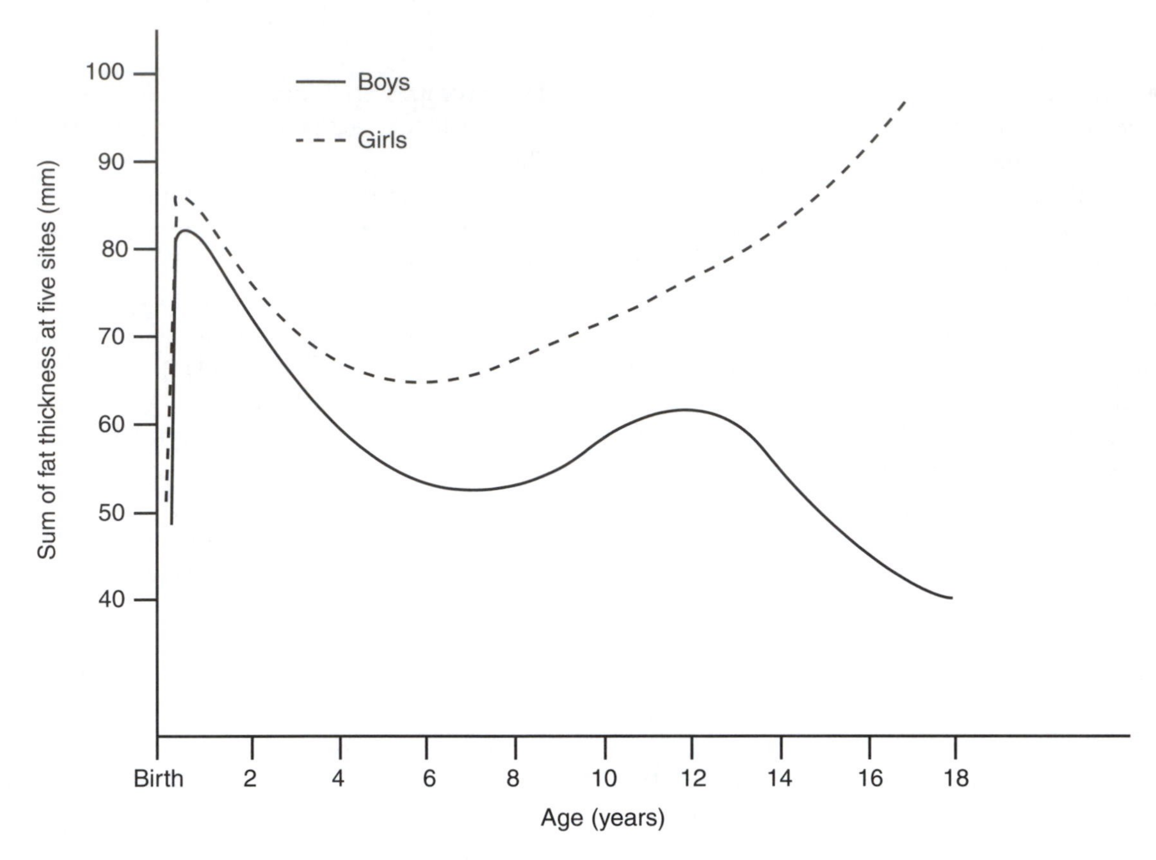

FIGURE 3.4 Changes in subcutaneous fat with age as estimated by skinfold measurements of boys and girls (from reference 80).

Reprinted by permission from R.M. Malina and C. Bouchard 1991.

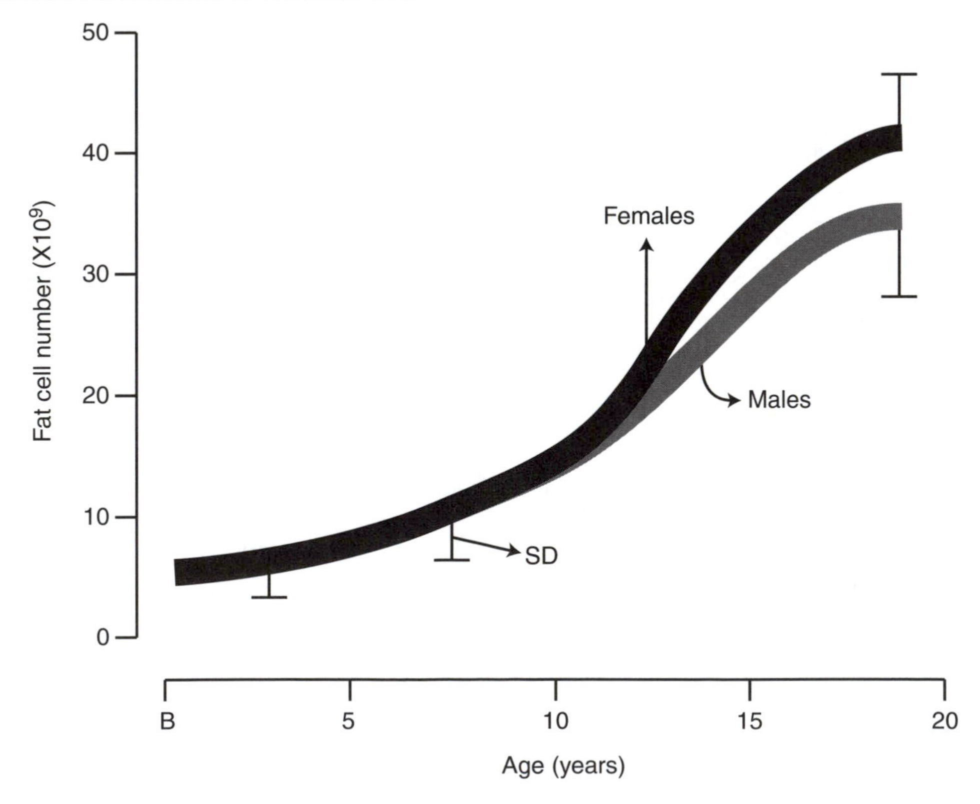

FIGURE 3.5 Increases in fat cell number in males and females during childhood and adolescence (from reference 80).

Reprinted by permission from R.M. Malina and C. Bouchard 1991.

Muscle Strength

Abundant evidence supports the time-honored hypothesis that testosterone increases skeletal muscle bulk and strength (see reference 17 for a review). Changes in muscle strength at puberty are closely associated with a rise in testosterone levels. Men who have depressed gonadal function characteristically also demonstrate low fat-free mass. Males who were given GnRH experimentally to suppress testosterone production demonstrated a fall in muscle protein synthesis and fat-free mass. The decline in testosterone as men age parallels a fall in muscle strength.

The mechanisms by which testosterone augments muscle protein synthesis and increases muscle size and strength are still under investigation. One possibility is that testosterone stimulates the anabolic effect of IGF-I in the muscle cell (120). Testosterone might also improve strength by some means other than increasing muscle mass. Neuromuscular transmission may be facilitated by testosterone, and the cognitive effects of androgens might alter motivation to train or perform at a higher intensity (17; discussed later in this chapter).

Cardiac Size and Function

Animal studies indicate a strong anabolic effect of testosterone on the myocardium. Scheuer et al. reported that gonadectomy (i.e., removal of sex hormone effect) decreased ventricular contractility in both pre- and postpubertal male and female rats (112). These effects were prevented by testosterone replacement in males and by both estrogen and testosterone treatment in females.

Koenig et al. reported myocardial hypertrophy, increased cell RNA and protein, and more cytochrome oxidase in animal heart muscle after testosterone administration (67). Total DNA did not increase, indicating that myocardial growth involved hypertrophy rather than hyperplasia. They concluded, "It is clear from present findings that endogenous androgens exert an important regulatory influence over the metabolism of the ventricular myocardium. Thus androgens would appear to merit inclusion among those hormones, e.g. growth hormone, thyroxine, insulin, which regulate protein turnover in the heart" (67, p. 785).

These myocardial changes may be associated with alterations in plasma volume that can also affect cardiac output. Broulik et al. reported that cardiac output of mice fell by an average of 13% after castration (22). Three weeks after castration the blood volume of the animals had decreased from 90 ± 3 ml · kg^{-1} to 82 ± 2 ml · kg^{-1}. Both of these effects were reversed with testosterone replacement. Gardner et al. studied the effects of biweekly pharmacologic doses of intramuscular testosterone on 15 old male veterans of the Spanish-American War (44). Treatment resulted in an increase in mean plasma volume from 2,056 ml to 2,306 ml and in total circulating albumin from 75.8 g to 89.9 g.

In humans, the size of the left ventricle in males increases at a faster rate during puberty than it does in females (53). As noted by both Janz et al. (59) and Daniels et al. (36), heart mass does not become any greater during the pubertal years than can be accounted for by increases in lean body mass (an effect of testosterone).

Erythropoiesis

Testosterone stimulates the production of erythrocytes and raises hemoglobin concentration, hematocrit, and red blood cell volume (115). Castrated adult male animals become anemic, and this is reversed by administration of androgens. The same phenomenon is seen in human patients with hypogonadism, while those with diseases characterized by high androgen levels (Cushing's syndrome, congenital adrenal hyperplasia) are often polycythemic. In the past (i.e., before the commercial availability of erythropoietin) androgens were used therapeutically for their erythropoietic effect in patients with conditions such as aplastic anemia, myeloid metaplasia, and chronic renal failure.

Before the onset of puberty, blood hemoglobin concentration is similar in boys and girls, with a slow rise from about 12.6 g · dl^{-1} at age 2 years to 13.7 g · dl^{-1} at age 12 (35). At puberty, hemoglobin levels continue to rise in males in association with the secretion of testosterone, while concentrations in females remain stable. The expected value for a 16-year-old boy (15.2 g · dl^{-1}) is 10.9% greater than that of a girl the same age (13.7 g · dl^{-1}). The average hematocrits in postpubertal male and female adolescents are 47% and 42%, respectively.

In vivo studies indicate that androgens can directly stimulate erythropoietic stem cells (94). Others have demonstrated, however, that testosterone may act indirectly on red cell production by enhancing production of erythropoietin by the kidney, as the effects of androgens on hemoglobin and red cell production are eliminated in nephrectomized rodents (85).

Cognitive Function, Pain Perception, and Mood

Estrogen exerts a number of effects in the central nervous system, and changes in estrogen levels are related to cognitive function and mood. Perhaps most importantly for exercise scientists, estrogen is recognized to alter the activity of serotonin, which can alter perception of pain (14).

Riley et al. performed a meta-analysis of 16 studies that examined the threshold for pain across the menstrual cycle (95). Creators of discomfort in these reports included pressor stimulation, cold pressor pain, heat, and muscle ischemia. These studies revealed a higher pain threshold during the early follicular phase of the menstrual cycle, when circulating levels of estrogen are lowest. This implies that estrogen decreases tolerance to pain. It is intriguing to infer from these data that estrogen-induced changes in motivation and persistence in exercise in the teen years might contribute to the decline in physical performance observed in females at this time.

Testosterone, on the other hand, has commonly been associated with aggressive behavior, criminal activity, and excessive sexual drive. Positive correlations are observed in men between testosterone levels and feelings of elation and joyfulness but negative correlations with depression and anxiety (28).

Some interesting studies relevant to this concept have been conducted in athletes. Scaramella and Brown found that testosterone concentrations in male hockey players correlated with their aggressive play in competition (111). Confidence in one's abilities as an athlete may be related to testosterone levels. Booth et al. reported that concentrations of androgens increased 15 minutes before a tennis match if the player had been victorious in the previous match and expected to win again (18).

These limited observations of the role of sex hormones in athletic play are intriguing, particularly considering the critical importance of psychologic factors in sport performance, particularly at the elite level.

Bone Development

Accumulating experimental evidence indicates that estrogen is the primary determinant of normal bone maturation in both males and females. Humans who are hypoestrogenemic (e.g., those with Turner's syndrome) have a significant delay in bone development, as do the rare males who lack estrogen but have normal levels of androgenic hormones (30). A deficiency of estrogen causes bone loss and increased susceptibility to fractures. The severity of menstrual dysfunction in athletes has been related to changes in bone density (93).

Osteoclasts and osteoblasts possess estrogen receptors, but the principal mechanism by which estrogen stimulates bone changes is inhibition of resorptive processes. A state of estrogen deficiency, then, results in excessive osteoclastic activity, with remodeling and resorption of bone tissue (50).

Normal bone development during puberty is considered important in preventing osteoporosis in later life. By age 18 years, more than 90% of peak skeletal mass has developed, and skeletal mass at this age accounts for most of the variability in bone mass at older ages (30).

Schoenau et al. provided evidence that estrogens promote "excessive" bone growth at puberty, perhaps in relation to the need for calcium for later gestation and lactation (113). They tested 318 healthy youths ages 6 to 22 years. Computed tomography was used to measure cortical area of the radius (CA) and muscle cross-sectional area (MA) as indicators of bone and muscle strength, respectively. Since muscle mass and strength are principal stimulants of bone growth, the ratio of CA to MA is expected to remain constant during puberty.

In this study, as expected, a close relationship was observed between CA and MA ($r^2 = .77$). In prepubertal children no sex differences were observed in the relationship of CA to MA. Once into the pubertal years, however, CA:MA was greater in girls than in boys. Analysis of covariance (ANCOVA) indicated that while muscle mass was the strongest predictor of CA, pubertal stage and sex both had significant effects.

Exercise-Induced Muscle Damage

Vigorous exercise, particularly with unaccustomed muscles performing eccentric contractions, produces muscle damage, increases circulating creatine kinase (CK), and causes delayed-onset muscle soreness. There is reason to expect that estrogen might have a protective effect against exercise-induced muscle damage. Estrogen acts as an antioxidant, stabilizes cell membranes, and diminishes inflammatory response, all factors related to muscle cell damage with exercise (62).

The evidence supporting this idea, however, is limited. Compared with men, women have lower levels of CK with exercise (100). And after removal of the ovary, female rats' CK responses to exercise were similar to those of males (12). However, histologic markers of muscle damage from exercise do not differ by sex (121), and no differences have been observed in exercise-induced muscle damage between pre- and postpubertal females (23). Rinard et al. demonstrated that muscle soreness and decreased strength following exercise are no greater in adult males than in adult females (96).

Substrate Utilization

Several lines of evidence suggest that estrogen may influence substrate utilization during exercise (10). Studies in animals have indicated that estradiol enhances utilization of fats by stimulating lipolysis and thereby increases fatty acid oxidation. This would spare glycogen utilization and improve exercise capacity. Kendrick et al. demonstrated that rats receiving estradiol had greater run times to exhaustion than a control group (63).

Studies in humans, however, have not been entirely convincing. Most have compared physiologic variables at different times in the menstrual cycle (i.e., in respect to varying levels of circulating estrogen). Some authors have reported changes in respiratory exchange ratio (RER), indicating alterations in lipid utilization, that correspond to different phases of the menstrual cycle, while others have not (10). Most reports have described no differences in RER during sustained submaximal exercise between amenorrheic and eumenorrheic runners.

Estrogen has also been suggested to inhibit catecholamine uptake, which might affect substrate utilization. Weise et al. described resting plasma catecholamine levels in healthy children (37 boys and 43 girls) in relationship to pubertal development (128). Concentrations of epinephrine fell with advancing stages of puberty. This decline was inversely related to increases in estradiol, testosterone, and insulin.

Skin Blood Flow and Thermoregulation

The threshold body temperature that triggers cutaneous vasodilatation is influenced by phase of the menstrual cycle (25). When estrogen and progesterone are both elevated (luteal phase), body temperature increases by 0.3 to 0.5°C, and this rise is associated with an increase in the threshold for skin vasodilatation. (Administered estrogen, however, causes the opposite effect.) These thermal effects of female sex hormones may represent a central neurologic function as well. Both estrogen and progesterone influence temperature centers in the hypothalamus.

One cannot fail to be impressed by this information about the remarkable, far-ranging effects on physiologic function created by the onset of sex hormone production at puberty. This biochemical metamorphosis is intended to permit sexual reproduction, but in the meantime the sex hormones influence anatomic and physiologic changes in virtually every body system. The outcome of these changes distinguishes male from female and, particularly important for the developmental exercise physiologist, child from adult. In later chapters we reconsider many of these hormonal influences, when we discuss differences between children and adults in various aspects of exercise physiology.

Pubertal Effects on Physical Fitness

From the broad spectrum of these effects of estrogenic and androgenic hormones, it is reasonable to expect that the increased production of these hormones would influence physiologic and performance fitness during puberty. Moreover, these alterations should be different in males and females. If this is true, (a) individual measures of exercise capacity should be expected to accelerate or "spurt" at some point during puberty, as height does, and (b) changes in physiologic factors should correlate closely with markers of pubertal development such as Tanner staging and blood levels of sex hormones.

The following discussion addresses these issues in regard to the various forms of physical fitness. Exercise variables have frequently been compared between prepubertal children and adults, and findings indicate maturation effects. In the following sections, however, only studies that assess fitness variables relative to stages of puberty or relative to changes in sex hormones—more valid indicators of maturational influence—are considered.

Maximal Aerobic Power

Longitudinal changes in absolute values for $\dot{V}O_2$max indicate that puberty strongly affects aerobic fitness, at least in boys. Values rise at a steady rate during the mid-childhood years, and no important differences are observed between boys and girls (8). At about age 11 years, however, the developmental curves for $\dot{V}O_2$max in boys and girls diverge. A continually increasing rate of rise is observed in males, while values plateau in females, such that at age 16 years mean $\dot{V}O_2$max is 60% greater in males (3.2 vs. 2.0 $L \cdot min^{-1}$).

In males the rise in $\dot{V}O_2$max is closely related to increases in body mass. Average $\dot{V}O_2$max expressed relative to body mass is stable at about 52 $ml \cdot kg^{-1} \cdot min^{-1}$ between the ages of 6 and 16 years, when it begins to decline. In girls, however, $\dot{V}O_2$max per kilogram declines progressively, beginning almost at the age when it can be first measured. Mean $\dot{V}O_2$max per kilogram is typically 50 $ml \cdot kg^{-1} \cdot min^{-1}$ in the 8-year-old girl but 40 $ml \cdot kg^{-1} \cdot min^{-1}$ at age 16. Changes in body composition that occur with puberty (i.e., augmented body fat) as well as low levels of habitual physical activity have traditionally been thought to account for these changes in females.

Improvements in $\dot{V}O_2$max as children grow are mediated by increases in maximal stroke volume, which, in turn, is a reflection of progressive left ventricular diastolic enlargement (see chapter 6). Increases in heart size with greater aerobic fitness should occur in concert with increases in other volume measures, such as lung size and skeletal muscle mass, as well as size-independent factors, such as muscle capillarization and muscle cell enzymatic aerobic capacity. As reviewed by Krahenbuhl et al., growth rates of these factors do parallel development of $\dot{V}O_2$max, but specific changes in these determinants relative to level of sexual maturation have not been delineated (71).

Some information is available regarding the influence of sexual maturation on development of the heart during puberty. In a five-year longitudinal study, Janz et al. described relationships between cardiac mass (determined by echocardiography), aerobic fitness, level of sexual maturation, and anthropometric variables in 125 healthy children (59). At the end of the study, subjects had advanced at least one Tanner stage, and over 80% were in late puberty or postpuberty. Spearman correlation coefficients of longitudinal changes of left ventricular mass with peak $\dot{V}O_2$, Tanner stage, and serum testosterone concentration were $r = .77$, .60, and .60, respectively, in the boys. Correlations were lower but still statistically significant in the girls ($r = .39$ for peak $\dot{V}O_2$ and $r = .32$ for Tanner stage).

Daniels et al. studied echocardiographic findings in relation to body composition and sexual maturity in subjects 10 to 17 years old (36). The variance in left ventricular mass was explained largely by growth of lean body mass, and multiple regression analysis revealed no influence of sexual maturation.

Malina et al. described longitudinal relationships between $\dot{V}O_2$max and maturation in 47 boys and 40 girls enrolled in sport schools (79). Anthropometric measurements, Tanner staging, and maximal exercise testing were performed annually for three years beginning at age 11. Boys were divided into early, average, and late maturers based on the slope of height velocity during the study period, while maturational age in the girls was based on the timing of menarche. Among the boys, early maturers demonstrated higher absolute $\dot{V}O_2$max at all testing sessions, while early- and average-maturing girls had greater $\dot{V}O_2$max than late maturers. These findings suggest that the time pattern, or tempo, of the development of aerobic fitness during puberty parallels that of sexual maturation.

Beunen and Malina reviewed studies that assessed the influence of puberty on aerobic fitness in boys by examining the temporal relationship between peak height velocity (PHV) and adolescent spurt of $\dot{V}O_2$max (15). Data were thought to be too scant to warrant such an analysis in girls.

In general, the age of maximal increase in $\dot{V}O_2$max very closely approximated the age of PHV. Estimated velocity of $\dot{V}O_2$max at PHV was 0.41 $L \cdot min^{-1} \cdot yr^{-1}$, compared with 0.17 $L \cdot min^{-1} \cdot yr^{-1}$ between the ages of 9 and 10. This analysis indicated that the spurt of development of $\dot{V}O_2$max with age was broader than that of height velocity. In boys, average height velocity rises quite abruptly at age 12, and after peaking declines to prepubertal levels by age 15, then falls to zero by age 17 or 18. The acceleration of $\dot{V}O_2$max, on the other hand, starts about five to six years before PHV, and the rate of rise continues to increase until the late teen years. That is, the acceleration of $\dot{V}O_2$max persists while height velocity is declining.

These data suggest that puberty influences improvements in aerobic fitness by increasing body size, particularly the dimensions of the heart, lungs, muscles, and circulatory system. Experimental evidence supports the concept that sexual maturation has no influence, other than the changes it induces in body size and composition, on the development of $\dot{V}O_2$max. First of all, $\dot{V}O_2$max expressed relative to body mass remains stable during puberty in males and actually declines in females. Fahey et al. found no significant relationship between peak $\dot{V}O_2$(expressed relative to body mass) and testosterone levels when the effect of age was eliminated (39). Welsman et al. reported that serum testosterone levels were not related to peak $\dot{V}O_2$ in 12- to 16-year-old boys when age, stature, and mass were considered (129). They therefore concluded that sexual maturation independent of increases in body size has a minimal role in the development of $\dot{V}O_2$max in males.

Others, however, have argued that improvements in $\dot{V}O_2$max at puberty are greater than can be accounted for by somatic changes alone. This interpretation is based on findings of studies that have avoided the use of ratio standards as a potentially misleading means of assessing changes in aerobic fitness over the pubertal years.

Williams et al., for example, compared the relationship of peak $\dot{V}O_2$ and body mass by both ratio standard ($\dot{V}O_2$max per kilogram of body mass) and linear regression analysis in pre- and postpubertal boys (131). No significant differences in average mass-relative $\dot{V}O_2$max were observed between the groups (49 and 50 $ml \cdot kg^{-1} \cdot min^{-1}$). Analysis of covariance with linear regression indicated no difference between the groups in the slopes of the lines that related mass to peak $\dot{V}O_2$, but the elevations of the lines were significantly different (the *y*-intercepts were 0.67 and 1.40 in the pre- and postpubertal groups, respectively). These findings indicate that peak $\dot{V}O_2$ relative to body size is, in fact, greater after the age of puberty than before. This implies that sexual maturation has a separate effect on peak $\dot{V}O_2$ beyond simply increasing body size.

Armstrong et al. used log-linear ANCOVA to demonstrate a significant influence of sexual maturation (determined by Tanner staging) independent of body size on peak $\dot{V}O_2$ in both boys and girls (7). The authors considered the possibility that changes in hemoglobin concentration (and thus blood oxygen-carrying capacity and arterial-venous oxygen difference) might contribute to these findings, at least in the boys. In the males, hemoglobin concentration increased 6.1% as adjusted peak $\dot{V}O_2$ rose 14.4% with advancement from Tanner stage I to IV. These comparisons were not so persuasive in females, however, who showed a 2.3% increase in hemoglobin concentration with the same increase in sexual maturation but an 11.8% rise in peak $\dot{V}O_2$.

In a subsequent longitudinal study using multilevel regression modeling, Armstrong and Welsman demonstrated that peak $\dot{V}O_2$ increases between the ages of 11 and 17 years above the effects of body size alone (6). No influence of blood hemoglobin concentration on this growth in aerobic fitness could be detected.

Statistical analyses other than ratio standards thus indicate that sexual maturation during puberty improves $\dot{V}O_2$max not only by increasing body size but by some other factor. What could this be? Consider the usual suspects from the Fick equation: $\dot{V}O_2$ = heart rate × stroke volume × arterial-venous oxygen difference. Maximal heart rate can be discounted since values are constant until late adolescence and are thus independent of level of sexual maturation. No information is available regarding the effects of sexual maturation on maximal stroke volume. Cross-sectional studies show no differences between prepubertal and adult subjects when values are related to body surface area (i.e., maximal stroke index; 89, 107, 109).

In males, an expected pubertal effect would be increased maximal arterial-venous oxygen difference caused by the rise in hemoglobin concentration in response to increased testosterone levels. In the same adult-to-child comparison studies mentioned earlier, variations in calculated values for maximal arterial-venous oxygen difference (AVO_2diff) suggest maturational and sex influences consistent with this difference in blood hemoglobin concentration and secondary rise in arterial oxygen content. Rowland et al. (107, 109) reported maximal values of 13.9 ± 3.0 and 17.2 ± 4.5 $ml \cdot dl^{-1}$ in boys and men, respectively ($p < .05$), but no differences between premenarcheal girls (12.2 ± 1.7 $ml \cdot dl^{-1}$) and adult women (12.1 ± 2.2 $ml \cdot dl^{-1}$). Similarly, Nottin et al. described a calculated maximal AVO_2diff of 13.8 ± 2.1 $ml \cdot dl^{-1}$ in boys and 16.9 ± 3.9 $ml \cdot dl^{-1}$ in men (89). As noted earlier, however, the multilevel regression study by Armstrong and Welsman did not support this effect of hemoglobin on the development of peak $\dot{V}O_2$ during puberty (6).

Janz et al. thought that the improvements in peak $\dot{V}O_2$ that occurred across the pubertal years and were independent of increases in body mass could be accounted for by changes in body composition

(58). Allometric analysis of their five-year longitudinal data from children indicated that peak $\dot{V}O_2$ increased with level of sexual maturation in both sexes, even with body mass considered. However, when lean body mass was substituted for body mass in the analysis, the influence of sexual maturation on the development of peak aerobic power was eliminated.

Improvements in $\dot{V}O_2$ max following a period of endurance training in prepubertal children (about 5%) are smaller than in adults (typically 15% to 20%). The explanation for this difference is not known. Katch suggested that "there is one critical time period in a child's life (termed the 'trigger point') which coincides with puberty in most children, . . . below which the effects of physical conditioning will be minimal or will not occur at all. It is suggested that this trigger phenomenon is the result of modulating effects of hormones that initiate puberty and influence functional development and subsequent organic adaptations" (61, p. 241).

Investigators have attempted to test this hypothesis by examining changes in $\dot{V}O_2$ max response to endurance training in subjects of varying levels of sexual maturation or longitudinally across the pubertal years. Unfortunately, these studies were handicapped by weaknesses such as small number of subjects (66) or high pretraining fitness (126), and a true experimental assessment of the trigger hypothesis has not yet been performed. A discussion of the factors that might be responsible for the diminished trainability of children is postponed until chapter 11.

It is difficult to determine if a spurt in endurance performance occurs at puberty, because normative values for a large number of subjects have been provided relative to chronologic age. This results in a smoothing of developmental curves in performance that might obscure individual accelerations. This considered, no clear-cut improvements are evident in endurance performance during puberty. Treadmill endurance times with the Bruce treadmill protocol tend to increase linearly with age in males until age 18 years, then decline (34). The peak in girls, however, is at age 12 years.

One-mile run time in both sexes improves until age 13 and then tends to plateau in boys. In girls, however, a decrement is observed during adolescence. These observations provide no good information regarding the biologic influence of puberty on aerobic performance. They do underscore the multifactorial, sex-related nature of physical performance, in this case the potential effect of increased body fat and social unacceptability of athleticism during adolescence for females (104).

Strength

Studies attempting to dissect the relative importance of pubertal versus hormone-independent factors on muscle strength, particularly as they relate to sex, have provided no consistent findings. Neu et al. studied serial changes in grip strength in 185 females and 181 males ages 6 to 23 years (88). Their findings indicated that the patterns of maximal isometric grip force reflected two components: growth in muscle size and grip force per muscle cross-sectional area (CSA). As expected, increases in muscle size and strength accelerated at puberty in males but not females, indicative of pubertal hormonal influences. However, when normalized to forearm length, grip strength per muscle CSA was not significantly different between sexes and was independent of the influence of sex steroids.

De Ste Croix et al. studied the effects of age, body size, and sexual maturation on the development of isokinetic knee extension and flexion using multilevel modeling (37). Forty-one subjects (20 boys and 21 girls) were studied on eight occasions over a four-year period, beginning when they were 10.0 ± 0.3 years old. Both stature and mass were found to be significant predictors of both peak knee extension and flexion, but once these variables were accounted for in the analysis, age and sexual maturation did not contribute to peak knee strength.

In a similar mixed longitudinal study of 50 boys and 50 girls from the ages of 8 to 17 years also using multilevel modeling, Round et al. determined changes in elbow flexor (biceps) and knee extensor (quadriceps) strength (103). In the girls, increases in body size fully accounted for improvement in muscle strength. In the boys, however, a size-independent contribution to strength improvements was observed that was fully explained by increases in blood testosterone concentration. The authors suggested that some other action of testosterone besides increasing muscle mass (possibly bone growth) was responsible.

A clear-cut spurt in explosive strength, such as that seen in performance of the long jump or vertical jump, occurs with puberty and appears to coincide with peak height velocity. When measured by flexed-arm hang time, an adolescent spurt is seen only in boys (80). In the Oakland Adolescent Growth Study, boys and girls between the ages of 11 and 18 years were grouped according to tempo of sexual maturation as early, average, or late maturers (80). Across this age span, the boys who were early maturers performed better on tests of grip strength and shoulder pushing strength than the other two groups of boys. No differences were seen in the strength tests of girls in the three groups.

Similar findings were observed in Belgian boys classified in the same manner (15, 16; figure 3.6). When body size was controlled for, the early maturers scored higher on static strength, explosive strength, and muscular endurance.

Anaerobic Fitness

Anaerobic fitness is more obviously influenced by size-independent factors (glycolytic metabolism, muscle architecture, neurologic input) as well as muscle bulk. One might expect, then, to see a greater influence of factors beyond body size in pubertal changes in anaerobic function. The extent to which these size-independent factors are influenced by the hormonal changes of puberty, however, is unclear.

Falk and Bar-Or employed a mixed cross-sectional/longitudinal design to assess the effects of sexual maturation on anaerobic fitness as measured by the Wingate test (41). They found significant increases in mean and peak anaerobic power (in watts per kilogram of body mass) in prepubertal, midpubertal, and late pubertal boys. The effect of sexual maturation per se was problematic, however, since similar improvements in anaerobic fitness in excess of increases in body mass are seen in the prepubertal years.

Falgairette et al. found close correlations between salivary testosterone levels and peak and mean anaerobic power (r = .45 and .47, respectively), both related to body mass, in boys ages 6 to 15 years (40). Armstrong et al. evaluated the effect of sexual maturation on Wingate test performance in 100 boys and 100 girls ages 12.2 ± 0.4 years (8). Sexual maturation was determined by Tanner staging (pubic hair) plus salivary testosterone levels in the boys. Allometric exponents were b = 1.02 for peak power and b = 0.82 for mean power. ANOVA indicated a main effect of sexual maturation on peak and mean power even when controlled for body mass. These findings indicate a size-independent influence associated with sexual maturation on anaerobic fitness.

In their study of 12- to 16-year-old boys, Welsman et al. could find no relationship between testosterone levels and submaximal blood lactate responses (a metabolic marker of anaerobic fitness; 129). Similarly, Williams and Armstrong reported no maturational influence on changes in peak $\dot{V}O_2$ percentage at blood lactate levels of 2.5 and 4.0 mmol · L^{-1} in groups of 11- to 16-year-old boys divided by Tanner staging (130). Mero, however, described a significant correlation between salivary testosterone levels and blood lactate concentrations in boys after 30 to 60 s of all-out cycling (86).

Influence of Exercise on Sexual Maturation

The anatomic and physiologic changes in response to the hormonal influences of puberty result in widespread effects on exercise fitness. But again the arrow

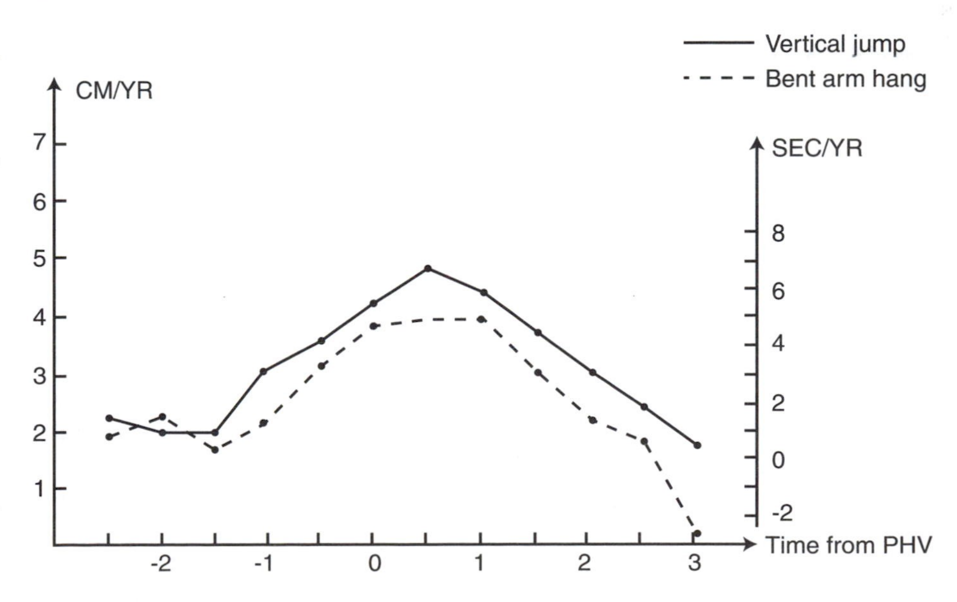

FIGURE 3.6 Velocity of improvements in bent-arm hang and vertical jump performance in boys (from reference 15).
Reprinted by permission from G. Beunen and R.M. Malina 1988.

can be turned around: Both acute and chronic muscular activity can profoundly alter the hypothalamic-pituitary-gonadal axis. These hormonal responses to exercise are expressed most obviously by the alterations in normal menstrual function in highly trained female athletes. Primary or secondary amenorrhea in such adolescent athletes was once considered a simple curiosity but is now viewed with concern because the associated hypoestrogenemia may increase risk of impaired skeletal mineralization. Current research suggests that disturbance in reproductive function with sport training in females is more likely due to caloric imbalance rather than the physical stress of training itself. Increasing evidence suggests a role for leptin as an intermediary in this process.

Hormonal Responses to Acute Exercise

An acute bout of exercise stimulates the release of sex hormones. Viru et al. measured serum estradiol responses to exercise in a three-year longitudinal study of 34 girls who were ages 11 to 12 years at the onset of the study (122). Resting levels rose from 240 pmol · L^{-1} at Tanner breast stage II to 350 pmol · L^{-1} at stage V. Estradiol concentrations after 20 minutes of cycling at 60% $\dot{V}O_2max$ rose to 305 pmol · L^{-1} at stage II and to almost 500 pmol · L^{-1} at stage V. Relative to resting levels, however, these responses represented a rise of 27% and 43% at breast stages II and V, respectively. The authors raised the possibility that greater relative response in the more sexually mature girls reflected a higher level of functional maturation of the pituitary-gonadal system.

Fahey et al. showed a small (about 10%) increase in serum testosterone levels in boys during a progressive cycle test to exhaustion at pubertal stages II, III, and IV (39). No increase was seen at either stages I or V. The small number of subjects in this study (three to seven per stage) makes it difficult to draw conclusions, however. Increases in testosterone with incremental and sustained submaximal exercise in adult men is well documented (52). Initial rises may be followed, however, by a more sustained suppression (3).

The mechanism for these hormonal responses is not known. A general stimulation of the entire hypothalamic-pituitary-gonadal axis seems likely in the case of testosterone, since FSH does not change with acute exercise and alterations in LH are variable (52). Other possible contributors include alterations in hepatic clearance, sympathetic stimulation, changes in protein binding, and hemoconcentration. The action of acute exercise on (a) the rate of androgen uptake by muscle cells and (b) the androgen sensitivity of muscles would also be of interest yet remains unknown (73).

The biologic meaning of this phenomenon—why reproductive hormones should rise in response to muscular work—remains equally uncertain. As noted in the previous sections, estrogen and testosterone might influence mental status, substrate utilization, or thermoregulation during an acute bout of physical activity.

Resting Hormone Levels After Training

Whereas acute exercise stimulates production of sex hormones, repetitive exercise (i.e., physical training) causes a significant decline in resting levels of both estrogen and testosterone. In the studies reviewed by Hackney, serum testosterone levels in endurance-trained men were 60% to 85% of those in nonathletic controls (52). Strauss et al. reported that the amount of weight lost was directly related to the decrease in serum testosterone concentrations seen in a group of collegiate wrestlers (116). At the peak of the competitive season, the correlation between serum testosterone level and percentage body fat was r = .72. After an analysis of research data regarding this relationship between body composition and testosterone level and other possible causes of depressed testosterone levels with training (changes in prolactin, cortisol, catecholamines, or opioids), Hackney concluded that "the available research examining the potential mechanisms for the finding of low resting testosterone in exercise-trained men is limited and confusing" (52, p. 187).

Little information is available about male adolescents. Roemmich and Sinning found that a season of high school wrestling caused a decrease in testosterone levels by approximately 20% without changes in serum LH concentrations (99). During the season the wrestlers lost an average of 4% of initial body weight. However, Rowland et al. could demonstrate no changes in total or unbound (free) testosterone during an eight-week cross country season in 15 male runners ages 14 to 17 years (108).

Highly trained female athletes often demonstrate low levels of circulating estrogen in association with menstrual dysfunction. Boyden et al., for example, studied the hormonal status of 19 women during training for marathon running (19). Resting estradiol concentration fell from 70 ± 14 pg · ml^{-1} at baseline to 34 ± 5 pg · ml^{-1} after each subject's weekly mileage had increased by 50 mi.

Reports indicate that athletic training can similarly induce hypoestrogenemia in adolescent girls. Baer found reductions in estradiol, LH, and FSH in young female athletes (11; figure 3.7). In that study, plasma estradiol levels were 113 ± 21 mmol · L^{-1} in young amenorrheic runners, 247 ± 15 mmol · L^{-1} in eumenorrheic runners, and 251 ± 31 mmol · L^{-1} in sedentary controls. Creatsas et al. studied the endocrine profile of 17-year-old ballet dancers (33). Estradiol levels were 15 ± 4 pg · ml^{-1} in those oligomenorrheic girls who had been training since they were 10 years old (i.e., premenarcheal), 23 ± 5 pg · ml^{-1} in oligomenorrheic dancers who had been training only since age 13, and 55 ± 14 pg · ml^{-1} in a group of athletes "with normal menstrual cycles and without strenuous exercise." Weimann described estradiol levels in 13-year-old elite German gymnasts (127). The expected pubertal rise was not observed: Concentrations were 17 ± 4 pg · ml^{-1} in the prepubertal girls and 24 ± 13 pg · ml^{-1} in those who were postpubertal ($p > .05$).

These findings of depressed sex hormone levels in trained athletes have generated particular concern. The low concentrations of estrogen may place athletes at risk of diminished bone maturation, with associated risks of fracture and later osteoporosis. At present there is no information to indicate that this observed hormonal status in athletes has

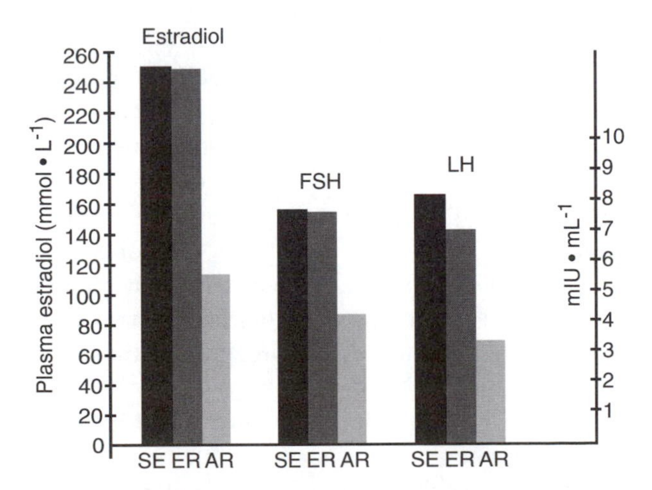

FIGURE 3.7 Blood levels of follicle-stimulating hormone (FSH), luteinizing hormone (LH), and estradiol in sedentary eumenorrheic girls (SE), eumenorrheic runners (ER), and amenorrheic runners (AR; from reference 11).

any implications for gamete production and future fertility.

This phenomenon, however, may not be entirely sinister. Allen considered the hormonal changes in response to training influences on the hypothalamic-pituitary axis to be physiologic and adaptive rather than pathologic (3). "A unifying principle is that frequent and persistent episodes of endurance activity exert a central dampening effect on the HP axis response to exercise, by inducing physiologic and biochemical adaptations that lower the degree of sympathetic system activation in response to a given absolute workload. . . . Sharing many similarities with cardiovascular and musculoskeletal training, this process appears to represent hypothalamic conditioning, which enhances an individual's ability to respond to future stressors of more prolonged [duration] and/or greater severity" (3, p. 43).

Pubertal Progression

Malina summarized the existing research literature that addresses the rate of development of secondary sexual characteristics (excluding menarche, discussed later) in young male and female athletes (78). In males from 11 to 16 years old, advanced maturation was typical of those playing baseball, football, basketball, ice hockey, and track and field. Only in gymnasts was later development observed. Overall, the intervals for progression from one pubertal stage to another were the same as in nonathletic youth.

Baxter-Jones et al. demonstrated that growth curves for testicular size in British male athletes were all within the normal range expected for nonathletes (13). Testicular dimensions relative to these norms, however, were sport specific. Tennis and soccer players followed the 50th percentile, swimmers were generally above the mean, and gymnasts were below the average.

Intense sport training in boys, then, does not appear to influence the timing or tempo of pubertal progression, at least as gauged by the appearance of secondary sexual characteristics. Differences in level of pubertal development reflect preselection by sport; early maturers who are bigger and stronger will excel in sports such as football and hockey.

A similar trend is observed in female athletes. No deviations from normal pubertal progression of secondary sexual characteristics are seen in most sports, including basketball, volleyball, running, and swimming (78). Only gymnasts and dancers show

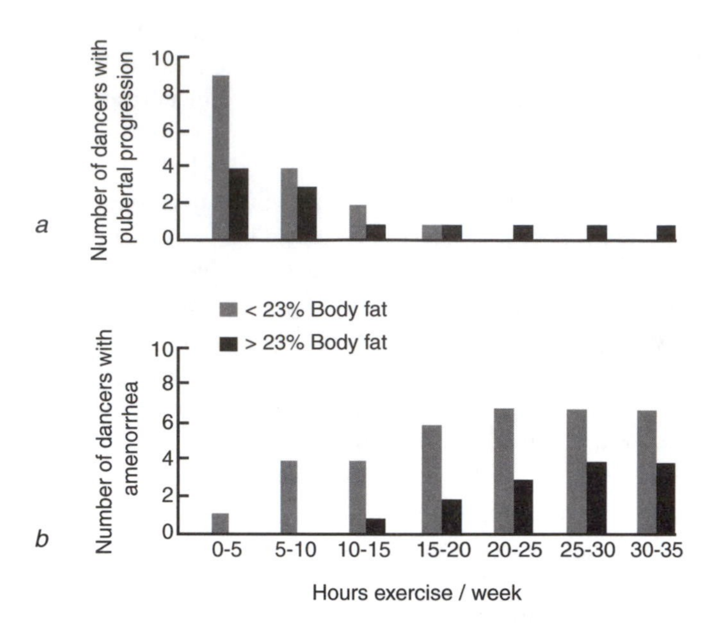

FIGURE 3.8 Effect of increasing training volume on *(a)* pubertal progression (Tanner breast stage) and *(b)* menstrual status in 15 adolescent ballet dancers (from reference 124).
Reprinted by permission from Warren et al. 1980.

later development. Delays in both breast and pubic hair development in a group of adolescent rhythmic gymnasts were reported by Georgopoulos et al. (47). Warren described delayed breast (but not pubic hair) development in 13- to 15-year-old ballet dancers (124; figure 3.8).

In other sports no sexual delay is typically seen. Geithner et al., for example, followed 23 female athletes (running, rowing, swimming) and 26 nontraining controls longitudinally from ages 11 to 18 years and compared markers of sexual development (45). Both average age of menarche and average peak height velocity occurred somewhat later in the athletes, but the differences were statistically insignificant when compared with nonathletes. Similarly, no significant differences were seen between athletes and nonathletes in age of attaining pubic hair and breast stages III, IV, and V or the estimated intervals between these stages.

Malina contended that patterns of growth and maturation in gymnasts and dancers were consistent with those of late-maturing youth: shorter stature and slower skeletal and sexual maturation (78). He concluded these findings were (a) to "a major extent familial and not the [direct] result of intense training" and (b) restricted to this group of athletes because of preselection for body characteristics important for competitive success and because of their "marginal caloric status" (78).

Amenorrhea

Girls involved in intensive athletic training generally experience menarche approximately one to two years later than girls in the general population (78). (As Loucks (76) pointed out, menarche that occurs as late as age 16 years is still considered by clinicians to be normal, meaning that this *primary amenorrhea* in female athletes does not usually indicate a pathologically delayed onset of menses.)

Considerable controversy has surrounded this observation. On one hand, Malina argued that girls who are more likely to be successful in sport—those with narrow hips, slender physiques, long legs, and low body fat—are normally more likely to experience later menarche, irrespective of participation in sports (78). That is, playing sports tends to preselect girls with later onset of menses, with no implication of a training effect. Loucks agreed, concluding that "it is correct to say that the average age of menarche is later in athletes than nonathletes, but there is no experimental evidence that athletic training delays menarche in anyone" (75, p. 276).

Others have considered the late age of menarche in athletes to represent a true delay (74). They see this trend as a manifestation of either stress effects of training or caloric inadequacy (i.e., a training-induced "caloric drain"), both of which are recognized to affect the hypothalamic-pituitary-gonadal axis and influence reproductive function (discussed later). The critical weight or body fat hypothesis of Frisch described earlier is an example (43).

It would not be unreasonable to suggest that both of these mechanisms might contribute to athletes' later menarche. Preselection for late menses (particularly in some sports, such as gymnastics) clearly occurs. At the same time, considering the etiology that underlies *secondary amenorrhea* (lack of menses in postmenarcheal females; described next), there is no question that intense training can inhibit normal endocrine function and disturb normal menstrual patterns.

Intense exercise training by previously menstruating females often results in a reversible disruption of the function of the hypothalamic-pituitary-gonadal axis, causing a depression of circulating estrogen levels and impairment of normal reproductive capacity. Disturbance in normal menstrual patterns has been described in up to half of elite runners and professional dancers (125). In these athletes the usual 24-hour pulsatile pattern of LH is minimal and disorganized, which reflects dampened activity of the GnRH pulse generator. The outcome is cessation of oogenesis by the ovary and reduced production of estrogen.

Two principal theories have been advanced to explain the endocrinologic basis of secondary amenorrhea from sport training. The exercise stress hypothesis holds that the physical stress of sport training is responsible for loss of normal menstrual function. Under this hypothesis, cortisol and other stress hormones triggered by exercise provide negative feedback on the hypothalamus and lower production of GnRH (figure 3.9*a*). The result is depression of LH production, estrogen levels, oogenesis, and menses.

An alternative explanation, the energy availability hypothesis, states that a negative caloric balance created by the energy demands of sport training signals the GnRH pulse generator to shut down (figure 3.9*b*). This theory is consistent with observations that malnutrition in nonathletes is often accompanied by secondary amenorrhea.

Loucks et al. (77) examined these two possible mechanisms in a study of three groups of young women (mean age 21 ± 0.2 years). LH responses to four days of intensive exercise were compared between subjects who had normal (group A) and decreased caloric intake (group B). They were also compared with another group of women who underwent caloric deprivation only (group C). LH pulse amplitude and frequency were no different in groups B and C, indicating no effect of exercise stress itself. However, pulse frequency in group A was 10% lower than that in group B. These findings indicate that LH pulsativity is disturbed by energy unavailability rather than by exercise stress itself. From a practical standpoint, this suggests that the female athlete might manage or prevent secondary amenorrhea by improving caloric intake rather than lessening the intensity of her training regimen.

Leptin, Exercise, and Reproductive Function

Earlier in this chapter we examined the evidence suggesting a role for leptin in the initiation of puberty. We return now to look at leptin in an associated capacity, as a possible link between exercise and reproductive function. In their description of the energy availability hypothesis to explain secondary amenorrhea with exercise training, Loucks et al. stated that this theory "holds that the GnRH pulse generator is disrupted by *an as-yet unidentified signal* [italics mine] that dietary intake is inadequate for the energy costs of both reproduction and locomotion" (77, p. 38). So is leptin that signal?

The concept is intuitively attractive. Leptin has been considered to act primarily as an antistarvation hormone (54), as its production by adipose cells is

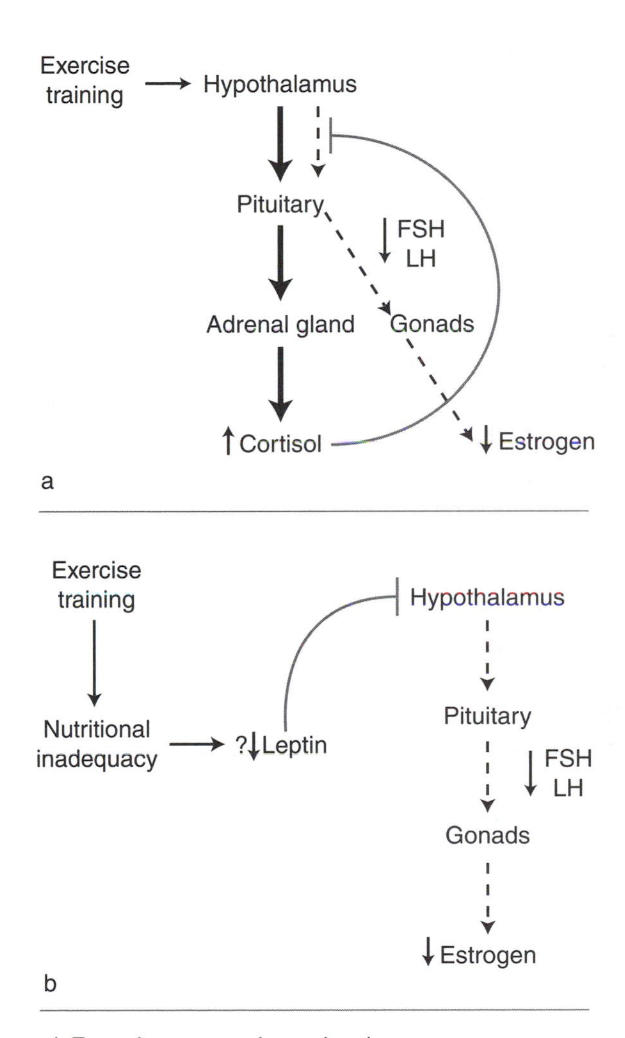

FIGURE 3.9 Two theories of the causal mechanisms of secondary oligomenorrhea and amenorrhea in young athletes. *(a)* The exercise stress hypothesis holds that stress hormones such as cortisol produced in response to intense training regimens block the hypothalamic-pituitary-gonadal axis and depress sex hormone production. *(b)* The energy availability hypothesis suggests that nutritional inadequacies (negative caloric balance) created by exercise training inhibit the hypothalamic-pituitary-gonadal axis, possibly with leptin as an intermediary.

markedly depressed during periods of energy starvation. Leptin "appears to be involved in regulating the physiologic adjustments to starvation, which defend the organism from excess energy expenditure in the face of limited availability of energy intake" (54, p. 584).

Intense exercise training can certainly result in such a negative caloric state, and both training and undernutrition are associated with secondary (and maybe primary) amenorrhea. Thus, it is a reasonable hypothesis that falling leptin levels might act as a signaling agent to shut down the hypothalamic GnRH pulse generator and LH production in the face of undernutrition and exercise training.

Does the research literature support this idea? Cross-sectional studies in adults, as well as investigations involving acute and chronic exercise, have reported a decrease in leptin with increasing exercise levels or increasing physical fitness (54). In almost all cases, however, this effect was found to be mediated by changes in body fat. That is, once adiposity was considered, no effects of exercise on leptin concentrations were observed. Leptin levels are depressed in amenorrheic adult female athletes, and hypoleptinemia is accompanied by low caloric intake and low estradiol levels (118). These observations, then, are consistent with the concept that changes in leptin concentration with exercise serve as a signal of caloric status.

Hilton and Loucks demonstrated that leptin concentrations do not change with energy intake or expenditure but rather decline in respect to the *difference* between the two (i.e., "energy availability"; 55). In their study, exercise stress itself had no influence on 24-hour leptin concentrations. Their data suggested, as well, that leptin responses may be regulated by carbohydrate nutritional state rather than body fat.

Some data are available in the pediatric age group. Weimann studied the link between energy stores, leptin levels, and patterns of pubertal development in elite 12- to 13-year-old female and male gymnasts (127). The girls demonstrated hypoestrogenemia, low leptin levels, reduced body fat content, inadequate caloric intake, and late menarche. However, pubertal development in the boys remained normal. Leptin levels correlated with fat mass (r = .60 in the girls and r = .44 in the boys; figure 3.10).

Kraemer et al. found no significant changes in resting serum leptin levels measured three times during a competitive season in adolescent female runners (70). Gutin et al. examined the effect of a four-month training program on leptin concentrations in 34 obese children ages 7 to 11 years old (51). At baseline, leptin levels were positively correlated with body fat mass. During the training period, leptin concentrations fell significantly, and the greatest decline was observed in subjects with the highest pretraining concentrations and who gained the least weight.

Salbe et al. assessed the relationship between fasting plasma leptin concentrations and energy expenditure (using isotopic water dilution) in 153 five-year-old Pima Indian children (110). Plasma leptin levels were directly correlated with body fat (r = .84). Leptin concentrations were also related to total and physical activity energy expenditure (r = .37 and r = .26, respectively, $p < .05$), even when body fat was taken into account.

Our understanding of the role of leptin as a factor linking exercise training, nutritional status, and reproductive function is clearly in its infancy. Still, there are both conceptual and experimental reasons to believe that leptin might play a key role in these interactions. The data suggest that exercise itself may have no direct effect on leptin production. Rather, it seems that states of caloric imbalance created by exercise training trigger a fall in leptin, which may serve to inhibit hypothalamic release of GnRH and interrupt normal reproductive function (127).

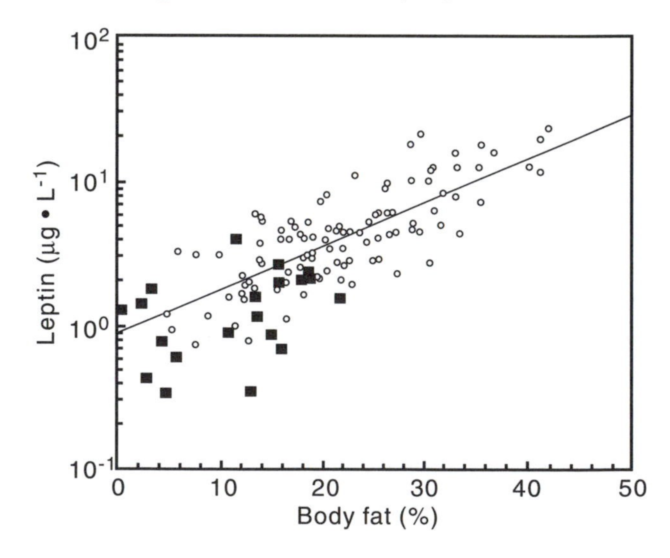

FIGURE 3.10 Serum leptin levels in girls related to percentage body fat. Open circles are healthy nonathletic females 5 to 16 years old. Black squares are elite female adolescent gymnasts (from reference 127).

Reprinted by permission from E. Weinmann 2002.

Conclusions

The increased secretion of sex hormones at the time of puberty influences a wide variety of physiologic functions that bear on exercise performance. These demarcate not only male and female physiologic responses to exercise but also child and adult responses. The specific actions of these hormones, then, need to be considered when making judgments regarding maturational differences in exercise physiology.

Exercise training can be accompanied by dramatic inhibition of the hypothalamic-pituitary-gonadal axis. This influence appears to be, at least in part, sex specific. The outcomes of this process are more obvious in females (e.g., amenorrhea) than in males. There is reason to believe that this impact of regular exercise on reproductive function is largely mediated by changes in nutritional state (caloric balance and body composition). While this endocrinologic response to exercise may be to some extent physiologic and adaptive, certain long-term outcomes, particularly lack of bone development in the face of hypoestrogenemia, are of concern.

Discussion Questions and Research Directions

1. What causes increased insulin resistance at puberty? How does this affect glucose homeostasis and substrate utilization during exercise?
2. How does leptin affect the initiation of puberty? Does it play a role in the exercise-induced alterations in reproductive function in athletes?
3. What exercise-related factors that are independent of changes in body size are altered at the time of puberty? How do hormonal changes influence these size-independent variables?
4. What is the importance of sex-related hormonal influences on psychologic function in respect to exercise performance?
5. What are the threshold levels beyond which sport training or caloric imbalance create health hazards for females (diminished bone mineralization, fractures)? How can these adverse outcomes best be prevented?
6. Are there reproductive or health implications of the diminished testosterone levels observed in male athletes?

The Metabolic Machinery

Morphology is not only a study
of material things
and of the form of material things,
but has its dynamic aspects
under which we deal
with the interpretation,
in terms of force,
of the quantities of energy.

—D'Arcy Thompson (1917)

In this chapter we discuss

- the metabolic processes that support muscular contraction during exercise,
- the means of assessing aerobic and anaerobic cellular metabolism,
- the patterns by which these biochemical processes evolve during growth in youth, and
- how biochemical changes might influence the development of physical fitness.

As children grow, their performance of motor tasks steadily improves. A 12-year-old boy can run a mile faster than when he was 6. The pubertal girl can perform more sit-ups than she could five years earlier. All the developmental curves of physical performance—at least until adolescence—go up. What we are witnessing here is a progressive increase in the *exercise fatigue threshold*. With increasing age, something (or some things) elevates the amount of motor work that can be performed before fatigue becomes limiting. The defining factors responsible for exercise fatigue and how these change with biologic development, then, become a critical issue for understanding this process.

In adults, investigations of the variables that define fatigue threshold have focused on explaining (a) the increased ability to tolerate exercise stress following a period of physical training and (b) the features that differentiate the performance of trained athletes and nonathletes. It may not necessarily be correct to assume that factors affecting changes in fatigue threshold as children grow are identical to these adult models. Nonetheless, it may be helpful to consider these determinants for their potential influence on maturational changes in exercise tolerance. Conversely, an understanding of the factors that affect the fatigue threshold during the growing years may provide insights into the basic mechanisms for exercise tolerance at all ages.

Exercise fatigue has generally been considered multifactorial, with potential contributions by factors such as neuromuscular innervation, actin–myosin contractile function, cellular calcium flux, and central brain drive. However, energy insufficiency has traditionally been the most commonly accepted model for exercise fatigue. The function of the exercise "motor" (i.e., skeletal muscle contraction) is contingent upon adequate fuel (carbohydrates, fat) and, for endurance events, a means of oxidation (oxygen delivery via the circulatory system). In this model, "the rate of ATP production by oxidative sources becomes inadequate, high rates of anaerobic glycolytic ATP production produce metabolites, particularly H^+, which interfere with energy production and cross-bridge cycling, causing fatigue and a failure of muscle contraction" (69, p. 133). [The reader is referred to the intriguing article in which Noakes describes the weaknesses of this concept, urging that "the modern generation of exercise physiologists challenge old dogma and so approach more closely the unattainable truth" (69, p. 142).]

In this chapter we begin our assessment of maturational changes in exercise fatigue where it should start, with a consideration of the development of the cellular metabolic machinery itself. Later chapters address the potential influences of other contributors, such as oxygen delivery, submaximal exercise economy, and thermal stress. Boisseau and Delamarche provided a comprehensive review of the metabolic responses to exercise in youth, to which the reader is referred for additional information (7).

Our understanding of metabolic processes within the skeletal muscle of children is clearly limited. Most information in adults is derived from needle biopsy studies performed before and after bouts of acute exercise or surrounding a period of physical training, an approach usually considered inappropriate for healthy children. Pediatric exercise scientists are, in fact, provided only a narrow peek at these processes through indirect measures of blood markers (lactate) or expired gases (respiratory exchange ratio, accumulated oxygen deficit). Newer methods, such as nuclear magnetic resonance spectroscopy (which we discuss in more detail later in the chapter), may provide a broader window through which metabolic events in children can be better observed.

These limitations notwithstanding, we will see in the course of this discussion that there is, in fact, good evidence for distinct changes in cellular energy metabolism as children grow. The general theme is that as metabolic function increases with age, children exhibit, relative to their increasing body size, a decline in cellular aerobic metabolic rate while glycolytic capacity improves. In some cases, possible explanations for these trends can be offered. For example, the fall in resting metabolic rate relative to body size with increasing age may reflect the need to maintain thermal neutrality as the ratio of body surface area to body mass declines. The reason for other changes, such as the apparent relative improvement in glycolytic capacity with age, are more obscure. Some evidence suggests that the biochemical *interactions* between energy systems may be influential.

Differences in metabolic capacities in children may be reflected in a different pattern of substrate utilization. Specifically, limitations in glycolytic function might be expressed by a greater reliance on fatty acid oxidation during endurance exercise in children. Some, but not all, experimental data support this concept. Patterns of increasing anaerobic and declining aerobic metabolism in children might also have a bearing on athletic trainability in young subjects. However, the extent to which alterations in cell metabolism during sport training by children mimic those observed in adults is problematic.

Basic Concepts in Exercise Physiology

It is useful to begin with a description of the current understandings of metabolic activity during exercise as determined by studies in adult humans and animals. We can then see how children and adolescents fit in to this scheme. The following information is summarized from reviews by Coyle (14), Gastin (35), and Spriet et al. (93).

The energy needed by the contractile elements of skeletal muscle is supplied by the high-energy phosphate bonds of adenosine triphosphate (ATP). ATP, in turn, is generated from adenosine diphosphate (ADP) through the transfer of these high-energy phosphates from phosphocreatine (PC). The body stores of phosphocreatine and ATP are limited, however (about 25–80 mmol per kilogram of skeletal muscle), and without a means of replenishing ATP, a person would become exhausted during physical activities lasting more than a few seconds. Interestingly, however, even during severe exercise, total depletion of ATP does not occur. Of the energy released by hydrolysis of ATP, only about 25% is effectively utilized by the actin–myosin contractile mechanism for external and internal work, the rest being released as heat.

Resynthesis of ATP occurs either by energy derived from anaerobic glycolysis or within mitochondria by chemical reactions in the Krebs cycle and subsequent utilization of oxygen in the electron transport chain (aerobic metabolism). In glycolysis, a series of 12 chemical reactions convert carbohydrate to pyruvate, utilizing either stored muscle glycogen or circulating glucose as the substrate. In the absence of oxygen, pyruvate is converted to lactate. The glycolytic pathway is capable of resynthesizing ATP at very rapid rates and is critical for high-burst physical activities lasting less than two to three minutes. The amount of energy that can be released in a single bout of exercise by glycolysis, however, is limited.

The key rate-limiting enzymes in the glycolytic cascade are glycogen phosphorylase, which converts glycogen to glucose-6-phosphate, and phosphofructokinase (PFK), which converts fructose-6-phosphate to fructose-1,6-diphosphate in the third step. Both of these enzymes are sensitive to and regulated by factors that signal the intensity of muscle contraction (i.e., epinephrine, calcium, and ATP demand; 93).

Aerobic metabolism, on the other hand, can generate much more energy, but that energy is released at a slower rate in response to exercise. Aerobic metabolism is limited by oxygen delivery via the cardiorespiratory system and by the rates of oxidative phosphorylation in the Krebs cycle and the electron transport chain. During aerobic metabolism, oxidative phosphorylation and the electron transport chain produce ATP by combining ADP and inorganic phosphate (P_i), provided by phosphocreatine. During this oxidative process, 1 mol of glucose produces 38 to 39 mol of ATP, compared with the 2 to 3 mol provided by the anaerobic metabolism of glucose or glycogen to lactate in the glycolytic pathway.

Substrate for aerobic metabolism is derived from two possible sources: circulating free fatty acids released by body fat stores after enzymatic breakdown of triacylglycerol (fat oxidation) and the same pyruvate produced by carbohydrate metabolism in the glycolytic pathway. Glycolysis thus can serve as a precursor to anaerobic or aerobic metabolism, depending on whether its end product pyruvate is converted to lactate (in the absence of oxygen) or enters the aerobic Krebs cycle (with oxygen).

The appearance of muscle or blood lactate is indicative of anaerobic glycolysis and has long been supposed to reflect conditions of cellular oxygen "starvation." This does not appear to be the case (48). Considerable evidence indicates that cellular oxidation and mitochondrial oxidation states are unaffected by exercise conditions in which lactate levels increase (91).

Substrate utilization during prolonged physical activity is determined by exercise duration and intensity as well as by diet and level of physical conditioning. Low-intensity exercise is characterized by oxidation of fats in the Krebs cycle. During exercise at an intensity of 25% $\dot{V}O_2max$, 60% to 85% of energy is derived from fat. When the intensity is raised to 65% $\dot{V}O_2max$, fat oxidation contributes only about half of the required energy. Duration of exercise also affects substrate utilization during endurance events. At any sustained intensity of 65% to 75% $\dot{V}O_2max$, the contribution of fat rises slightly, while that of carbohydrate falls. More importantly, with increasing duration, utilization of glycogen (stored muscle carbohydrate) progressively declines. Depletion of glycogen stores, then, appears to serve as a limiting factor for endurance exercise performance (e.g., marathon running). Depletion of glycogen stores with fatigue, though, occurs even when large fat stores remain unutilized as energy substrate. In fact, in adults, total carbohydrate stored as glycogen in the liver and muscle amounts to about 375 g, equivalent to 2,000 kcal of energy (enough to run about 20 mi). Fat stores, on the other hand, amount to as much as 70,000 kcal (106, pp. 91-96).

Considerable interplay is observed between substrate utilization patterns. Increased carbohydrate utilization through glycolysis is accompanied by a decrease in fatty acid oxidation (15). And infused lipid increases fat oxidation and diminishes carbohydrate metabolism (25, 41). Endurance training in adults improves utilization of fats as an energy source during exercise. This glycogen-sparing shift to greater levels of fatty acid oxidation has been associated with improved endurance performance.

Resting ATP Stores

Limited research data consistently indicate that body ATP stores at rest are equal in children and adults and that ATP-PC activity is independent of maturation. Lundberg et al. reported biopsy findings of the vastus lateralis muscle in 25 healthy children ages two months to 11 years (60). Mean ATP concentration was 4.47 ± 0.92 mmol · kg^{-1}, and no correlation was observed between ATP concentration and age. The resting ATP levels reported by Karlsson et al. (51) in a group of 14 military conscripts (mean age 20 ± 0.8 years) was somewhat lower (3.8 ± 0.2 mmol · kg^{-1}), however.

Eriksson and Saltin (30) observed similar ATP concentrations at rest in boys ages 11.6, 12.6, 13.5, and 15.5 years (4.8 to 5.1 mmol · kg^{-1}). Resting values of phosphocreatine (PC), the enzyme responsible for transferring high-energy phosphate to form ATP, showed a tendency to increase in these age groups (14.5 mmol · kg^{-1} in the youngest and 23.6 mmol · kg^{-1} in the oldest). Values in adult men are typically 15 to 23 mmol · kg^{-1}. In this study, biopsies of the quadriceps femoris muscle were performed at each work stage of a progressive exercise test. ATP concentrations remained essentially unchanged, while PC levels gradually fell. The decline in the oldest group was approximately 50% greater than that observed in the youngest boys.

Berg et al. (6) compared the activity of creatine kinase in the vastus lateralis tissue of three groups of healthy subjects ($n = 33$, mean ages 6.4, 13.5, and 17.1 years). No significant differences were observed in the three cohorts (5,208 ± 774, 6,886 ± 2,251, and 6,351 ± 3,063 μmol · min^{-1} · g^{-1}, respectively), and no correlation was evident between enzyme activity and age ($r = .24$, $p > .05$).

Among adult animals no relationship has been observed between body size and activity of mitochondrial ATP synthase, an enzyme important in ATP synthesis. Rouslin found that activity per milligram of heart mitochondrial protein was similar in mammals ranging in size by a factor of 10^5 (83).

Resting ATP concentrations, then, appear to be a biologic constant, irrespective of body size, level of maturation, or capacity for anaerobic or aerobic metabolism. The effect of training on resting ATP levels is not clear. Eriksson et al. (28) and Gollnick et al. (36) indicated that ATP levels in boys and men, respectively, rose following training. However, Saltin and Gollnick found no such response in adult men (87).

Glycolysis

A number of lines of evidence indicate that the functional capacity of the glycolytic pathway is greater in adults than in children. Moreover, glycolytic capacity appears to improve steadily throughout the growing years. In this section we examine the evidence for this premise and address the question of why such changes might occur. How limitations in glycolysis might be expressed in the performance of short-burst, anaerobic activities is addressed in chapter 9.

Is Glycolytic Metabolism Diminished in Children?

Evidence that the capacity for glycolytic metabolism is less in immature subjects comes from a diverse set of data obtained by measurement of cellular enzyme activity and lactate production and also by noninvasive methods such as nuclear magnetic resonance spectroscopy and the accumulated oxygen deficit technique.

Body Size

Indicators of glycolytic function seem to reflect progressive biochemical development during childhood. Studies in biologically mature adult animals suggest, in addition, that size per se may contribute to these changes. Emmett and Hochachka determined the catalytic activities of glycolytic enzymes in the gastrocnemius muscle of adult animals varying in size from the shrew to the cow (26). A direct relationship was observed between activity of the glycolytic enzymes (pyruvate kinase, lactic dehydrogenase, and glycogen phosphorylase) and animal size, with allometric mass exponents ranging from +0.09 to +0.15.

Somero and Childress reported the same findings in their study of enzymes in the white muscle of fish (92). Lactic dehydrogenase (LDH) and pyruvate kinase (PK) activity per gram of muscle was significantly greater in the larger fish. Enzyme activity of LDH was related to fish mass by an average exponent of +0.35, while the mean mass exponent for PK was +0.22. No relationship was observed between enzyme activity in brain tissue and body mass. This led the authors to conclude that scaling of glycolytic enzymes in muscle is related to selected factors associated with locomotion.

Emmett and Hochachka suggested that this factor might be the power required for short-burst activities, which depend on anaerobic metabolism (26). They noted that power needs for this type of burst work scale differently from the energy demands of sustained performance. From the data of Davies et al. in children (19), they calculated that $\dot{V}O_2max$ was related to $M^{1.06}$, while anaerobic power scaled to $M^{1.76}$. In other words, in children, "being big and running very fast anaerobically requires relatively more power than being big and running aerobically, a result predictable from our data and requiring that glycolytic enzyme potentials scale directly with size" (26, p. 271).

Somero and Childress tested this concept by estimating the scaling of muscle power needed to maintain identical burst-swimming abilities (in body lengths per second) in large and small fish (92). Mass exponents ranged from +1.22 to +1.53, agreeing closely with those observed for glycolytic enzyme activity.

These data, then, suggest that glycolytic activity is matched allometrically to the energy required to perform the types of short-burst muscle activity it supports. This, of course, provides a mathematical link but tells us nothing about the mechanism. It also raises the question of which comes first (à la the chicken and the egg). Does some factor that increases glycolytic metabolic activity in bigger animals permit greater anaerobic performance relative to body size? Or is the glycolytic pathway more active because such performance is greater with large body size for some reason?

Lactate Production

During low-intensity exercise, pyruvate produced as the end product of glycolysis is largely converted by pyruvate dehydrogenase (PDH) to acetyl CoA in the mitochondria, where it enters the Krebs cycle for aerobic metabolism. During high-intensity exercise, however, the demand for ATP rises, increasing the activity of the glycolytic pathway and rate of pyruvate production. When this production exceeds the capacity of PDH to divert pyruvate to aerobic pathways, pyruvate is converted to lactate by lactic dehydrogenase (LDH). The rate of lactate production is therefore dictated by (a) the rate of pyruvate production, (b) the extent that pyruvate is utilized by oxidative phosphorylation, and (c) the activity of LDH.

Muscle lactate spills into the bloodstream, and serum concentrations have therefore been used as a readily available marker of anaerobic glycolysis. It is important, however, that a particular level of lactate in the blood may reflect more than simply glycolytic function. Differences between muscle and serum lactate concentration, the balance between lactate production and removal, and oxidation of lactate or conversion to glucose all can significantly influence blood concentrations.

This point notwithstanding, the observation that children often demonstrate lower serum lactate levels than those observed in adults during exercise has traditionally served as a starting point for the argument that immature subjects have depressed glycolytic capacity. This position is supported by the single study that has been conducted to assess maturational effects on maximal muscle lactate concentrations, which showed a consistent increase in males between the ages of 11 and 23 years (29).

Welsman and Armstrong (104) and Pfitzinger and Freedson (73) have comprehensively reviewed information regarding lactate levels during exercise in children. These authors emphasized that when studies of child and adult lactate responses are compared, it is critical to carefully consider the variables than can influence serum concentrations. For example, the mode, duration, and intensity of the exercise testing protocol affect glycolytic response and lactate concentrations. Mero reported that peak lactate levels in children during a 15-s Wingate test and treadmill $\dot{V}O_2max$ test were 61% and 68% as high, respectively, as those measured during a 60-s Wingate test (66). Meanwhile, Fellman et al. demonstrated that peak lactate levels in 10- to 12-year-old boys were similar in a $\dot{V}O_2max$ test and a 30-s Wingate test (34). This finding suggests that neither test provoked a total lactate response (because the 30-s test was too short in duration to maximize glycolytic function, while the $\dot{V}O_2max$ test relied more on aerobic metabolism). Pfitzinger and Freedson thought that maximal lactate production in children demands a test of one to six minutes in duration (73). Other factors that can influence lactate concentrations, including site of blood sampling, timing of measurements, and method of assay, need to be considered in study comparisons (104).

A truly definitive study that longitudinally assesses lactate production during maximal exercise with respect to level of biologic maturation or chronologic age in children has not yet been performed. Almost all studies are cross-sectional and usually limited to a fairly narrow age range. And, as Pfitzinger and Freedson pointed out (73), the findings in these studies have not always been consistent regarding trends during the growing years (e.g., reference 105). However, when we consider the total accumulated information, a clear trend of levels of lactate during maximal exercise increasing with age is observed in both boys and girls. While recognizing that peak lactate levels are lower in children than adults, Welsman and Armstrong concluded that "it is impossible to infer a true pattern of change with increasing age because of the many methodological differences among the various studies" (104, p. 146).

Figures 4.1 and 4.2 indicate cross-sectional findings from selected studies of maximal blood lactate levels by age in boys and girls. With the preceding considerations in mind, the data from the studies used in this analysis suggest that levels increase by approximately 50% in boys between the ages of 6 and 14 years, with perhaps a slightly shallower upward slope in girls. Two more-recent studies support maturational changes in maximal lactate production. Ratel et al. reported lactate concentrations of 8.5 ± 2.1 and 15.4 ± 2.0 mmol · L^{-1} in 11 boys (mean age 9.6 ± 0.7 years) and 10 men (age 20.4 ± 0.8 years), respectively, after ten 10-s sprints (79). Hebestreit et al. described mean postexercise lactate levels of 5.7 and 14.2 mmol · L^{-1} after a 30-s all-out cycle bout in prepubertal boys and men, respectively (42).

Figures 4.1 and 4.2 lead us to suspect that lactate levels during maximal exercise might tend to be greater in females. In their study of 100 boys and 91 girls ages 11 to 16 years, Williams and Armstrong demonstrated just that (105). Mean peak blood lactate levels were 6.1 mmol · L^{-1} for the girls and 5.8 mmol · L^{-1} for the boys ($p < .01$). However, others have reported no effect of sex on maximal lactate levels with exercise (19, 90, 101).

Level of sexual maturation probably has no effect on lactate responses to exercise (71, 105), despite theoretical support for such an effect. Welsman and Armstrong reviewed data indicating that testosterone levels are linked to glycolytic capacity in animals, and correlations between postexercise lactate and testosterone levels have been reported in boys (104). These researchers concluded, however, that "research evidence . . . for the view that testosterone and glycolysis are causally linked . . . in children and adolescents is equivocal" (p. 147).

Consistent with these data on lactate, blood hydrogen ion concentration is greater during maximal exercise in adults than in children. Bar-Or presented data demonstrating such maturational differences in pH after cycling and running exercise (3). More recently, the study of Ratel et al. described acid–base balance after a series of ten 10-s cycle sprints in prepubertal boys and young men (79). After the 10th sprint, hydrogen ion concentrations were 43.8 ± 1.3 and 66.9 ± 9.9 nmol · L^{-1} in the boys and men, respectively.

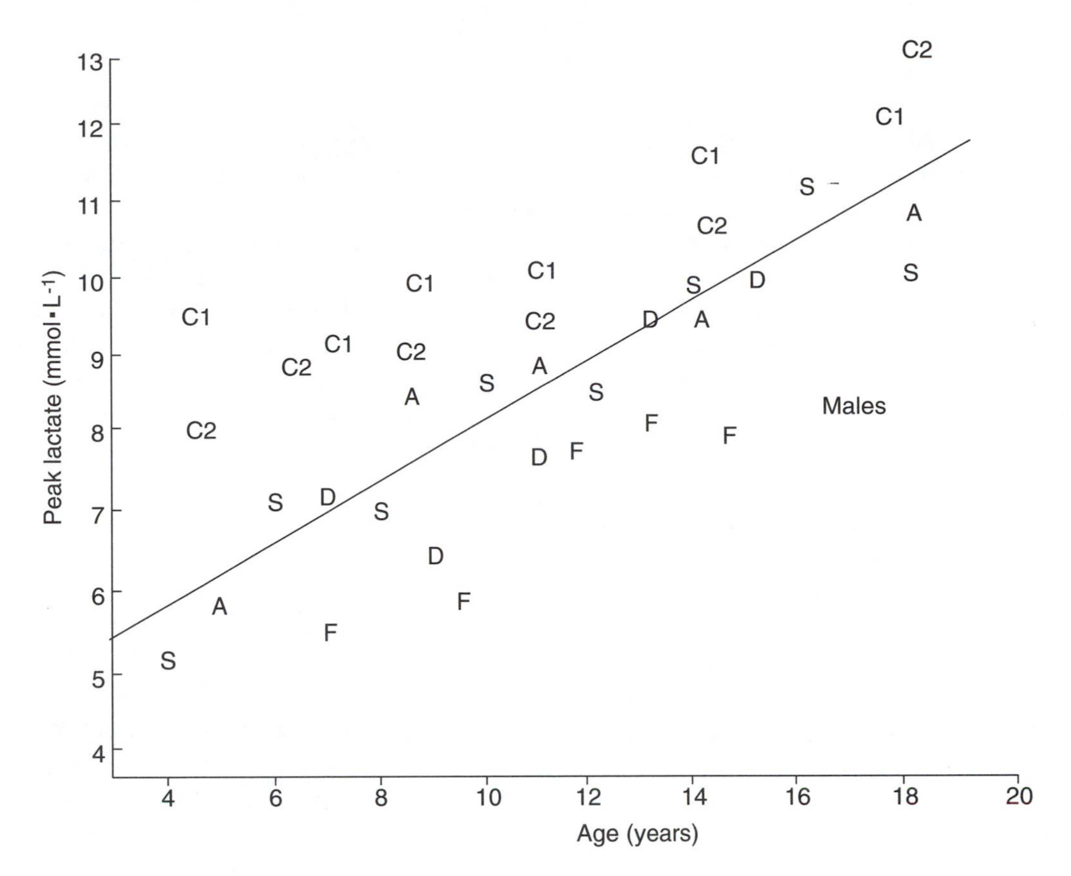

FIGURE 4.1 Cross-sectional studies of maximal blood lactate concentrations by age in boys. Data are from various sources: A (2), C1 (16), C2 (17), D (19), F (32), S (88).

Reprinted by permission from T.W. Rowland 1996.

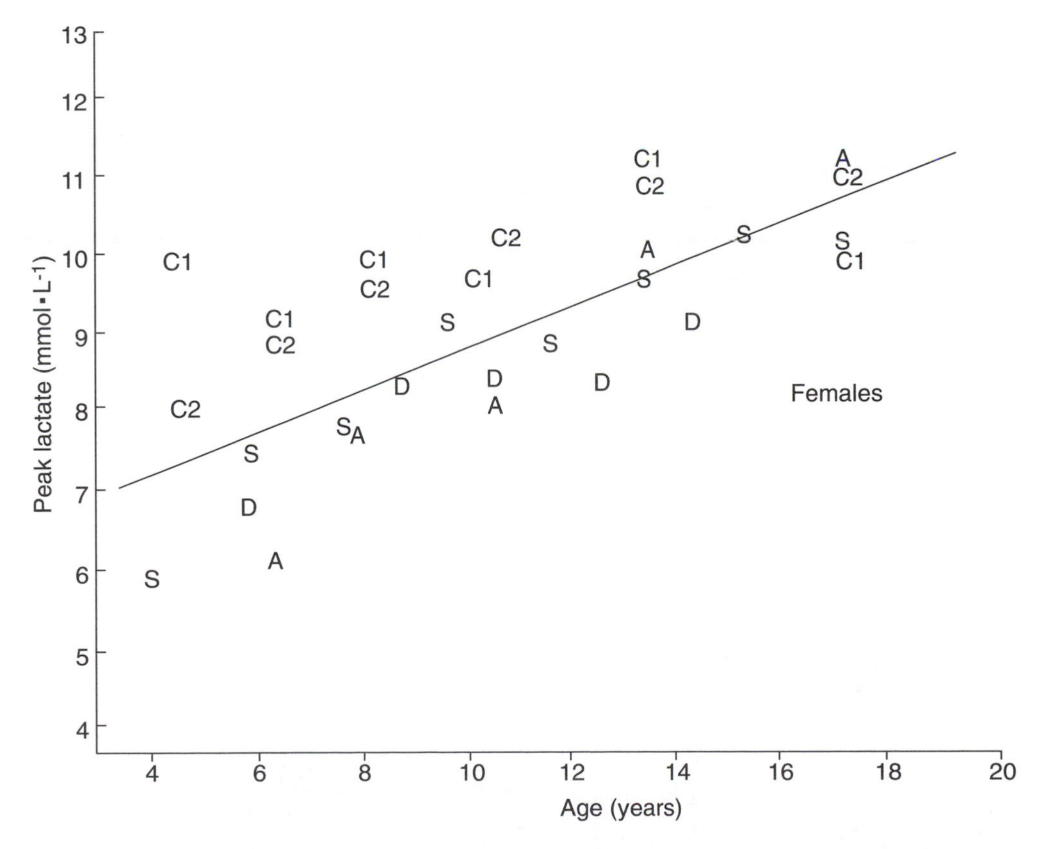

FIGURE 4.2 Cross-sectional studies of maximal blood lactate concentrations by age in girls. The key to references is the same as for figure 4.1.

Reprinted by permission from T.W. Rowland 1996.

As noted earlier, this apparent developmental change in lactate production with exercise might not necessarily imply depressed glycolytic function in younger children. Another explanation, for example, might be that lactate removal during exercise occurs at different rates in children and adults, perhaps by developmental differences in hepatic blood flow (73). There are currently no experimental data to assess this possibility.

Glycolytic Enzyme Activity

Direct measurement of cellular enzyme activity would obviously be invaluable in defining developmental changes in glycolytic capacity. Ethical constraints, though, have largely obstructed our view of the cellular milieu in pediatric subjects. The limited data that are available, however, do generally support the concept that the glycolytic machinery turns at a slower pace in growing children than in adults.

We have already seen that animal studies suggest that size itself may influence glycolytic capacity: Larger animals exhibit greater glycolytic enzyme activity than smaller animals. The limited information available indicates the same trend in children, but it is not possible to distinguish the relative contributions of size versus other factors involved in biologic maturation.

The closest parallel to the animal data in children is the study of Berg et al., in which vastus lateralis muscle biopsy samples were obtained during orthopedic or traumatic surgery (6). The 33 subjects were divided into three age groups (6.4 ± 2.1 years, n = 8; 13.5 ± 1.3 years, n = 12; and 17.1 ± 0.8 years, n = 13). These subjects were therefore clearly distinct in age, body size, and level of sexual development. Activity of the glycolytic enzymes pyruvate kinase and aldolase increased with age (r = +.45 and r = +.35, respectively; figure 4.3). Between 6 and 17 years of age, these two enzymes increased in activity by 45% and 47%, respectively. Lactic dehydrogenase activity rose until age 12 but then declined. Changes in hexose phosphate were more equivocal. The overall picture, though, was of a progressive rise in glycolytic activity as children become larger and older.

Haralambie also reported significantly lower activity of various glycolytic enzymes in prepubertal children (38). A rise in the activity of 3-phosphoglycerate kinase, enolase, and pyruvate kinase was found until the age of 12 to 13 years, when values became similar to those of adults. In subsequent studies in adolescent subjects, however, no differences from adults were observed. Lactic dehydrogenase activity was actually greater in girls (mean age 12.9 ± 1.4 years) than in women (36.2 ± 7.0 years old; 40). Values were 165 ± 40 U · g^{-1} in the girls and 126 U · g^{-1} in the adult women. The authors felt this "simply suggests a daily physical activity directed towards more intensity than endurance in this group [of girls]" (p. 265).

In another study, Haralambie found that the activity of nine glycolytic enzymes in 13- to 15-year-old pubertal subjects did not differ from that in adults (39). Phosphofructokinase (PFK) levels were 18% higher in the adults (45.5 ± 10.9 U · g^{-1} vs. 38.6 ± 4.4 U · g^{-1}), but the difference was not statistically significant.

Great attention has been paid to a report by Eriksson and Saltin in 1974, which described low levels of PFK in 11-year-old boys (30). This has been considered a particularly important observation, as PFK serves as a rate-limiting step in glycolytic metabolism, and action of this enzyme should reflect

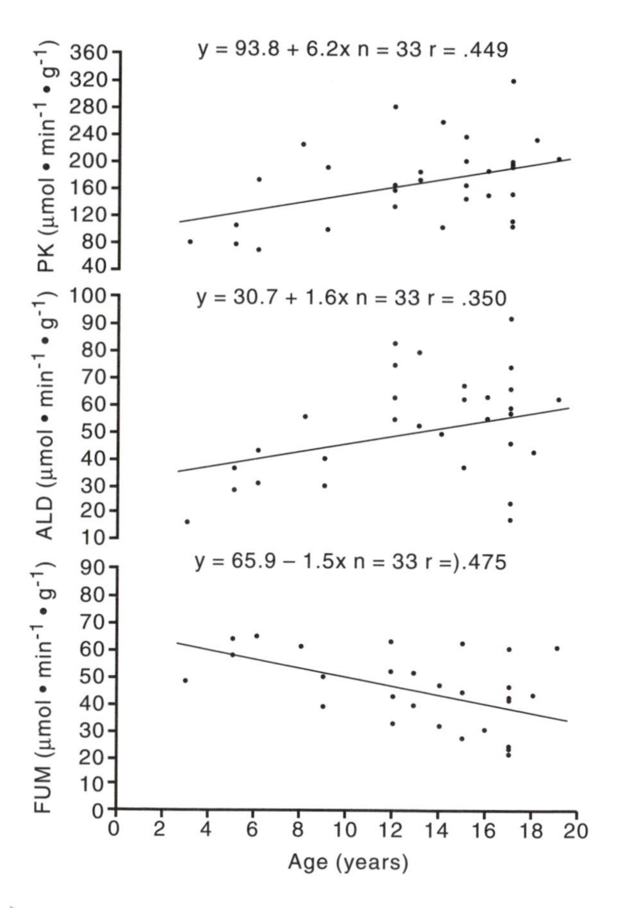

FIGURE 4.3 Correlation between age and activity of the enzymes pyruvate kinase (PK), aldolase (ALD), and fumerase (FUM) in vastus lateralis muscle biopsies (from reference 6).

Reprinted by permission from Berg et al. 1986.

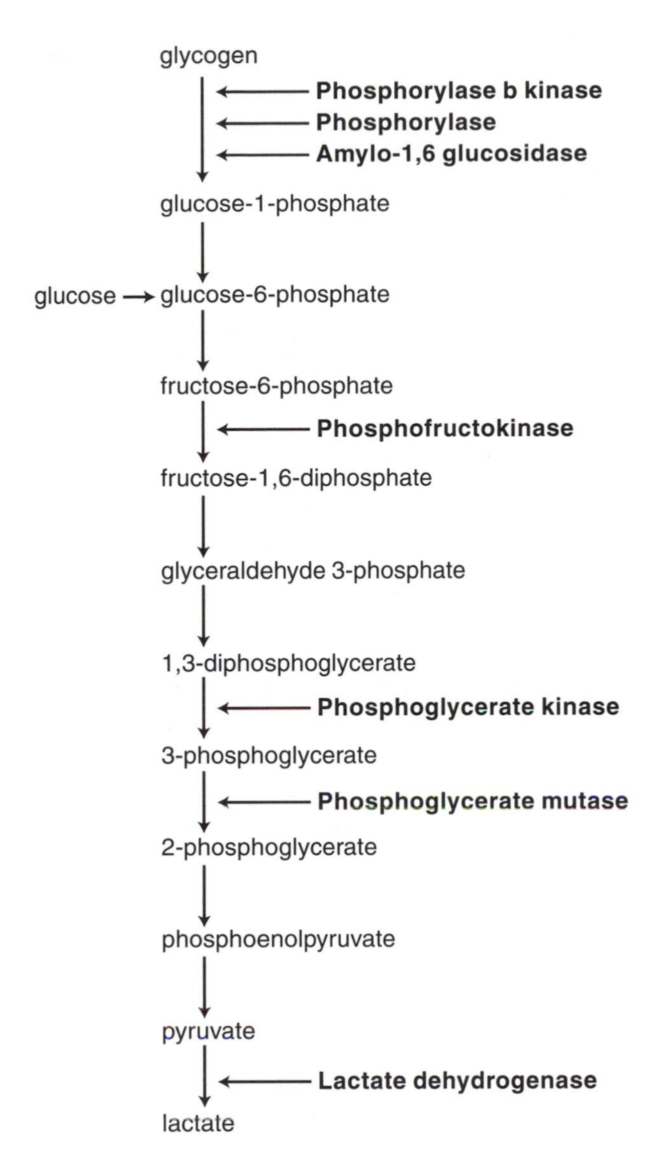

FIGURE 4.4 The anaerobic glycolytic pathway (from reference 58).

Reprinted by permission from S.F. Lewis and R.G. Haller 1990.

that of the entire glycolytic pathway (figure 4.4). This study, however, involved only five subjects, who had a wide range of PFK values in vastus lateralis muscle biopsies (6.0 to 19.8 μmol · min^{-1} · g^{-1}; mean 8.4). No direct comparisons were made with adults in this study, but the boys' values were approximately 30% of those previously described in untrained adult men (36).

Dunaway et al. confirmed that PFK activity rises markedly during development in young mice (22). They found that this increased activity was accompanied by an alteration in enzyme isoform, which they explained as "very likely associated with the [glycolytic] energy requirements of the muscle."

This evidence, tenuous as it might be, has been interpreted as indicating a slowed glycolytic chain in young children that is inhibited by enzymatic deficiency. It may be useful to see if children do, in fact, show the biochemical features of PFK deficiency observed in people who have an inborn absence of this enzyme. These individuals have a metabolic block at the third stage of the glycolytic chain and cannot utilize either glycogen or circulating glucose for exercise energy but must depend on oxidation of fatty acids via the Krebs cycle. They are characterized by marked exercise intolerance, lack of lactate elevation with muscular work, hyperglycemia, and increased levels of circulating free fatty acids (58, 102). $\dot{V}O_2$max is approximately 35% to 50% of normal. This is due to a lower peak arterial-venous oxygen difference, as maximal cardiac output is not affected.

With the exception of low exercising lactate levels, the physiologic characteristics of children during exercise neither quantitatively nor qualitatively mimic this picture. That is, it seems unlikely that children operating at 30% of adult glycolytic capacity would show no clinical manifestations of PFK deficiency. The scant information that we have, however, does suggest that prepubertal subjects have lower glycolytic enzyme activity than mature individuals.

Accumulated Oxygen Deficit

The oxygen deficit is a reflection of the energy used during exercise that is not supplied by aerobic metabolism, and thus it measures the contribution of anaerobic metabolic processes. The deficit is the difference between actual oxygen uptake during supramaximal work and that predicted by the relationship of work and oxygen uptake during submaximal exercise. A plateau in accumulated oxygen deficit with increasing supramaximal work indicates true anaerobic capacity. Naughton and Carlson have reviewed this technique and its applications in the pediatric age group (10, 68).

Only one study has evaluated differences in accumulated oxygen deficit relative to biologic maturation. Stear et al. (94) demonstrated a significant increase in values for accumulated oxygen deficit with age and pubertal status: prepubertal (mean age 10.5 ± 1.43 years), 36 ± 9 ml · kg^{-1} · min^{-1}; adolescent (mean age 13.6 ± 1.6 years), 58 ± 16 ml · kg^{-1} · min^{-1}; and postadolescent (mean age 16.6 ± 0.7 years), 62 ±

21 ml · kg^{-1} · min^{-1}. This increase was not linear, causing Naughton and Carlson to suggest that factors other than level of sexual maturation, particularly athletic training, might affect anaerobic capacity (68).

Nuclear Magnetic Resonance Spectroscopy

Nuclear magnetic resonance (NMR) spectroscopy is a noninvasive technique that provides information regarding intracellular metabolic processes. When the subject is placed in a magnetic coil, the nuclei of atoms align with the magnetic field. A second oscillating magnetic field is then applied, and spectroscopic analysis of the subsequent nuclear transitions reveals changes in molecular dynamics.

While this technique offers a safe means of examining metabolism during exercise in children, use of NMR spectroscopy is constrained by the need for muscle contractions to occur within a magnetic tube. Exercise studies by this method, then, have typically involved foot force against a pedal apparatus (treadle ergometer). Subjects in the supine position perform repeated dorsal flexions of the ankle joint (typically 30–60 repetitions per minute).

NMR spectroscopy permits estimation of changes in cellular phosphocreatine (PC) and inorganic phosphate (P_i), as well as intracellular hydrogen ion concentration (pH). During a progressive exercise test, the ratio of P_i to PC initially rises linearly, an expression of aerobic metabolism. The subsequent appearance of a second, steeper slope in the P_i/PC relationship is thought to reflect anaerobic glycolysis (12).

NMR studies comparing cellular metabolic responses to exercise in children and adults have found an inferior glycolytic capacity in immature subjects. Zanconato et al. (107) measured calf-muscle high-energy phosphate metabolism during exercise in 10 children (7–10 years old) and 9 adults (ages 20–42 years). There were no differences in pH or P_i/PC at rest between the two groups. With exercise, the adults exhibited a greater increase in P_i/PC and a more substantial fall in pH than the children. During the initial phase of exercise, the slopes of the slow rise in P_i/PC were similar, but in the accelerated phase, the slope was significantly greater in the adults (23.6 ± 9.8 vs. 10.7 ± 2.5). At that time, too, the slope of the decline in pH was steeper in the adults (–60 ± 1.9 vs. –3.7 ± 1.2).

Using NMR spectroscopy, Kuno et al. also found lower a P_i/PC ratio during maximal exercise in children than in adults (56). Mean intracellular pH was

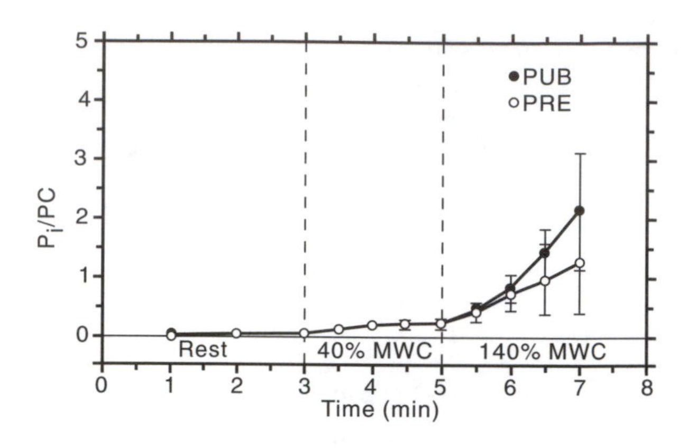

FIGURE 4.5 Changes in the ratio of inorganic phosphate to phosphocreatine (P_i/PCr) at rest, during submaximal exercise (40% maximal work capacity), and during supramaximal exercise (140% maximal work capacity) in prepubertal (PRE, n = 9) and pubertal (PUB, n = 9) female swimmers (from reference 72).

Reprinted by permission from Peterson et al. 1999.

6.71 in 12-year-olds and 6.58 in 25-year-old adults. Taylor et al. showed that 6- to 12-year-old children had higher pH levels with exercise than 20- to 29-year-old adults (97).

Peterson et al. elected to study trained prepubertal and pubertal female swimmers because they expected that these subjects would be better able to tolerate high exercise intensities (72). Their exercise protocol called for a two-minute bout of leg exercise at 40% maximal work capacity followed by two minutes of supramaximal work (140% maximal work capacity). At the end of exercise, muscle pH was 6.76 ± 0.17 and 6.66 ± 0.11 in the prepubertal and pubertal girls, respectively, and the P_i/PC ratio was 70% greater in the pubertal girls (figure 4.5). These differences, however, were not found to be statistically significant, causing the authors to conclude that "these findings do not support the concept that glycolysis is attenuated in prepubertal children" (p. 2154). However, considering the magnitude of differences seen between the two groups and the small number of subjects (n = 9), one wonders if these differences do, in fact, have biologic meaning and would have been statistically significant given a larger study sample.

Why Is Glycolytic Activity Lower in Children?

Each of these pieces of evidence regarding glycolytic metabolism in children by itself would have to be considered meager. But we cannot escape that, collectively, they all indicate the same trend: a continual improvement in glycolytic metabolic capacity throughout the growing years. We now move on to the question of how this phenomenon might be explained. The performance ramifications of this development of glycolytic function are examined in chapter 9, where we assess developmental changes in short-burst physical activities.

Glycogen Stores

Muscle glycogen is the primary substrate for glycolysis. Glycogen is the dominant fuel in exercise performed at an intensity above 50% $\dot{V}O_2max$; during sustained exercise above 70% to 80% $\dot{V}O_2max$, endurance is related to the initial glycogen content of the muscle, and fatigue corresponds to its depletion. Maturational differences in glycogen stores might therefore be expected to influence exercise metabolism.

Eriksson and Saltin described increasing stores of glycogen in the quadriceps femoris with age in boys in the age groups 11.6, 12.6, 13.5, and 15.5 years (30). Mean glycogen contents were 54, 70, 69, and 87 mmol · kg^{-1}, respectively. During a progressive exercise test, the rate of glycogen utilization was directly related to age group. Utilization rate was three times greater in the oldest than in the youngest subjects (figure 4.6). An identical pattern was observed in muscle lactate levels.

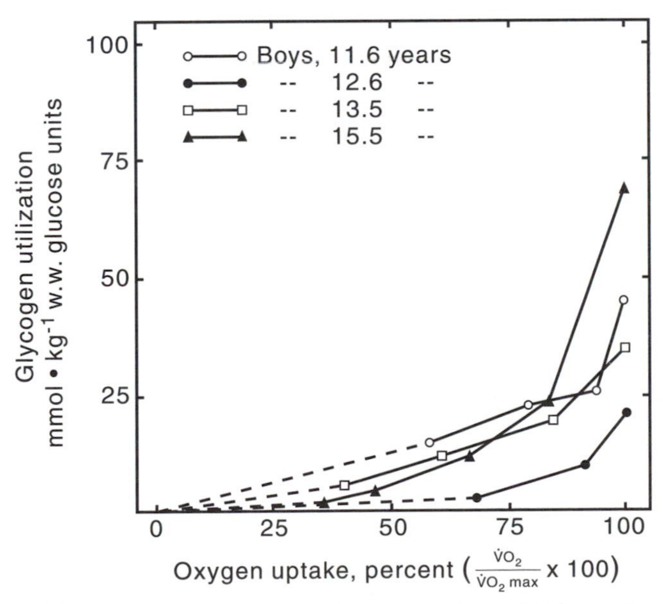

FIGURE 4.6 Glycogen utilization with maximal exercise relative to age in boys (from reference 30).
Reprinted by permission from Eriksson and Saltin 1974.

Lundberg et al. described glycogen content in needle biopsies of the vastus lateralis muscle of 25 healthy children ages two months to 11 years (60). Mean concentration was 61 ± 16 mmol · kg^{-1}. This study failed to suggest developmental change, since no variation in glycogen stores was seen in respect to age.

Differences in Epinephrine Secretion

Circulating epinephrine stimulates both glycolysis (by enhancing glycogen breakdown and PFK activity) and free fatty acid oxidation (by promoting lipolysis). Developmental differences in adrenal secretion of epinephrine might then be expected to influence both glycolytic and aerobic metabolic rate. Data examining maturational differences in serum epinephrine levels is too scant, however, to permit any conclusions.

Weise et al. reported resting values of epinephrine in 80 healthy children and adolescents (43 girls and 37 boys), ages 5 to 17 years (103). Plasma concentration progressively fell with increasing age and increasing pubertal stage. Mean values in the prepubertal subjects were almost three times those in subjects at Tanner stage V. The males demonstrated higher levels than the females. The authors felt that actions of the sex hormones, known to suppress adrenal secretion of epinephrine, might be responsible for these changes. They acknowledged, however, that a possible effect of age-related changes in catecholamine clearance could not be dismissed.

Analysis of the few studies of maturational differences in epinephrine responses to exercise is hampered by the wide interindividual variability in catecholamine values (7). Moreover, values can vary markedly, depending on measurement methodology as well as extrinsic factors such as change in body position, food intake, and placement of intravenous catheters (75).

During a maximal treadmill test, peak epinephrine levels were not significantly different in eight boys and seven men, as reported by Lehmann et al. (57). Average levels rose from a mean resting value of 0.115 ng · ml^{-1} in the boys to 0.970 ng · ml^{-1} during maximal exercise. Respective values for the men were 0.092 and 1.028 ng · ml^{-1} ($p > .05$).

Delamarche et al. had 17 children ages 8.5 to 11.0 years perform a 60-minute bout of cycle exercise at 60% $\dot{V}O_2max$ (20). During exercise, epinephrine levels rose and were always higher in the boys. Values increased 2.7-fold in the boys and by a factor of 1.9 in the girls. In a similar study of eight adult men and women, Hoelzer et al. observed a 2.8-fold rise at the end of 60 minutes of exercise (45).

Berg and Keul (5) suggested that a reduction in adrenal catecholamines might be responsible for the reduced anaerobic capacity of children based on the tight inverse relationship observed by Lehmann et al. (57) between lactate levels and total plasma catecholamines during exercise. Rowland also found a close association between serum lactate and epinephrine levels in a progressive cycle test in 11 healthy 10- to 12-year-old boys (unpublished data). Venous blood specimens were drawn at rest, during exercise at 59% and 73% $\dot{V}O_2$max, and during maximal exercise. The patterns of the rises in serum lactate and epinephrine were virtually identical. Correlation coefficients for lactate versus epinephrine at 73% and 100% $\dot{V}O_2$max were $r = .59$ ($p = .07$) and $r = .81$ ($p = .0025$), respectively (figure 4.7).

Change in Fiber Type Distribution

The distribution of slow-twitch oxidative and fast-twitch glycolytic muscle fibers has generally been considered fixed at birth or very shortly thereafter (65, pp. 243-245). However, any change in their distribution during biologic development would affect relative oxidative and glycolytic capacities in children (12). As one might expect, few data exist to help determine whether such a change occurs. Therefore, no firm conclusions can be drawn.

In an autopsy study, Oertel could find no differences in the percentages of slow- and fast-twitch fibers between the ages of 2 and 20 years (70). Lexell et al. (59), on the other hand, reported that the percentage of fast-twitch fibers increased significantly from age 5 (about 35%) to age 20 years (about 50%). Biopsy studies in children by Bell et al. (4), Lundberg et al. (60), and Eriksson (27) indicated slow-twitch percentages of 59%, 59%, and 60%, respectively. Komi and Karlsson reported 50% and 56% slow-twitch fibers in vastus lateralis biopsies of 15- to 24-year-old

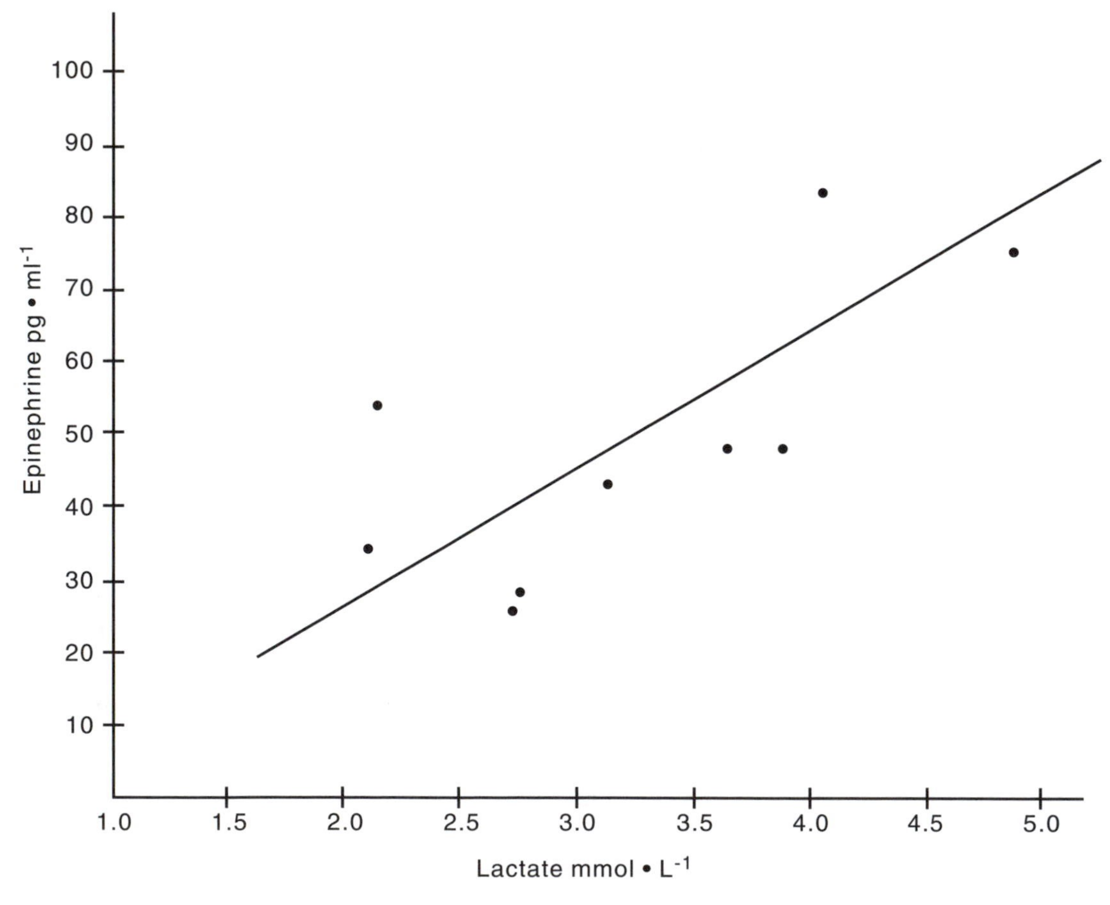

FIGURE 4.7 Correlation of serum epinephrine and lactate levels at an exercise intensity of 70% $\dot{V}O_2$max in children (Rowland, unpublished data).

males and females, respectively (53). Typical values in sedentary adults are 45% to 55% (65).

The Glucose–Fatty Acid Cycle

Exercise scientists have traditionally believed that fat metabolism regulates carbohydrate metabolism. This idea was derived from a study by Randle et al. in the 1960s that indicated that increases in fatty acids in rats reduced glycolysis by the inhibitory action of acetyl CoA on the activities of pyruvate kinase and PFK, a process termed the glucose–fatty acid cycle (78). This matter has been the subject of some debate. Coyle et al. disagreed with this concept, which they felt had little support in human studies (15).

Holloszy argued, though, that the conclusion of Coyle et al. "did not seem compatible with the extensive evidence that elevation of plasma FFA [free fatty acids] reduces carbohydrate utilization" (47, p. 323). He cited studies indicating that raising FFA level slows depletion of muscle glycogen in both humans and animals. Dyck et al., for instance, showed that men given a lipid infusion demonstrated a significant increase in circulating FFA and a 20% to 76% sparing of muscle glycogen during a 15-minute cycling bout (25). Indeed, glycogen sparing by increased utilization of FFA and fat oxidative metabolism during exercise is considered a beneficial effect of endurance training.

We shall see later in this chapter that (a) the aerobic pathways appear to be more active in small children than in large and (b) FFA levels and fat metabolism during exercise may be more active in children than adults. Traditionally, this possibly greater reliance of children on fatty acid oxidation has been considered a default phenomenon because of a primary inhibition of glycolytic pathways in the young. Is it possible, however, that the low levels of glycolysis in children might simply reflect an inhibition imposed by their superior aerobic metabolism? That is, do children rely less on anaerobic metabolism because they don't *need* to? Are aerobic and anaerobic metabolism reciprocal in children? If so, which is the driving factor?

Pianosi et al. reasoned that if a superior level of aerobic metabolism was the primary factor, adults should be expected to show higher pyruvate levels in the blood during exercise with a stable lactate-to-pyruvate ratio (74). That is, a greater bottleneck for entry into the Krebs cycle would be evident in adults than in children. The researchers examined blood lactate and pyruvate levels during exercise in three groups of subjects ages <7 years, 11 to 14 years, and 15 to 17 years. Values were determined at rest, immediately after 6 minutes of exercise at one third and two thirds of maximal work capacity, and 20 minutes postexercise. Lactate levels with exercise increased with age, while pyruvate concentrations were stable. This resulted in an increase in the lactate-to-pyruvate ratio with increasing age. These findings suggest that the greater increase in lactate during exercise in mature subjects is related to a primary improvement in glycolytic function rather than a decrease in aerobic metabolic capacity.

Rowland and Cunningham studied the relationship of aerobic ($\dot{V}O_2max$ per kilogram) and anaerobic fitness (50-yd sprint time) with ventilatory anaerobic threshold (VAT) during a progressive treadmill-walking protocol in children (84). VAT expressed as percentage of $\dot{V}O_2max$ is typically higher in children than adults, which might be explained by either the child's greater aerobic fitness or inferior anaerobic capacity. In this study, VAT as percentage of $\dot{V}O_2max$ was found to be inversely correlated with both $\dot{V}O_2max$ per kilogram ($r = -.77$) and sprint performance ($r = -.63$). This suggests that children's greater VAT as a percentage of $\dot{V}O_2max$ is a reflection primarily of their inferior anaerobic capacity, since a greater level of aerobic fitness would be related to a lower, not higher, relative VAT.

As Cooper and Barstow pointed out, too, an inhibitory effect of aerobic fitness on glycolytic capacity might occur at submaximal exercise levels (12). But no influence would be expected on markers of glycolysis (such as peak lactate) at exhaustive levels of exercise. These data, then, support the hypothesis that the lower levels of glycolytic function observed in children represent a primary rather than a secondary metabolic phenomenon.

Aerobic Metabolism

While glycolytic capacity appears to increase during the course of biologic development, the aerobic metabolic machinery seems to slacken. This conclusion is based on animal studies demonstrating the effect of body size on metabolic rate and is supported by reports of declines in cellular aerobic enzyme content and possibly mitochondrial density with age in children. Identifying changes in aerobic metabolism that limit endurance performance as children grow, however, is more problematic.

Resting Metabolic Rate

Resting or basal metabolic rates are assumed to represent entirely aerobic pathways of energy transfer via fatty acid oxidation (i.e., a respiratory exchange ratio of approximately 0.70). Studies of basal metabolic rate (BMR) in relationship to body size have ascertained differences in aerobic metabolism in the resting state.

We introduced in chapter 1 the concept that resting metabolic rate relative to body size is greater in small mature animals than in large ones. The horse has a metabolic rate at rest that is about 1,200 times that of the mouse, but mice have a metabolic rate relative to body mass that is 10 times that of the horse (76). Over the span of mammalian size, resting metabolic rate relates to body mass by the scaling exponent 0.75. This mathematically indicates that as body mass increases, so does resting metabolic rate, but at a relatively slower rate. In other words, resting metabolic rate expressed relative to body mass decreases as body mass becomes larger. Expressed allometrically, BMR per kilogram relates to $M^{-0.25}$.

The snag here is that by both dimensionality theory and mechanistic hypothesis, resting metabolic rate is expected to relate to body mass by the exponent 0.67, not 0.75. This difference, long the source of considerable consternation by biologists and a multitude of attempts at explanation, remains unreconciled (see reference 89 for full discussion).

Equally obscure is the physiologic meaning of the 0.75 exponent itself. In 1982, Heusner commented, "Thirty years of experimental work with this approach have failed to offer any direction for understanding the physiological significance of the allometric equation [for BMR] within or beyond the realm of energy metabolism, a situation which illustrates Claude Bernard's statement: 'Empiricism may serve to accumulate facts, but it will never build science'" (43, p. 15).

He and others have discounted the traditional explanation that animals need to generate heat to match that lost by surface area to maintain a stable body temperature (the surface law). This explanation implies that, in a circular fashion, an animal's body surface area becomes the determinant of both metabolic rate and the heat lost from the body. This, Heusner argued, violates the second law of thermodynamics, which holds that heat is a consequence of metabolism and not its cause.

We are discussing phylogenetic scaling exponents here, scaling variables among adult mammalian species. As discussed in chapter 1, we expect that ontogenetic exponents, those describing the relationship between physiologic variables and body size during growth, to be different. And this is indeed the case when we examine the allometric scaling of resting metabolic rate to body mass in children.

From several studies, Holliday et al. derived curves for the relationship of basal metabolic rate with increasing body mass in children (46). The mass exponent was found to be 1.02 (i.e., the ratio standard) up to approximately 10 kg, but the exponent changed to 0.58 between 10 and 20 kg. Beyond that size, the allometric relationship between mass and BMR has not been worked out but can be assumed to be similar. As evidence, if BMR is expressed relative to body surface area, which relates to $M^{0.67}$, a progressive fall is seen with age. BMR declines from 52 cal $\cdot$ m^{-2} $\cdot$ h^{-1} at age 6 in males to 43 cal $\cdot$ m^{-2} $\cdot$ h^{-1} at age 18. Respective values in females are 50 and 36 cal $\cdot$ m^{-2} $\cdot$ h^{-1} (52).

Clearly, then, the BMR of small children differs from that of large ones in the same way as is seen in animals. An 8-year-old boy exhibits a significantly higher resting metabolic rate relative to his body size than a 13-year-old. There are two possible explanations for this developmental pattern: Either (a) the aerobic metabolic processes within individual cells are more intense in the small boy (i.e., cellular aerobic metabolic rate is inversely proportionate to body size), or (b) his metabolically active organs represent a greater proportion of his body mass (in which case no age or mass differences in the rate of cellular oxidative metabolism would be assumed).

The various organs do, in fact, have different metabolic rates, and it is well recognized that the percentage of body mass contributed by those that are most metabolically active does decrease as body size becomes greater. In a mouse, for instance, the liver constitutes 6% of body mass, while in the elephant it is closer to 1.6% (89). But is this difference sufficient to explain the lower body mass–relative metabolic rate in the pachyderm?

To answer this question, Schmidt-Nielsen examined the allometric equations for the mass-specific metabolic rates of most important metabolic organs themselves in respect to body mass (89). Mass exponents were –0.15 for kidney, –0.30 for brain, –0.13 for liver, –0.02 for heart, and –0.01 for lungs. Given that the mass-specific metabolic rate for the entire animal is related to $M^{-0.25}$, it is evident that only the brain has a similar exponent. Schmidt-Nielsen concluded, "We can therefore immediately see that the observed decrease in specific metabolic rate cannot be explained by the decreases in the relative sizes of the metabolically most active organs" (89, p. 91).

In growing children, however, the picture may be different. Holliday et al. found that the decline in the combined weights of brain, liver, kidneys, lungs, and heart relative to body weight in growing children directly paralleled the decline in mass-relative metabolic rate with age (46). They reported a mass scaling exponent of 0.53 for the metabolic activity of these organs in respect to mass between 10 and 60 kg, similar to the 0.58 exponent observed for BMR (figure 4.8). These authors concluded that "upward from 10–12 kg, BMR per kilogram declines during growth because its principal source (the internal organs) becomes a smaller proportion of body weight as growth progresses" (46, p. 192).

In animals, however, there is also abundant evidence that individual cells become increasingly hypermetabolic with decreasing body size. Maximal respiratory rate per unit of mitochondrial volume is relatively constant across animal sizes (3–5 ml $O_2 \cdot cm^{-3} \cdot min^{-1}$), but mitochondrial density is inversely related to body size (~$M^{-0.10}$; 89, 95). That is, individual mitochondria in the mouse and horse operate at the same metabolic rate, but the concentration of mitochondria in the cell is greater in the mouse.

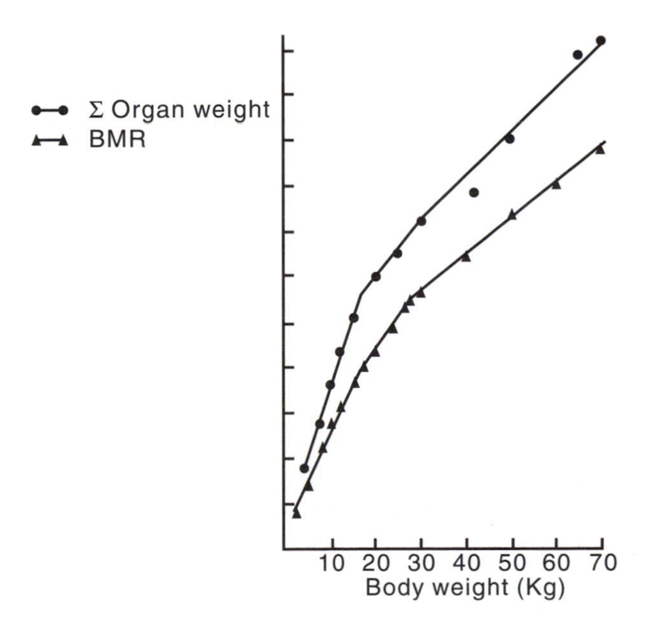

FIGURE 4.8 Log-log plots of basal metabolic rate (BMR) and the combined mass of brain, liver, kidney, lungs, and heart relative to body mass in children (from reference 46). Note: Units on the ordinate (y axis) are arbitrary to allow comparison of curve slopes.

Similarly, an inverse relationship is observed between the activities of mitochondrial oxidative enzymes and body mass in mature animals. Porter and Brand determined the metabolic rates of hepatocytes isolated from nine species of animals (76). The rate of oxygen consumption per unit mass of cells related to body mass by the exponent –0.18. This demonstrated, then, that cells from larger animals consume less oxygen than those from smaller ones. Others, including Krebs (54) and Couture and Hulbert (13), demonstrated similar findings.

Muscle content of cytochrome oxidase, an important enzyme in the aerobic electron transport chain, has been reported to relate negatively to mass by the exponent –0.24 in animals (55). When Emmett and Hochachka analyzed gastrocnemius tissue in 10 mammalian species, the activity of enzymes critical to oxidative metabolism decreased with greater body size (26). The overall mass scaling exponents for the activity of citrate synthase, beta-hydroxybutyryl-CoA dehydrogenase, and malate dehydrogenase were between –0.07 and –0.21 (well below the allometric exponent –0.25 for mass-specific total-body resting metabolic rate).

Does this trend hold up in children? It is unclear if mitochondrial density changes as children grow. Van Ekeren et al. described mitochondrial density ranging from 2.6% to 9.8% (mean 5.8%) in quadriceps femoris biopsy specimens in 21 children and adolescents 1.2 to 20.7 years old (99). This is consistent with the findings of Bell et al. in 13 children (mean age 6.4 years), who had an average mitochondrial density in vastus lateralis muscle of 5.5% (4). These numbers are somewhat higher than those reported in adults. Hoppeler et al., for example, described a mitochondrial density of 3.92 ± 0.16% in five men and 4.74 ± 0.30% in five women (50).

Researchers have described decreases in cellular aerobic enzyme activity with increasing age in children that mimic those reported in respect to size in animals. Haralambie demonstrated a decline in citrate synthase activity in vastus lateralis muscle from approximately 30 $U \cdot g^{-1}$ at age five to seven years to 20 $U \cdot g^{-1}$ in mature subjects (38).

Berg et al. (6) reported mean citrate synthase values of 28.3 ± 6.8 and 22.1 ± 8.2 µmol · min^{-1} · g^{-1} in muscle specimens of children 6.4 and 17.1 years old, respectively ($p > .05$). In that study, fumerase activity was inversely related to the child's age. The value at age 6.4 was 57.6 ± 9.5 µmol · min^{-1} · g^{-1} but 41.0 ± 13.4 at age 17.

Eriksson and Saltin reported mean succinic acid dehydrogenase activity of 5.4 µmol · min^{-1} · g^{-1} in muscle of five healthy 11-year-old boys (30). Values of 3.6 µmol · min^{-1} · g^{-1} were reported in 24- to 30-year-old untrained men by Gollnick et al. (36).

Haralambie compared oxidative enzyme activity in vastus lateralis tissue in 14 adolescents ages 13 to 15 years and 14 adults ages 22 to 42 years (39). Of the six enzymes studied, all but one had significantly greater activity in the adolescents. In another study, the same author found significantly higher activity of the oxidative enzymes fumerase (by 56%) and isocitrate dehydrogenase (by 85%) in muscles of 11- to 14-year-old girls and young adult women (40).

The mechanism and the biologic explanation for this link between aerobic enzyme activity and body size are problematic. Enzyme production is a phenotypic expression of gene control, and the contribution of variations in factors such as gene transcription by messenger RNA and binding of mRNA to ribosomes in respect to body mass is unknown (8).

As with PFK, there is a clinical model of aerobic metabolic dysfunction that demonstrates the importance of this pathway to physical fitness. Products of the Krebs cycle become oxidized in the electron transport chain, where the reduction of molecular oxygen is coupled to the phosphorylation of ADP to form ATP. Children who have a congenital deficiency of components of the aerobic chain (i.e., cytochromes) or enzymes within the Krebs cycle demonstrate abnormal skeletal muscle function, and in some cases the myocardium and central nervous system are involved as well. As would be expected, these children characteristically exhibit exercise intolerance, with high levels of circulating lactate. $\dot{V}O_2max$ is typically about 40% of normal, a reflection of limited maximal arterial-venous oxygen difference. Maximal cardiac output in those without myocardial involvement is normal (37).

From these data, then, it would be reasonable to conclude that the decline in mass-relative resting metabolic rate as children grow reflects *both* a decrease in cellular oxidative metabolic rate and the lesser contribution of highly metabolically active organs to body mass with growth.

Exercise Metabolism: A Different Scaling Factor

Allometric analyses of oxygen uptake relative to body mass during maximal exercise in children have generally indicated an exponent that is greater than that expected at rest ($M^{0.58}$) but less than the ratio standard ($M^{1.00}$). In fact, in studies of children the values for the exponent b in the equation $\dot{V}O_2max \sim M^b$ have ranged considerably, from as low as 0.37 to as high as 1.22. These differences are thought to reflect small sample sizes as well as variations in factors such as body composition, athleticism, and sex.

On the average, however, most reported mass scaling exponents for $\dot{V}O_2max$ have been between 0.70 and 0.90 (with the ratio standard within 95th percentile confidence limits in many studies). This indicates that the relationship between aerobic metabolism and mass during peak exercise is different from that at rest. More specifically, it tells us that body size has more of an effect on total-body aerobic metabolism at high levels of exercise than at rest.

This shift in mass scaling exponent for aerobic metabolic rate may be explained in the following way. At rest, changes in body metabolic rate with increasing body size reflect both (a) a fall in cellular rate of oxidative metabolism and (b) a decrease in the size of the organs most metabolically active at rest—liver, heart, lungs, kidney, and so on—relative to body size. During maximal exercise, however, things are different. Peak aerobic metabolic rate still is influenced by the limits of cellular aerobic machinery, which we have seen has a small but real inverse relationship with body size. But the tissue responsible for metabolic rate has shifted: It is now skeletal muscle. At rest, the combined contribution of the metabolic rate of brain, heart, kidney, liver, and lung to total-body rate is about 80%, while that of skeletal muscle is 20%. During maximal exercise, the percentages reverse, and locomotor muscle may represent as much as 85% of the metabolic rate.

While the size and metabolic contribution of the visceral organs relative to body mass decline with age, the mass of skeletal muscle relative to body mass *increases.* At age 5 years, muscle mass is 42% of the body mass in males, and this increases to 50% at age 15 (63, pp. 126-128). The values for females are 40%

and 43%, respectively. It has been estimated, too, that the muscles of the lower extremity constitute 40% of the total muscle mass at birth but 55% at the time of puberty.

Changes in the relative size of the most metabolically active tissue (i.e., skeletal muscle), then, should serve to increase the mass scaling exponent for $\dot{V}O_2max$ with growth (in fact, it should be greater than 1.00). In the meantime, a decrease in cellular aerobic metabolism with increasing body size has the opposite effect. The balance of these two factors is expected to determine the mass scaling exponent for $\dot{V}O_2max$ in any group of children. Interindividual variability in these two influences might explain the wide variability in this exponent observed empirically in pediatric studies.

Substrate Utilization

We have seen that good evidence exists that the capacity for glycolysis improves as children age, while cellular aerobic function appears to decline. Given these reciprocal changes, we would expect that younger children rely more on fatty acid oxidative metabolism for energy during exercise than older children and adults. Some research observations have supported this idea, while others have not.

Assessment of substrate utilization during exercise in children has been based on determination of serum levels of free fatty acids and glycerol and on measurement of gas exchange variables. The relative contribution of fat and carbohydrate as energy substrates can be estimated by the respiratory exchange ratio (RER; equivalent to $\dot{V}CO_2/\dot{V}O_2$) at the mouth, which, in steady-state exercise at an intensity below the anaerobic threshold, reflects the cellular respiratory quotient. An RER of 0.70 indicates 100% fat utilization, while 1.00 indicates that carbohydrates are serving entirely as the metabolic fuel.

Martinez and Haymes found significantly lower RER values in 8- to 10-year-old girls during a 30-minute treadmill run at 70% $\dot{V}O_2max$ than in 20- to 32-year-old women (64). However, Rowland and Rimany found no significant differences in RER between young adult women and premenarcheal girls during 40 minutes of cycling at 63% $\dot{V}O_2max$ (86). Mean RER values at the end of 60 minutes of submaximal exercise were not different between men and boys in the studies of Asano and Hirakoba (1) and Macek and Vavra (61).

Riddell et al. found that the fat contribution to the energy requirements of sustained exercise at 55% peak $\dot{V}O_2$ was approximately 50% in 10 to14-year-old boys (81). This finding contrasts with an earlier study in older boys, ages 14 to 17 years, whose fat contribution was approximately 30% while cycling at a slightly higher intensity of 60% $\dot{V}O_2max$ (80).

In their study of prepubertal boys cycling for 60 minutes at 60% $\dot{V}O_2max$, Delamarche et al. found a greater turnover of free fatty acids than had been reported previously in studies of adults (20). They suggested that since "the intervention of this metabolic pathway results in an accumulation of citrate in the tricarboxylic cycle, which inhibits glycolysis, the greater utilization of the FFA by the children would be a means to limit the decrease of blood glucose during prolonged and moderate exercise" (p. 70).

No obvious differences in blood glucose levels during exercise have been reported between children and adults (7). In particular, there has been no indication of a hyperglycemic response as has been described in patients with congenital PFK deficiency (100).

Among adults, men have greater glycolytic function than women. Women demonstrate lower activity of glycolytic enzymes and rely to a greater extent on fatty acid oxidation. At the same time, adult women exhibit a greater relative resistance to muscle fatigue than men. This has led to the suggestion that estrogen might have glycogen-sparing properties (44).

Riner and Boileau examined this issue in prepubertal children (82). They compared substrate utilization in 10 prepubertal girls and boys during 60 minutes of continuous treadmill walking at workloads of 50% (low) and 75% (high) $\dot{V}O_2max$. The fuel source was mainly carbohydrates at both levels of work in boys and girls. At the low workload, carbohydrate constituted 54% and 56% of the substrate utilized in the girls and boys, respectively. At the high workload, however, carbohydrate utilization was greater in the girls (66% versus 59%), a finding that the authors concluded was "not readily explained."

Training Effects

The effects of endurance training on energy metabolism have been assessed by examining cellular enzyme activity as well as alterations in substrate utilization.

What scant information is available suggests that the responses might be similar in children and adults.

Enzyme Activity

Studies in adults indicate a clear-cut enzymatic response to physical training. The highly trained adult male endurance athletes studied by Gollnick et al. had twice the activity of succinic acid dehydrogenase of nontrained individuals (36). A two- to three-month period of endurance training in adults (30–60 minutes at 70–80% $\dot{V}O_2$, three to five times per week) caused a 40% to 50% increase in muscle content of oxidative enzymes and a similar rise in mitochondrial density (49).

In adults, 20% to 35% increases in glycolytic enzymes following sprint training have been reported (98). However, Gollnick et al. could find no significant differences in mean levels of PFK activity in weightlifters, canoeists, or cyclists when compared with untrained adult men (36).

In the one study of children, Eriksson et al. examined the effects of four months ($n = 8$) and six weeks ($n = 5$) of exercise training on muscle metabolic markers in boys 11 to 13 years old (28). In the first group, significant increases were observed in mean blood lactate at maximal exercise following training (4.7 ± 0.6 to 5.6 ± 0.7 mmol · L^{-1}). Similarly, muscle lactate concentration was 56% higher at peak exercise as a result of training. In the second group, resting levels of succinic acid dehydrogenase and PFK rose by 30% and 83%, respectively. No changes were seen in distribution of fiber types.

Based on these few data, of course, it is impossible to draw any conclusions regarding maturational changes in skeletal muscle enzyme responses to training. The single study of five circumpubertal boys suggests that young subjects do respond with enzymatic changes similar to those seen in adults.

Substrate Utilization: The Crossover Concept

At the beginning of a progressive exercise test, utilization of fat as an energy substrate is high. As intensity of work rises, the relative contribution of fat falls, and by the time of the subject's exhaustion, fat oxidation is minimal. The opposite trend is seen with carbohydrates, which provide an increasing proportion of energy substrate as exercise intensity rises. At some time in the course of such exercise, the curves of the percentage contribution of fats and carbohydrates as substrate intersect; this is termed the crossover point. Below this point, fats predominate as the energy source, while above it, carbohydrates prevail.

A tenet of our understanding of endurance training in adults has been a shift to increasing fatty acid oxidation during exercise, sparing glycogen stores (which seem to be the energy-limiting factor for sustained exercise) and thereby improving endurance fitness. The crossover concept, then, indicates that the intersection of the fatty acid–carbohydrate utilization curves should shift to the left with endurance training (9).

This concept might be useful in understanding substrate utilization and energy flux with exercise and might help identify differences in substrate utilization related to maturation, diet, intensity and forms of exercise, type of training, and sex.

Duncan and Howley tested this concept in children (23, 24). They studied substrate utilization during a progressive cycle exercise test before and after a four-week cycling training program in 10 children compared with nontrained controls. A significant decrease in RER was observed with training at all levels of exercise except the lightest, indicating an effect of fat utilization. The curves of fat and carbohydrate utilization at the different exercise intensities before and after training are presented in figure 4.9. First, it is clear that the pattern of substrate utilization (decline in fat and increase in carbohydrate with greater intensity) is the same as that observed in adults. Second, the intersection point of the carbohydrate and fat utilization curves has shifted from 38% $\dot{V}O_2max$ before training to 43% after training, an effect of greater reliance on fatty acid oxidation.

This study, then, confirms that reliance on fat as energy substrate is enhanced following exercise training in children. It suggests a tool by which future studies might identify any maturational differences in this response.

Are Children Metabolic Nonspecialists?

Bar-Or first presented the idea that children, as compared with adults, might be "metabolic nonspecialists" in respect to sports performance (3). He observed that adult athletes are typically highly specialized in their sports skills: Compare the long,

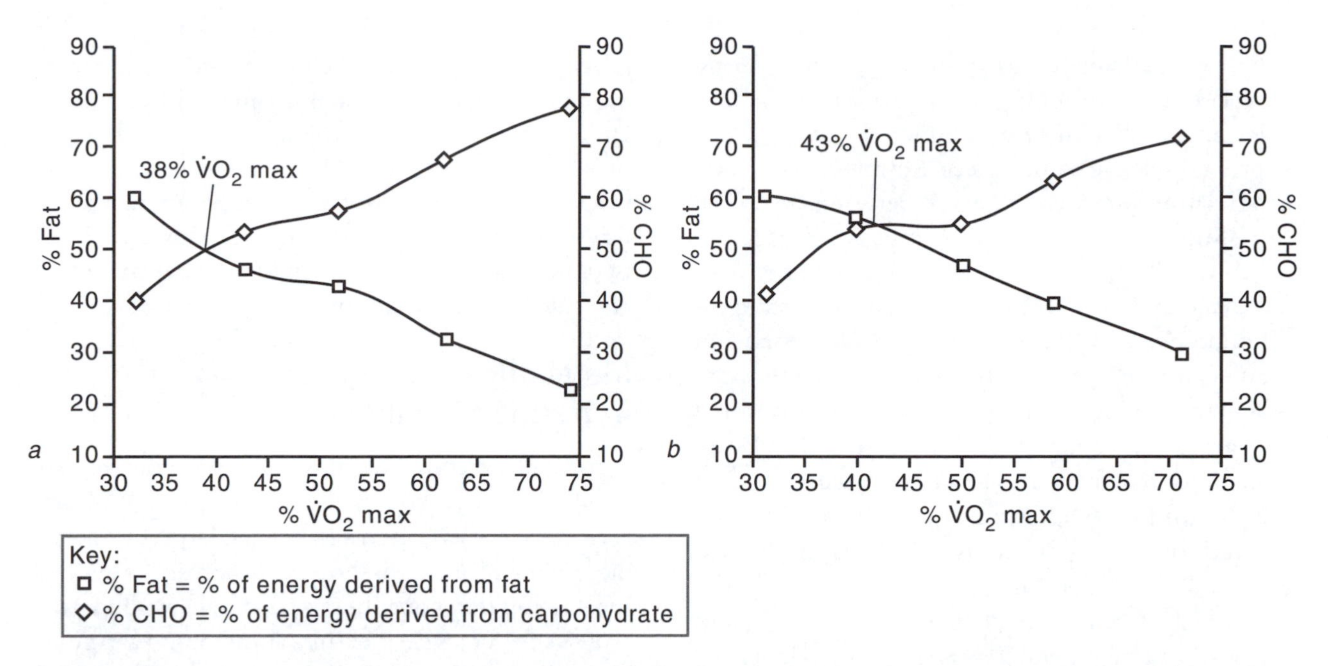

FIGURE 4.9 The crossover point *(a)* before and *(b)* after four weeks of cycle ergometer training in children (from reference 24).

Reprinted by permission from G.E. Duncan and E.T. Howley 1999.

thin distance runner with a high $\dot{V}O_2$max with the muscular weightlifter with great strength but low aerobic fitness and the sprinter who is very fast in a 50-yd dash but possesses only average aerobic power. In children, on the other hand, "the sprinting star of his class is often also above average in long-distance running and successful in a variety of team sports. [This is also evident] in the laboratory, as children who possess a high maximal O_2 uptake also perform above average anaerobically" (3).

Some of the studies that tested this concept over the years have been supportive, others not. These investigations have all been cross-sectional and have been unable to definitively resolve the issue.

In studies that have involved strictly prepubertal subjects, a link has been observed between the different forms of physical fitness (3, 31, 77). Some have shown a significant relationship between Wingate anaerobic performance and $\dot{V}O_2$max (although in some reports the subjects were trained athletes). Rowland and colleagues studied the relationships between performances on vertical jump, 50-yd sprint, and treadmill $\dot{V}O_2$max in 20 children ages eight to nine years (unpublished data). Maximal aerobic power was expressed relative to body mass, but vertical jump and sprint times were examined in absolute terms, since no significant correlation with body weight was observed in these variables. Moderate correlations were seen between all three measures (r = .51 to .57).

Docherty and Gaul described a similar study of 52 children ages 10 to 11 years (21). Mass-relative maximal aerobic power during Wingate testing was significantly related to total work per kilogram of body mass (r = .62) but not to peak power (r = .47). The researchers observed no relationship between $\dot{V}O_2$max per kilogram and maximal knee flexor or extensor angular velocity, which was considered to indicate leg muscle power. These authors stated that "the strength of the relationship between different metabolic systems was weak and unable to account for a large portion of the variance between scores" (p. 530). They felt that their results did not support the theory that children are metabolic nonspecialists.

Other investigators have compared associations between these types of fitness in prepubertal versus postpubertal subjects. Falk and Bar-Or compared the association of aerobic and anaerobic power in pre-, mid-, and postpubertal subjects (33). Correlation coefficients were typically higher in the pre- and midpubertal children, but these comparisons were made with absolute rather than size-adjusted values. And, as the authors observed, "correlations were based on unequal and small samples, as well as a small range of data for the late pubertal group" (p. 328).

Suei et al. described similar findings in a comparison of explosive power (vertical jump), isometric strength, and anaerobic power in subjects at Tanner stages I, II–IV (considered as a single group), and

V (96). When values were expressed in absolute terms, correlations were significant between the three variables and higher in the prepubertal group. When expressed relative to body size, however, no significant relationships were observed between any of the fitness measures regardless of maturational stage.

Murphy (67) examined the relationship between $\dot{V}O_2max$ and anaerobic power (Wingate testing) in 20 untrained girls (mean age 10.2 ± 0.7 years) and women (mean age 22.1 ± 2.9 years). When maximal aerobic power and peak and mean anaerobic power were adjusted allometrically for body size, moderate but significant correlations were seen within each age group (r = .40 to .56). The strength of these associations was similar in the two groups, leading the author to conclude that "the phenomenon of 'metabolic specialization' is evident for both untrained children and adults. When correlations between aerobic and anaerobic power are examined as they develop naturally [i.e., without training], they appear to be similarly related regardless of level of maturation."

There are a number of possible explanations for the linkages observed between the various forms of physical fitness. These need to be considered when interpreting correlation studies that test the question of maturational influences on metabolic specialization.

Parallel but Noncausal Relationships

Many physiologic variables that have nothing to do with each other increase (or decrease) in parallel during the childhood years. Simply demonstrating an association between two factors as children grow does not therefore imply a causal relationship between the two. Muscle strength, anaerobic power, and $\dot{V}O_2max$ all are closely associated with changes in body size during the pediatric years. In any group of young people, the child who is larger or sexually more mature can be expected to outperform the smaller, less mature child in all areas of strength, aerobic, and anaerobic fitness, and a linkage can be expected between these different forms of fitness.

Obviously, then, assessing relationships between fitness variables at different levels of maturation using absolute values is treacherous. And if we are considering a variable that changes with age even relative to body size, such as anaerobic power, the same problem exists. How would we explain an association between peak anaerobic power per kilogram on Wingate testing and performance on a mile run test in a group of 10-year-old boys? Would this mean that children are metabolic nonspecialists? Or could the interpretation simply be that this association between anaerobic fitness and endurance performance is based on level of physical and biologic maturation, independent of any true biologic relationship between the two?

Metabolic Crossover in Testing Modality

A particular testing modality is often considered an indicator of a certain type of fitness (e.g., Wingate cycle testing tests "anaerobic" fitness; 1-mi run, "aerobic"). This is not entirely true, however. Anaerobic metabolism may contribute to endurance events, and a 30-s all-out cycle test has an aerobic metabolic contribution that is not negligible. Only about 25% of the variance in 1-mi run times in children can be attributed to $\dot{V}O_2max$ (85), and evidence exists that anaerobic fitness may contribute in such events (18, 62). Chia et al. (11) found a 19% to 44% aerobic contribution to a Wingate anaerobic test (depending on mechanical efficiency). Van Praagh et al. described values of 60% to 70% $\dot{V}O_2max$ during Wingate testing of children (100). This "cross-contamination" of metabolic sources used in different testing modalities might affect correlations between different forms of fitness.

It is also important to recognize that subjects' motivation is important for peak performance on all these different tests of fitness. Correlation between fitness test results in different groups could therefore be a reflection of psychologic factors that allow some subjects to withstand discomfort at high levels of exercise.

Influence of Training

The original idea of metabolic nonspecialization in children was based on observations of children and adults who were participants in sports. Adults tend to be highly skillful in a particular type of athletics, while child athletes seem to demonstrate competence in a variety of sports. This observation might be explained if adult athletes are capable of enhancing strength, speed, power, and aerobic fitness with training to a greater degree than children can. That is, what appears to be metabolic specialization in the adults might simply reflect the mature person's greater capacity to improve a particular form of fitness with training. Maturity-related differences in metabolic or performance specialization might only indicate differences in trainability between children and adults.

There are some data to support this idea. Prepubertal children demonstrate a dampened response of $\dot{V}O_2max$ with endurance training compared with adults (about a 5% increase in the child compared with 25–30% in the young adult). The relative increases in muscle strength with a short-term program of resistance training are similar in children and adults, but it is obvious that children cannot develop the massive strength seen in adult powerlifters who have been training for years. No information is yet available regarding maturational differences in training effects on anaerobic fitness.

Somatotype Specialization

Differences in body composition and somatotype can dictate performance in events described as aerobic or anaerobic. The person who is highly muscular and heavy-boned is more likely to be successful in weightlifting than the asthenic, lean individual who can excel in distance running. Many of the somatotypic and anthropometric features that define athletic capabilities in adults are developed at the time of puberty under the influence of the sex hormones. Particularly obvious are the effect of testosterone on muscle mass development in males and the changes in body fat in females. It may be, then, that children, rather than lacking metabolic specialization, are actually "somatotype nonspecialists." By this explanation, the physical characteristics that permit expertise in certain kinds of physical activities in adults are developed at the time of puberty.

Metabolic nonspecialization in children versus adults is a complex question. From the preceding discussion, it is evident that an understanding of this concept might provide some interesting insights into the maturational changes that affect physical performance.

Conclusions

The view we have of cellular metabolic activity in children is extremely limited. Nonetheless, the available information suggests a progressive rise in anaerobic glycolysis as children age and a reciprocal decline in aerobic metabolic capacity.

The explanation for these patterns of change is not clear. The only clues we have are that (a) these changes appear to progress smoothly through the pediatric years, suggesting that influences of puberty and sex hormones are not primary, and (b) these patterns mimic those observed in adult mammals in relation to body size. The latter point might indicate that the alterations in metabolic function seen during development in children reflect biologic fundamentals associated with changes in body dimensions.

How these patterns of metabolic change contribute to alterations in exercise fatigue threshold and define the development of anaerobic and aerobic fitness in children remains to be established. It is obvious that a true understanding of these dynamics requires improvements in noninvasive research methodologies.

Discussion Questions and Research Directions

1. What mechanisms explain developmental changes in anaerobic and aerobic cellular metabolism? Does change in one influence the other?
2. Do children, as might be predicted, rely more on lipid oxidation during sustained exercise? If so, is this a primary function, or is it secondary to diminished glycolytic capacity?
3. Does the theoretically greater reliance of children on fat oxidation during exercise have a glycogen-sparing and endurance-promoting effect?
4. Do children show the same qualitative and quantitative shifts in substrate utilization with endurance training as adults?
5. How are shifts in cellular metabolic function with age translated into alterations in physical fitness?

5

Aerobic Fitness

La respiration est donc une combustion.

—Lavoisier (1780)

In this chapter we discuss

- the normal development of maximal aerobic power and the proper means of expressing values of $\dot{V}O_2max$ with respect to body size,
- factors influencing interindividual and sex differences in $\dot{V}O_2max$, and
- the relationships of $\dot{V}O_2max$ with endurance fitness and physical activity.

Aerobic fitness can be defined physiologically as maximal oxygen uptake ($\dot{V}O_2max$), the highest rate at which skeletal muscle cells can utilize oxygen in the provision of energy for locomotion. Or it can be described in functional terms as endurance performance: the time it takes to bike, run, walk, or swim a particular extended distance, or the distance one can achieve in a certain time. While the former contributes to the latter (i.e., $\dot{V}O_2max$ correlates with endurance performance), the factors that define the limits of oxygen utilization and those establishing field performance are not the same. $\dot{V}O_2max$ has generally been thought to be limited by components of oxygen delivery, most particularly stroke volume. Endurance performance, on the other hand, is influenced by a multitude of factors, including not only $\dot{V}O_2max$ but also submaximal economy, cellular aerobic enzyme activity, substrate stores, anaerobic fitness, motivation, and environmental conditions.

All the determinants that distinguish the aerobic fitness of one person from another appear to be no different in children than in adults. The added dimension during childhood is the *growth* of aerobic fitness, both physiologic and functional, and the determinants of ontogenetic change that may differ from those that define interindividual variations. That, then, is the theme of this chapter.

We propose that the principal determinants of the development of $\dot{V}O_2max$ in respect to body mass as children grow are (a) the increase in size of the exercise "motor," skeletal muscle, and (b) the relative decrease in oxidative enzyme activity within those skeletal muscle cells. At the same time, according to the concept of developmental symmorphosis, all the factors responsible for defining the changes in $\dot{V}O_2max$ with growth should develop in concert. That is, there is no biologic rationale for one part of the system to possess a greater or more advanced functional capacity than the other members of the system during biologic development. We shall see, in fact, that cardiac size and circulatory dimensions grow at the same time to match oxygen delivery to the increasing cellular oxygen consumption of exercising muscle.

Improvements in endurance fitness during growth, on the other hand, are not influenced by $\dot{V}O_2max$. The performance of events such as a 1-mi run is instead affected by developmental enhancement of submaximal energy economy and changes in the threshold of exercise intensity that can be sustained for long periods of time. The latter most likely has a metabolic basis, reflecting the functional capacity of aerobic metabolism to utilize pyruvate, prevent its conversion to lactate, and thus prevent resulting increases in hydrogen ion concentration.

The Development of $\dot{V}O_2max$

The course of childhood is marked by progressive enlargement of the system components that determine $\dot{V}O_2max$—lungs, heart, skeletal muscle—as well as improvement in endurance performance. Consequently, absolute values of maximal aerobic power increase as the child grows. Between the ages of 6 and 12 years, the $\dot{V}O_2max$ of a boy more than doubles (from 1.2 $L \cdot min^{-1}$ to 2.7 $L \cdot min^{-1}$). Average values for females are about 200 $ml \cdot min^{-1}$ lower than those of males at the same chronologic age. At puberty the rise in $\dot{V}O_2max$ accelerates in males as a result of the anabolic influences of testosterone, while values in females plateau. The normal development of maximal aerobic power in children has been extensively reviewed by Krahenbuhl et al. (54), and Armstrong and Welsman (6, 8).

When $\dot{V}O_2max$ is expressed relative to body mass, no substantial changes are observed across the pediatric years in boys; their typical values during treadmill testing are 50 to 52 $ml \cdot kg^{-1} \cdot min^{-1}$. In girls, however, a progressive decline is observed in mass-relative $\dot{V}O_2max$ from age 8 years on, and the 15-year-old girl has a $\dot{V}O_2max$ that is 20% lower (42 $ml \cdot kg^{-1} \cdot min^{-1}$) than her male counterpart.

When $\dot{V}O_2max$ is related to body mass by allometric analysis, a wide range of cross-sectional scaling exponents have been observed, from 0.23 to 1.23. Most, however, are within the range of 0.70 to 0.90, and few exceed 1.00 (although it may be assumed that the 95th percentile confidence limits of many include the ratio standard). This is consistent with the mass scaling exponent for $\dot{V}O_2max$ of 0.81 observed in phylogenetic studies of mature mammals (97, pp.106-108). It has been suggested that factors such as small sample size, variations in body composition and somatotype, athleticism, and sex may be responsible for the wide differences in reported scaling factors for $\dot{V}O_2max$ in children (8).

When $\dot{V}O_2max$ is related allometrically to mass, some interesting differences appear in sex-related development of aerobic fitness. Welsman et al. compared the relationships of peak $\dot{V}O_2$ with body mass using both the ratio standard and allometric scaling in subjects 11 to 23 years of age (106). By the ratio standard, as expected, mass-relative peak $\dot{V}O_2$ showed no change with age in the males and a decrease in the females. However, when peak $\dot{V}O_2$ was adjusted by allometric scaling exponents for body mass, values increased in males. In the females, peak $\dot{V}O_2$ rose into puberty but then remained stable.

Longitudinal studies provide a better estimate of ontogenetic changes in peak $\dot{V}O_2$. Armstrong and Welsman used multilevel modeling to assess serial changes in peak $\dot{V}O_2$ in children between ages 11 and 17 years (9). Lean body mass was the major factor influencing growth of peak $\dot{V}O_2$; no effect of blood hemoglobin concentration was observed. The multilevel regression models indicated that even when body mass and composition were considered, peak $\dot{V}O_2$ increased with age and maturation in both sexes.

We still await a full understanding of how aerobic fitness develops with age in respect to body size in youth. That is, from a physiologic standpoint, it is not yet clear whether children are more or less aerobically fit than adults. What is clear, however, is that the means by which maximal aerobic power is adjusted for body dimensions has a significant impact on how such changes are interpreted.

Ontogenetic Scaling Exponents for $\dot{V}O_2max$

In the previous chapter it was proposed that the interplay between two factors is responsible for defining the mass scaling exponent for $\dot{V}O_2max$ in children and adolescents: (a) the limitation of cellular aerobic enzyme activity and (b) the size of the tissue that is the major consumer of energy during exercise (i.e., skeletal muscle) relative to body mass.

Aerobic Enzyme Activity

Cellular aerobic enzyme function defines the limits to which the aerobic metabolic machinery can run. In both adult animals and human children, the activity of these enzymes (expressed per gram of muscle tissue) in the resting state is observed to decrease in respect to increasing body mass (see chapter 4). While allometric information is not available for children, mass scaling exponents for oxidative enzyme activity and mitochondrial density among adult mammals range from −0.07 to −0.24. That is, the cellular metabolic "fires" of larger animals burn less intensely than those of smaller ones. It follows that the maximal rate of cellular aerobic metabolism becomes relatively less as children grow.

As discussed in chapter 4, just *why* aerobic enzyme activity should be influenced by body dimensions per se remains a mystery. The speed of oxidative reactions should be governed by rate of ATP depletion, substrate availability, and enzyme concentration. Limited research data indicate no maturational differences in ATP depletion or in availability of substrate at maximal exercise.

Skeletal Muscle Mass

Skeletal muscle is by far the most metabolically active tissue during exercise, in contrast to the resting state, when liver, lungs, heart, kidney, and brain create the major energy demand. As described in the previous chapter, the progressive decline in the relative size of these internal organs in respect to body mass during the growing years has been used to explain the mass exponent for resting metabolic rate ($\sim M^{0.58}$) in children.

At the same time, the mass of body muscle constitutes a progressively *increasing* percentage of muscle mass. From a number of cross-sectional sources, Malina and Bouchard (60) calculated that average muscle mass as percentage body weight in males increases from 42% at age 5 years to 53% at age 17. However, no appreciable change is observed in females across the same age span (41% and 42%, respectively). Consistent with these observations, Alexander et al. (1) reported that proximal leg muscle mass in mammals relates to body mass by the exponent 1.10, and the same scaling exponent was reported by Nevill (67) for the relationship of leg volume and body mass in adolescent boys. Welsman et al. (105) reported a correlation coefficient of .84

between total thigh muscle volume (determined by MRI) and peak $\dot{V}O_2$ in prepubertal girls, and Zanconato et al. (112) found that peak $\dot{V}O_2$ related allometrically to calf muscle cross-sectional area by the scaling exponent 1.04.

The switch in the major energy-consuming tissue from the visceral organs (which decrease relative to increasing body mass) to skeletal muscle (which rises in proportion to increasing body mass) accounts for the higher mass-scaling exponent for $\dot{V}O_2$ during maximal exercise (0.70–0.90) than in the resting state (0.58). The mass scaling exponent for $\dot{V}O_2$max in any given child or adolescent should be expected, then, to represent the combined influences of these two factors: one (aerobic enzyme activity) that declines with greater body mass and the other (relative size of skeletal muscle mass) that increases. It may be expected, too, that individual variations in the relationship of these two dynamic trends might contribute to the marked variability reported in scaling exponents for $\dot{V}O_2$max.

It is apparent, as well, that scaling exponents varying by sex or body composition might be similarly explained. Since skeletal muscle mass does *not* contribute an increasing percentage of body mass with growth in girls, their mass scaling exponents for $\dot{V}O_2$max should be lower. In most (but not all) studies this sex difference has been observed (10). For example, the mass exponents for $\dot{V}O_2$max in 51 females studied by Cooper et al. were 0.79 and 0.91, compared with 1.01 and 1.02 in males (24). Pettersen et al. found an average scaling exponent of 0.82 for peak $\dot{V}O_2$ in boys and 0.69 in girls 8 to 17 years old (74). In the study by Welsman et al. of 29 males and 34 females, scaling exponents were 0.92 and 0.84, respectively (106).

If this concept is correct, we should expect to see that components of the oxygen supply chain also relate to muscle mass in the same way as $\dot{V}O_2$max. In animals, heart and lung size often correlate with skeletal muscle volume. The proverbial heart of the lion, for instance, is 1.36 times bigger relative to body mass than the average mammal, corresponding to a muscle mass that is 1.31 times greater and a lung capacity that is 1.67 times larger (19). In a group of 201 children and adolescents ages 6 to 17 years, Daniels et al. found that lean body mass explained 75% of the variance in left ventricular mass (29). In a group of premenarcheal girls, Rowland et al. (87) reported that maximal stroke volume and $\dot{V}O_2$max both related to body mass by the same exponent (0.55). Scaling exponents for $\dot{V}O_2$max and stroke volume with respect to lean body mass were 0.82 and 0.89, respectively. In a group of boys (mean age 12.0 ± 0.4 years), the respective values were 0.79 and 0.97 (86).

There is convincing evidence, then, that $\dot{V}O_2$max in children is closely associated with muscle volume. We have seen that if this were the only factor influencing the relationship of $\dot{V}O_2$max to increasing body size, however, the mass scaling exponent for $\dot{V}O_2$max would be greater than 1.00. This is not the case: In most reports, mass-scaling exponents for $\dot{V}O_2$max are in fact less than 1.00. The added effect of a progressive decline in aerobic metabolic activity per unit of muscle as a child grows may explain these lower scaling exponents.

Is $\dot{V}O_2$max (or Peak $\dot{V}O_2$) Really $\dot{V}O_2$max? Physiology and Semantics

The traditional paradigm holds that oxygen uptake during a progressive exercise test rises linearly until the functional limits of the oxygen supply chain are reached. At exercise intensities above this point, defined as $\dot{V}O_2$max, values of $\dot{V}O_2$ plateau. Energy utilization then must rely on anaerobic metabolism, and, with the accumulation of lactate and other byproducts, fatigue and termination of exercise rapidly ensue.

This viewpoint has not been without detractors. Noakes contended that "the belief that oxygen delivery alone limits maximal exercise performance has straightjacketed exercise physiology for the past 30 years" (68). He proposed that "an alternative mechanism may need to be evoked to explain exhaustion during maximal exercise . . . [and] that the factors limiting maximal exercise performance might be better explained in terms of a failure of muscle contractility . . . which may be independent of tissue oxygen delivery."

Observations of the behavior of $\dot{V}O_2$ in pediatric subjects at the end of what appears to be a truly exhaustive effort during exercise testing seem to support this. A true plateau (or flattening) of oxygen uptake is hardly ever witnessed, and criteria that have been used to identify a plateau (which, in reality, is instead a tapering) have been satisfied in only a minority of cases. (These criteria have included changes in $\dot{V}O_2$ in the final stage of exercise of less than 150 ml · min^{-1}, less than 2.1 ml · kg^{-1} · min^{-1}, and less than two standard deviations of mean changes between submaximal stages.)

In their review of exercise studies in children, Rowland and Cunningham noted that from 21% to 60% of children have demonstrated a $\dot{V}O_2$ plateau by one of these criteria during a progressive treadmill exercise test (84). When maximal heart rate, RER, and serum lactate are considered, there is no evidence that a child's failure to demonstrate a plateau is related to motivation, level of aerobic fitness, or anaerobic capacity. Lack of a $\dot{V}O_2$ plateau was once considered unique to children, but a close scrutiny of the adult literature reveals similar low percentages of subjects who demonstrate a plateau by standard criteria (26, 65).

This issue of nonplateau of $\dot{V}O_2$ during exercise testing in children has two implications, one pragmatic and the other conceptual. In defining maximal values of physiologic variables during exercise, it is necessary to rely on a criterion that accurately indicates that the performance of a given subject on a progressive exercise test is truly an exhaustive effort. The question thus is, Can a test in which no $\dot{V}O_2$ plateau is observed, but in which other markers of a "maximal" test are satisfied (e.g., RER > 1.00, heart rate over 190 bpm), be safely assumed to be truly exhaustive?

The second issue relates to understanding the relationship between the maximal capacity of the oxygen delivery chain and exercise performance (at least as determined by a progressive treadmill or cycle test). The absence of plateau might lead one to suspect that the true limits of oxygen delivery had not been reached, even with a true exhaustive effort. This is in line with the argument set forth by Noakes (68). Perhaps the oxygen uptake when the subject reaches a peak exercise effort just happens to be the oxygen uptake when performance is being limited by another, oxygen-independent factor (such as limits of skeletal muscle enzyme activity or fatigue of the contractile apparatus). After all, $\dot{V}O_2$ per kilogram of body mass during submaximal exercise is interpreted as a marker of exercise economy (how much energy it takes to move a certain number of kilograms). Why can't the $\dot{V}O_2$ at peak exercise simply be interpreted as the economy at the time of exhaustion rather than as an indicator of the limits of the oxygen delivery chain? The absence of a plateau at the end of a progressive exercise test lends credence to this idea.

One approach to both of these issues would be to determine if $\dot{V}O_2$ can increase above the peak $\dot{V}O_2$ observed on a standard progressive test when supramaximal loads are imposed. Two such studies have been performed in children, in fact, and both demonstrated the same results.

In the study by Armstrong et al., 18 girls and 17 boys (mean age 9.9 ± 0.4 years) performed three treadmill running tests to exhaustion (12). The first was a standard discontinuous incremental protocol at 7 km · h^{-1} with increasing slope. An exhaustive effort was defined by a leveling of heart rate near or above 200 bpm, peak RER over 1.00, and subjective signs of exhaustion (hyperpnea, facial flushing, unsteady gait, sweating). The second and third tests were performed at treadmill slopes that were 2.5% and 5.0% greater, respectively, than the highest slope achieved on the first test. Thirteen of the subjects (37%) demonstrated a $\dot{V}O_2$ plateau (<2 ml · kg^{-1} · min^{-1}) during the first test. Average peak $\dot{V}O_2$ was no different on tests 2 and 3 than on the first. In the girls, peak $\dot{V}O_2$ values were 51 ± 6 ml · kg^{-1} · min^{-1} on test 1, 52 ± 7 on test 2, and 52 ± 7 on test 3, while the respective values in the boys were 62 ± 6, 63 ± 8, and 64 ± 7.

Rowland performed a similar study in nine children ages 10 to 13 years (79). Three (33%) satisfied the criteria for a plateau of $\dot{V}O_2$ on the first test. Follow-up tests were performed at grades 2.5%, 5.0%, and 7.5% steeper than that achieved on the initial test. Values for peak $\dot{V}O_2$ were 53.9 ± 4.3 ml · kg^{-1} · min^{-1} on test 1, 55.0 ± 3.8 on test 2, 54.0 ± 4.9 on test 3, and 53.1 ± 3.2 on test 4. In both of these studies, peak RER rose markedly as slope increased (from 0.99 ± 0.03 in test 1 to 1.14 ± 0.08 in test 4 in the study by Rowland).

These two studies both indicated, then, that $\dot{V}O_2$ cannot increase with supramaximal loads above values observed in a standard progressive test in which a plateau is not observed. This implies that, despite lack of a plateau, $\dot{V}O_2$ at peak exercise does, in fact, reflect the limits of oxygen delivery in children. Accordingly, peak $\dot{V}O_2$ can be considered to reflect $\dot{V}O_2$max if certain subjective and objective criteria (RER, heart rate) are met.

Many pediatric exercise scientists prefer to use the label peak $\dot{V}O_2$, feeling that the term $\dot{V}O_2$max implies that a plateau has been observed. As noted in the preface, this book respects each author's choice of term when specific literature is cited. Otherwise $\dot{V}O_2max$ is used, since, as indicated by these two studies, the two terms have the same physiologic meaning.

These two reports do not, however, necessarily imply that the concepts presented by Noakes are incorrect. We shall see in the next chapter that peripheral factors, particularly skeletal muscle contractile function and endurance, may serve to regulate and limit circulatory responses to exercise—and thus oxygen delivery.

The Meaning of Physiologic Aerobic Fitness

According to one current viewpoint, aerobic capacity should be considered a "distributed property" (45). That is, $\dot{V}O_2max$ is not defined by a single limiting variable but rather by multiple factors, each with their own limits and controlling influences, which then interact as a system to define the boundaries of physiologic aerobic fitness. This concept is consistent with the principle of symmorphosis, which holds that the function of no single component in a system should exceed that of the other components (97).

Although compelling, the idea of $\dot{V}O_2max$ as a distributed system proves somewhat dissatisfying to the biologic scientist who seeks the mechanisms that serve as primary determinants of aerobic capacity as a child grows. There is evidence that certain elements do, in fact, serve as primary limiting factors in biologic systems. Rate-limiting enzymes such as PFK govern the rate of the glycolytic chain of biochemical reactions. In an acute bout of exercise, peripheral factors (reviewed in the next chapter) such as arteriolar dilatation and skeletal muscle pumping function, not the heart, appear to regulate and limit circulatory flow during exercise.

In this section we discuss two mechanisms that might determine physiologic fitness ($\dot{V}O_2max$) among children: (a) a ceiling for oxygen delivery and (b) constraints created by the limits of aerobic enzyme activity. Data in adults and animals suggest that the former is more critical than the latter. Enzyme concentration and activity appear to relate more to limits of endurance performance than of oxygen utilization. This hypothesis has not, however, been tested in children.

Aerobic Enzyme Activity

Data presented in the previous chapter for both children and animals indicate that the activity of aerobic metabolic enzymes diminishes with increasing body size. As Hochachka has discussed, this decline in activity could be accounted for by a reduction in either enzyme catalytic efficiency or cellular enzyme content (43). The first option is unlikely, since the oxygen consumption of a given volume of muscle mitochondria (4–5 ml · min^{-1}) is similar among animals regardless of size. It can be assumed, then, that the mass-relative decline in aerobic enzyme activity with increased body size reflects a fall in cellular enzyme concentration.

The rate of activity of the Krebs cycle and electron transport chain during exercise is driven by the downstream depletion of ATP as the energy demands of contracting muscle rise. According to the principles of classic enzyme kinetics, the peak velocity (V_{max}) of these enzymatic reactions is determined by the concentration of the enzyme if sufficient substrate is present (20). If the supply of oxygen to the oxidative machinery becomes limiting before V_{max} is reached, oxygen delivery is the limiting factor for $\dot{V}O_2max$. If, on the other hand, V_{max} is reached before oxygen supply becomes limiting, the ceiling of enzyme activity determines $\dot{V}O_2max$. (Devotees of the principle of symmorphosis would suggest that *both* oxygen delivery and V_{max} are reached simultaneously.)

Another scenario, however, needs to be considered in which aerobic enzyme activity might limit $\dot{V}O_2max$. According to the traditional paradigm, the rate of glycolysis increases at high work intensities as the cell becomes starved for oxygen. That is, anaerobic metabolism becomes a forfeiture replacement for aerobic metabolism as oxygen availability becomes limited. Accumulation of the by-products of anaerobic glycolysis, lactate and hydrogen ions, then lead to muscle fatigue.

An alternative, more contemporary viewpoint, however, holds that the increased rate of glycolysis during a progressive exercise test does not reflect a lack of available oxygen. Other factors, such as recruitment of type 2 fibers with increasing exercise intensity, may be responsible. The greater the capacity of the aerobic metabolic pathways, the more pyruvate produced by this glycolytic process can enter the Krebs cycle for oxidative metabolism. As a consequence, less pyruvate is converted to lactate, and pH falls more slowly. The result is increased tolerance for higher exercise intensities and consequently, at least theoretically, a greater $\dot{V}O_2$max.

Whether the maximal capacity of cellular aerobic enzyme activity or oxygen delivery is limiting to $\dot{V}O_2$max has long been a source of debate among exercise physiologists. Most data support the idea that oxygen delivery is the decisive factor, including information indicating that the amount of muscle mass that needs perfusion during exercise far exceeds the heart's pumping capacity (96). Moreover, animals studies indicate that muscle oxidative capacity is closely linked to endurance performance but not to $\dot{V}O_2$max (30). As noted earlier, no research specifically addresses this issue of cellular metabolism versus oxygen delivery in children. We shall return to this question in chapter 11, where we examine variations in enzyme concentration responses to aerobic training.

Oxygen Delivery: Reflections on the Fick Equation

For optimal operation, each of the components of the oxygen delivery chain, from the upper airways to the terminal conversion of ADP to ATP, must be functioning normally. With a breakdown in any component—lung disease, anemia, abnormal hemoglobins, myocardial dysfunction—oxygen delivery becomes impaired. In this sense, *any* of the parts of the oxygen delivery chain can, in a state of abnormal function, serve as a limiting factor for oxygen delivery to the contracting skeletal muscle.

In the normal, healthy individual, though, constraints on oxygen delivery during maximal exercise have been sought within the mathematical confines of the Fick equation, which indicates that $\dot{V}O_2$max is equal to the product of cardiac stroke volume, heart rate, and arterial-venous oxygen difference (AVO_2diff) at the end of an exhaustive bout of progressive exercise. The potential for each of these components to contribute to $\dot{V}O_2$max in children has been assessed in low- and high-fit youth before and after a period of endurance training and in highly trained child endurance athletes. The findings in these studies are outlined later. They indicate no differences from adult subjects. That is, there appear to be no maturational differences in the determinants of maximal oxygen delivery to exercising muscles. We shall see, however, that quantitative differences need to be considered in any child-to-adult comparisons.

Heart Rate and Arterial-Venous Oxygen Difference

Maximal heart rate can be immediately dismissed as the defining determinant of $\dot{V}O_2$ max: values in children and adolescents are clearly independent of level of aerobic fitness, sex, and chronologic or biologic age (78, 104). Similarly, AVO_2diff during maximal exercise changes little during childhood, and values are similar in unfit, fit, and trained populations of boys and girls. AVO_2diff values during maximal exercise that were calculated from cardiac output estimated by the same Doppler echocardiographic technique in multiple studies from two laboratories are presented in table 5.1. Values are remarkably similar for prepubertal untrained boys and girls and highly trained male child cyclists.

Obert et al. (72) studied the effects of 13 weeks of aerobic training on the cardiovascular responses to exercise of 10 girls and 9 boys (mean age 10.5 ± 0.3 years). $\dot{V}O_2$max rose by 15% in the boys and 9% in the girls. However, no significant change was seen in maximal AVO_2diff in either sex. Pre- and posttraining values were 13.0 ± 2.1 and 13.1 ± 1.9 ml · dl^{-1} in the boys, respectively, and 13.2 ± 1.6 and 13.0 ± 1.9 ml · dl^{-1} in the girls, respectively. These findings mimic those of an earlier study by Eriksson and Koch in which 16 weeks of aerobic training failed to increase maximal AVO_2diff in nine boys ages 11 to 13 years, despite a 16% improvement in $\dot{V}O_2$max (34).

In their study of 39 prepubertal boys, Rowland et al. (89) compared cardiovascular findings in eight subjects with low fitness ($\dot{V}O_2$max 38.8 ml · kg^{-1} · min^{-1}) with those with high fitness (54.8 ± 1.2 ml · dl^{-1} · min^{-1}). Mean maximal AVO_2diff was 12.9 ± 3.0 and 12.4 ± 1.1 for the two groups, respectively ($p > .05$).

The explanation for these similarities in maximal AVO_2diff is straightforward. Arterial-venous oxygen difference represents the difference between the arterial oxygen content of blood approaching the muscle cell and the venous oxygen content as it leaves. It therefore represents the amount of oxygen extracted from a given volume of blood passing the cell. Arterial oxygen content is determined principally by blood hemoglobin concentration. As for venous oxygen content, studies indicate that values are very low in nontrained adults; that is, oxygen extraction

TABLE 5.1 Maximal stroke index and maximal arterial-venous oxygen difference in children

	Source	Stroke index ($ml \cdot m^{-2}$)	AVO_2 diff ($ml \cdot dl^{-1}$)
PREPUBERTAL BOYS			
	Rowland et al. (91)	58±9	13.9±3.0
	Rowland et al. (88)	61±11	13.0±2.5
	Rowland and Blum (83)	62±12	13.1±2.2
	Nottin et al. (69)	59±5	13.8±2.1
	Obert et al. (72)	56±5	13.0±2.1
	Nottin et al. (70)	52±8	13.4±1.2
PREMENARCHEAL GIRLS			
	Rowland et al. (90)	55±9	12.2±2.2
	Obert et al. (72)	47±7	13.2±1.9
MALE CHILD CYCLISTS			
	Rowland et al. (93)	76±6	13.1±0.8
	Nottin et al. (70)	63±5	14.0±1.5

Calculated from cardiac output determined by the Doppler echocardiographic technique and oxygen uptake.

is essentially maximized. (Such studies have not been done in children.) The only means by which AVO_2diff can be increased, then, is by (a) increasing hemoglobin concentration (i.e., increasing arterial oxygen content) or (b) altering the distribution of blood flow during exercise. Since these factors are not appreciably altered by level of aerobic fitness or athletic training, maximal AVO_2diff is not altered from that of untrained subjects.

Hemoglobin levels do increase at the time of puberty in males, and this causes an increase in arterial oxygen content. Between the ages of 12 and 16 years, the average hemoglobin concentration in males rises from 13.7 $g \cdot dl^{-1}$ to 15.2 $g \cdot dl^{-1}$. A gram of fully saturated hemoglobin holds 1.34 ml of oxygen. The typical arterial oxygen content therefore increases during this age span, from 18.4 $ml \cdot dl^{-1}$ to 20.4 $ml \cdot dl^{-1}$, an 11% rise.

This corresponds to differences observed in calculated maximal AVO_2diff between adult men and prepubertal boys. Maximal values for men (21.2 ± 2.7 years old) and boys (11.7 ± 0.6 years old) in the study of Nottin et al. were 16.9 ± 3.9 $ml \cdot dl^{-1}$ and 13.8 ± 2.1 $ml \cdot dl^{-1}$, respectively (69). These findings are similar to the maximal AVO_2diff values of 17.2 ± 4.5 $ml \cdot dl^{-1}$ in young men and 13.9 ± 3.0 $ml \cdot dl^{-1}$ in prepubertal boys reported by Rowland et al. (91). As expected, similar differences are seen in maximal AVO_2diff between untrained adult men and women, since the former have a hemoglobin concentration that is 15% greater than the latter (90, 91).

It can be expected from these observations, then, that maximal AVO_2diff should not contribute to differences in oxygen supply to exercising muscle between children and adolescents of the same sex and pubertal status. Maximal AVO_2diff *does* influence ontogenetic increases in $\dot{V}O_2$max in males at puberty, but not in females.

Stroke Volume

From a Fickian perspective, the principal factor that differentiates level of physiologic aerobic fitness between children is maximal stroke volume. As indicated in table 5.1, values for stroke index (stroke volume related to body surface area) during maximal exercise are greater in boys than in girls, and in child athletes than in nonathletes. In the training study of Obert et al. (72), the increase in $\dot{V}O_2$max in both boys and girls was entirely due to a rise in maximal stroke index (52 ± 8 to 60 ± 7 ml · m^{-2} in the boys and 47 ± 7 to 52 ± 4 ml · m^{-2} in the girls). Similar findings were also reported by Eriksson and Koch, who found a 20% increase in stroke volume but no change in maximal AVO_2diff in nine boys ages 11 to 13 years after 16 weeks of training (34).

As discussed in chapter 6, the typical stroke volume response to progressive exercise performed in the upright position is an initial rise of about 30% to 40% and then a plateau to the point of exhaustion. Figure 5.1 depicts the increases observed in stroke index during cycle testing in fit and unfit boys and highly trained child cyclists (89, 93). Two observations are pertinent. First, the pattern of rise is identical in all three groups, with an upward offset in those with greater fitness. This implies that the cardiovascular adaptation to endurance training is the same as that to the effects of daily habitual activity or inherent genetic aerobic capacity.

Second, the differences in maximal stroke volume among the three groups reflect variations in stroke index at rest. From this observation it follows that the factors that define resting stroke volume are those that define stroke volume at maximal exercise as well as, by extension, maximal cardiac output and maximal oxygen uptake. To find the underlying determinants of $\dot{V}O_2$max, then, we need to decipher the factors that influence resting stroke volume.

Resting stroke volume can be influenced by filling volume (ventricular end-diastolic dimension), myocardial contractility, and afterload. Table 5.2 shows a comparison of these factors between untrained prepubertal boys and highly trained child cyclists from studies using two-dimensional echocardiography. In these reports, the left ventricular shortening fraction (a one-dimensional surrogate of the ejection fraction) has been used as an indicator of myocardial contractility (although it can be influenced by pre- and afterload as well). In the single study that

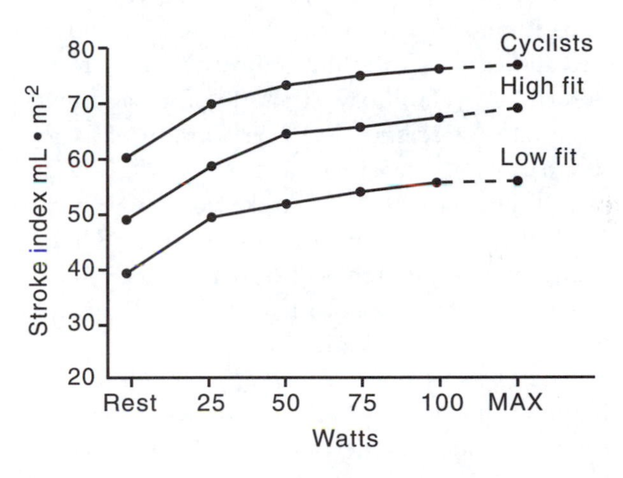

FIGURE 5.1 Stroke index responses to progressive upright cycle exercise in unfit, highly fit but untrained, and highly trained (cyclists) boys. (Data from references 89 and 93.)

TABLE 5.2 Echocardiographic measures of left ventricular size and function at rest in untrained boys and highly trained male child cyclists

	Untrained boys	Male child cyclists
SHORTENING FRACTION (%)		
Nottin et al. (70)	41±4	37±4
Rowland et al. (93)	29.5±4	35±3
LEFT VENTRICULAR END-DIASTOLIC DIMENSION (MM · $BSA^{-0.50}$)		
Nottin et al. (70)	36.7±2.2	39.7±2.8
Rowland et al. (93)	36.4±3.1	41.0±2.3

calculated resting systemic vascular resistance (considered a marker of afterload), Nottin et al. (70) found values of 19.6 ± 6.1 U in untrained boys and 17.3 ± 3.0 U in child cyclists ($p > .05$). From these data, it is clear that only the left ventricular end-diastolic dimension (filling volume) distinguishes the different groups of subjects.

Given the preceding information, an allometric analysis of the relationship of either resting stroke volume or ventricular end-diastolic dimension to increasing body mass should be consistent with the relationship of oxygen uptake to increases in body size as children grow. Some data are available on this point. De Simone et al. (31) found that the mass exponent for resting stroke volume (by two-dimensional echocardiography) in children and adolescents was 0.57, almost identical to that described by Holliday et al. (44) for basal metabolic rate (i.e., resting $\dot{V}O_2$). In a group of 24 girls in a narrow age range (12.2 ± 0.5 years), Rowland et al. found that maximal stroke volume related to $M^{0.55}$, the same exponent as for $\dot{V}O_2max$ (87). In a group of 904 children ages 6 to 16 years, echocardiographic left ventricular mass scaled to body mass by the exponent 0.75 (59).

The explanation for fitness-related differences in filling volume is not known. A number of candidates can be considered. Interindividual differences in left ventricular end-diastolic dimension may be largely under genetic control. Primary cardiac differences may be defined by hormonal influences (growth hormone, IGF, testosterone, insulin). Or the factors defining resting ventricular volume may be extracardiac (plasma volume, parasympathetic tone). A further discussion of these factors that might affect aerobic fitness in children is postponed until chapter 11, when maturational differences in training influences on resting left ventricular dimensions are considered.

To summarize, maximal stroke volume defines interindividual differences in peak oxygen delivery during exercise in childhood as it does in adulthood. As left ventricular end-diastolic dimension—the determinant factor of maximal stroke volume—increases as a child grows, increasing stroke volume during maximal exercise is also the sole factor responsible for ontogenetic changes in the limits of oxygen delivery until puberty. In males, the rise in peripheral oxygen extraction (AVO_2diff) from pubertal hormonal changes contributes to this development of maximal aerobic power.

Does $\dot{V}O_2max$ Reflect the Development of Endurance Fitness?

In any heterogeneous group of individuals, $\dot{V}O_2max$ per kilogram serves as a good indicator of an individual's ability to perform endurance exercise. This is true in children as well as adults. Rowland et al. examined the relationship between 1-mi run performance and $\dot{V}O_2max$ determined during cycle testing in 36 sixth-grade boys from the same school (88). To ensure a wide range of fitness, 10 subjects were selected from each quartile of finishers in a 1-mi run in earlier testing (four subjects did not complete the study). The correlation coefficient between $\dot{V}O_2max$ (per kilogram) and mile run velocity was $r = .77$, similar to that observed in comparable studies in adults. Multiple regression analysis indicated that percentage body fat and $\dot{V}O_2max$ accounted equally for 60% of the variance in run performance.

When $\dot{V}O_2max$ is employed to assess longitudinal changes in endurance fitness in youth, however, a different picture emerges. And here is our first clue that development of a child's maximal ability to deliver oxygen to exercising muscle may not be the best means of assessing his or her changes in aerobic performance: During the growing years, endurance performance progressively improves, while $\dot{V}O_2max$ per kilogram does not (figure 5.2).

Take 1-mi run performance again as an example. Results of mass health-related fitness testing in American schools indicate that the average five-year-old boy can run a mile in 13:46 (3). Ten years later, at age 15, he can cover the same distance in 7:14, an almost 100% improvement in running speed. Between ages 5 and 15, however, his $\dot{V}O_2max$ relative to his body mass will not have budged, remaining stable at about 50 $ml \cdot kg^{-1} \cdot min^{-1}$.

In girls this discrepancy is even more marked. While average 1-mi run time decreases from 15:08 at age 5 to 10:05 at age 15, $\dot{V}O_2max$ per kilogram actually *decreases* from about 50 $ml \cdot kg^{-1} \cdot min^{-1}$ to 40 $ml \cdot kg^{-1} \cdot min^{-1}$ (54). Clearly, then, $\dot{V}O_2max$ per kilogram is capable of giving us information regarding endurance performance capacity in cross-sectional comparisons but is of no value in predicting developmental (ontogenetic) changes in fitness as a child grows. This longitudinal "uncoupling" of

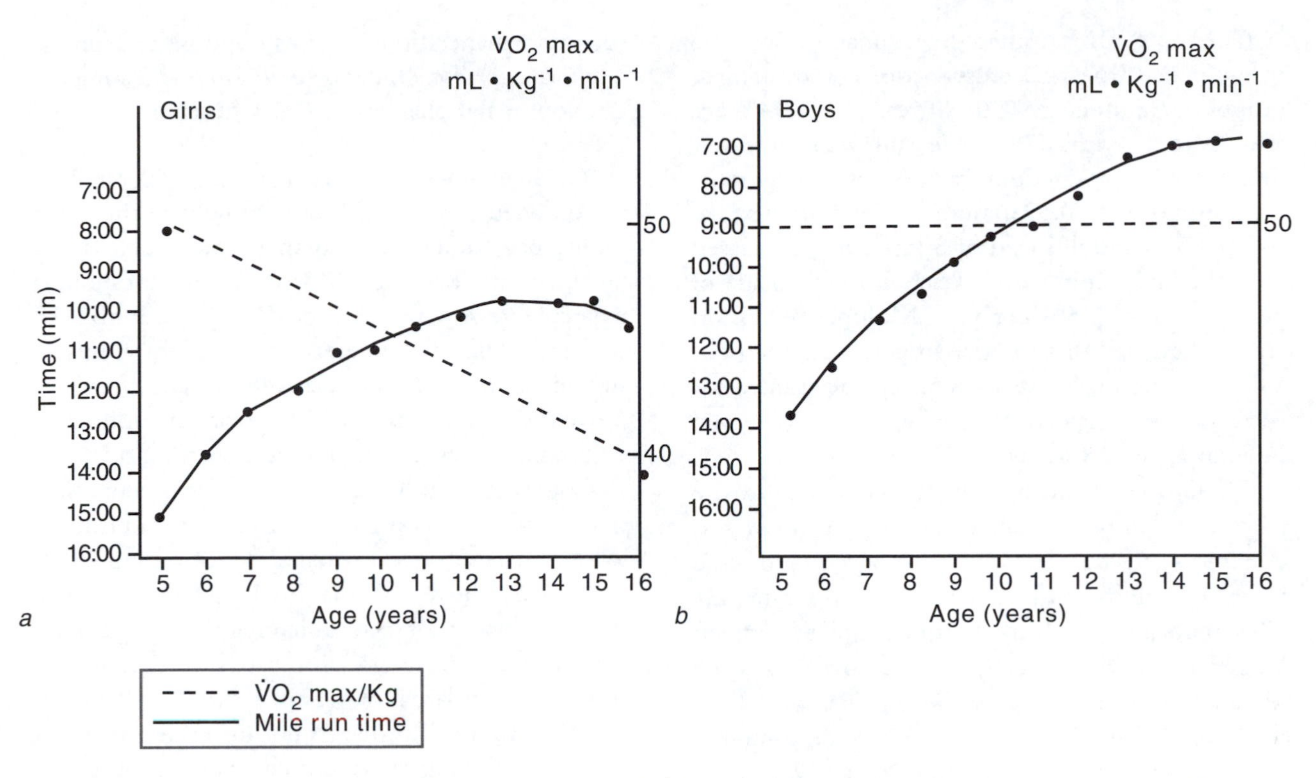

FIGURE 5.2 Comparison of developmental changes in $\dot{V}O_2$max per kilogram and endurance performance (1-mi run velocity) in *(a)* girls and *(b)* boys. (Data from references 3 and 54.)

mass-relative aerobic power and endurance fitness leads also to the conclusion that improvements in endurance performance with growth are not caused by increased capacity of the oxygen delivery chain (relative to the child's size).

This suggests that models other than $\dot{V}O_2$max need to be considered to explain the development of endurance fitness as children grow. A number of alternative conceptual approaches have been proposed, but none has been extensively tested experimentally. As discussed in the following sections, however, these models promise better understanding of the factors that stimulate improved endurance fitness during growth.

Running Economy and Percentage of $\dot{V}O_2$max

The metabolic demands relative to body mass of weight-bearing activities (running, walking) decline as a child grows (see chapter 8). This developmental improvement in exercise economy with age has been linked to better endurance performance, since the relative intensity of a given work rate decreases with age. That is, $\dot{V}O_2$max per kilogram remains stable as a child grows, while $\dot{V}O_2$ per kilogram while running flat at 5 mi · h^{-1} progressively decreases. As a result, the child at age 10 years runs at a significantly lower relative intensity (percentage of $\dot{V}O_2$max) than the 5-year-old (figure 5.3).

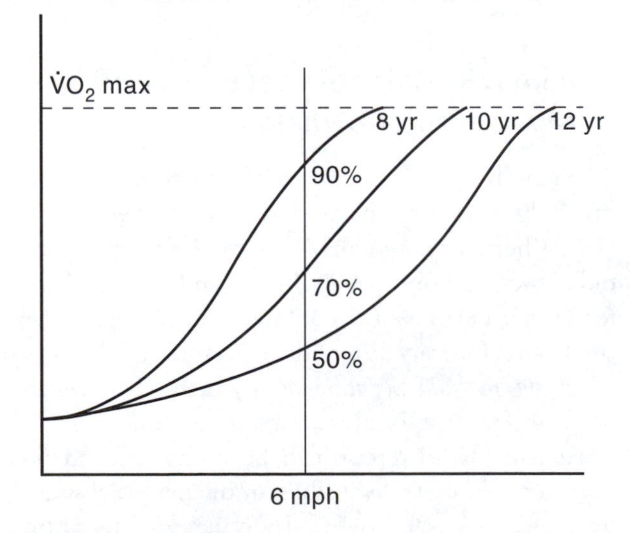

FIGURE 5.3 Improvements in running economy with age mean that the older child works at a relatively lower intensity (percentage of $\dot{V}O_2$max) at the same speed.
Reprinted by permission from T.W. Rowland 1989.

Cureton et al. examined determinants of the rise in endurance fitness in a cross-sectional study of three groups of children ages 7 to 10 years, 11 to 14 years, and 15 to 17 years (28). Mile run/walk time was determined on an outdoor track. Across this age span, times improved by 0.52 minutes per year, submaximal $\dot{V}O_2$ (at a treadmill speed of 8 km · hr^{-1}) decreased 1.0 ml · kg^{-1} · min^{-1} each year, and percentage of peak $\dot{V}O_2$ rose 1.5% per year. Multiple regression analysis revealed that the rise in percentage of peak $\dot{V}O_2$ and improvements in economy accounted for 41% and 31%, respectively, of the increase observed in endurance performance.

These data fit nicely with the concept that the progressive improvement in running economy as children grow translates into a progressive decline in percentage of $\dot{V}O_2$max at a given work rate (63). This trend, in turn, should permit a child to perform longer at the same speed, or faster over the same distance, as he or she grows. This is just what Krahenbuhl et al. showed in their study of nine-minute run performance and running economy in boys who were initially tested at mean age 9.9 years and then again 7 years later (53). $\dot{V}O_2$max per kilogram did not change over that time span, but distance run performance increased on the average by 29%. Running economy improved by 16%, and estimated $\dot{V}O_2$max percentage fell by 13%. These observations invite a discussion of the factors responsible for the improvements in economy with age (and, presumably, for increases in endurance performance), which is postponed until chapter 8.

Maximal Sustainable Level of Muscular Performance

The length of time that exercise can be sustained is inversely related to the workload, or power output (75). Whereas heavy work rates can be tolerated for only several seconds, lesser loads can be maintained for longer periods of time. There exists a particular work rate for each individual, termed the *limit of sustainable muscular performance,* at which he or she can perform exercise for an extended duration.

At this load, exercise can be continued indefinitely or—more realistically—until energy substrate stores are exhausted or until temperature elevation or fluid depletion becomes limiting. Understanding this threshold of exercise intensity involves (a) understanding its metabolic determinants and (b) defining performance fitness and its response to training. This concept also should be useful in delineating the developmental changes in fitness during children's growth.

The highest level of work that can be tolerated for a prolonged time should relate closely to the more traditional markers of endurance fitness such as distance run performance. Moreover, this threshold is expected to increase with age during childhood. An understanding of the determinants of the limits of sustainable muscular performance should therefore prove useful in identifying the factors that influence endurance fitness during biologic maturation.

Conley et al. outlined the metabolic mechanisms that might explain and define the sustainable level of muscular performance (22; figure 5.4). At workloads above the maximal sustainable level, the by-products of anaerobic glycolysis (hydrogen ions, lactate, phosphate) inhibit ADP concentration, limit oxidative phosphorylation, and lead to muscle fatigue. This occurs because the process of glycolysis during submaximal exercise produces an amount of pyruvate that exceeds the demands of the aerobic metabolic pathways. It is now generally agreed that this process occurs independently of oxygen availability. That is, hydrogen ions and lactate accumulate from increased glycolysis even when the oxygen supply is adequate for oxidative metabolism.

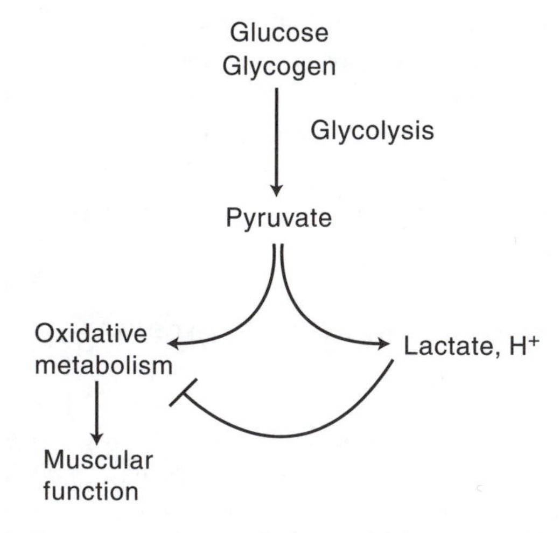

FIGURE 5.4 The metabolic model for sustainable level of muscular performance. Increased work intensity accelerates glycolysis and production of pyruvate. Given sufficient oxidative metabolic activity, conversion of pyruvate to lactate and H$^+$, with resulting fatigue, is limited.

The limit of sustainable muscular performance, then, is defined by the highest work intensity that fails to produce a progressive rise in hydrogen ions, lactate, and other anaerobic glycolytic by-products. At this level, the pyruvate generated by glycolytic metabolism is shunted primarily into aerobic metabolic pathways. As a consequence, ATP supply is maintained, and the inhibitory actions of these by-products of anaerobic glycolysis on oxidative metabolism are prevented. In this model, cellular metabolic processes, rather than oxygen supply, limit oxidative metabolism during steady-state exercise at intensities above the maximal sustainable level. The fit child possesses a greater cellular oxidative capacity than the unfit, and this permits a higher sustainable level of muscle performance.

This concept predicts a certain pattern of change in the limits of sustained exercise—and endurance performance—during the growth of children. We have seen previously that, at least metabolically speaking, children become progressively more anaerobic and less aerobic as they age. These trends are manifest in the relationship between anaerobic power and $\dot{V}O_2max$ during the growing years. Blimkie et al. demonstrated this when they combined data from three studies to show that the anaerobic-to-aerobic power ratio between the ages of 8 and 19 years increased from 2.0 to 3.0 in both males and females (18; figure 5.5).

Endurance fitness increases with age in children. But the concomitant decrease in aerobic capacity and increase in anaerobic capacity, according to the preceding mechanistic construct, should cause the limits of sustainable endurance performance relative to body size to steadily decrease as children grow. This is, in fact, the trend observed in endurance performance relative to body size.

Rowland estimated the mass scaling exponent for changes in mile run performance in children between the ages of 5 and 13 years, using data from the AAHPERD test battery and body mass norms of healthy children in the United States (81). In boys, mile run velocity related to body mass by the scaling exponent 0.66, while the exponent in the girls was 0.45.

Between the ages of 5 and 15 years, the velocity of a boy performing his best on a mile run test doubles, but his weight can be expected to increase by a factor of 3.4. Relative to body size, then, endurance performance declines during the childhood years. This observation, given the concomitant trends in aerobic and anaerobic capacities, is consistent with the metabolic explanation for the limits of sustainable muscular performance set forth by Conley et al. (22). It also provides evidence that developmental changes in metabolic function contribute to the pattern of changes in endurance fitness in growing children.

The limits of sustainable muscular performance obviously cannot be defined by direct observation of these metabolic processes. There are, however, two means by which this threshold can be operationally approached: maximal lactate steady state and critical power. Limited data are available in the pediatric age group for these measures, but both hold promise for insights into developmental changes in endurance fitness.

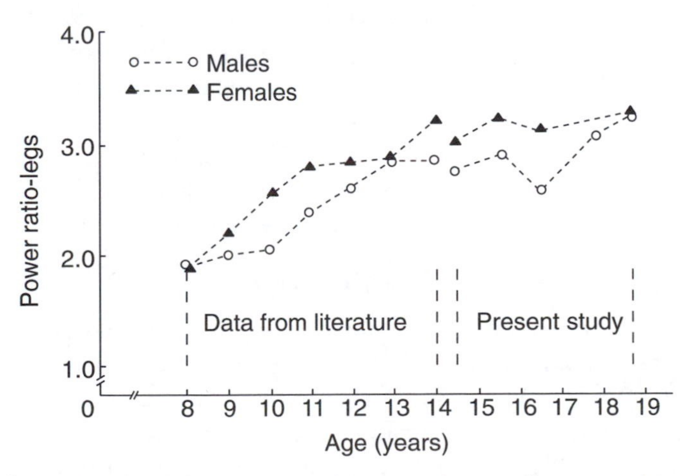

FIGURE 5.5 Changes in the ratio of anaerobic to aerobic power (by Wingate testing and measurement of $\dot{V}O_2max$) during childhood and adolescence (from reference 18).

Reprinted by permission from C.J.R. Blimkie, P. Roche, and O. Bar-Or 1986.

Maximal Lactate Steady State

Maximal lactate steady state (MLSS) is the greatest exercise intensity that can be sustained without a rise in blood lactate level (16). MLSS, then, is expected to reflect the maximal sustainable level of muscular performance and, by extension, endurance performance capability. From the foregoing findings, values of MLSS expressed relative to body size should decrease as children grow. This effect would be consistent with the observation that endurance performance does not improve proportionally with body mass as children grow (e.g., relative to her body size, the 6-year-old girl can a run a mile faster than her 16-year-old counterpart).

In adult subjects, MLSS has been estimated to occur at about 4.0 mmol · L^{-1}. This value may be lower in children, but the few available studies are conflicting. Williams and Armstrong (110) demonstrated that MLSS occurred at levels of 2.1 and 2.3 mmol · L^{-1} in 13-year-old boys and girls, respectively (an exercise intensity equivalent to 77% peak $\dot{V}O_2$). MLSS was defined as the level of blood lactate that corresponded to the highest exercise intensity that could be sustained with a lactate accumulation less than or equal to 0.5 mmol · L^{-1} over the final 5 minutes of a 10-minute bout of exercise. The authors noted that MLSS and percentage of peak $\dot{V}O_2$ were similar to those described in highly trained adults.

Mocellin et al. (62) found a much higher MLSS (5.0 mmol · L^{-1}) in their study of 11- to 12-year-old boys, a value that represented 88% peak $\dot{V}O_2$. Billat et al. calculated running velocity at MLSS using only two 15-minute running stages at 60% and 75% $\dot{V}O_2$max separated by a 40-minute rest period (17). Blood lactate values at MLSS were 3.9 ± 1.0 mmol · L^{-1} for both boys (n = 6) and girls (n = 7). Velocity at MLSS was 8 to 8.5 km · h^{-1} (about 5 mi · h^{-1}, equivalent to a running pace of 12 minutes per mile), which corresponded to 65% $\dot{V}O_2$max. The authors concluded that "the knowledge of this value [is] of particular interest to begin an endurance training program without overtraining of [12-year old] children" (p. 71).

When Benecke et al. studied 34 males 11 to 20 years old, MLSS (figure 5.6) and percentage $\dot{V}O_2$max were independent of age (14). The mean MLSS value was 4.2 ± 0.7 mmol · L^{-1}, and mean intensity at MLSS was 66.5 ± 7.7% $\dot{V}O_2$max. These authors concluded that these findings "support the theory that with physical maturity neuromuscular factors may contribute to the changes in response to selected exercise more than changes in oxidative metabolism and/or glycolysis" (p. 1474).

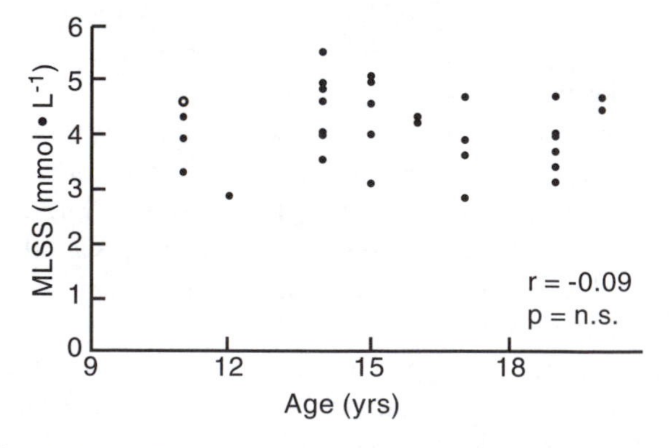

FIGURE 5.6 Maximal lactate steady state (MLSS) in relation to age (reference 14).
Reprinted by permission from R. Benecke et al. 1996.

Considering this information, Armstrong and Welsman felt that "differences in methodology preclude the combining of data from several studies to form a composite picture of children's submaximal blood lactate responses during development. Furthermore, the predominance of adolescent boys in the subject population restricts extrapolation to other groups, for whom information is notably sparse" (6, p. 460). Nevertheless, this approach seems conceptually promising as a means of identifying maturational trends in endurance fitness, given a standard means of (a) measuring MLSS and expressing values relative to body size and exercise intensity and (b) studying lactate responses at different stages of biologic development in boys and girls.

Closely aligned to the concept of MLSS is the ventilatory anaerobic threshold (VAT), the point in an incremental exercise test when minute ventilation diverges from oxygen uptake. VAT is considered to reflect the onset of lactate accumulation in the blood, which, according to the concepts outlined by Conley et al. (22), would indicate the intensity above which anaerobic glycolytic pathways overtake aerobic metabolism.

The idea that the mass-relative value of maximal sustainable muscular performance decreases as children grow is supported by the observation that VAT as a percentage of $\dot{V}O_2$max declines during the childhood years. Reybrouck et al. described a decline in relative VAT from 70% $\dot{V}O_2$max at age 6 years to 55% $\dot{V}O_2$max at age 16 years in both boys and girls (76). Similar findings were reported by Cooper et al. (24), Weymans et al. (108), and Kanaley and Boileau (48). In a longitudinal study, Vanden Eynde et al. tested 30 boys serially for six years (102). The average value of VAT as a percentage of $\dot{V}O_2$max fell from 68.4% $\dot{V}O_2$max at 1.5 years prior to peak height velocity to 59.4% $\dot{V}O_2$max at 1.5 years after peak height velocity.

The Critical Power Concept

When a subject performs a series of exercise bouts at varying work rates, power is inversely related to time until exhaustion by a hyperbolic curve (41). This curve asymptotically approaches a "critical power" (CP) that defines the highest work rate that can be sustained for long periods of time (figure 5.7). The critical power, then, provides an assessment of aerobic endurance capacity, while the degree of curvature

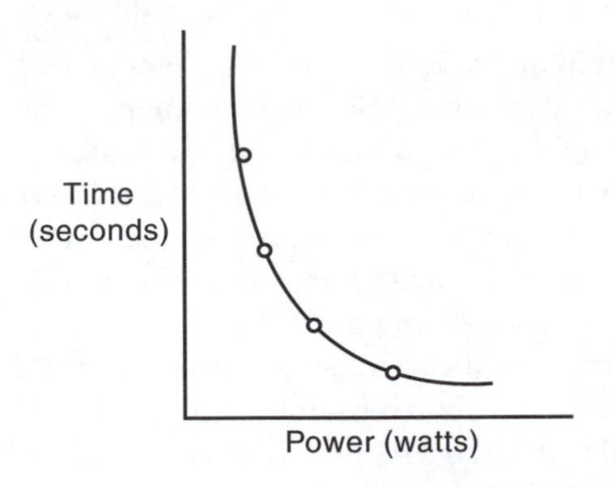

FIGURE 5.7 The critical power concept. A hyperbolic curve is obtained when exercise time to exhaustion is plotted against power output. Critical power is the asymptote. The same relationship can be expressed as a linear function, in which critical power becomes the *y*-intercept.
Reprinted by permission from S.G. Fawkner and N. Armstrong 2002.

of the relationship has been considered to reflect anaerobic work capacity (AWC). The critical power corresponds conceptually to maximal sustainable muscular performance and should be expected to relate to MLSS if the metabolic hypothesis is correct. And all three should be markers of field performance on endurance events. [The findings of Poole et al. in adults did, in fact, suggest that the work rate at MLSS coincides with the asymptote of the critical power relationship (75).]

The curve can be expressed as $t = AWC/(P - CP)$, where t is the time to fatigue and P is power output. This equation can be arranged to give a linear model: $P = CP + (AWC \cdot t^{-1})$, where AWC is the slope and CP is the y-intercept.

Conducting a test to determine critical power is simple, requiring only the timing of a series of cycle bouts at various workloads with a stopwatch. The difficulty lies in defining an exhaustive performance, which is particularly uncertain at lighter work rates. The number of trials, proper pedal cadence, periods of rest between tests, and the range of exercise loads or duration all need to be considered in designing such an investigation (see reference 4 for discussion). We also clearly need to be aware of these factors when comparing critical power findings from one study to another.

Fawkner and Armstrong studied different methods of assessing critical power in eight boys and nine girls (10.3 ± 0.4 years old) using an electronically braked cycle ergometer (35). Critical power was calculated from three tests ranging in duration from 2 to 25 minutes with a cycling cadence of 70 rpm. No significant differences in critical power were seen in values from three tests on one day (86.2 ± 18.1 W), five tests on separate days (87.5 ± 14.6 W), and three tests on separate days (88.7 ± 16.0 W). These represented 79.3%, 72.6%, and 73.0% of peak $\dot{V}O_2$, respectively. These values correspond to 75% and 82% of peak $\dot{V}O_2$ in similar studies of adults using similar protocols and analytical methods (38, 61).

In adults, critical power is high in endurance athletes (47) and has been demonstrated to increase with aerobic training (46). When the critical power concept is translated into a velocity–time relationship in running and swimming, the critical velocity is related to performance as well as lactate accumulation.

Hill and his colleagues (42) investigated this approach in swimmers grouped by age (8–18 years old). Three all-out time trials were conducted over distances individually chosen based on age and swimming performance (23–457 m). Calculated critical velocity was correlated with performance in a long time trial (183–2,286 m), which was not used to determine critical velocity. Correlation coefficients ranged from $r = .89$ to .99 depending on age group, and values were about 33% greater in the older than the younger swimmers. Critical velocity did, however, overpredict the sustainable velocity for the distance swim, causing the authors to caution that "while critical velocity is an index of endurance ability that can be used to evaluate and monitor swimmers, the parameter estimate should be interpreted with care if is to help determine training velocities or predict endurance performance" (42, p. 290).

Critical velocity or power should be expected to rise with growth of children in the same way (and for the same reasons) that absolute values for $\dot{V}O_2max$ increase. By the metabolic explanation of sustainable muscle performance, however, it is expected that the rise of CP with age should be less than that expected by increases in body mass. That is, relative to body size, the CP in small children should be greater than in older ones. In the study by Hill et al. (42), this probably held true, since weight (not described) would be expected to have been about twice as great in the older swimmers, while critical velocity rose by about 33%.

Oxygen Uptake Kinetics

Oxygen uptake rises during the first one to two minutes of a low to moderate workload to reach a steady-state

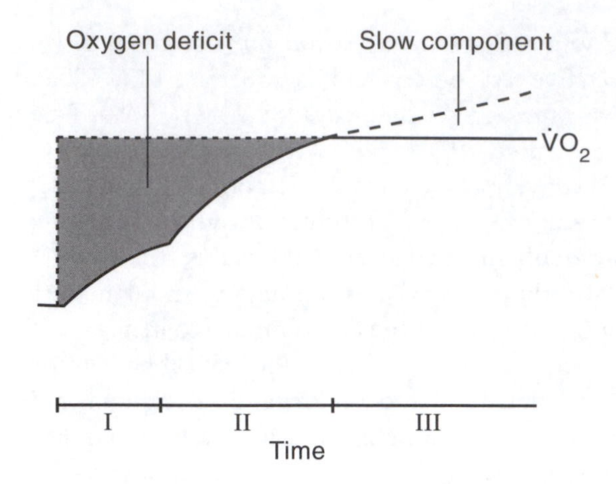

FIGURE 5.8 Oxygen uptake kinetics at the onset of exercise.

plateau value. Since the energy requirement is present from time 0, the oxygen deficit—or energy required before oxygen uptake kicks in—must be provided by anaerobic metabolic sources. The rate of rise in oxygen uptake and the extent of the initial oxygen deficit, then, are considered to reflect the relative capacities of the aerobic and anaerobic metabolic pathways.

At the onset of moderate-intensity exercise, the oxygen kinetics are best described by a three-phase model (109). Phase I indicates the rise in $\dot{V}O_2$ resulting from increases in muscle blood flow, phase II reflects augmented oxygen extraction, and phase III is steady state (figure 5.8). In the analysis of oxygen kinetics at the onset of exercise, it is considered important to exclude phase I from the model, since only phase II corresponds to the exponential rise in oxygen consumption at the level of the muscle cell.

When exercise is more intense (i.e., above the anaerobic threshold), $\dot{V}O_2$ does not plateau at the end of phase II. This increase in oxygen uptake beyond the third minute of constant-load exercise, termed the *$\dot{V}O_2$ slow component,* may approach the level of $\dot{V}O_2$max. However, in some cases, highly intense exercise can be tolerated for such a short duration that there is insufficient time for this slow component to develop. The means by which this slow component is considered in mathematical models of oxygen kinetics can confound comparisons between studies. Its physiologic meaning remains uncertain. Factors postulated to contribute include lactate, epinephrine, cardiac and ventilatory work, temperature, potassium, and recruitment of less-efficient fast-twitch fibers (37).

From our considerations of developmental metabolic patterns, we expect that children would demonstrate more rapid $\dot{V}O_2$ kinetics (i.e., more aerobic and less anaerobic capacity) than adults. Whether this is true is open to some debate, as different methodological and analytical approaches to this question have produced varying results.

Armon et al. (5) studied oxygen uptake dynamics at work rates above anaerobic threshold (AT) in 22 healthy children ages 6 to 12 years and seven adults ages 27 to 40 years (figure 5.9). At exercise intensities well above AT, $\dot{V}O_2$ increased for six minutes in each of the adults (i.e., all demonstrated an oxygen slow component), but this rise was observed in fewer than half of the children. The magnitude of the slow component was much greater in the adults than the children (1.76 ± 0.63 vs. 0.20 ± 0.42 ml · kg^{-1} · min^{-1}). The children demonstrated faster oxygen kinetics, conforming to the concept that children are more aerobic and less anaerobic than adults. Some researchers have considered the $\dot{V}O_2$ slow component to reflect metabolism of lactate, and the smaller increase observed in children in this study fits this concept. Similar findings of faster oxygen kinetics in children were described by Macek and Vavra (57) and Sady (94).

Other studies, however, have failed to verify this maturational difference in $\dot{V}O_2$ kinetics. Cooper et al. found no difference in rate of rise in $\dot{V}O_2$ at onset of exercise in a comparison of 7- to 9-year-old subjects with 15- to 18-year-old subjects exercising at 75% of anaerobic threshold (23). Average $t_{1/2}$ values (time

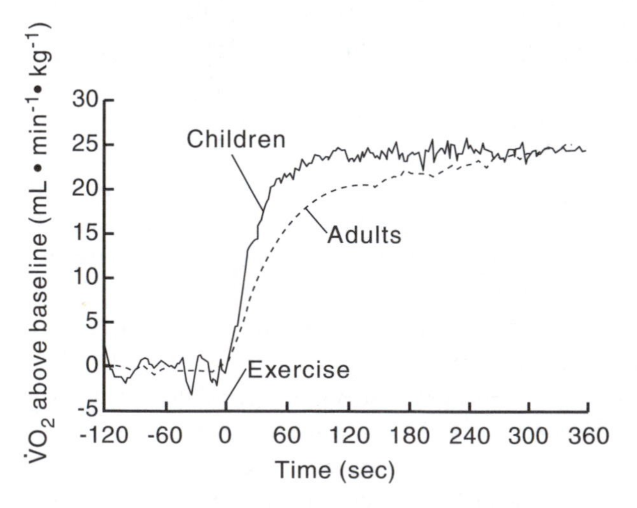

FIGURE 5.9 Mean $\dot{V}O_2$ responses to the same work rate in children and adults, demonstrating faster oxygen kinetics in the children (from reference 5).
Reprinted by permission from Armon et al. 1991.

required to reach a $\dot{V}O_2$ halfway between the value at onset and at steady state) for children and adults were similar (24.8 and 23.0 s, respectively) during exercise at intensities above and below AT in the study of Zanconato et al. (111).

More recently, Hebestreit et al. reported that $\dot{V}O_2$ kinetics of boys and men were similar during the first one to two minutes of exercise ranging from 50% to 130% peak $\dot{V}O_2$ (40). In this study, phase I was excluded from the analysis. The respective values for time constants for boys versus men were 22.8 ± 5.1 s versus 26.4 ± 4.1 s at 50% peak $\dot{V}O_2$, 28.0 ± 6.0 s versus 28.1 ± 4.4 s at 100% peak $\dot{V}O_2$, and 19.8 ± 4.1 s versus 20.7 ± 5.7 s at 130% peak $\dot{V}O_2$. The authors concluded that "these findings are in contrast to common beliefs that children rely less on anaerobic energy turnover early in exercise."

These authors felt that the difference between their findings and others showing faster time constants in children might be explained by differences in data analysis. In some previous studies a single exponential equation was used to model the entire $\dot{V}O_2$ response, an analysis that failed to differentiate between phase I and phase II and did not consider the influence of the $\dot{V}O_2$ slow component. In fact, when Hebestreit et al. reanalyzed their data using the approaches used in previous studies, faster time constants were revealed in the children.

Riner and his colleagues studied oxygen kinetics in 25 children and adolescents during uphill treadmill walking (3.5 mi · h^{-1}) at 85% and 100% $\dot{V}O_2max$ (77). Linear and exponential regression analyses of the time course of the increase in $\dot{V}O_2$ were conducted for the first 120 s of exercise. The $t_{1/2}$ was significantly shorter in the premenarcheal girls and prepubertal boys than in the older adolescents.

Results from these studies suggest that when the entire $\dot{V}O_2$ time response at onset of exercise is modeled, the expected maturational difference in time constants between children and adults is observed. But when phase I is excluded, no differences in $\dot{V}O_2$ kinetics are seen between children and adults. However, Fawkner et al. recently described faster oxygen kinetics in children even when phase I was excluded from the analysis (36). She and her colleagues observed 30 healthy 11- to 12-year-old children and 30 adults exercising for six minutes at 80% of $\dot{V}O_2$ at anaerobic threshold. Multiple transitions from unloaded to loaded pedaling were performed to obtain a confidence interval of ≤5 s. Mean response time was calculated from a single exponential model following phase I, which was considered to be 15 s in duration. The children were found to have a significantly faster response than the adults, both males (19.0 ± 2.0 and 27.9 ± 8.6 s, respectively) and females (21.0 ± 5.5 and 26.0 ± 4.5 s, respectively). No sex differences were seen within the child and adult groups.

Fawkner et al. considered it most likely that the more rapid $\dot{V}O_2$ kinetics in children reflected a greater mitochondrial capacity for oxidative phosphorylation than in adults (36). This is consistent with other evidence pointing to the same conclusion presented in this and the previous chapter.

While the explanation for the slow component is unclear, the magnitude of this rise in $\dot{V}O_2$ in children does not appear to be related to level of aerobic fitness. Obert et al. found that the slow component contributed about the same amount of the total $\dot{V}O_2$ response in 12 highly trained child swimmers (10–13 years old) and untrained controls during exercise above and below the anaerobic threshold (71). This finding is not concordant with studies in adults, which have indicated an attenuation of the slow component with endurance training. Obert et al. noted, however, that this discrepancy might be explained by the fact that posttraining measurements in those adult studies were performed at the same absolute rather than relative exercise intensities. An interesting finding in the study by Obert at al. was the lack of relationship between the magnitude of the $\dot{V}O_2$ slow component and the change in blood lactate levels. This suggests that lactate did not play a critical role in the genesis of the rise in $\dot{V}O_2$ in those subjects.

Gaesser and Poole differentiated the short-term $\dot{V}O_2$ slow component from the gradual rise in oxygen uptake seen during more extended exercise (40–60 minutes), which has been termed *oxygen drift* (37). The mechanics of oxygen drift are presumably different, given that this rise in $\dot{V}O_2$ is not typically accompanied by an increase in blood lactate level as is usually observed with the $\dot{V}O_2$ slow component. The degree of oxygen drift in children and adults when cycling at the same relative intensity (60–65% $\dot{V}O_2$) has consistently been demonstrated to be similar (13, 21, 58, 92).

It seems appropriate to conclude that the issue of maturational differences in $\dot{V}O_2$ kinetics is far from settled. Progress in this area will be hastened by identifying a standard means of measuring and analyzing $\dot{V}O_2$ responses at the onset of exercise. How to interpret any differences between children and adults in terms of metabolic capacities remains the next challenge.

The Relationship Between Aerobic Fitness and Physical Activity

Physical fitness and physical activity are two entirely different operational and mechanistic constructs. Physical fitness describes how well one can perform an exercise task, while physical activity denotes the amount of movement an individual engages in on a daily basis. The first is a personal *attribute,* governed by factors such as muscle strength, oxygen delivery, or neuromotor coordination, which are largely genetically based. The second is a *behavior,* determined for the most part by psychosocial influences. Fitness is measured by markers of performance (time to run a mile, number of sit-ups), while activity is assessed in terms of caloric expenditure. A link between the two can occur, however. Anatomic and physiologic responses to sufficient activity (i.e., physical training) cause fitness to improve.

The distinction between fitness and activity has come under scrutiny in the context of the role of exercise in preventive health care. The research literature on adults now clearly indicates that physical activity or fitness can play a salutary role in a wide range of diseases processes, including atherosclerosis (coronary artery disease), osteoporosis (bone fractures), accumulation of body fat (obesity), and hypertension (stroke).

Considerable attention has been focused in adult populations on the question of which is more important—physical fitness or activity—in achieving these health benefits. As Paffenberger et al. concluded: "Studies [in adults] not only of physical activity but also of physical fitness have revealed important relationships to health. Surely this is not an either/or situation. These questions are often asked: Can there be fitness without physical activity? Can there be health without fitness? Can there be health without physical activity? The answer is probably not" (73, p. 38).

In children the issues of the exercise–health connection and the relative impact of fitness and activity are less clear. Since the diseases influenced favorably by exercise occur in the adult years, connections between exercise in childhood and eventual health outcomes are much more difficult to document. Still, a reasonable case for promoting exercise in children to diminish future health risks can be made from the rationale that (a) activity and fitness may track from childhood into the adult years and (b) the disease processes in question—atherosclerosis, osteoporosis, obesity, and hypertension—often have their origins in the pediatric years (82).

Whether physical activity or fitness is more important for the health outcomes of exercise in the pediatric population is not clear. Fifty years ago the focus was entirely on what was perceived as a "fitness crisis," a reaction to reports of American children scoring unacceptably low on motor performance tests. With a more comprehensive picture of the mechanisms underlying the exercise–health link, that view has changed. Presently it is generally accepted that establishing habits of regular physical activity of different types is equally or more important than fitness for the present and future health of children (25, 82).

The answer to this question of fitness versus activity is important for those physical educators and public health professionals who are devising strategies for improving healthy exercise habits in children. Improving fitness requires formal training programs, while improving activity relies on different interventions aimed at increasing daily exercise. An important question in this regard, then, is how much daily physical activity by children contributes to their physical fitness. This issue has almost always been considered in the context of physiologic aerobic fitness (i.e., how much daily activity influences $\dot{V}O_2$max). The effects of a sedentary versus an active lifestyle in childhood on muscle strength, short-burst (anaerobic) fitness, and endurance performance have not been well addressed.

This distinction between health effects from fitness and from activity might be considered unnecessary quibbling, since it might be assumed that active individuals are more fit, and vice versa (82). In adults, this seems to be the case (15, 32), but the connection between level of habitual activity and physical fitness (at least as defined by $\dot{V}O_2$max) is much more tenuous in children.

In 1994, Morrow and Freedson reviewed 10 published studies that dealt with the association of habitual physical activity and aerobic fitness in young people (64). In only approximately half of these reports was a significant relationship observed, and even in those studies the correlations were low ($r \cong .20$). (It is perhaps pertinent to point out that the opposite relationship—a negative association between activity and fitness—was not described in a single study.) The authors suggested three possible explanations for their findings: (a) Physical activity has not been accurately measured in these studies, (b) youth have a high level of aerobic fitness, or (c) there truly exists no—or at least a very limited—relationship between physical activity and aerobic fitness in youth.

More recent studies have provided no more compelling evidence of an aerobic fitness–activity link in children. Loftin et al. examined the association of $\dot{V}O_2$ peak and physical activity levels in 16 elementary and 16 high school students (55). Aerobic fitness was assessed with treadmill testing, whereas physical activity patterns were measured by heart rate over two 12-hour weekdays. A median correlation of $r = .27$ was observed from eight independent comparisons of activity and peak $\dot{V}O_2$.

Allor and Pivarnik contended that the low relationship between aerobic fitness and physical activity in children could be explained by the use of imprecise field methods for estimating activity (2). They studied the relationship between $\dot{V}O_2max$ (with treadmill testing) and physical activity (by heart rate monitor, accelerometer, and activity recall) in 46 sixth-grade girls. A significant association was found between $\dot{V}O_2max$ and activity when the latter was estimated by heart rate ($r = .44$, $p < .01$). However, no relationship of fitness with activity was observed using accelerometers ($r = .13$) or by recall ($r = .09$).

Armstrong et al. examined the relationship between heart rate estimates of physical activity over a period of three school days and both performance on a Wingate anaerobic test and $\dot{V}O_2$ peak measured by treadmill running in 12-year-old British children (11). No significant relationships were observed for either aerobic or anaerobic fitness with daily activity. Correlation coefficients ranged from −.13 to +.16 in the boys and from −.02 to .04 in the girls.

Kemper et al. investigated the association of habitual physical activity and $\dot{V}O_2max$ longitudinally in the 15 years between ages 13 and 27 among 181 participants in the Amsterdam Growth and Health Longitudinal Study (51). They found a significant relationship between daily physical activity (by standardized interview) and $\dot{V}O_2max$ (treadmill running test) in the same individuals over that 15-year time span. However, the authors noted that the practical implication of this finding is small, since an increase in physical activity score corresponded to only a 2% to 5% rise in aerobic fitness.

From these data it is difficult to muster an effective argument that the range of regular daily activities in the childhood population has much effect on aerobic fitness. There are good reasons, in fact, to expect this lack of an effect. Sustained exercise of the duration and intensity needed to improve $\dot{V}O_2max$ in children is rarely observed in the typical daily activities of children. Moreover, activity levels relative to body size decrease during childhood in boys, while mass-relative $\dot{V}O_2max$ is stable (both, however, diminish in girls). Moreover, factors other than a training response, particularly body fat content, could explain any weak associations that have been observed between $\dot{V}O_2max$ per kilogram and amounts of daily physical activity in youth. A more definitive answer to this question, however, awaits more accurate methods for assessing levels of habitual activity in pediatric subjects.

Can we estimate how much the normal daily activities of children contribute to $\dot{V}O_2max$? There are no direct studies addressing this question, but we can get some idea of the magnitude of such an effect by examining the extremes of physical activity in young people. If a prepubertal child is place in a standard endurance training program of sufficient duration, intensity, and frequency, improvements in $\dot{V}O_2max$ are not dramatic, typically 5% to 10% (see chapter 11). Little information is available at the other end of the activity spectrum. In one study, average $\dot{V}O_2max$ during treadmill walking was reported to be 37.2 $ml \cdot kg^{-1} \cdot min^{-1}$ in five children who had been confined to complete bed rest for nine weeks due to a femoral fracture (80). Testing was repeated monthly for four months and again at six and nine months. Peak $\dot{V}O_2$ rose with successive tests until the third month, when values subsequently plateaued. The difference between average peak $\dot{V}O_2$ on the initial test and that at the plateau (43.1 $ml \cdot kg^{-1} \cdot min^{-1}$) was considered an estimate of loss during the prolonged bed rest (i.e., about 13%).

This scant information suggests that the plasticity of $\dot{V}O_2max$ in children from the extremes of extended bed rest to endurance training is approximately

20% (i.e., +7% to −13%). Since the normal variations in activity among a population of school-age children is far less than this, it can also be inferred that such differences in activity do not have a large impact on $\dot{V}O_2max$.

It is interesting to compare these findings with those in studies of adults, where data on the decline in aerobic fitness with bed rest are more plentiful (95). Endurance training increases $\dot{V}O_2max$ by an average of about 20% in previously sedentary adults. Bed rest of two to three weeks' duration results in a similar decline in aerobic fitness. The plasticity in adults, then, is about 40%, twice that of children. This suggests that variations in daily physical activity in adults might be more influential on $\dot{V}O_2max$, and this is consistent with the observation that higher correlations exist between aerobic fitness and habitual physical activity in the adult population.

Sex Differences in $\dot{V}O_2max$

Among young adults, average values of $\dot{V}O_2max$ expressed relative to body mass are 20% to 25% higher in males than females. A number of factors help explain this discrepancy, including sex-related differences in lean body mass and body fat content, hemoglobin concentration, habitual activity, and ratio of muscle to body mass (33). In a meta-analysis of 13 studies in men and women, Sparling found that expressing $\dot{V}O_2max$ in liters per minute, milliliters per kilogram per minute, and milliliters per kilogram of lean body mass per minute resulted in sex differences of 56%, 28%, and 15%, respectively (99). When a similar analysis was performed in studies of athletically trained men and women in the same sports, the differences were 52%, 19%, and 9%. Although hemoglobin levels are higher in males, this appears to account for only a small part of the sex differences in $\dot{V}O_2max$ in adults (27).

These data indicate, then, that a number of factors combine to account for sex differences in maximal aerobic power in adults. At the end of such analyses, a small percentage of differences in $\dot{V}O_2max$ between the sexes goes unexplained. We will see that this seems to be true in prepubertal children as well.

The principal determinants of sex differences in $\dot{V}O_2max$ in adults (muscle mass, body fat, hemoglobin concentration) reflect influences of the sex hormones testosterone and estrogen. Accordingly, these variations between males and females become overt at the time of puberty. Prior to this age, when no influence of sex hormones is expected, sex differences in $\dot{V}O_2max$ are smaller but still evident. In this section we address the determinants of these sex differences in children. Are sex-related variations a reflection of the same factors observed in postpubertal subjects? Or do additional determinants play a role in differentiating aerobic power between boys and girls? What explains these influences prior to the hormonal events of puberty, however, remains a mystery.

We make this analysis stepwise, beginning with differences in absolute $\dot{V}O_2max$ and ending with more specific ultimate determinants of maximal aerobic power. It will be apparent that this progression moves from data that are more or less indisputable to those that must be considered entirely speculative.

Absolute $\dot{V}O_2max$

Absolute $\dot{V}O_2max$ rises during the prepubertal years in a curvilinear fashion in both boys and girls, and while there exists considerable overlap, mean values for boys are consistently greater than for girls. In their analysis of 32 studies in the literature, Krahenbuhl et al. found that the average sex difference in the age range of 6 to 11 years was approximately 0.200 $L \cdot min^{-1}$, a 15% to 19% difference (54). At age 12, values of absolute $\dot{V}O_2max$ were observed to be 25% greater in boys than in girls.

In a similar literature review, Armstrong and Welsman examined sex differences in peak $\dot{V}O_2$ with age in 3,050 boys and 2,167 girls (6). Using linear regression, they found that mean peak $\dot{V}O_2$ determined by treadmill testing is approximately 12% higher in boys than in girls at age 10 years, increasing to 23% at age 12 and 37% by age 16.

In summary, the ability of the average prepubertal boy to utilize oxygen during maximal exercise testing is about 12% to 15% higher than that of the average girl the same age.

$\dot{V}O_2max$ per Kilogram of Body Mass

Comparisons of maximal aerobic power between groups necessitate expressing these values in relation to body size. When $\dot{V}O_2max$ is related to body mass, the small mean sex differences seen in absolute values during the prepubertal years are amplified. And instead of increasing in parallel, the curves for boys and girls diverge. In boys there is, in fact, no curve at all: $\dot{V}O_2$max per kilogram remains stable at approximately 52 ml · kg^{-1} · min^{-1} throughout the growing years. Values in girls, on the other hand, decline steadily almost from the age when $\dot{V}O_2$max can first be measured, from about 50 ml · kg^{-1} · min^{-1} at age 8 years to 40 ml · kg^{-1} · min^{-1} at age 15 (6, 54).

The use of the ratio standard in these studies may cloud the true picture of how $\dot{V}O_2$max develops with increasing body size. Nonetheless, these data indicate a distinct sex difference in the relationship of aerobic fitness and body dimensions. The sex discrepancy is small at age 6 (about 1.5%) but rises progressively to 32% at age 16 (54).

In summary, when adjusted for body mass, sex differences in maximal aerobic power in children persist. Moreover, the magnitude of these differences between boys and girls becomes greater with age. The differences in these patterns of change need to accounted for.

Body Composition

$\dot{V}O_2$max is closely related to lean body mass (LBM) or, more specifically, skeletal muscle mass. Sex differences in these measures is therefore expected to contribute to variations in absolute $\dot{V}O_2$max. In fact, the developmental curves of estimated muscle mass with age in boys and girls and those for absolute $\dot{V}O_2$max are essentially identical (figure 3.3 in chapter 3, figure 5.10a). Sex differences in absolute $\dot{V}O_2$max in the prepubertal years can therefore be accounted for largely by variations in the development of body skeletal muscle mass (or LBM) in boys and girls.

Differences in skeletal muscle mass relative to body mass may also contribute to sex differences in mass-specific $\dot{V}O_2$max. Malina and Bouchard noted that muscle mass in boys increases from approximately 42% to 54% of body weight between the ages of 5 and 17 years (60). In girls the increase is less: 40% at age 5 years rising to 45% at age 13. The explanation for these sex differences in muscularity before the age of puberty is unknown. Testosterone levels during childhood are very low and equal in boys and girls (60). Differences in muscle growth in the prepubertal years between boys and girls may reflect differential sex-related actions of the GH/IGF-I axis.

While absolute $\dot{V}O_2$max is closely associated with lean body mass, adipose tissue reduces the value of $\dot{V}O_2$max per kilogram, acting as an inert component that inflates the denominator. The fall in mass-relative $\dot{V}O_2$max during childhood in girls has therefore traditionally been explained by the accumulation of body fat during the later prepubertal years.

The numbers fit nicely to support this idea. By skinfold measurements, percentage of body fat increases progressively in the average girl from 13.5% at age 6 years to 18.5% at age 12 (56). By the age of 16 years this has increased to 23.0%. As we have already seen, a 25% reduction in $\dot{V}O_2$max per kilogram is observed during these 10 years. Over the same period of time, percentage body fat—and $\dot{V}O_2$max per kilogram—remain stable in boys.

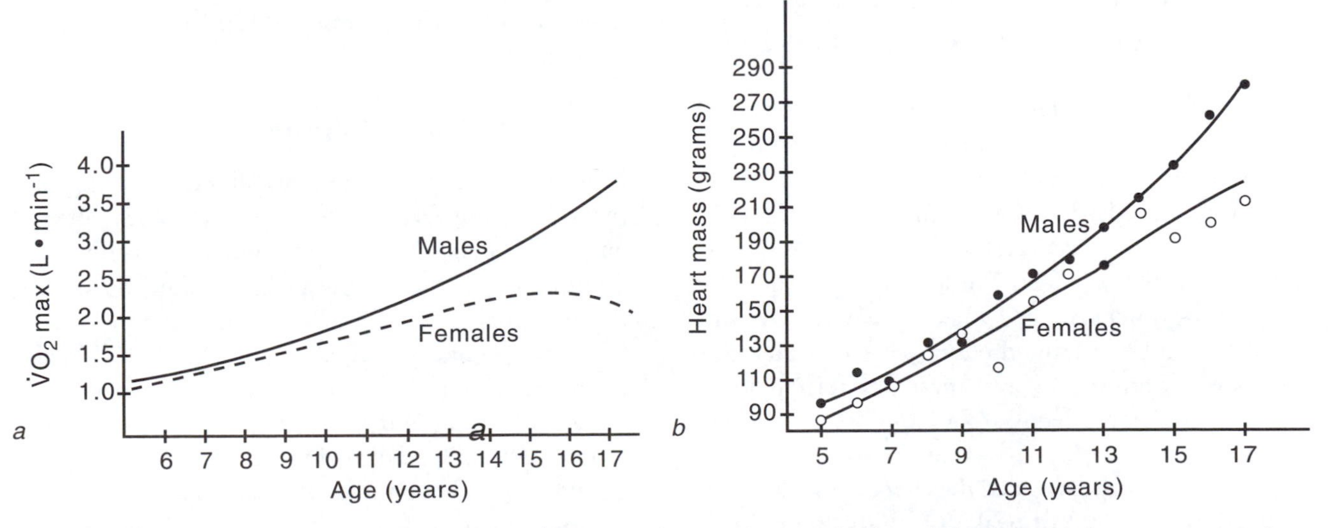

FIGURE 5.10 *(a)* Development of absolute $\dot{V}O_2$max with age in boys and girls and *(b)* change in heart mass over the same years (from references 54 and 50). The two curves are similar to that for increases in skeletal muscle mass (figure 3.3 in chapter 3).

Most studies have indicated that when $\dot{V}O_2max$ is expressed relative to body composition, sex differences in children are reduced but not entirely eliminated. Rowland et al. (86) compared $\dot{V}O_2max$ values during cycle testing between 25 boys (12.0 ± 0.4 years old) and 24 girls (ages 11.7 ± 0.5 years). Mass-relative $\dot{V}O_2max$ values for the boys and girls were 47.2 ± 6.1 and 40.4 ± 5.8 ml · kg^{-1} · min^{-1}, respectively (a 17% difference). When expressed relative to lean body mass (calculated from skinfold measures), the difference fell to 6%. By these data, then, differences in lean body mass accounted for approximately two thirds of the sex difference in $\dot{V}O_2max$. These numbers mimic those described earlier in adults.

These findings are similar to those described by Sunnegardh and Bratteby in eight-year-old Swedish children (100). These researchers found a 14.8% difference between the boys and girls in mass-relative $\dot{V}O_2max$ (52.7 and 45.9 ml · kg^{-1} · min^{-1}, respectively). When $\dot{V}O_2max$ was adjusted for LBM, the difference narrowed to 6.8%. In a longitudinal study of children 12 to 17 years, Kemper et al. reported that $\dot{V}O_2max$ per kilogram was approximately 20% greater in males at all ages (52). When values of $\dot{V}O_2max$ were expressed relative to LBM, a 6% sex difference remained.

The findings in these reports, which used skinfold thicknesses to assess body composition, are very consistent. However, when Vinet et al. estimated LBM by dual X-ray absorptiometry in their study of sex differences in $\dot{V}O_2max$, the findings were different (103). The boys exhibited a higher $\dot{V}O_2max$ per kilogram than the girls (47.9 versus 40.9 ml · kg^{-1} · min^{-1}), but when adjusted for LBM (expressed allometrically to the exponent 1.33), the sex difference in maximal aerobic power totally disappeared. When $\dot{V}O_2max$ was related to LBM by the ratio standard ($LBM^{1.0}$), the boys had a 6.2% greater value on average, but the difference was not statistically significant. These authors concluded that differences in $\dot{V}O_2max$ between boys and girls could be explained entirely by sex differences in body composition, and these sex-related variations were "not reflective of a more basic functional gender difference."

Welsman et al. examined the effect of children's sex on the relationship of peak $\dot{V}O_2$ with thigh muscle volume determined by MRI (107). All but 2 of the 32 subjects were in Tanner stage I according to pubic hair indexes. No significant difference in mean thigh muscle volume was observed between the boys and girls, but the boys had a significantly greater mass-relative peak $\dot{V}O_2$. Thigh muscle volume was highly correlated with absolute peak $\dot{V}O_2$ in both sexes (r = .80). No significant sex differences were seen when peak $\dot{V}O_2$ was expressed in respect to thigh muscle volume.

In summary, sex differences in absolute $\dot{V}O_2max$ in the prepubertal years can be accounted for largely by variations in the development of body skeletal muscle mass (or LBM) in boys and girls. Increases in percentage body fat, on the other hand, are responsible for the decline in maximal aerobic power expressed relative to body mass during childhood in girls. Even with these influences considered, however, some studies have indicated that a small sex-related difference in $\dot{V}O_2max$ (about 5–6%) remains.

Hemoglobin Concentration

No significant differences are observed in hemoglobin levels between boys and girls during the prepubertal years, and hemoglobin concentration is therefore expected to play no role in sex differences in $\dot{V}O_2max$. The similarity of hemoglobin concentration is reflected in similar values of maximal arterial-venous oxygen difference in boys and girls (discussed earlier in this chapter). Armstrong and Welsman found that blood hemoglobin concentration was a nonsignificant explanatory variable in their multilevel analysis of $\dot{V}O_2max$ changes in 11- to 13-year-olds (7). Blood hemoglobin concentration, then, does not influence sex differences in $\dot{V}O_2max$ in the prepubertal years.

Maximal Stroke Volume

Most studies have indicated that after considering all of the preceding factors, a small (about 5%) sex difference in $\dot{V}O_2max$ still needs explaining. In an effort to do so, we can return to the Fick equation and assess the potential roles of the usual suspects: maximal heart rate, stroke volume, and AVO_2diff. Maximal heart rate during the pediatric years is independent of sex (86, 103), and we have seen earlier in this chapter that AVO_2diff during exhaustive exercise is no different in boys and girls. Again, maximal stroke volume is identified as the key element defining differences in $\dot{V}O_2max$ between individuals, in this case between boys and girls.

A number of studies have examined sex-related differences in stroke volume in children during submaximal and maximal exercise. In these analyses, submaximal stroke volume (measured at moderate exercise intensity) can be assumed to represent stroke volume at peak exercise, since submaximal values are highly predictive of maximal stroke volume (85).

A number of studies have described significantly higher stroke volumes and lower heart rates at a given submaximal work rate in boys than in girls (4, 7, 39, 49, 101). Turley and Wilmore (101) compared cardiovascular variables between 12 boys and 12 girls eight to nine years old at three steady-state cycle workloads (20, 40, and 60 W) and treadmill speeds (3, 4, and 5 mi $\cdot$ h^{-1}). Absolute values were compared, since no significant group differences were observed for body size, fat-free mass, or estimated lean leg volume between the two groups. There were no statistically significant differences in stroke volume between the two groups in any of these comparisons. However, in all six comparisons, the values were greater in the boys than in the girls. The average from all trials was 61 ml for the boys and 57 ml for the girls (amounting to a 7% sex difference, consistent with the studies reviewed earlier that related $\dot{V}O_2max$ to body dimensions and composition).

Armstrong and Welsman examined this question from a longitudinal perspective (7). They used the multilevel regression modeling approach to assess sex differences in cardiac responses to submaximal treadmill exercise (8 km $\cdot$ h^{-1}) in children tested annually from ages 11 to 13 years. Cardiac output was estimated by the CO_2 rebreathing technique. Heart rate was significantly greater in the girls on all three tests. Stroke volume was greater in the boys, but the difference was statistically significant only in the second test. These differences were observed even when body size and composition were controlled for.

Two studies have looked at sex influences on stroke volume during maximal exercise. Among 24 premenarcheal girls and 25 prepubertal boys, Rowland et al. (86) found that average maximal stroke index measured by Doppler echocardiography was 62 ± 9 ml $\cdot$ m^{-2} in the boys and 55 ± 9 ml $\cdot$ m^{-2} in the girls ($p < .05$). When stroke volume was expressed relative to lean body mass, sex differences diminished, but values remained greater by 5.2% in the boys. No significant differences were observed between boys and girls in maximal heart rate or arterial-venous oxygen difference.

Vinet et al. could not confirm this result in a similar study that used dual X-ray absorptiometry instead of skinfolds for body composition measurement (103). Sex differences in maximal stroke volume were found to disappear when values were related to the allometrically derived $LBM^{0.79}$.

Other Factors

Are there other factors that might explain the small unexplained sex difference in $\dot{V}O_2max$ observed in most studies once body fat and lean body mass are considered? Are there sex differences in the ratio of skeletal muscle mass to lean body mass? Might there be a contribution by differences in habitual physical activity between the sexes, usually considered to be unimportant in sex differences in $\dot{V}O_2max$? Is the small unexplained difference in $\dot{V}O_2max$ between the sexes simply a reflection of variations in methods of estimating body composition? Could there be minor differences in cellular aerobic enzyme activity related to sex? No data are available to address these possibilities.

There is evidence to indicate that heart size is smaller in girls, which would be reflected in a smaller left ventricular end-diastolic dimension and thus a smaller stroke volume (50, 66, 98). Figure 5.10b demonstrates cross-sectionally derived heart weights by age from autopsy studies. Values are consistently greater in males than females, and the developmental curve is remarkably similar to those of skeletal muscle mass (figure 3.3) and $\dot{V}O_2max$ (figure 5.10a). In terms of explaining the small residual sex difference in $\dot{V}O_2max$, however, we would have to establish that heart size is *disproportionately* smaller in respect to lean body mass in females, and that information is not available.

After all these factors are considered, the functional capacity of the oxygen transport and utilization system relative to the demands of skeletal muscle is not substantially different between prepubertal boys and girls. At most, a 5% difference that is not accounted for by variations in body size and composition may be observed.

Conclusions

Ontogenetic improvements in $\dot{V}O_2max$ in respect to body mass may be determined by increases in the size of the exercising tissue (skeletal muscle) and changes in the functional capacity of its energy system (aerobic enzyme capacity). Cardiac size increases in concert with this development to match oxygen delivery to cellular oxygen consumption in exercising muscle.

In the Fick equation, factors influencing resting left ventricular size are responsible for a child's resting stroke volume, which in turn determines maximal exercise stroke volume → maximal cardiac output → maximal oxygen uptake. From a cardiac standpoint, then, physiologic aerobic fitness can be defined by variables that affect ventricular diastolic dimension, such as plasma volume, heart rate (vagal tone), and hormonal effects on the heart. (We shall see in the next chapter, however, evidence that indicates that peripheral factors, rather than the heart, may be critical in controlling and defining the limits of blood circulation during exercise.)

Increases in endurance performance as children grow are independent of $\dot{V}O_2max$ and are instead linked to improvements in submaximal exercise economy. The limit of sustainable muscular performance, defined by the balance of aerobic and anaerobic cellular enzyme function, may be a useful concept in understanding factors that define field performance during growth.

Level of habitual physical activity in children, in contrast to adults, is not a useful predictor of $\dot{V}O_2max$. Variations in exercise from the usual daily activities of young people therefore cannot be expected to substantially alter their maximal oxygen utilization.

Once body composition and size are considered, little difference is observed between boys and girls in maximal aerobic power. It is not clear whether the small remaining sex variations in $\dot{V}O_2max$ (about 5%) represent true biologic differences between boys and girls.

Discussion Questions and Research Directions

1. How does $\dot{V}O_2max$ truly relate to body dimensions as children grow? Do sex differences exist in this trend independent of body size and composition?
2. What factors cause sex differences in body composition in the prepubertal years that influence physical fitness?
3. Are the factors that influence interindividual differences in $\dot{V}O_2max$ among children similar to those responsible for ontogenetic changes?
4. What factors are responsible for determining resting left ventricular volume, which appears to be the critical determinant of $\dot{V}O_2max$?
5. How does the limit of sustainable muscular performance change during growth? What practical testing approaches can be used to measure it? What determines this threshold?
6. How much do daily activities contribute to the aerobic fitness—physiologic and functional—of children?
7. What is the most appropriate means of analyzing oxygen uptake kinetics? Do maturational differences really exist? If so, how do they reflect developmental trends in the aerobic and anaerobic capacity of children?

6

Cardiovascular Responses to Exercise

The animal's heart
is the basis for its life,
its chief member,
the sun of its microcosm;
on the heart all its activity depends,
from the heart all its liveliness
and strength arise.

—William Harvey (1642)

In this chapter we discuss

- the means by which the cardiovascular system responds to exercise,
- how these responses change with age during childhood, and
- the appropriate means of adjusting cardiac variables to body size.

There have been concerns—persistent even today—that the child's heart might not be able to tolerate exercise as well as that of the adult. In the late 1800s, Beneke and others brought to attention a perceived discrepancy between the growth of the heart and the size of the aorta and pulmonary artery (see reference 43). While the volume of the heart increases proportionally with body mass, the circumference of the great vessels grows in respect to body height (i.e., linearly). This "natural disharmony" was considered to predispose the child to hypertension and to "diminish the child's vigor at this period" (43).

More recently, another issue has arisen: Exercise studies in children have consistently demonstrated that at a given level of oxygen uptake, cardiac output tends to be smaller in the child than in the adult (8). At a $\dot{V}O_2$ of 1.4 $L \cdot min^{-1}$, for instance, the typical cardiac output for an adult is about 14 $L \cdot min^{-1}$, but in a child it is 11 $L \cdot min^{-1}$. This hypokinetic response has been considered important in limiting children's adaptations to high-intensity exercise, particularly in the heat (31).

Both concerns seem to be equally spurious. In the first case, Beneke's interpretations expose his failure to grasp the fundamental principles of hydrodynamics. As Karpovich pointed out, the blood flows in the aorta and pulmonary area are not related to the circumference of these vessels but rather to their cross-sectional area (43), and the cross-sectional area has been shown to rise in parallel with heart volume as a child grows. For many decades, school children had been restricted in their activities because of this geometric error.

The second issue might be considered biologically irrelevant, since children and adults do not typically exercise at the same absolute oxygen uptake. At a given $\dot{V}O_2$ the child has a lower stroke volume than the adult because his or her heart is smaller; stroke volume expressed proportionally to body size, however, is the same in the two groups. As we shall see later in this chapter, all available information indicates that no maturational differences in cardiac functional reserve during exercise exist when variables are adjusted appropriately for body dimensions. During exercise, cardiac and stroke index, pattern of stroke volume response, myocardial function, relative ventricular dimensional changes, systolic and diastolic filling periods, cardiovascular drift, and exercise factor (ratio of change in cardiac output to increase in oxygen uptake) are all similar in children and adults (77).

Several noninvasive measurement techniques have permitted these insights into cardiac responses to exercise in children, particularly Doppler and two-dimensional echocardiography, thoracic bioimpedance, carbon dioxide rebreathing, and the acetylene rebreathing technique. The reader is referred to several recent reviews that have assessed the accuracy and feasibility of these methods (7, 88, 111, 112, 113). The results of these studies, besides failing to demonstrate any uniqueness in the cardiac responses of youth, have supported traditional—but often ignored—concepts regarding the control and limitation of blood circulation during exercise. This chapter begins with an overview of current knowledge regarding children's cardiovascular responses to exercise. In light of these empiric observations, evidence about the factors that define these responses are then considered. Before beginning this analysis, however, it is important to address the question of the proper means of expressing cardiac variables in relation to body dimensions.

Relating Cardiac Variables to Body Size

The size-dependent cardiac variables, stroke volume and cardiac output, have traditionally been related to body surface area (BSA) when individuals of different body dimensions are compared. This practice grew out of a consideration of dimensionality theory and the concept that basal metabolic rate (BMR) should be linked with body heat loss (i.e., the surface rule). However, whether cardiac output and stroke volume are truly best adjusted by BSA, particularly during exercise, is controversial (17). BSA is not, in fact, a proper means of normalizing BMR in children, since BMR per square meter of BSA falls with age, decreasing by about 18% between ages 6 and 14 years. Error is introduced because BSA is never actually measured but rather is estimated from equations of dubious quality. It has been troublesome, too, that BSA does not take into account differences in body composition (i.e., percentage of body fat). These concerns notwithstanding, many empiric observations have supported the use of BSA for adjusting cardiac variables during exercise for body size in children.

Heart Dimensions

According to dimensionality theory, heart mass or volume (three-dimensional constructs) should relate to BSA (two-dimensional) by the exponent 1.5. Gutgesell and Rembold confirmed this when they combined data from multiple echocardiographic and angiographic studies to estimate the growth of the heart in respect to BSA (35). Across the age span from newborn to adult, the relationship between cardiac volume and BSA was nonlinear but became so with BSA expressed to the 1.5 power. They noted that when subjects under the age of five years were excluded, the relationship of left ventricular volume and mass with BSA approximated linearity.

In the Muscatine Study of 904 children ages 6 to 16 years, echocardiographic left ventricular mass related to height to the 2.6 power, BSA to the 1.3 power, and body mass to the 0.75 power (56). Nidorf et al. came to the same conclusion in their echocardiographic study of 268 normal people ages 6 days to 76 years (61). Since left ventricular volume relates to the cube of the diameter, these studies indicate that height cubed would be an appropriate normalizing variable for resting ventricular volumes.

In 201 boys and girls between ages 6 and 17 years, Daniels et al. (25) found close relationships of left ventricular mass (LVM; measured by echocardiography) with weight (r = .84), height (r = .81), BSA (r = .87), and lean body mass (r = .86). In multiple regression analysis, lean body mass ($LBM^{1.0}$) was the most important of these, accounting for 75% of the variance in ventricular mass. In another report, Daniels et al. indicated that left ventricular mass to height cubed, compared with BSA or other exponents of height, was the closest relationship to LVM/LBM (24).

After reviewing this literature, Batterham et al. concluded that "the current consensus is that an estimate of fat-free mass (FFM) provides the most appropriate body size variable [and] that cardiac dimensional data should be scaled by appropriate power of FFM, derived from allometric modeling" (10, p. 500). The description by Blimkie et al. of relationships between $\dot{V}O_{22}$max, heart size, and lean body mass in 117 boys ages 10 to 14 years supports this (15). Left ventricular end-diastolic diameter, resting stroke volume, and calculated left ventricular mass were all closely correlated with $\dot{V}O_2$max (r = .75 to .84). Multiple regression analysis indicated, however, that all these relationships could be explained by their shared association with lean body mass.

Cardiac Output and Stroke Volume

Again by dimensionality theory, volume flow rates (i.e., volume over time, such as cardiac output in liters per minute) should be expressed as length cubed divided by length (which is considered equivalent to time), or length squared. This means that body surface area, which relates to the square of body height, should be an appropriate denominator for the size adjustment of cardiac output.

De Simone et al. examined the association of stroke volume and cardiac output at rest (measured by M-mode echocardiography) with anthropometric measures in children and adolescents (27). Across the pediatric population, resting stroke volume was related to BSA by the scaling exponent 0.82, to body weight by the exponent 0.57, and to height by the exponent 1.45. Respective exponents for cardiac output were 0.53, 0.38, and 0.92. Scaling exponents were different for cardiac output and stroke volume because of the influence of the fall in resting heart rate with age. The authors cautioned that the use of any anthropometric measure to normalize cardiac measurements needs to take into account the influence of body fat.

This study reminds us that the proper size adjustment for cardiac variables at rest in children is expected to be different from that during maximal exercise. Allometric analysis indicates that $Q^{x+y} = SV^x \times HR^y$. At maximal exercise, the exponent y for heart rate (HR) in children is zero (i.e., maximal heart rate is independent of body size across the age range of 6 to 16 years). Therefore, maximal values of cardiac output (Q) and stroke volume (SV) should relate to other anthropometric measures by similar exponents. At rest, however, heart rate is inversely related to mass and age. The exponent in children has not been determined, but in animals $HR_{rest} \sim M^{-0.25}$ (94). Consequently, at rest, allometric exponents for stroke volume and cardiac output relative to any given anthropometric measure are different.

Limited data obtained in children during exercise suggest that body surface area is an appropriate means of adjusting children's cardiac variables for body size. Armstrong and Welsman studied the influence of body size on cardiac output and stroke volume during a submaximal treadmill run (8 km · h^{-1}) in boys and girls between ages 11 and 13 years (4). Changes in cardiac output (determined by CO_2 rebreathing) in both sexes increased with age in direct proportion to body surface area.

However, Turley and Wilmore reported that cardiac index (Q/BSA) and stroke index (SV/BSA) were not size-independent values in their study of 7- to 9-year-old children during submaximal exercise (107). At a 60-W workload, a correlation of $r = -.82$ was seen between Q/BSA and BSA, and $r = .29$ between SV/BSA and BSA.

Rowland et al. (84) found that BSA served as a valid means of normalizing values of maximal cardiac output and stroke volume for body size in a group of 24 premenarcheal girls (mean age 12.2 ± 0.5 years). The equations were

$$Q_{max} = 1.02 BSA^{1.08}$$
$$SV_{max} = 1.73 BSA^{1.05}$$

The correlation coefficient between Q/BSA and BSA was .04. Other appropriate anthropometric variables for adjusting maximal cardiac output in this study were mass, $M^{0.59}$, and height, $H^{1.76}$. In a similar study in prepubertal boys, the same authors found that that maximal stroke volume related to BSA by the exponent 1.03 (83). Vinet et al. reported that maximal stroke volume and cardiac output related to lean body mass by exponents 0.79 and 0.76, respectively, in 35 prepubertal children (110).

In summary, the current literature suggests that exercise values of cardiac output and stroke volume might be properly adjusted for body size by $BSA^{1.0}$, $H^{3.0}$ or $LBM^{1.0}$. However, allometric analysis of the specific subject population at hand is advised to ensure that the 95% confidence limits of the allometric exponents fit these predictions.

Circulatory Responses to Exercise: The Basics

While lacking a precise tool for noninvasively estimating cardiac output, researchers have provided a reasonably consistent description of how circulatory variables respond to exercise during the pediatric years. The essential trends involve changes in size-related (stroke volume) and size-independent (heart rate) variables. Interestingly, the balance of these two developmental patterns in establishing cardiac output is different at rest than during maximal exercise.

The following information regarding cardiac variables at rest and during exercise in children is summarized from two recent reviews (77, 104). The reader is referred to these sources for further discussion and references.

The Heart at Rest

The resting metabolic rate in small children is greater relative to their body mass and surface area than that in older children. Consequently, it is expected that cardiac output should follow the same trend. Katori established this in an investigation of resting cardiac output by the earpiece dye dilution method in 151 subjects ages 4 to 78 years (44). Cardiac output per kilogram of body mass declined from approximately 240 ml · kg^{-1} · min^{-1} in the youngest subjects to about 120 ml · kg^{-1} · min^{-1} by mid adolescence. When expressed relative to body surface area, similar decreases were observed. This curve of declining mass-relative resting cardiac output is essentially identical to the pattern of fall in BMR. Studies using cardiac catheterization, impedance cardiography, and Doppler echocardiography have all revealed a consistent mean value for resting cardiac index in children in the supine position of 4.1 to 4.4 L · min^{-1} · m^{-2} (49, 54, 98).

The progressive decrease in resting cardiac output relative to body mass as children grow must reflect concomitant changes in resting stroke volume, heart rate, or both. Values for stroke volume parallel heart size and body dimensions. Krovetz et al., for example, found that resting stroke volume during cardiac catheterization in 29 young people ages 0 to 20 years correlated with body mass and BSA by coefficients of $r = .87$ and $r = .90$, respectively (49). Similar findings were described by Sproul and Simpson (98). In these two studies the supine stroke index values averaged 44 and 42 ml · m^{-2}, respectively.

These data indicate that the size-related change in resting cardiac output that parallels BMR must be accounted for by the decline in resting heart rate with age. Heart rate in the basal state falls 10 to 20 bpm between the ages of 5 and 15 years. Mean basal heart rate is about 80 bpm at age 5 and 62 bpm at age 15. After age 10 years, the basal heart rate is approximately three to five beats faster in girls than in boys.

As expected, the rate of decline in resting heart rate closely approximates the decline in mass-specific $\dot{V}O_2$. BMR (expressed in calories per square meter of BSA per hour) falls about 23% in girls between the ages of 6 and 16 years, while basal heart rate falls by about 20%. In mature animals, resting heart rate (f_h) relates to body mass (M) by the allometric equation $f_h \sim M^{-0.25}$, which is identical to the equation relating weight-relative metabolic rate to body mass (94).

Obviously, then, the rate of sinus node discharge at rest is tied closely with metabolic energy expenditure. The mechanism for this association is not known. Autonomic blocking studies indicate that the fall in resting heart rate with age is probably independent of maturational differences in parasympathetic or sympathetic input. Intrinsic changes in the rate of sinus node depolarization seem more likely, such as changes in sinoatrial node membrane ion flux or permeability, or alterations in location of the predominant pacemaker cells within the node.

A number of studies have failed to indicate any maturational differences in myocardial contractility as children grow. Contractile function is age independent, as demonstrated by measurement of systolic time intervals, echocardiographic ventricular shortening fraction and wall stress, and ejection fraction determined by radionuclide angiography.

Systemic blood pressure at rest progressively increases as a child grows. A healthy, full-term newborn infant has a blood pressure of about 70/55 mmHg. By age 10 years the expected value will have risen to 110/62 mmHg, and by age 15 to 115/65 mmHg.

Cardiac Variables With Exercise

The relationships of heart rate, stroke volume, cardiac output, and metabolic rate at maximal exercise in a progressive incremental treadmill or cycle test can be expected to differ from those at rest. While the relationship between $\dot{V}O_2max$ and body mass changes from that at rest, the scaling exponent is still typically <1.00. Since maximal arterial-venous oxygen difference is size independent (at least until puberty), the allometric relationships of maximal cardiac output to anthropometric measures should be similar to that with $\dot{V}O_2max$. Heart rate at maximal exercise in childhood is unchanged, so all of the increase in cardiac output can be attributed to the growth of stroke volume, which in turn follows heart diastolic size.

Heart rate rises steadily with progressive exercise in children as it does in adults. At high work intensities, however, heart rate begins to taper as load increases. In one treadmill walking study, all 13 children demonstrated such a taper when exercise intensity reached 60% $\dot{V}O_2max$ (81). About one third of the subjects demonstrated a plateau in heart rate (defined as less than a three-beat increase in the final stage).

Heart rate at a given work intensity declines as children grow. This reflects the growth in heart size and stroke volume. Since the oxygen demand and cardiac output needed for a particular workload generally stays the same regardless of body size, the larger child satisfies those requirements by a greater reliance on stroke volume and less reliance on heart rate than the younger, smaller child.

As noted previously, maximal heart rate during treadmill or cycle testing remains stable across the pediatric years, and values are similar in boys and girls. This means that formulas used for estimating a target maximal heart rate in adults (such as 220 minus age) are not appropriate for children and young adolescents. Peak heart rate does depend on testing modality, though. The highest rates are achieved during treadmill running (200–205 bpm), while those during walking and cycling are usually about 5 to 10 beats less.

Maximal stroke index in boys is typically 58 to 63 $ml \cdot m^{-2}$, and values in girls are about 5 $ml \cdot m^{-2}$ less. As reviewed previously and addressed again in later sections of this chapter, interindividual differences in stroke volume are related to ventricular diastolic filling size rather than intrinsic contractility or afterload.

No dramatic differences have been observed in maximal cardiac output in relation to body surface area during growth. Comparisons of studies using a variety of measurement techniques have provided reasonably consistent values of 10 to 12 $L \cdot min^{-1} \cdot m^{-2}$. Consistent with measurements of $\dot{V}O_2max$, maximal cardiac output values are typically smaller in girls than in boys. The rate of rise in cardiac output with exercise in respect to change in oxygen uptake (the exercise factor) in children (5.7 to 6.5) is similar to that observed in adults.

An acute bout of endurance exercise is a hypertensive event in children as well as adults. During an incremental treadmill or cycle test, systolic pressure rises to values that are typically 40% greater than at rest, with little change in diastolic pressure. There is some evidence to suggest that the magnitude of increase in systolic pressure from rest to peak exercise may be larger in older than in younger subjects. The steady decline of peripheral vascular resistance in children during such an exercise test (generally about 50%) is also not dissimilar to that observed in adults.

Cardiovascular Responses to Acute Progressive Exercise

Cardiac responses to a bout of incremental exercise have traditionally been considered in the context of the Fick equation: The oxygen consumption of exercising muscle reflects the product of oxygen delivery by cardiac output and its cellular extraction as indicated by the difference in arterial and venous oxygen content. This provides an accurate mathematical expression of $\dot{V}O_2max$. However, the Fick equation has also been used for defining factors that regulate and limit oxygen uptake, and from this perspective it has caused misconceptions regarding the control of blood circulation during exercise.

Most research has indicated that the limits of circulatory provision of oxygen to exercising muscle limits $\dot{V}O_2max$. From a Fickian viewpoint, then, this indicates that cardiac factors that limit output are central to understanding the ceiling of physiologic aerobic fitness. (We have already taken such a traditional approach when we identified factors accounting for maximal stroke volume as those defining $\dot{V}O_2max$.) This concept, however, is not consistent with a century of research that indicates that peripheral, not central, factors control and limit the circulation during progressive exercise. In particular, the sharp decline in peripheral vascular resistance coupled with the pumping action of skeletal muscle are considered most instrumental in augmenting blood flow during exercise. Indeed, most researchers have considered the heart a *responsive* rather than *responsible* factor in this circulatory response to exercise. Recent information in children indicates that circulatory response in immature subjects follows this same pattern.

It is important to examine this construct in some detail, for if the peripheral perspective is correct, the factors responsible for limiting $\dot{V}O_2max$ are different from those suggested by the Fick equation. This section begins with a brief historical overview of research data that has led to the focus on peripheral circulatory factors. We then examine empiric findings, most specifically in children, to determine if such observations are consistent with the model of peripheral control of circulatory responses to exercise.

Historical Perspective

Historically, uncertainty regarding the normal cardiac responses to increased physical activity has reflected "the difficulty in studying the subject during the violent motions of exercise" (100, p. 237). Not surprisingly, then, an understanding of circulatory dynamics has paralleled the development of techniques that permit measurement of the cardiovascular system in motion.

In the first part of the 20th century, information was gained from either isolated heart–lung preparations or anesthetized open-chest dogs. When in 1914 Patterson and Starling pronounced their law of the heart (69)—stretching of the myocardium causes an increase in muscle contractile force—it was concluded that circulatory flow during exercise occurred in the same way. Augmented venous return to the heart (from peripheral vasodilatation and action of the skeletal muscle pump) was thought to increase the left ventricular filling pressure and diastolic dimension (preload). This stretch of the myocardium triggered an increase in contractile force and stroke volume. Heart rate rose at the same time as a result of sympathetic stimulation from atrial stretching (the Bainbridge reflex, discussed later). The combined increases in heart rate and stroke volume were then responsible for the rise in cardiac output with exercise.

This viewpoint was widely accepted until the 1950s, when studies using dye dilution and cardiac catheterization techniques in both humans and dogs indicated that stroke volume in supine subjects did not, in fact, change with increasing exercise intensity (92). At the same time, left ventricular diastolic volume remained constant or decreased slightly (41). In their 1959 review article, Rushmer and Smith argued that the "greatly distorted conditions" of earlier research models had failed to adequately imitate physiologic changes in awake, intact subjects, "seriously retarding" an understanding of cardiac responses to exercise (92). (Indeed, this was not a specious argument, since Starling's heart–lung model had eliminated the effects of venous resistance, intrathoracic pressure, vascular volume, change in heart rate, neurogenic influence, and humoral factors.) Consequently, the Starling mechanism for explaining the increase in cardiac output with exercise fell into disrepute.

Still, the idea that local arteriolar dilatation in contracting muscle served as the primary factor increasing blood flow during exercise prevailed. In 1967, Guyton proposed that individual tissues autoregulated blood flow in respect to their metabolic need through increased vascular conductance and that in the case of vigorous exercise, "the primary cause of augmented cardiac output is believed to be local vasodilatation in the skeletal muscle" (36, p. 806). He concluded from the research to that date that "the heart has relatively little effect on the normal regulation of cardiac output."

This reasoning was based largely on the report of Donald and Shepherd (28), who observed no impairment in the normal rise in cardiac output during exercise in denervated dogs (i.e., animals lacking sympathetic stimulation of the heart). Heart rate response was dampened, but stroke volume was greater than before denervation, consistent with control of the circulation being a primary effect of increased venous return from peripheral nonautonomic factors. [Similar findings are observed today in patients who have undergone heart transplantation and are devoid of sympathetic stimulation (20)]. Supporting this concept, Guyton's own animal studies found that "an increase in efficacy of the heart as a pump cannot by itself increase the cardiac output more than a few percent, unless some simultaneous effect takes place in the peripheral circulatory system at the same time to translocate blood from the peripheral vessels to the heart" (37, p. 182).

In their classic 1967 review article, Bevegard and Shepherd agreed with this passive role of the heart: "The heart serves as a force-feed pump designed to discharge whatever volume it receives by increasing its rate or its stroke volume. Unless there is dilatation of resistance vessels in some systemic vascular bed, mediated by local, humoral, or nervous mechanisms, an increased rate will not result in an increase in cardiac output" (11, p. 180). Smith et al. (97) and Clausen (21) considered this local vasodilatation in skeletal muscle to be a graded response to metabolic need, indicating that the fall in peripheral vascular resistance was responsible not only for increasing blood circulation during exercise but also for controlling circulatory flow to meet metabolic needs.

More recent reviews of mechanisms surrounding circulatory responses to exercise have focused attention on another peripheral factor, the skeletal muscle pump, in augmenting blood flow during exercise. While it had long been recognized that skeletal muscle could propel blood back to the heart, Rowell et al. considered this function a primary one: "The muscle pump is a major determinant of venous return and ventricular filling pressure during exercise [and] can be viewed as a second heart on the venous return portion of the circuit, having capacity to generate blood flow rivaling that of the left ventricle" (76, p. 775).

During the past two decades, the research focus has shifted to the role of the autonomic nervous system in controlling cardiovascular responses to exercise. Specifically, a reflexive increase in sympathetic drive to augment heart rate, myocardial contractility, and blood pressure in response to various afferent inputs is now considered central to the circulatory responses to increased physical activity. Initial studies dealt with static exercise, but the concept of reflex autonomic control of the circulation during exercise has been generalized to dynamic activities as well (58, 75).

Empiric Observations

Recent observations in children support these earlier descriptions of normal cardiovascular dynamics during exercise. It is important to scrutinize these empiric findings closely, since any proposed construct of the mechanisms that control these responses must, by necessity, conform to these observations.

Stroke volume rises in the initial phase of upright exercise but then remains stable as exercise intensity increases. This pattern occurs while heart rate and cardiac output increase linearly with work rate. This plateau of stroke volume is one of the most consistent findings in cardiovascular physiology, having been demonstrated in children using thoracic bioimpedance (48, 89), Doppler echocardiography (64, 80), CO_2 rebreathing (9, 47, 105), dye dilution (30), and acetylene rebreathing techniques (23, 29). Values typically rise approximately 30% to 40% by the time work intensity reaches 50% $\dot{V}O_2max$, but no changes are observed beyond this point as work rate increases.

The clue to the meaning of this initial rise in stroke volume comes from the observation that when a child (or adult) exercises in the supine position, typically *no* significant change in stroke volume occurs with progressive exercise. Rowland et al. compared differences in cardiovascular responses to cycle exercise performed in the upright and supine positions in 13 boys ages 10 to 15 years (82). At rest, stroke index and cardiac index were significantly greater in the supine position (stroke index: 71 ± 15 vs. 51 ± 12 ml · m^{-2}; cardiac index: 4.18 ± 1.00 vs. 4.01 ± 1.39 L · min^{-1} · m^{-2}). While subjects pedaled supine at 50 rpm with incremental 25-W loads, no significant changes were seen in stroke volume (figure 6.1). In the upright position, a typically 29% rise was observed by the second workload, but stroke volume then remained unchanged at higher intensities. At and above the 50-W work rate, no significant differences were observed in stroke index in the sitting and supine positions. These findings are consistent with similar reports in adults (11, 41, 92) and with Rushmer and Smith's studies in dogs (92).

This information supports the concept that the initial rise in stroke volume during exercise reflects the mobilization of blood pooled by gravity in the dependent lower extremities. When an adult subject assumes the upright position, blood volume in the legs increases by as much as 500 to 1,000 cm^3, and the resulting decrease in central blood volume causes

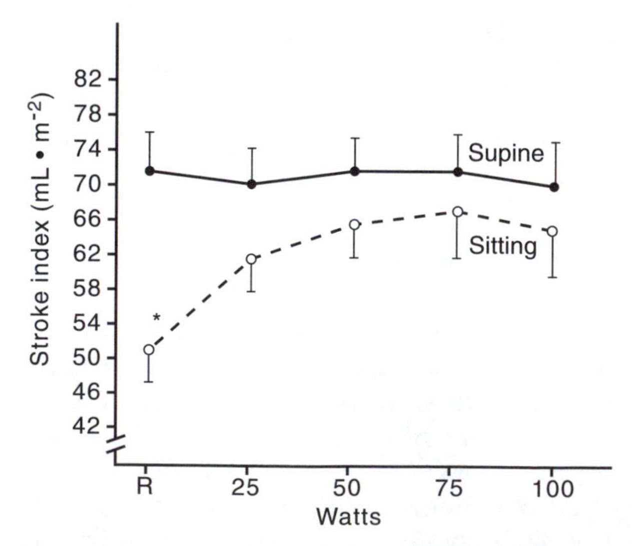

FIGURE 6.1 Stroke index response to supine and upright progressive cycle exercise in children (from reference 82).

Reprinted by permission from Rowland 2003.

cardiac output and stroke volume to fall by 20% to 40%. Similar changes have been observed in children (48). With the onset of upright exercise, contractions of the skeletal muscles in the legs mobilize this dependent blood, central volume increases, and stroke volume and cardiac output rise (63). Once this process is complete, stroke volume remains stable with increasing workload, just as it does during supine exercise, and values are similar during supine and upright exercise at higher work intensities. By this interpretation, then, the 30% to 40% rise in stroke volume observed at the onset of upright exercise should be interpreted simply as a "refilling" phenomenon—a mobilization of dependent blood volume in the legs—and not part of the fundamental process by which circulatory flow increases during exercise.

Two additional conclusions are suggested from these data. First, when the supine position is considered as the baseline, stroke volume provides *no* contribution to increases in cardiac output in response to exercise. And second, the ratio of maximal to resting stroke volume during upright exercise is not, as has been supposed, a marker of cardiac functional response to exercise. This ratio, in fact, reflects only the difference between stroke volume in the supine and upright postures. Individual differences between maximal and resting stroke volume while cycling upright indicate variations in postural influences (possibly autonomic function), not myocardial capacity.

Left ventricular end-diastolic dimension remains stable or gradually declines as exercise intensity increases. Left ventricular filling volume does not increase as heart rate, cardiac output, and oxygen uptake rise during progressive exercise to exhaustion. The most typical pattern seen in echocardiographic studies, in fact, has been a gradual decline in left ventricular end-diastolic dimension with increased workload. (The only exception to this is a small rise in diastolic size at the beginning of upright exercise, considered to reflect the transient increase in cardiac filling from blood mobilized from the lower extremities.)

Nottin et al. (64) demonstrated this pattern in boys (mean age 11.7 ± 0.6 years). Left ventricular end-diastolic dimension fell from 44 ± 4 mm at rest to 41 ± 5 mm during maximal exercise. The same pattern was seen in a comparison group of adult subjects. Using the same technique, Rowland and colleagues described an identical trend of changes in left ventricular diastolic dimension with exercise in

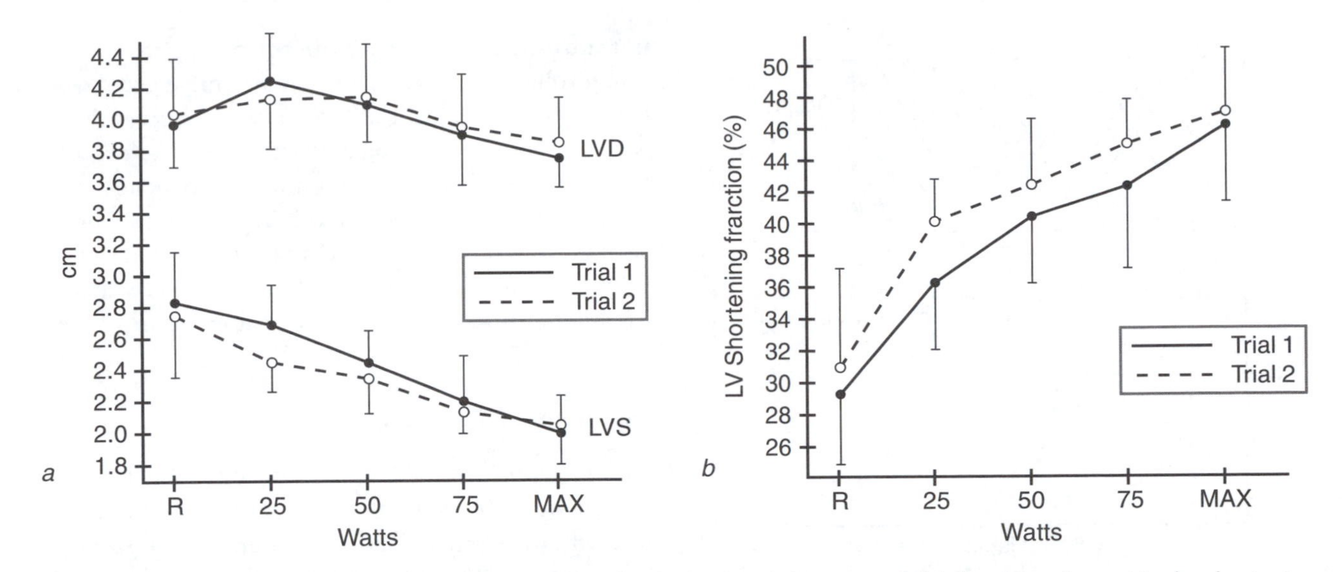

FIGURE 6.2 *(a)* Left ventricular systolic and diastolic dimensional changes and *(b)* alterations in ventricular shortening fraction during maximal upright cycle exercise in 10- to 12-year-old boys (from reference 80).

two separate studies of prepubertal children (80, 82; figure 6.2). In effect, then, studies of both children and adults indicate that left ventricular preload does not change or may even diminish slightly during an acute bout of progressive exercise.

We are thus faced with the need to explain how left ventricular diastolic size can remain stable during an exercise test in which systemic venous return increases fourfold. The only reasonable explanation is that the rise in heart rate must match the increase in venous return to maintain a constant, stable volume of blood crossing the mitral valve with each beat. The result of this heart rate response is a constant (or slightly diminishing) end-diastolic size and little change in atrial pressure.

That explanation begs two further questions: *How* does the sinus node firing rate match the volume of venous return to the heart, and *why* should this occur? The Bainbridge reflex, by which heart rate rises in response to increases in right atrial pressure, is an ideal mechanism to explain the close matching of heart rate to venous return. The existence of this reflex, however, has always been clouded by controversy (see reference 38 for discussion).

But why does the heart rate "defend" left ventricular size during exercise? Linden has pointed out that there are both physical and physiologic reasons that preventing ventricular chamber enlargement is advantageous for the heart (53). Most importantly, as expressed by the law of Laplace, an increased ventricular radius results in an increase in wall tension. Dilatation of the ventricle during exercise would cause a decrease in mechanical efficiency by increasing the workload of individual muscle fibers.

The contractile force of the heart becomes accentuated as work rate increases. With increasing work rate, the heart empties more completely during each beat, a phenomenon most obvious with two-dimensional echocardiography. While left ventricular end-diastolic size (LVED) gradually decreases, a more dramatic decline is seen in end-systolic dimension (LVES). As a result, the left ventricular shortening fraction, (LVED – LVES)/LVED × 100%, rises during progressive exercise.

In their comparison study, Nottin et al. found that left ventricular shortening fraction rose from an average of 37% at rest to 47% during maximal exercise in boys and from 36% to 49% in men (64). Rowland and Blum described an increase from 31% before exercise to 47% at peak exercise in 10- to 12-year-old boys (80).

This raises another issue: If the left ventricular diastolic size is stable as exercise intensity increases and the ventricular chamber empties more completely with each beat, how can the stroke volume (i.e., the amount of blood leaving the ventricle) remain unchanged? The answer must be, as illustrated in figure 6.3, that the increase in contractile force serves to eject the same volume of blood in a shorter ejection period as the heart rate increases

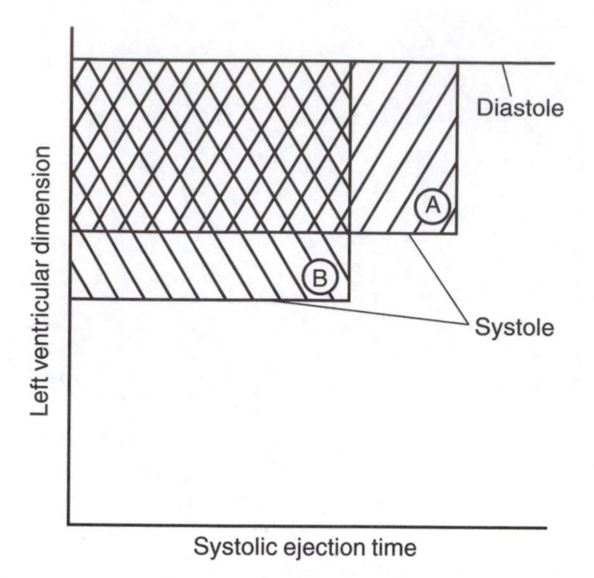

FIGURE 6.3 As systolic ejection time diminishes from low exercise intensity (stroke volume depicted as the area of rectangle A) to high intensity (rectangle B), the left ventricle empties more completely (systolic dimension decreases and shortening fraction rises in B compared with A). The stroke volume, however, does not change (the area of rectangle A equals that of rectangle B).

(16). In support of this concept, Rowland et al. (90) found a 20% increase in shortening fraction between the first workload and peak exercise in children, while their mean systolic ejection time fell 24% (and stroke volume remained stable). Increase in myocardial contractile force during progressive exercise, then, appears to *maintain* rather than increase stroke volume.

The onset of exercise is accompanied by a dramatic fall in peripheral vascular resistance, which continues to decline in a curvilinear fashion as work intensity increases. Typical results were observed in a study of 10 boys in whom mean calculated systemic vascular resistance fell precipitously at the onset of cycle exercise, from 13.9 ± 4.4 units preexercise to 8.0 ± 1.5 units with the first 25-W workload (80). Only minimal decreases were subsequently observed as exercise intensity rose (6.8 ± 1.1 units at 50 W and 6.0 ± 1.5 at 75 W). This fall in peripheral resistance reflects arteriolar dilatation in exercising muscle and is reflected in the minimal changes in mean systemic arterial blood pressure (typically about 20 mmHg) that occur in a maximal exercise test even while cardiac output rises about fourfold (figure 6.4).

From these empiric observations, it is apparent that what the left ventricle experiences during exercise is not substantially different from the resting state. The filling volume is the same, and the amount ejected per beat does not change. The alteration during exercise is that blood must be ejected more often and at a faster rate per beat. The physiologic challenge for the heart during exercise is to beat more frequently with greater force, the result being a stable ventricular filling volume (preload) and stroke volume. Studies in children suggest no maturational effect on these patterns.

Peripheral Factors

The scenario outlined in the preceding section is consistent with a central cardiac pump functioning in a responsive role to factors that define the systemic venous return. That is, the empiric observations support the concept that determinants of the circulatory response to exercise and its link with the metabolic requirements of muscle are peripheral rather than central in origin. In this section, we examine the peripheral factors that might contribute to this response.

Decreased Peripheral Vascular Resistance

Causing heart rate to increase in itself will not cause a rise in cardiac output. When the heart of an animal or human is artificially paced, the rise in heart rate is instead accompanied by a fall in stroke volume (74). These observations indicate that an increase in

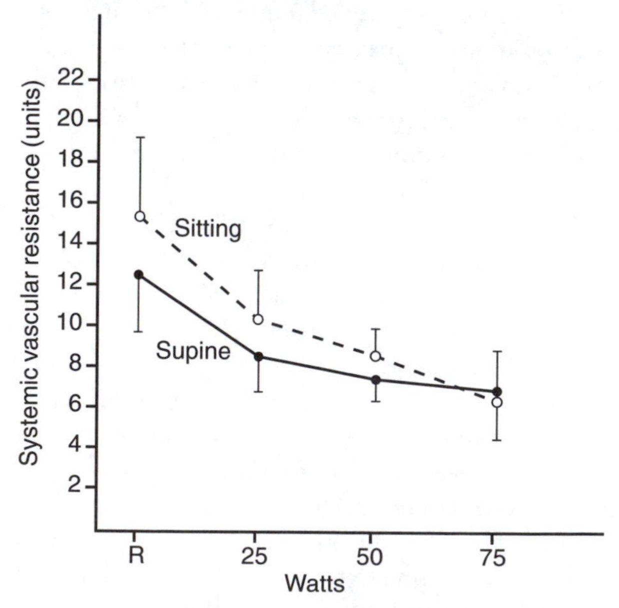

FIGURE 6.4 Decreases in systemic vascular resistance in children during progressive exercise are equal in the supine and upright positions (from reference 82).
From T.W. Rowland et al. 2003.

systemic venous return to maintain cardiac filling pressure is necessary to augment circulatory flow. This is accomplished by a fall in peripheral arterial resistance.

Arteriolar dilatation within skeletal muscle occurs immediately at the onset of exercise, with a rapid increase in muscle blood flow. Neural control mechanisms may be involved, but most attention has been directed to the local vasodilatory effect of chemical factors—such as potassium ions, hydrogen ions, adenosine, nitric oxide, and lactate—as well as tissue hypoxia (26).

The resulting fall in peripheral vascular resistance increases cardiac output, as predicted by the equation Pressure = Flow × Resistance, in which an increase in flow can be expected with a fall in resistance as long as the heart maintains pressure by increasing myocardial contractile force. This effect is seen in experimental, surgically induced, or naturally occurring arteriovenous fistulas (3, 14). These shunts produce an increase in cardiac output by causing blood to bypass the high-resistance arterioles but have no direct cardiac effects.

The influence of an arteriovenous fistula on peripheral resistance and, secondarily, blood circulation is not far removed from the conditions of dynamic exercise. Binak et al. (14) demonstrated this in a study of seven subjects with traumatic arteriovenous fistulas of the lower extremity who underwent exercise testing with and without the shunt occluded (by a blood pressure cuff). At rest, cardiac output rose from 3.9 to 7.9 $L \cdot min^{-1}$ when the fistula was opened, while peripheral resistance fell from an average initial of 1,183 $dyn \cdot s^{-1} \cdot cm^{-2}$ to 639 $dyn \cdot s^{-1} \cdot cm^{-2}$. Exercise with the fistula occluded resulted in a rise in cardiac output from an initial 3.9 $L \cdot min^{-1}$ to 4.9 $L \cdot min^{-1}$ and a fall in peripheral resistance from an initial 1,183 $dyn \cdot s^{-1} \cdot cm^{-2}$ to 918 $dyn \cdot s^{-1} \cdot cm^{-2}$.

Skeletal Muscle Pump

Like the heart, skeletal muscle is contractile tissue surrounding a vascular bed (multiple vascular channels rather than a ventricle), equipped with one-way (venous) valves, and a nerve supply to coordinate contraction. And like the heart, skeletal muscle has a defined "stroke volume" which is a reflection of preload (blood supply from the systemic arterioles), intrinsic contractility, afterload (systemic venous tone, ventricular diastolic function), and rate of contraction (e.g., pedaling cadence during cycle exercise).

Evidence of the function of skeletal muscle as a vascular pump is derived largely from studies of central shifts of blood volume and pressure away from the lower extremities during exercise (see reference 78 for a review). There are two mechanisms by which contracting skeletal muscle can serve as a circulatory pump. First, contraction of muscle compresses intramuscular veins, providing kinetic energy to blood and propelling it centrally (with the assistance of the one-way orientation of venous valves). Second, the effect of muscular contraction on reducing intramuscular venous pressure to low or even negative values causes an increase in the arterial-venous inflow gradient during the subsequent period of muscle relaxation (i.e., a "suction" effect).

A number of investigators have attempted to characterize features of skeletal muscle pump function during exercise. Gotshall et al. introduced the idea of skeletal muscle efficiency, defined as the ability of the muscle pump to satisfy the metabolic requirements of exercise (32). These authors viewed the magnitude of the cardiac output response to exercise as an indicator of muscle pump output and the arterial-venous oxygen difference as an inverse marker of pump efficiency. Their adult subjects pedaled at a constant workload of 200 W, at cadences of 70, 90, and 110 rpm. With increased pedaling cadence, oxygen uptake and cardiac output rose, but arterial-venous oxygen difference declined. These authors felt that the increased pumping rate of the skeletal muscle (i.e., pedaling cadence) at a constant contractile force (workload) was associated with improved skeletal muscle pump effectiveness.

Rowland and Lisowski performed a similar study in children (86). Hemodynamic responses were measured of subjects cycling at cadences of 41, 63, and 83 rpm under constant-load (50 W) and zero-load conditions. Energy expenditure and cardiac output rose but arterial-venous oxygen difference was unchanged with greater pedaling cadence in both the unloaded and constant-load conditions. Values of arterial-venous oxygen difference were significantly greater in the loaded versus the unloaded bouts, suggesting that skeletal muscle pump effectiveness in children is influenced by increased load but not contractile frequency (rate of pedaling).

Many questions remain to be answered regarding the functional characteristics of the skeletal muscle pump and its role in circulatory dynamics during exercise. Do rapid contractile rates prevent adequate filling time, thereby limiting function? Do high-intensity contractions have the same effect? What limits contractile function during a maximal exercise test? What is the interplay between rate and contractile force in defining pump effectiveness?

Ventricular Suction

While not a peripheral factor per se, the diastolic suction effect of ventricular contractility is important in controlling systemic venous return to the heart. Diastolic filling of the ventricle occurs only when a pressure gradient exists between the atrium and ventricle, causing the atrioventricular valve to open. This occurs at two separate times during a single ventricular diastole. The first, termed the *rapid-filling phase,* occurs early, when the pressure in the relaxing ventricle drops below that of the atrium. This phase accounts for approximately 80% of ventricular filling in the resting state. In late diastole the gradient created by atrial contraction accounts for the remainder. With the onset of light exercise and increased heart rate, the diminished diastolic filling time causes these two filling events to occur virtually simultaneously.

The extent to which ventricular suction serves to increase venous return and augment diastolic filling is determined by the nadir of ventricular pressure in early diastole, since this establishes the magnitude of the gradient between atrium and ventricle. Evidence suggests that the increased inotropic systolic function of the ventricle at high exercise intensities may be directly related to the extent of this pressure decline and widening of the atrial-ventricular pressure gradient. Udelson et al. reported that the adrenergic stimulation of an isoproterenol infusion in adult men lowered the nadir of left ventricular pressure in diastole, thereby increasing transmitral blood flow (108).

Rowland et al. estimated changes in the left atrium–left ventricle gradient using Doppler echocardiography in 10- to 14-year-old boys during incremental upright cycle exercise (87). The mean and peak transmitral pressure gradients increased fourfold from rest to maximal exercise. This was accompanied by a shortening of the average estimated diastolic filling period, from 0.48 to 0.14 s. Flow per beat across the mitral valve (equivalent to the ventricular stroke volume) rose by 40% at the onset of exercise and subsequently showed no change (figure 6.5). The authors interpreted these data as suggesting that the main effect of an increase in atrial-ventricular gradient with exercise is to increase the velocity of filling to adjust to the shortened diastolic filling time rather than increase the overall volume of systemic venous return. If so, the increased suction effect of the ventricle during exercise would permit but not define increases in circulatory flow.

Hemodynamic Responses to Exercise: Alternative Explanations

The idea that peripheral factors within the circulatory system govern the circulatory responses to exercise is well accepted by most exercise scientists. It is surprising, then, that in the efforts at explaining the factors that limit these responses—for example, defining $\dot{V}O_2$max—the importance of these peripheral determinants is often ignored. In fact, when we consider the variables that might influence hemodynamics during exercise from a peripheral rather than a central perspective, the defining factors of physiologic aerobic fitness can be viewed in an altogether different light. These alternative viewpoints may be important in explaining the developmental changes in aerobic power as children grow.

Limits of Oxygen Uptake

When considered from the standpoint of the Fick equation (as we did earlier in this chapter), the ability to generate maximal stroke volume is the factor that defines interindividual differences in $\dot{V}O_2$max.

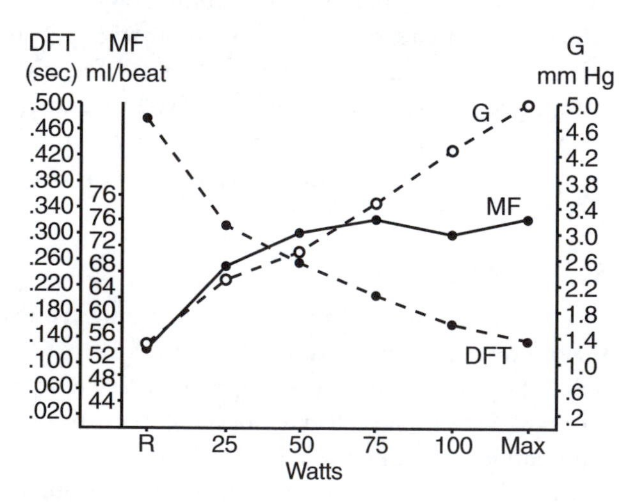

FIGURE 6.5 Comparison of changes in mean mitral flow gradient (G), flow volume per beat (MF), and estimated diastolic filling time (DFT) with progressive exercise in children (from reference 87).

Further analysis indicates that variables that influence left ventricular diastolic filling size at rest are key. This leads to a consideration of the importance of factors such as plasma volume, autonomic nervous control on the sinus node, and anabolic cardiac effects of hormones such as growth hormone, IGF-I, insulin, and testosterone.

From a peripheral, non-Fickian perspective, however, the list of potential variables that might influence $\dot{V}O_2$max is different. In particular, prime candidates are now thought to be factors that influence the magnitude of peripheral vasodilatation and the contractile capacity of skeletal muscle. Decreases in peripheral vascular resistance appear to be the critical factor in initiating and sustaining cardiac output with exercise (51, 103). However, peripheral resistance approaches its nadir midway through a progressive exercise test and changes little at high exercise intensities. Thus, it seems unlikely that factors affecting arteriolar dilatation limit circulatory responses to an acute bout of progressive exercise (although such factors clearly could define interindividual differences in $\dot{V}O_2$max).

Fatigue of the skeletal muscle pump is conceptually an attractive explanation for the limits of blood flow with exercise, because that would imply that the limits of both oxygen delivery and external work—the supply and demand of energy—are identical. This would be a handy means of explaining the close link observed between energy demand ($\dot{V}O_2$) and oxygen supply (cardiac output) as workload increases. But other factors affecting the skeletal muscle pump might also be involved. As work intensity increases, for instance, the force of muscle contraction increasingly obliterates arterial vessels and interrupts blood flow and oxygen supply (102, 111).

Another factor that might limit blood flow during exercise is actually a central one: As heart rate rises to match the increasing systemic venous return, diastolic filling time might begin to compromise blood flow to the heart. Because myocardial oxygen consumption is entirely dependent on rate of coronary blood flow (as opposed to oxygen extraction) during exercise, such an effect would create conditions of myocardial ischemia. The roles of these mechanisms in limiting circulatory flow with exercise are largely conjectural, but they might serve as plausible alternatives to those offered by the restrictive viewpoint of the Fick equation.

Physiologic Governor

There is an intriguing point regarding these potential limiting factors: At maximal, exhaustive exercise we observe *no* evidence of any these mechanisms (62). That is, there are no electrocardiographic or biochemical markers of coronary insufficiency, and the skeletal muscle is not in tetanic spasm. Moreover, systemic venous return, cardiac output, and heart rate are still rising at the time of exhaustion. Noakes (62) felt that these observations could be explained by the existence of a "physiologic governor"—initially proposed by Hill et al. (39)—that terminates exercise before dangerous myocardial ischemia or skeletal muscle anaerobiosis can occur.

Similar mechanisms may differentiate levels of $\dot{V}O_2$max and endurance fitness in child endurance athletes from those of nonathletes. These athletes, of course, are characterized by superior circulatory flow at maximal exercise. Studies of adult endurance athletes reveal findings that are consistent with superior function of peripheral circulatory variables. They exhibit (a) greater vascularization of skeletal muscle (i.e., increased "stroke volume" of the skeletal muscle pump; 93), (b) higher muscle aerobic enzymatic function (40), and (c) enhancement of factors such as muscle anaerobic power and neuromuscular capacity (67). These athletes may demonstrate greater potential for peripheral arteriolar dilatation (33), and improved diastolic ventricular filling may augment their systemic venous return (95). These issues have not been addressed in child athletes.

Prolonged Steady-Rate Exercise

An adult who performs constant-load exercise at about 60% $\dot{V}O_2$max for 45 to 60 minutes demonstrates hemodynamic changes called *cardiovascular drift*. These changes include (a) a progressive rise in heart rate to about 15% above baseline, (b) a simultaneous 15% decrease in stroke volume, and (c) small decreases in mean arterial blood pressure; there is little change in cardiac output (72). In several studies, children have demonstrated similar quantitative and qualitative patterns of drift as adults (5, 19, 91, 106).

Cheatham et al. studied cardiovascular responses to 40 minutes of cycle exercise at ventilatory anaerobic threshold (about 65% $\dot{V}O_2$max) in eight 10- to 13-year-old boys and ten 18- to 25-year-old men

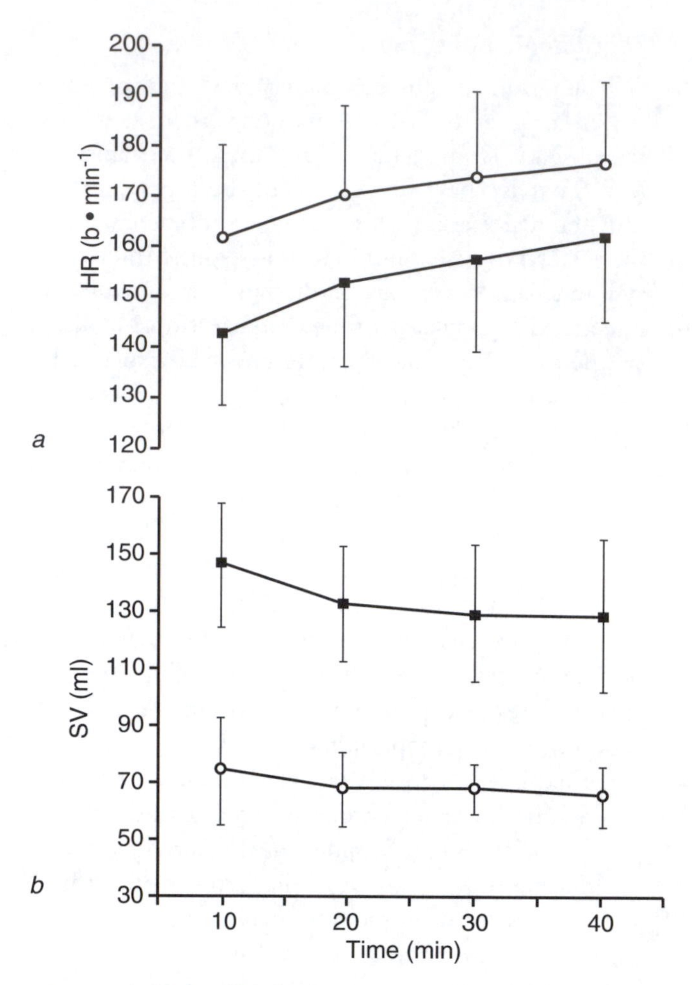

▶ FIGURE 6.6 Changes in *(a)* heart rate and *(b)* stroke volume during sustained submaximal exercise in boys *(open circles)* and men *(filled squares)*. (From reference 106.)

Reprinted by permission from C.C. Cheatham et al. 2000.

(19; figure 6.6). From 10 to 40 minutes, heart rate rose by 9.5% in the boys and 13.6% in the men. Meanwhile, stroke volume fell 8.8% and 11.6%, respectively. Mean arterial blood pressure fell by 4.2% in the men but remained stable in the boys. The decrease in plasma volume was greater in the men than in the boys (–10.2% vs. –5.7%, $p < .05$), but the fall in plasma volume was not related to the decline in stroke volume in either group.

While a number of mechanisms have been proposed to explain cardiovascular drift, most have centered around responses to increases in body temperature. Increased sweating with resulting dehydration and loss of plasma volume, compensated for by increased heart rate, have commonly been assumed. Cheatham et al.; however, considered that the primary cause for cardiovascular drift is the need for greater cutaneous blood flow to dissipate heat, causing decreases in central blood volume, diastolic filling pressure, and stroke volume (19). The rise in heart rate was explained as a compensatory response to maintain cardiac output.

Under the peripheral model of circulatory dynamics during exercise outlined earlier, however, another explanation is tenable. The constant cardiac output with time during steady-rate work indicates that systemic venous return is not changing. A primary effect of such exercise on sinus node firing rate would then account for all the changes seen with cardiovascular drift. There is, in fact, evidence that increases in catecholamine levels—associated with rises in body temperature, [H^+], and lactate concentration—parallels (and may cause) the rise in heart rate during prolonged exercise.

We return to this issue when maturational differences in temperature regulation during exercise are discussed in chapter 12. The point here is that viewing circulatory responses to exercise from a peripheral rather than central perspective can alter perceptions of the mechanisms that govern hemodynamic response during sustained exercise.

Recovery Cardiac Dynamics

A peripheral rather than a central perspective similarly affects interpretation of differences in heart rate recovery after exercise. Two trends have been identified. First, the rate of heart rate recovery is directly related to level of aerobic fitness; that is, those with higher $\dot{V}O_2$max show more rapid decrease in heart rate from maximal values after a progressive exercise test, which is believed to be due to their greater level of parasympathetic tone.

Second, a progressively slower rate of heart rate recovery is seen as children grow, irrespective of level of aerobic fitness—or of maximal heart rate, which remains stable (73, 114). Washington et al., for example, reported mean heart rates at one minute of recovery of 133, 138, and 148 bpm in boys grouped by BSA as <1.0 m^2, 1.00 to 1.19 m^2, and ≥ 1.2 m^2, respectively (114).

Baraldi et al. showed that heart rate recovery from one minute of high- and low-intensity exercise was faster in 7- to 11-year-old children than in adults ages 26 to 42 years (6). They hypothesized that lower levels of circulating catecholamines, associated with less lactate and lower [H^+], in the children were responsible for this difference.

Ohuchi et al. assessed the role of maturity-related changes in autonomic nervous system activity in determining differences in heart rate recovery between children and adults (65). Seven boys and two girls (ages 9–12 years) and six young adult men and two women (ages 17–21 years) performed a maximal treadmill test and a four-minute constant load test with measurement of heart rate at one-minute intervals during recovery. Assessment of autonomic nervous system activity was performed at rest by measurement of high- and low-frequency components of heart rate variability. A significant inverse correlation was observed between log of high-frequency variability and rate of decline in heart rate after exercise in both protocols. These findings suggest that a greater parasympathetic modulation of heart rate in children may be responsible for their more rapid decrease in heart rate after exercise compared with adults.

By the peripheral model, however, differences in heart rate recovery after exercise can be defined in terms of amount of systemic venous return, which should be governed by degree of arteriolar dilatation and—when active recovery is involved--actions of the skeletal muscle pump. Age-related variations in these factors upon cessation of exercise, then, could be responsible for developmental changes in heart rate recovery.

Heart rate recovery in children has generally been found to be more rapid in boys than girls. However, this might be explained by the typically higher levels of aerobic fitness in boys. Mahon et al. examined heart rate recovery after submaximal exercise at both the same relative and the same absolute intensity in boys and girls (55). At the same absolute workload (70 W), heart rate decline was faster in the girls than in the boys. At 85% to 90% $\dot{V}O_2$max, however, no sex differences were seen in rate of heart rate recovery.

Left ventricular dimensions during recovery from maximal exercise have been described in adults (46, 71) and children (79). They show identical trends. The left ventricular end-diastolic dimension, which fell slightly during exercise, reverts to preexercise values. End-systolic dimension, which fell more dramatically with progressive exercise, shows the same recovery trend. Consequently, ventricular shortening fraction slowly declines from maximal levels.

Rowland and Lisowski examined changes in systemic venous return immediately upon cessation of skeletal muscle pumping in children and adults (85). In study 1, 12 boys (age 12.0 ± 1.3 years) performed a standard cycle test to exhaustion. At maximal exercise, subjects immediately dropped their feet off the pedals and dangled their legs limply. Cardiac variables were recorded by Doppler echocardiography and electrocardiography in the first 15 s of recovery. During that time, cardiac output (equal to systemic venous return) fell slightly in 9 of the 12 subjects but remained essentially unchanged in the other 3, with an overall decline of 15.8% (figure 6.7).

In study 2, cardiac output was measured by changes in thoracic bioimpedance during recovery from a four-minute bout of exercise at approximately 60% $\dot{V}O_2$max in nine men (age 27 ± 3.7 years). In these subjects, cardiac output fell to 89% of maximal levels during the first 20 s after the activity of the skeletal muscle pump was eliminated. After a full minute of recovery, cardiac output was still 40% above resting levels.

The principal finding in these two studies was that after the action of the skeletal muscle pump is suddenly halted, little change in seen in cardiac output (or systemic venous return). This suggests that any developmental differences in heart rate recovery in children need to be explained, at least in the peripheral model, in the context of variations in systemic venous return. Moreover, the findings in these studies

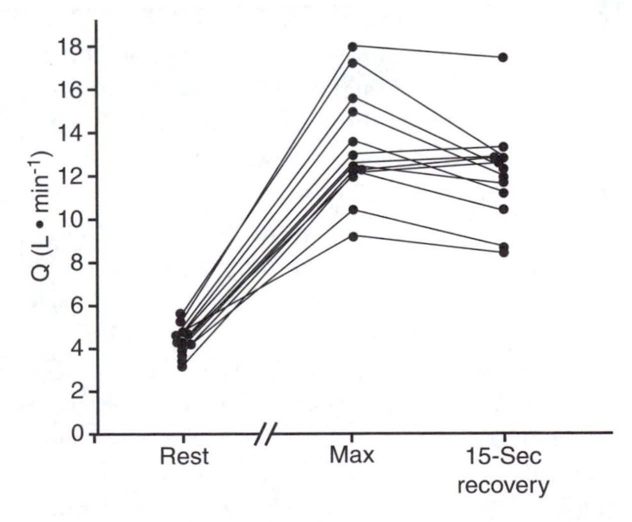

FIGURE 6.7 Individual subjects' curves for cardiac output at rest, during maximal cycle exercise, and during 15 s of passive recovery in 12-year-old boys (from reference 85).

Reprinted by permission from T. Rowland and R. Lisowski 2003.

imply that such variations would reflect differences in peripheral vasodilatation after exercise.

Myocardial Energetics

The heart's work at rest and during exercise is reflected by myocardial oxygen uptake ($M\dot{V}O_2$), which is directly related to rate of coronary blood flow. Because direct measurement of these variables is impractical, $M\dot{V}O_2$ can be predicted noninvasively in adult subjects by the product of systolic blood pressure and heart rate (the pressure–rate product), which has a correlation coefficient of $r = .85$ (60). This measure has not gained much attention in children, possibly because the relationship of $M\dot{V}O_2$ with the pressure–rate product has not been established in this age group.

Resting myocardial oxygen uptake relative to heart mass should not be expected to change as children grow. Between ages 6 and 15 years, the resting heart rate in males falls from 92 to 74 bpm, while systolic pressure increases from 94 to 105 mmHg. That is, the pressure–rate product in the resting state does not appear to be influenced by biologic maturation.

At maximal exercise the developmental story is different. Peak heart rate is stable throughout the pediatric years, while systolic pressure at maximal exercise steadily increases. Peak systolic pressure during a progressive cycle test in a child with a BSA of 1.00 m^2 is 130 mmHg, while in an older youngster with a BSA of 2.00 m^2, the maximal pressure is 170 mmHg (2). With these figures, the $M\dot{V}O_2$ relative to heart mass in the older child should be 30% greater at maximal exercise than that of the younger child.

The limited studies of pressure–rate product in children bear this out. Riopel et al. measured the pressure–rate product during maximal treadmill walking in 288 healthy young people ages 4 to 21 years (73). Subjects were divided by body size into groups I (0.72–1.09 m^2), II (Figure 1.10–1.39 m^2), III (1.40–1.89 m^2), and IV (1.90–2.31 m^2). No differences in pressure–rate product were observed between the groups at rest. With exercise, values rose and diverged among the four groups; the rate of increase was related to body size (figure 6.8). At maximal exercise the pressure–rate product was about 40% greater in the subjects in group IV than those in group I. Values rose to 3.8 times resting values in the former group and 2.8 times in the latter.

Pressure–rate products can also be calculated from the data reported by Washington et al. in 7- to 12-year-old children during cycle exercise (114). Subjects were divided into three groups by body surface area. Resting pressure–rate product was independent of body size. At maximal exercise, however, the product was higher in the largest subjects than in either group of smaller subjects (35.3×10^3 for the largest versus 29.2×10^3 for the smallest).

Colan et al. calculated resting ventricular wall stress, another index of myocardial oxygen uptake, using echocardiography in 256 normal children ages 7 days to 19 years (22). Wall stress expressed per beat rose significantly with age, but because of the concomitant decrease in heart rate, no significant change in total minute stress with age was observed. There were no sex differences in these findings.

These limited data demonstrate that $M\dot{V}O_2$ in the resting state does not change during childhood. However, myocardial oxygen consumption at maximal exercise appears to increase as children grow. The implications of this rise in myocardial metabolic rate at peak exercise are unknown.

The role of heart rate changes during childhood in cardiac energetics is suggested by the comparative animal study of Ianuzzo et al. (42). They contrasted glycolytic enzyme (phosphofructokinase, phosphorylase) and aerobic enzyme (citrate synthase,

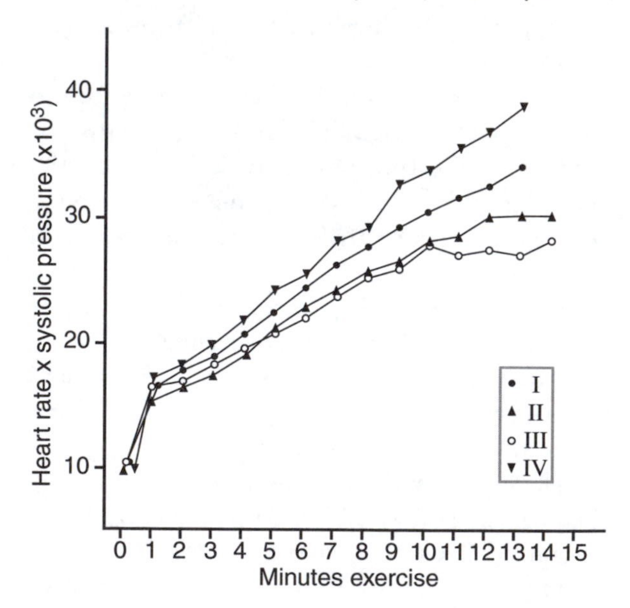

FIGURE 6.8 Pressure–rate product with treadmill walking exercise in children grouped by body surface area: I (0.72–1.09 m^2), II (1.10–1.39 m^2), III (1.40–1.89 m^2), and IV (1.90–2.31 m^2). (From reference 73.)

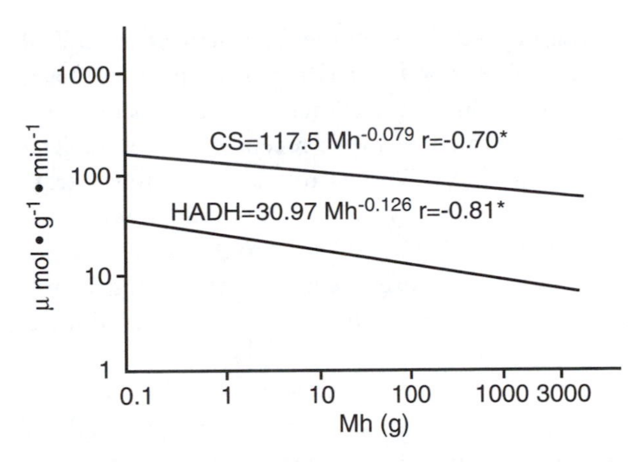

FIGURE 6.9 Mammalian heart aerobic enzyme activity is negatively related to heart mass (Mh). CS = citrate synthase; HADH = 3-hydroxyacyl-CoA dehydrogenase (from reference 42).

Reprinted by permission from S.F. Lewis and R.G. Haller 1990.

3-hydroxyacyl-CoA dehydrogenase) activities in the ventricular myocardium of mammals ranging in size from mice to cattle. No relationship was observed between heart mass and glycolytic enzyme capacity. On the other hand, the capacity of the aerobic enzymes was inversely related to heart mass (r = –.70 to –.81; figure 6.9) and was positively correlated with heart rate (r = .86). Citrate synthase activity, for instance, was 73 µmol · g^{-1} · min^{-1} in cattle and 181 µmol · g^{-1} · min^{-1} in the mice. This suggests that a causal relationship exists between the frequency of myocardial contractions and myocardial aerobic enzyme activity. (It should be noted that mean arterial pressure values were found to be similar in all these animals.)

To test this concept—that frequency of myocardial contractions defines aerobic biochemical capacity in cardiac muscle—the same authors artificially paced the heart of large animals (pigs) to match the rate of a smaller animal (180 bpm) for 35 to 42 days (42). No significant changes were observed in glycolytic function. However, the activity of Krebs cycle enzymes increased by 39%, and fat oxidation enzymes by 57%.

Heart rate, then, is the central determinant of cardiac metabolic rate when other hemodynamic influences (e.g., blood pressure) remain constant. Changes in stroke volume, on the other hand, only minimally affect $M\dot{V}O_2$. As Starnes pointed out, myocardial metabolic activity is not necessarily related to cardiac output (99).

This information suggests, too, that changes in $M\dot{V}O_2$ during childhood can be expected to reflect alterations in aerobic rather than glycolytic enzyme activity. This is consistent with the concept that the metabolic requirements of the heart can be satisfied by oxidative phosphorylation as long as coronary blood flow is adequate. Because the myocardium cannot pause to rest and recover, it has little tolerance for oxygen debt or reliance on nonaerobic metabolic pathways (42).

Heritability of Cardiac Size

Endurance fitness in children, as well as in adults, has generally been assumed to be linked to genetic rather than environmental factors. That is, performance by the star eight-year-old swimmer is expected to reflect principally her genetic endowment of superior cardiovascular capacity and muscular endurance. It is evident, too, that a defining characteristic of the talented endurance athlete—child or adult—is a larger left ventricular filling volume (i.e., end-diastolic size). It follows, then, that heritability studies should be expected to demonstrate a strong genetic influence on cardiac size. Quite surprisingly, this has not been the case.

Bielen et al. examined the influence of heredity on echocardiographic measures of left ventricular dimensions in 15 monozygotic and 19 dizygotic six- to eight-year-old twin pairs (13). This analysis revealed no significant influence of genetic endowment on left ventricular internal diameter or wall thickness. A significant genetic variance was observed, however, for calculated left ventricular mass. The authors concluded that, at least in the resting state, the cardiac characteristics associated with superior endurance fitness are not inherited. They suggested that the heritability of $\dot{V}O_2$max (about 50%) may be due to inheritance of noncardiac features or to cardiac factors that are expressed only during exercise (such as diastolic filling or myocardial contractility).

The latter idea was supported by another echocardiographic study by the same authors in 18- to 31-year-old twins (12). In this investigation, cardiac variables were assessed not only at rest but also during submaximal supine exercise at a heart rate of 110 bpm. Mean left ventricular end-diastolic dimensions increased with exercise by 1.1 to 1.8 mm. At rest, a significant genetic influence was observed for ventricular wall thickness but not for internal diameter. However, the magnitude of the increase in end-diastolic dimension with exercise showed a genetic component of 24%.

Adams et al. compared measures of cardiac size in college-aged monozygotic twins, dizygotic twins, and siblings of like sex (1). Average intrapair differences in ventricular size for monozygotic twins were no different from those in the other groups. However, differences for all three groups were significantly smaller than those for a group of random subjects. After 14 weeks of endurance training, average $\dot{V}O_2max$ increased by 9.2 ml · kg^{-1} · min^{-1}, and average left ventricular end-diastolic dimension increased by 1.9 mm ($p < .07$). However, the same findings in twin comparisons prevailed. These results were felt to indicate that cultural or familial influences are more critical to heart size than genetic influences.

Isometric (Static) Exercise

The cardiovascular responses to isometric exercise, in which the length of the muscle does not change, differ from those to dynamic exercise. Two reasons can be immediately suggested: (a) Static exercise occurs without the rhythmic contractions of the skeletal muscle pump, and (b) during isometric exercise, local arteriolar dilatation competes with the compression of blood vessels by contracting muscle. This latter effect should become progressively more influential as the force of contraction increases. The combined result of these effects is less response by cardiac output and heart rate to static exercise than to dynamic activities.

The hallmark of isometric exercise, though, is a significant increase in systolic and diastolic blood pressure (70). The extent of this rise depends on the absolute and relative load, duration of contraction, muscle mass involved, and joint angle. This hypertensive response has been viewed by some as critical for maintaining flow to vessels whose lumens are being compromised by muscular contractions.

As mentioned previously, neurologic reflexes are thought to be important in mediating sympathetic responses to static exercise. Two types of neural input during exercise have been described. In the first, termed *central command,* signals from the motor cortex of the brain trigger both motor efferent traffic to skeletal muscle and autonomic control centers of the medulla (115). Together, these augment sympathetic input to increase heart rate and blood pressure in parallel with the contractions of skeletal muscle. A second mechanism involves a feedback loop beginning with afferent signals arising from mechanical and metabolic receptors in the skeletal muscle itself (66). These cause a rise in reflex triggering of medullary autonomic centers, which in turn increase sympathetic drive to the cardiovascular system with rising exercise intensity. Studies supporting these two constructs have thus far been limited to adult subjects.

Blood flow to exercising muscle during static contractions does increase, but as the strength of the contraction increases (measured as a percentage of maximal voluntary contraction [MVC]), flow declines. Some have suggested that flow through blood vessels in totally occluded above 70% MVC.

Cardiac changes with isometric exercise in adults vary in different reports. Perhaps most typical are the findings in 16 female athletes ages 17 to 21 years described by Cassone et al. (18). With handgrip contractions at 75% MVC for one minute, no changes were seen in ventricular diastolic or systolic size, stroke volume, or myocardial contractility. Cardiac output rose modestly by 43%, which was entirely the consequence of an increase in heart rate. However, others have described decreases in stroke volume and left ventricular dimensions with high-intensity static contractions (45, 52).

Studies in Youth

A number of studies have examined cardiovascular responses to static exercise in children (34, 50, 101, 109). Strong et al. provided information regarding blood pressure responses to isometric exercise in 170 healthy black boys and girls (101). Blood pressure measurements were made after 30 s of handgrip exercise at 50% MVC. The increase in systolic pressure was 18 ± 9 mmHg for boys and 16 ± 8 mmHg for girls, corresponding to a 19% and 15% increase from the resting value, respectively. In this study an inverse relationship was observed between resting blood pressure and percentage change ($r = -.43$ for boys and $r = -.36$ for girls).

Gumbiner and Gutgesell described cardiovascular responses to three minutes of 33% MVC handgrip in 18 children ages 9 to 18 years (34). Mean systolic pressure rose from 115 ± 4 to 128 ± 6 mmHg, and diastolic pressure rose from 64 ± 3 to 76 ± 4 mmHg. Heart rate increased from 78 to 91 bpm. Echocar-

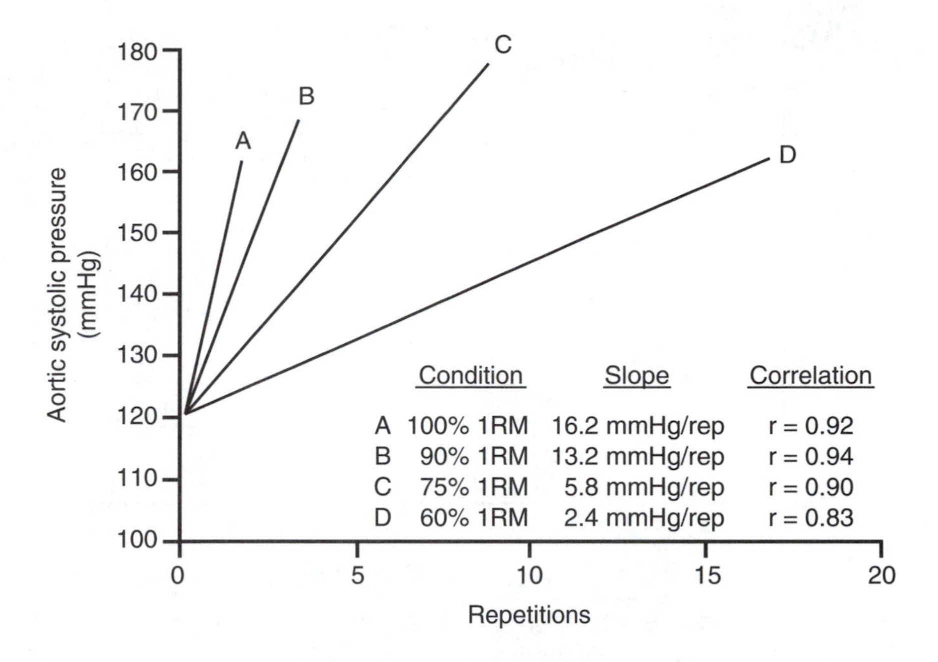

FIGURE 6.10 Aortic systolic pressure response in children to varying resistance conditions (from reference 59).
Reprinted by permission from Nau et al. 1990.

diographic measurements indicated no change in left ventricular shortening fraction nor in end-diastolic or end-systolic dimension.

Laird et al. performed a similar study with similar results in 32 healthy adolescents who performed contractions at 25% MVC (50). Again, left ventricular diastolic and systolic dimensions as well as stroke volume remained constant. Heart rate rose from 70 ± 9 to 88 ± 11 bpm. Mean arterial pressure rose from 78 ± 7 to 92 ± 7 mmHg.

Verhaaren et al. made the same measurements in 82 children (age 11.0 ± 0.9 years) (109). Subjects performed handgrip exercise for two minutes at 33% MVC. Cardiac output was measured by the Doppler echocardiographic technique. An inverse relationship was observed between stroke volume and heart rate response. In the boys, heart rate rose 12.7% while stroke volume fell 3.9%. Heart rate in the girls rose 12.7% with an unchanged stroke volume. Blood pressure changes were similar by in boys and girls. Systolic pressure rose from 108 ± 10 to 121 ± 14 mmHg in the boys and from 112 ± 11 to 126 ± 14 mmHg in the girls.

When a series of isometric exercises is performed, blood pressure rises progressively with each exercise. Nau et al. demonstrated this in eight children who were tested during cardiac catheterization for dysrhythmias (59). Subjects performed bench-press weightlifting at 60%, 75%, 90%, and 100% MVC to exhaustion. The peak intraarterial blood pressure observed was independent of lifting condition. The 60% MVC load was lifted an average of 16.9 repetitions, the 75% load for 8.9 repetitions, and the 90% load for 3.7 repetitions. At fatigue, peak blood pressures showed no significant differences among the submaximal loads and were similar to that observed with the 100% MVC (1RM) lift, in which aortic pressure rose on the average from 120/81 mmHg at rest to 162/130 mmHg at maximal exercise (figure 6.10). At the same time, mean heart rate rose from 86 to 139 bpm. When considered as percentage changes from resting values, these responses are similar to those observed in adult subjects.

Comparisons Between Children and Adults

Other studies have directly compared hemodynamic changes with static exercise between children and adults. Smith et al. found similar cardiovascular changes during three minutes of sustained handgrip at 30% MVC in a supine position in premenarcheal girls and college-aged women (96). Stroke volume and cardiac output were estimated using thoracic bioimpedance. Heart rate rose by 8% in the children and 13% in the adults, with no significant change in cardiac output in either group. Stroke volume decreased 13% in the girls and 12% in the college women.

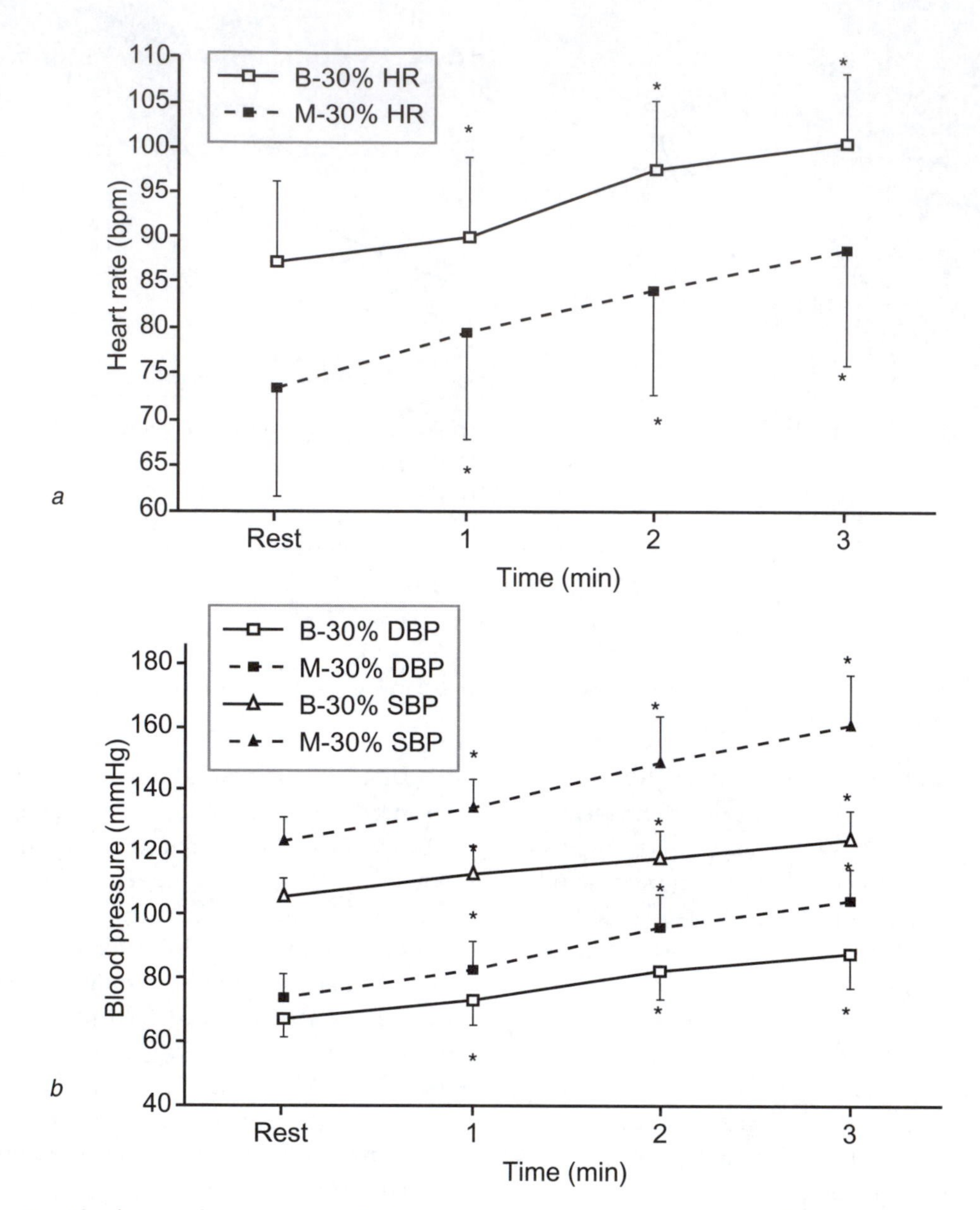

FIGURE 6.11 Absolute *(a)* heart rate and *(b)* blood pressure responses to 30% MVC handgrip exercise in men (M) and boys (B). (From reference 105.)

Turley et al. compared heart rate and blood pressure responses to static handgrip at 10%, 20%, and 30% MVC in 27 boys ages 7 to 9 years and 27 men ages 18 to 26 years (105). No differences were seen in heart rate response between the two groups at any of the intensities (at 30% MVC, the increase was 15 ± 9 % in boys and 21 ± 15 % in men). At 30% MVC, the rise in blood pressure was greater in the men (35 ± 11 %) than in the boys (24 ± 11 %; figure 6.11).

An important finding in this study was that test-retest reliability of mean blood pressure measurements was low at 10% and 20% MVC in the boys (intraclass correlation coefficients of −.04 and −.07) and at 10% MVC in the men (intraclass r = .29). Reproducibility of diastolic pressure responses was particularly suspect, with intraclass correlation coefficients ranging widely (from −.16 to .70). Heart rate reproducibility, on the other hand, was moderately high for both groups at all intensities (r = .35 to .87).

Matthews and Stoney (57) and Palmer et al. (68) found no differences between adults and children in blood pressure response to 30% MVC exercise.

However, these authors described greater heart rate responses in the children.

This information generally indicates that few or no maturational or sex differences, either qualitative or quantitative, occur in cardiovascular responses to isometric exercise. The extent to which blood pressure in children rises with high-resistance loads has not been studied, however, and the effects of this type of exercise on myocardial function, particularly in those with heart disease, needs to de determined.

Conclusions

The patterns and relative magnitude of cardiovascular responses to progressive and sustained dynamic exercise as well as to isometric contractions are similar between children and adults. This suggests that the determinants of circulatory adaptations to increased muscular work are not affected by biologic maturation. Evidence in both adults and children points toward peripheral factors, particularly arteriolar dilatation and the pumping action of skeletal muscles, as the controlling determinants that cause the circulation to rise in response to the metabolic demands of dynamic exercise.

Increases in stroke volume are responsible for matching cardiac output to body size, but changes in metabolic expenditure at rest relative to body size are accounted for by differences in heart rate. Myocardial contractility, however, is essentially independent of age or maturation.

Myocardial oxygen uptake relative to heart mass in the resting state remains constant during the childhood years. However, $M\dot{V}O_2$ during maximal exercise appears to increase as children grow, because of the rise in peak systolic pressure. The implications of this trend are unknown.

Studies of the heritability of cardiac size have demonstrated only a small genetic influence on left ventricular size. This is inconsistent with data indicating that $\dot{V}O_2max$ in children, which is at least moderately hereditary, is related to resting left ventricular filling dimensions.

Discussion Questions and Research Directions

1. How is heart rate linked to metabolic expenditure, both at rest and during exercise? Why do children have higher peak rates of sinus node firing during maximal exercise?
2. What mechanisms are responsible for the rise in systemic blood pressure during the course of childhood?
3. How do peripheral factors (skeletal muscle pump, arteriolar dilatation) define circulatory responses to exercise?
4. What factors cause local arteriolar dilatation in exercising muscle? Are there maturational changes in these responses?
5. What causes cardiovascular drift?
6. What is responsible for the age-related difference in rate of heart rate recovery after exercise?
7. What are the implications of the higher $M\dot{V}O_2$ at maximal exercise in older than in younger children?

7

Ventilation Responses

Control of ventilation is one
of physiology's unsolved problems.
Resting ventilation may be
compared to a jigsaw puzzle,
whereas exercise ventilation may be
compared to a three-dimensional puzzle
such as a Rubik's cube.
The numerous combinations
of possible factors
must be considered simultaneously.

—Erling Asmussen (1983)

In this chapter we discuss

- how the contributions to minute ventilation (tidal volume, breathing frequency) at rest and during exercise evolve during childhood,
- mechanisms by which the control of ventilation may differ in children and adults, and
- ventilatory responses to sustained exercise in youth.

As the first portion of the oxygen delivery chain, the lungs serve as the link between the ambient air and the blood circulation of oxygen to the exercising muscles. During exercise, providing this external respiration is no simple task. Arterial oxygen and carbon dioxide content must be allowed to deviate no more than a small degree, and blood [H^+] must be maintained within narrow tolerance limits. Moreover, this blood gas and acid-base homeostasis must be sustained while the demand for ventilation rises during maximal exercise to a value that is 20 to 30 times that in the resting state.

The rise in ventilation is accomplished through an increase in both rate of breathing (f_R) and tidal volume per breath (V_T). The relative contributions of each must be balanced to prevent inappropriate increases in airway resistance, impedance to airflow, or dead-space ventilation (V_D). Additionally, some mechanism must be in place that matches these ventilatory responses closely with the metabolic demands of exercising muscle. As Whipp and Ward concluded, "The appropriateness of the ventilatory response to exercise is best considered with respect not to the actual level of ventilation achieved, but to the degree of arterial blood gas homeostasis for moderate exercise and by the degree of compensatory hyperventilation at work rates that engender a metabolic acidosis" (49, p. 13).

The work of ventilation during exercise to distend the lung and overcome airway resistance to flow comes at a metabolic cost. At maximal exercise, the metabolic demands of breathing in a healthy adult may reach as much as 10% to 14% of the total body $\dot{V}O_2$. Still, ventilatory capacity does not reach its peak at exhaustive exercise. Minute ventilation ($\dot{V}_E$) at maximal exercise is typically about 60% to 70% of the maximal voluntary ventilation (MVV), the highest ventilation that can be achieved in a brief, all-out breathing test at rest.

It is tempting to draw parallels between the circulatory and ventilatory responses to progressive exercise, since in both cases these involve combined increases in (a) a volume that is related to body size (stroke volume in the case of circulation and tidal volume in the case of respiration) and (b) a size-independent, time-defined factor (heart rate and breathing frequency). However, the driving factors that cause cardiac output and minute ventilation to rise with exercise are different. Increases in cardiac output (Q) occur in direct response to metabolic demand; that is, a linear relationship is observed between Q and $\dot{V}O_2$ throughout a progressive test to exhaustion.

At lower intensities of exercise, ventilation also follows metabolic demand, or $\dot{V}O_2$, in the same linear fashion. At approximately 55% to 65% $\dot{V}O_2$max, however, excessive CO_2 is generated above that produced through skeletal muscle metabolism by the buffering of increasing lactic acid with bicarbonate. As a result $\dot{V}_E$ rises more steeply than $\dot{V}O_2$. At even higher levels of intensity, arterial pH begins to fall as a consequence of accumulating lactic acid, further driving $\dot{V}_E$ as an acid-base compensatory mechanism. At this point, the increase in $\dot{V}_E$ diverges from $\dot{V}CO_2$, and arterial PCO_2 falls (figure 7.1).

Maximal cardiac output and maximal minute ventilation thus reflect two different physiologic constructs. The former reflects the limits of aerobic metabolic activity during exercise. The latter, on the other hand, represents the ventilatory responses not

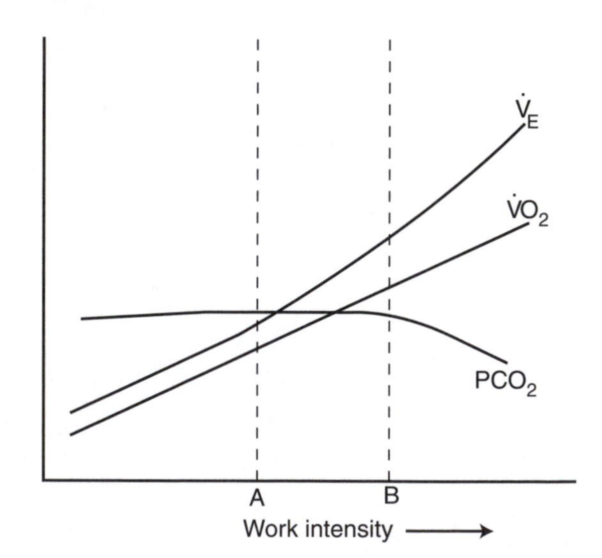

FIGURE 7.1 Changes in minute ventilation ($\dot{V}_E$), oxygen uptake ($\dot{V}O_2$), and arterial PCO_2 with increasing work intensity. $\dot{V}_E$ parallels $\dot{V}O_2$ until intensity A, when increasing lactate buffering produces excessive CO_2 and $\dot{V}_E$ accelerates. At intensity B, metabolic acidosis further stimulates $\dot{V}_E$ as a compensatory mechanism, and PCO_2 falls.

only to the metabolic demands imposed by exercising muscle but also to the increasing needs to eliminate the added carbon dioxide produced by the buffering of lactic acid and to compensate for metabolic acidosis. It is impressive that the pulmonary system accomplishes this at approximately two-thirds its maximal functional capacity.

Ventilatory responses in a healthy individual are assumed to be capable of sustaining both acid-base and gas exchange homeostasis during even exhaustive levels of exercise. Arterial PO_2 is maintained within 5 mmHg of resting values, and declines in pH at maximal exercise are minor. Thus, maximal ventilation is not considered a limiting factor of aerobic fitness in either children or adults. Dempsey et al., however, have pointed out that "this pulmonary system is far from perfect" (13, p. 114). Studies in both children and adults indicate that while alveolar hyperventilation during exercise compensates for rising $[H^+]$, the fall in arterial PCO_2 to 25 to 32 mmHg does not fully correct the metabolic acidosis. At the same time, the ventilation-to-perfusion ratio becomes less uniform, causing a threefold increase in the alveolar-arterial oxygen difference at maximal exercise. While arterial oxygenation remains stable in untrained people, the demands of high-level maximal exercise in trained adult athletes outstrips ventilatory capacity to maintain normal oxygen saturation, causing arterial PO_2 to decline.

The basic principles of ventilatory responses to exercise are as true in children as in adults. However, a number of maturational differences in pulmonary adjustments to exercise have been recognized. The central difference is that, at any given level of metabolic work, children hyperventilate in comparison with adults. This is most obviously expressed by their greater ventilatory equivalent for oxygen ($\dot{V}_E/\dot{V}O_2$) and lower levels of arterial PCO_2 during exercise. That is, factors influencing exercise hyperpnea are set to a lower PCO_2 threshold in prepubertal subjects. The explanation is unknown, but the answer presumably lies among the anatomic, metabolic, and mechanical features of ventilation that are unique to children.

In this chapter we explore these maturational issues. Following a description of age differences in ventilatory responses, variations in respiratory mechanics for achieving adequate minute ventilation are explored. How these differences affect acid-base and blood gas homeostasis is considered, particularly maturational differences in factors responsible for ventilatory drive. Finally, variations between children and adults in how ventilation responds to prolonged exercise are examined. Studies addressing the question of how these factors might be influenced by athletic training are postponed until chapter 11.

Developmental Changes in Ventilatory Components

The changes that occur in the components of ventilation as children grow, both at rest and during exercise, have been well described. Much of this development parallels increases in lung size. Between the ages of 5 and 14 years, the total capacity of the lung increases from approximately 1,400 to 4,500 cm^3. Lyons and Tanner reported that lung volume correlated closely with both height and height cubed, irrespective of sex (28).

Resting Values

Vital capacity (VC), the greatest amount of air that can be expelled in a single maximal expiratory effort, increases with age in direct proportion to body mass. Vital capacity per kilogram, however, is greater in males than females at all ages. Armstrong et al., for example, described a mean forced VC of 2.50 L in 11-year-old boys and 2.19 L in 11-year-old girls when body mass and stature were accounted for (3). Taussig et al. thought that this finding might be explained by sex-related differences in the mechanical properties of the lung (47).

In the resting state, tidal volume (V_T) increases with lung growth, but values decrease in respect to both body mass and surface area during the course of childhood. Cassels and Morse found average V_T values of 321, 297, and 242 $ml \cdot m^{-2}$ of body surface area in girls ages 6 to 8, 8 to 12, and 12 to 17 years, respectively (11). This means that the proportion of vital capacity used for tidal volume declines as children age. Robinson found a mean V_T/VC value of 0.23 in 6-year-old boys and 0.13 in 17-year-olds (40).

Breathing rate at rest steadily falls during childhood. Robinson found that between the ages of 6 and 17 years, breathing frequency in boys declined from an average of 24 to 13 breaths per minute (40). The fall in resting breathing rate with increasing body size reportedly can be expressed by the scaling exponent for mass of -0.53 and for height of -1.17 (6).

Considering that both breathing frequency and tidal volume (expressed relative to body size) decline with growth, it is not surprising that resting minute ventilation likewise does so. Recognizing, too, that ventilation in the resting state should be adjusted to match metabolic rate, we would expect similarities to be observed in the relationship between size-related $\dot{V}_E$ and $\dot{V}O_2$.

This is true in phylogenetic studies of mammals, in which minute ventilation at rest relates to body mass by the exponent 0.80, an exponent similar to that for oxygen uptake (44). An allometric analysis of resting minute ventilation in relationship to body size in children has not yet been performed. Morse et al. reported that resting minute ventilation in a 10-year-old boy is expected to be about 200 ml · kg^{-1} · min^{-1}, while that in a 16-year-old boy would be 158 ml · kg^{-1} · min^{-1} (33).

Maximal Exercise

It should be remembered that values of $\dot{V}_E$ at maximal exercise, unlike those for maximal cardiac output, are not expressions of the limits of ventilatory capacity. $\dot{V}_E$max reflects not only a response to peak metabolic rate but also the influences of excessive CO_2 produced by the buffering of lactate and compensatory needs to reduce metabolic acidosis. Consequently, $\dot{V}_E$max is not an expression of aerobic metabolic fitness alone.

From an ontogenetic standpoint, $\dot{V}_E$max becomes greater as lung volume increases, but developmental curves for $\dot{V}_E$max may also be influenced by maturational changes in lactate production and factors governing cellular acidosis.

Studies examining values of $\dot{V}_E$max in children relative to body size have produced varying results. Rowland and Cunningham reported $\dot{V}_E$max during treadmill walking in 20 children studied annually for 5 years between the ages of 9 and 13 years (41; figure 7.2). Average $\dot{V}_E$max expressed per kilogram of body mass was relatively stable in the boys, ranging over the five years from 1.82 to 1.89 L · kg^{-1} · min^{-1}. The girls, however, showed a steady decline in values from 1.90 L · kg^{-1} · min^{-1} at the start of the study to 1.70 L · kg^{-1} · min^{-1} at the end. Other authors have reported stable values for $\dot{V}_E$max per kilogram across the male childhood years (33, 40).

On the other hand, Mercier et al. found that $\dot{V}_E$max scaled to mass by the exponent 0.68, indicating that $\dot{V}_E$max increased more slowly than body mass during growth (32). Åstrand described similar findings, noting a mean $\dot{V}_E$max of 1.94 L · kg^{-1} · min^{-1} in 4- to 6-year-old boys and 1.59 L · kg^{-1} · min^{-1} in 14- to 15-year-olds (7).

Armstrong et al. performed a cross-sectional study of ventilatory responses to treadmill exercise in 101 boys and 76 girls ages 8 to 11 years (3). The allometric exponent relating $\dot{V}_E$max to stature was 0.69, and to mass 0.48. They concluded that "the conventional approach of accounting for body size differences in the interpretation of children's ventilatory data, i.e., by dividing by either body mass or stature, is inappropriate" (p. 1558). In that study, $\dot{V}_E$max was significantly greater in the boys than in the girls, even after adjusting values for body size.

Definitive norms for $\dot{V}_E$max in children, then, are not available. Most studies have described values for maximal ventilation in children of 1.60 to 1.90 L · kg^{-1} · min^{-1} in relation to body mass and about 0.50 L · m^{-1} · min^{-1} when adjusted for body height. As Armstrong et al. pointed out, much of the variability in $\dot{V}_E$max values and their relationship to body size might be explained by the influence of different exercise modalities and protocols (3). Boileau et al. found that $\dot{V}_E$max was 10% lower during cycle testing than during treadmill testing of the same children (8, pp. 162-168). Paterson et al. described mean $\dot{V}_E$max values of 67.6 and 73.3 L · min^{-1} during a walking and a running treadmill protocol, respectively (37).

Breathing frequency at maximal exercise slowly declines with age during childhood, and values are independent of body size. Rowland and Cunningham found that maximal rate during serial treadmill walking tests over five years in the same children fell

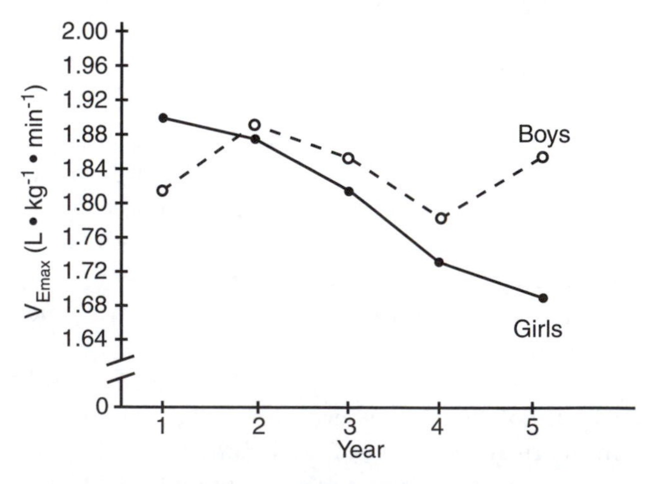

FIGURE 7.2 Longitudinal changes in $\dot{V}_E$max per kilogram in boys and girls (from reference 41).
Reprinted by permission from T.W. Rowland 1997.

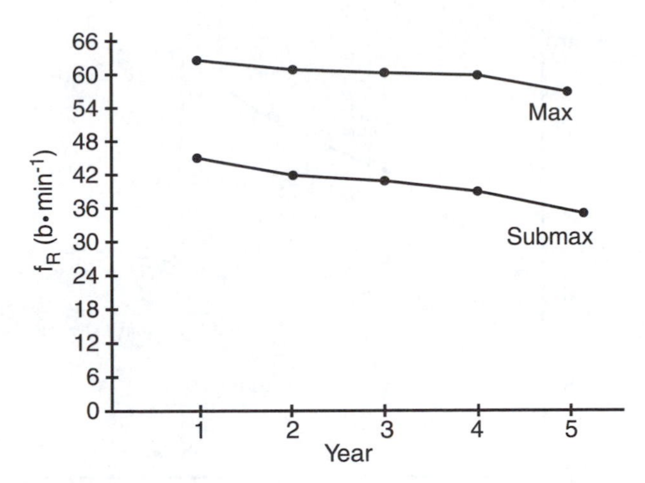

FIGURE 7.3 Longitudinal changes in maximal and submaximal breathing frequency (f_R) with age in children (from reference 41).

Reprinted by permission from T.W. Rowland 1997.

from 65 to 57 breaths per minute on average in the boys and from 63 to 57 in the girls (41; figure 7.3). Robinson reported that 6-year-old boys took 62 breaths per minute at peak exercise, compared with 46 in 18-year-olds (40). Mercier et al. (32) calculated that maximal breathing rate ($f_R$max) in children could be expressed relative to body mass (M) by the allometric equation

$$f_R\text{max} = 137M^{-0.27}$$

It should be noted that this mass scaling exponent is similar to others that reflect biologic time (see chapter 1).

No sex differences in either $f_R$max or rate of decline in breathing rate with age have been observed in children. In the study by Armstrong et al., breathing frequencies at peak exercise were 52 ± 10 and 53 ± 12 breaths per minute for the boys and girls, respectively (3).

Absolute values of tidal volume at maximal exercise increase with growth in accord with greater lung dimensions. The ratio of f_R to V_T at peak exercise, then, falls as a child grows. Rowland and Cunningham found that this ratio fell from 60.7 to 29.2 in boys and from 58.9 to 32.6 in girls between the ages of 9 and 13 years (41). Contrary to these findings, Armstrong et al. (3) observed that the average f_R/V_T ratio at peak exercise was greater in girls (60.0) than in boys (48.0).

Armstrong et al. found allometric scaling exponents for $V_T$max of 1.71 for stature and 0.27 for mass (3). When adjusted for stature and mass, values for $V_T$max were significantly higher in boys than girls. Other studies have found that tidal volume at peak exercise is closely linked to body mass (7, 32, 41). Rowland and Cunningham reported that $V_T$max per kilogram was essentially unchanged over their five-year longitudinal study (41). The mean value, about 30 ml · kg^{-1}, was similar in boys and girls. Mercier et al. found that $V_T$max related to body mass by the scaling exponent 0.96 and to height by the exponent 2.90 in children (32).

In summary, most studies indicate that the absolute increase in $\dot{V}_E$max with age during childhood is a reflection of greater tidal volume with growth. Differences in $\dot{V}_E$max related to body size, however, may be more influenced by variations in f_R.

Ventilatory Mechanics

The lungs are essentially elastic sacs that expand through the action of the diaphragm and intercostal muscles. The resultant negative intrapulmonary pressure draws air through the conducting airways and expands the alveolar air spaces. There, diffusion of gases occurs across the alveolar-capillary membrane, an interface that is large (about 125 m^2 in the average adult) and thin (1/50 the thickness of a piece of airmail stationery; 13).

The energy expenditure of the breathing muscles to achieve minute ventilation is influenced by the both the compliance, or distensibility, of the lung and chest wall and the resistance to flow offered by the conducting airways. Compliance is defined by the change in lung volume in response to a given transmural pressure gradient. Resistance conforms to the predictions of both Ohm's and Poiseuille's laws, changing in response to airway radius as well as pressure gradients driving airflow. Energy efficiency in respiration calls for a maximization of airflow conductance from optimal compliance and optimal limitation of resistance.

These aspects of lung mechanics and flow dynamics change as children grow. The compliance of the lung improves during childhood, with most significant changes in the early years. At the same time, airway resistance progressively diminishes, again most prominently in the preschool years. Beyond the age of five years, changes in both compliance and resistance are relatively small.

Lanteri and Sly described changes in breathing mechanics during general anesthesia (for urological surgery or repair of inguinal hernias) in 63 children ages three weeks to 15 years who were considered to have normal lungs (26). Compliance increased with body size, scaling to height by the exponent 1.76. Airway resistance decreased with respect to body height by the exponent −1.29. These cross-sectional findings indicate that compliance and resistance change as children grow but that these factors do not evolve at the same rate.

Whether these observations apply to the effects of compliance and resistance on ventilatory work at high work intensities is unknown. The findings suggest that the metabolic expenditure of the muscles of respiration to meet the ventilatory demands of exercise might be greater in small children, with improvements in energy efficiency during the growing years. The extent to which compliance changes during exercise in adults is controversial. Decreased lung compliance has been described and attributed to the increase in pulmonary blood volume. Others have found no changes or even an increase in compliance with exercise (49).

Patterns of Breathing Frequency and Tidal Volume

Another aspect of breathing mechanics that can affect the efficiency of ventilation is the relative contributions of breathing frequency and tidal volume to achieve a given minute ventilation. A greater reliance on f_R (i.e., a higher f_R/V_T ratio) increases the ventilatory work needed to overcome airway resistance. On the other hand, a large V_T is generated at the cost of a greater effort to overcome lung and chest wall elastic forces. The elastic work performed in accomplishing a tidal volume of 1 L is four times that required for a tidal volume of 0.5 L (49). As suggested by Whipp and Ward, the body adjusts the relative contributions of f_R and V_T "in such a way that the work necessary will be minimized" (49, p. 20).

These considerations dictate the relative contributions of breathing frequency and tidal volume in the course of a progressive exercise test. Observations in adults indicate that in order to limit the work needed to overcome elastic forces, tidal volume rises to only about 60% of vital capacity at moderate work loads and then plateaus. At higher intensities, increases in ventilation are provided by a greater breathing rate.

Some data suggest that these patterns of the responses of f_R and V_T are different in children than

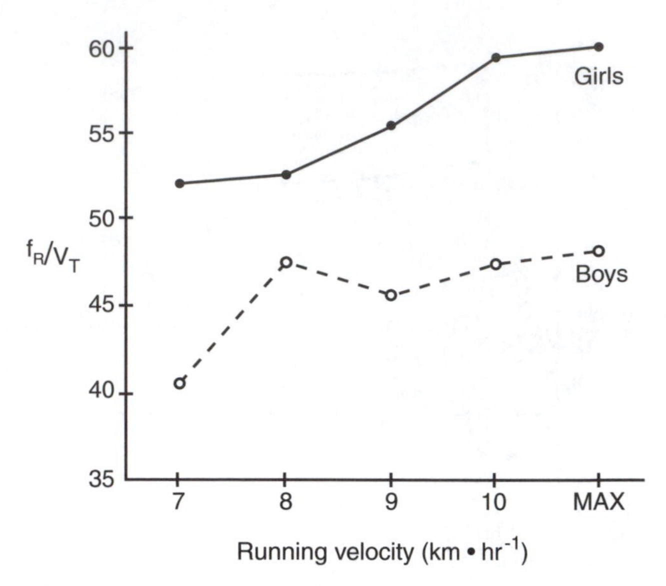

FIGURE 7.4 Increases in mean values of f_R/V_T during progressive exercise in boys and girls (data from reference 3).

in adults; if so, this would imply that the relative influences of changes in airway resistance and lung compliance during exercise are affected by biologic maturation. Boule et al., for example, found that V_T rose linearly during maximal cycle testing in children ages 6 to 15 years, while f_R plateaued at 67% of peak exercise (9). Similarly, Rowland and Green found that the ratio of f_R to V_T fell at high treadmill intensities in 10- to 14-year-old boys (42).

Other information indicates a typical adult pattern. Armstrong et al. described a rise in both V_T and f_R during a progressive treadmill test in 8- to 11-year-old boys and girls (3). As exercise intensity rose, a greater reliance on f_R to create $\dot{V}_E$ was indicated by an increase in the f_R/V_T ratio (figure 7.4). Between 67% and 85% $\dot{V}O_2$max in the boys, for example, the ratio rose from 40.5% to 47%. During this rise in work intensity, V_T as a percentage of vital capacity remained relatively stable (38.2% and 41.1% at the start and end of this range of intensities).

Dead-Space Ventilation

The other aspect of ventilatory volumes that needs consideration is a possible change in relative anatomic or physiologic dead-space ventilation (V_D) in growing children. Alveolar ventilation is $V_E - V_D$, so changes in unventilated airways with respect to increases in body size might be expected to influence gas exchange. The data are quite clear, though, that

V_D/V_T in a given breath remains constant during the pediatric years.

Resting dead-space ventilation is closely related to body weight throughout childhood, with a value of approximately 1.0 $cm^3 \cdot lb^{-1}$ (38). Shephard and Bar-Or found no significant differences in calculated relative dead space between preadolescent children and adults exercising at 80% $\dot{V}O_2$max (46). Similarly, Gadhoke and Jones described no age-related differences in V_D/V_T in boys ages 9 to 15 years who were cycling at identical work loads (17). Average values were 0.21, 0.18, and 0.14 at 400, 600, and 800 $kp \cdot m \cdot min^{-1}$, respectively.

Control of Ventilation

Not long ago, ventilatory responses to exercise were considered to be under the control of a chemoreceptor feedback system, sensitive to changes in arterial PO_2 (hypoxic drive), PCO_2 (hypercapneic stimulus), and $[H^+]$. It is now recognized that this concept is overly simplistic and that "exercise PCO_2 and PO_2 seem to be regulated functions more than primary regulating factors" (5, p. 25). The principal means by which exercise stimulates hyperpnea, however, remains uncertain. Humoral mechanisms seem improbable, given the very rapid rise in $\dot{V}_E$ with the onset of exercise. Most researchers favor a neurologic explanation, either a primary descending drive from the brain that is proportional to stimuli for locomotor activity or ascending reflexes from contracting skeletal muscle (13, 14). Forster concluded that "the mechanisms of exercise hyperpnea remain controversial because investigators have yet to devise an ideal preparation to study the phenomenon. This failure is at least partly because the basic ventilatory control system is extremely complex and poorly understood" (16, p. 134).

Considerable evidence exists that ventilatory drive has a lower PCO_2 set point in children than in adults. This conclusion is based primarily on observations that the slope of the relationship between $\dot{V}_E$ and $\dot{V}CO_2$ becomes flatter as children grow (2, 12). Also, end-tidal PCO_2 (an indicator of arterial PCO_2) during exercise is lower in children than in adults (2). Gratas-Delamarche et al. compared ventilatory responses to inhaled CO_2 in prepubertal boys and young adults (20). The children exhibited a lower CO_2 sensitivity threshold (the value of end-tidal PCO_2 when ventilation increased above a steady-state level). The children also had a steeper slope in the linear relationship between minute ventilation and end-tidal PCO_2. Gaultier et al. found that the mouth pressure generated 0.1 s after airway occlusion decreased with age between the ages of 4 to 16 years (18). These authors felt that the greater pressures observed in the younger children were indicative of a higher level of neural ventilatory drive.

More recently, Ohuchi et al. compared ventilatory and blood gas responses to maximal treadmill exercise between seven boys ages 8 to 11 years and six young males ages 14 to 21 years (36). Maximal respiratory rate was higher in the boys (66 ± 6 vs. 46 ± 4 breaths per minute in the older males), but no significant differences were observed in maximal tidal volume with respect to body mass. At peak exercise, arterial pH was greater in the boys than in the men (7.33 ± .01 vs. 7.28 ± .01), but arterial PO_2 values were similar in the two groups. At both the anaerobic threshold and peak exercise, however, arterial PCO_2 was significantly lower in the boys (figure 7.5). The boys demonstrated a smaller improvement in the ratio of physiologic dead space to tidal volume as exercise progressed, but there were no significant differences between the two groups in total alveolar ventilation. These accumulated data indicate that the homeostatic set point for arterial PCO_2 is lower in children than adults. But why this should be so is problematic.

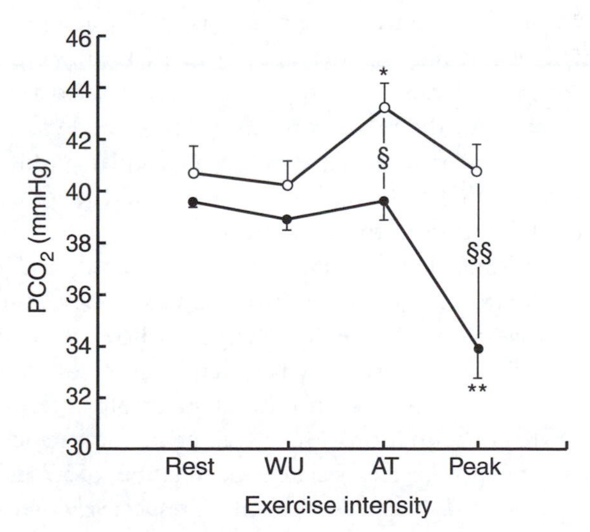

FIGURE 7.5 Changes in PCO_2 with increasing exercise intensity in children *(solid circles)* and postpubertal subjects *(open circles)*. (From reference 36.) WU = work below anaerobic threshold; AT = anaerobic threshold; * = $p < 0.05$; ** = $p < 0.01$ vs. rest; § = $p < 0.05$ between two groups; §§ = $p < 0.01$ between two groups.

Reprinted by permission from Ohkwwa et al. 1999.

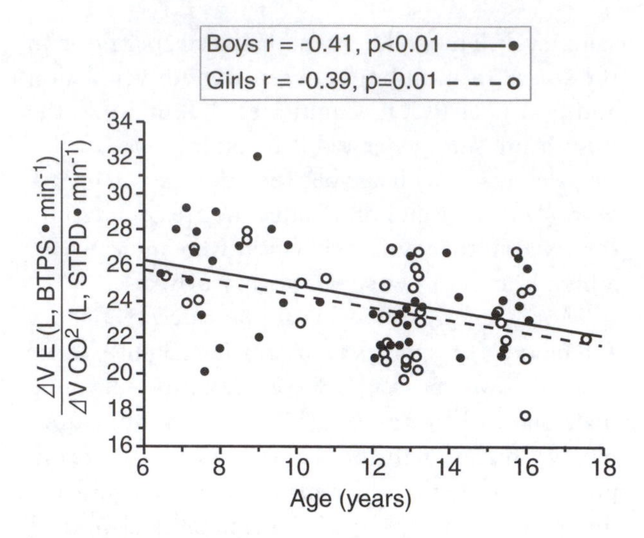

FIGURE 7.6 Age-related changes in the relationship between $\dot{V}_E$ and $\dot{V}CO_2$ (from reference 34).
Reprinted by permission from Nagano et al. 1998.

The findings of Nagano et al. support the idea that young children breathe more during exercise to eliminate a given amount of CO_2, and this results in a lower arterial PCO_2 set point (34). These authors examined the relationship between age and ventilatory control during treadmill exercise in 80 children ages 6 to 17 years. Both at rest and at peak exercise, no correlation was observed between age and either ratio of $\dot{V}_E$ to CO_2 production, ratio of effective alveolar ventilation to CO_2 production, ratio of lung dead space to tidal volume, or arterial PCO_2. At the ventilatory anaerobic threshold, arterial PCO_2 increased with age, while the ratio of alveolar PCO_2 to $\dot{V}CO_2$ decreased. Linear regression analysis indicated that the slope of the relationship between $\dot{V}_E$ and $\dot{V}CO_2$ (figure 7.6) as well as that between alveolar ventilation and $\dot{V}CO_2$ decreased with age. There were no sex differences in these findings.

Children demonstrate lower lactate levels and a smaller fall in arterial pH during a maximal exercise test than adults do. Evidence suggests, however, that the mild decline in pH with exercise in children is not consistent with their more dramatically inferior lactate concentrations. Hebestreit et al., for instance, reported postexercise lactate concentrations of 5.7 and 14.2 mmol · L^{-1} in children and men, respectively, after a 30-s all-out cycle test (23). Average venous pH values were 7.32 in the children and 7.18 in the adults.

Ratel et al. hypothesized that this limited decline in pH might be due to a different time course of regulation of arterial PCO_2 by ventilation in children (39). To investigate this question they compared acid-base findings in 11 boys (9.6 ± 0.7 years) and 10 men (20.4 ± 0.8 years) during several repeated bouts of short-term, high-intensity exercise. After the 10th sprint, mean lactate concentration was 8.5 ± 2.1 mmol · L^{-1} in the boys and 15.4 ± 2.0 mmol · L^{-1} in the men, with respective $[H^+]$ of 43.8 ± 11.3 and 66.9 ± 9.9 nmol · L^{-1}. At a given lactate level, $[H^+]$ was lower in the boys than in the men (figure 7.7). Also, arterial PCO_2 values determined at a given $[H^+]$ were less in the boys than in the men (figure 7.8). The authors concluded that children regulate their blood $[H^+]$ better than adults, and this may be due to a greater ventilatory compensatory response.

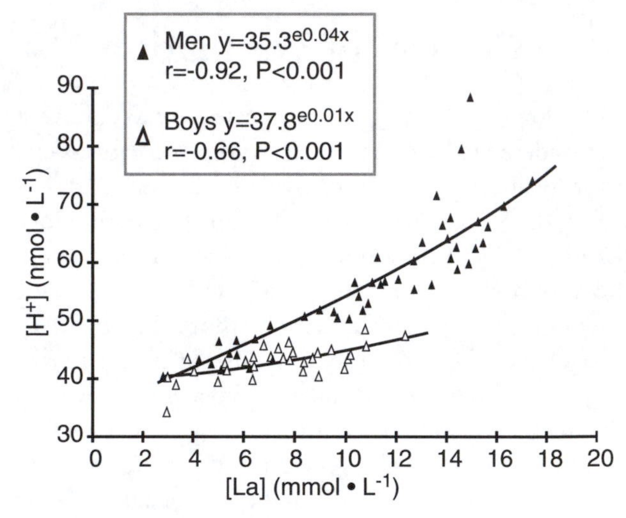

FIGURE 7.7 Relationship between $[H^+]$ and lactate concentration, [La], during 10 sprint exercises in boys and men (from reference 39).
Reprinted by permission from Ratel et al. 2002.

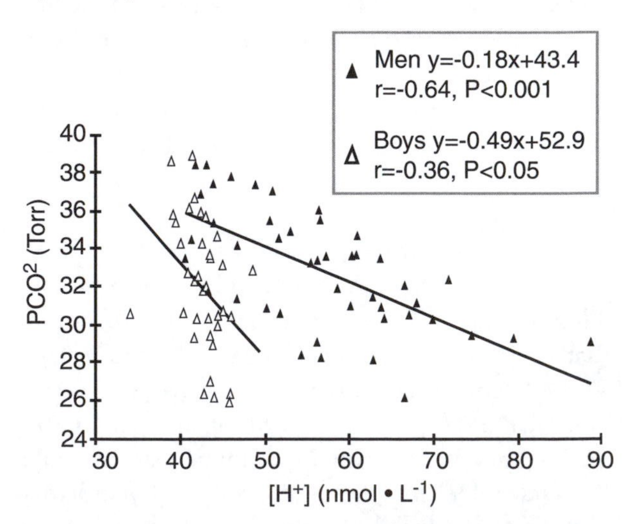

FIGURE 7.8 Relationship between arterial PCO_2 and blood $[H^+]$ during 10 sprint exercises in boys and men (from reference 39).
Reprinted by permission from Ratel et al. 2002.

Ventilatory drive was found to be significantly influenced by genetic factors in a study comparing sensitivity to hypercapnia between pairs of adolescent monozygotic and dizygotic twins (25). There is evidence, too, that sports training may influence ventilatory responses to CO_2. Some studies in adult athletes have shown a lower ventilatory drive than in nonathletes (10), but others have been unable to confirm this finding (30). More specifically, studies have suggested that adult endurance athletes have a low ventilatory response to CO_2, while those excelling in anaerobic sports (e.g., sprinting) have high responses (35).

McGurk et al. examined this issue in a group of 12-year-old swimmers who were involved in regular practice but were not considered heavily trained (31). Based on ventilatory responses to inhaled CO_2, subjects were divided into two groups, high and low responders. They then underwent two sprint tests (50-m run, alactic peak power test) and two endurance tests (1.6-km run, cycle PWC_{170}). Results were mixed. Sprinting ability in the high responders was significantly better than in the low responders on the alactic power test, but no differences were observed in 50-m run time. There were no differences between groups in the endurance tests. The authors could only conclude that these findings were inconclusive and that further research is required.

Ventilatory Equivalent for Oxygen

Children also hyperventilate in relation to the aerobic metabolic requirements of physical work. This is manifested by their greater ventilatory equivalent for oxygen ($\dot{V}_E/\dot{V}O_2$) at all levels of exercise (29). A gradual decline in $\dot{V}_E/\dot{V}O_2$ occurs continuously through the childhood years, as demonstrated by the cross-sectional study of Andersen et al. (1). They reported a gradual decline in $\dot{V}_E/\dot{V}O_2$ at maximal exercise between the ages of 8 and 16 years, with values falling from about 34 to 24. No differences were seen in these changes between boys and girls.

Similarly, Armstrong et al. could find no sex-related differences in $\dot{V}_E/\dot{V}O_2$ during treadmill running (3). Values for boys and girls at 70% to 75% peak $\dot{V}O_2$ were 23.3 ± 2.9 and 24.1 ± 2.4, respectively, and 29.8 ± 3.2 and 29.5 ± 3.1 at peak exercise. However, in the longitudinal study of Rowland and Cunningham, boys demonstrated a lower submaximal $\dot{V}_E/\dot{V}O_2$ than girls, but with a similar rate of decline over five years (41). At maximal exercise, mean $\dot{V}_E/\dot{V}O_2$ declined from 37.2 to 34.1 in the boys, but no changes were observed in the girls (39.6 to 39.4). Godfrey also reported that girls have a greater ventilatory equivalent for oxygen relative to metabolic requirements at maximal exercise than boys do except at very young ages (19).

In any such comparisons at high exercise intensities, it must be remembered that $\dot{V}_E/\dot{V}O_2$ is influenced by degree of lactate production. As Godfrey pointed out, any age or sex comparisons of $\dot{V}_E/\dot{V}O_2$ at maximal exercise might be influenced by group differences in anaerobic metabolism (19).

Ventilation Kinetics at Exercise Onset

Other evidence for maturational differences in ventilatory control comes from studies examining respiratory dynamics at the onset of exercise. At the start of constant-load exercise in adults, the time constant of change for $\dot{V}O_2$ is typically 20 to 40 s, and steady state is reached by about three minutes. $\dot{V}CO_2$ rises more slowly, and $\dot{V}_E$ is delayed even more, not reaching steady state for four to five minutes.

Welsman et al. showed that children's rate and magnitude of hyperpneic response were enhanced in comparison with adults' (48). These authors studied ventilatory adjustments at onset of exercise in 12 boys and 10 girls (11–12 years old) and 12 men and 9 women (19–26 years old). Subjects completed a ramp protocol with no less than four square-wave transitions. The time constants for $\dot{V}O_2$, $\dot{V}CO_2$, and $\dot{V}_E$ were all significantly shorter for the children than for the adults. Increase in $\dot{V}CO_2$ lagged behind that of $\dot{V}O_2$ by a longer period in the adults than in the children (in boys by 23.8 ± 5.9 s; girls, 22.9 ± 8.4 s; men, 44.9 ± 14.5 s; women, 37.1 ± 7.6 s). The slope of the change in $\dot{V}_E$ with $\dot{V}CO_2$ was significantly greater in the children than in the adults and in the women compared with the men.

This study supports previous findings. Cooper et al. found a 30% faster rate of rise in $\dot{V}_E$ and $\dot{V}CO_2$ in 7- to 10-year-old children compared with 15- to 18-year-old adolescents (12). Because CO_2 production relative to a particular level of metabolic work should be age independent, the authors suggested that their finding might be explained by a lower storage of CO_2 in prepubertal children.

Armon et al. reported no significant differences between adults and children in whole-body CO_2 stores at rest measured using bicarbonate tracer techniques (2). However, the findings of Zanconato et al.

suggest that this might not be true during exercise (50). They measured CO_2 stores in boys ages 6 to 10 years and young adults ages 21 to 39 years. No differences were seen between the two groups at rest. During exercise, however, CO_2 stores increased by 31% in the adults but remained unchanged in the children. These authors suggested that adults may store more CO_2 because of a greater amount of body fat (which dissolves CO_2) or higher levels of hemoglobin (which binds CO_2).

Maintaining Normoxemia

At rest, the alveolar PO_2 in adults is approximately 100 mmHg; the drop from 160 mmHg in the ambient atmosphere results from humidification of air as it passes through the upper airways. Because of ventilation–perfusion inequalities in the lung, the arterial PO_2 is about 90 mmHg. During exercise, increases in minute ventilation serve to replenish oxygen in the alveoli as diffusion of oxygen across the alveolar-capillary membrane accelerates. As a result, in the untrained adult the arterial PO_2 is maintained at 90 mmHg, even at high exercise intensities. This occurs despite an increase in the alveolar-arterial O_2 difference to 20 or 30 mmHg, an effect of poorer matching of ventilation and perfusion.

It generally takes about 0.3 s for PO_2 equilibrium to occur between the alveoli and pulmonary capillaries. In a resting adult there is plenty of time to accomplish this, since mean red cell transit time in the pulmonary capillaries is approximately 0.75 s. With severe exercise, transit time shortens by approximately one half, so time for oxygen exchange does not normally limit arterial oxygenation at maximal exercise.

In highly trained adult athletes, the demands of higher metabolic and cardiovascular activity exceed the ability of ventilatory responses to maintain normoxemia at high exercise intensities (14). At maximal exercise a typical athlete has a pulmonary blood flow that is up to 50% greater than that of a nonathlete. A shortened red cell transit time may therefore limit the time available to accomplish full oxygen exchange. In addition, the alveolar-arterial gradient increases to 34 or 45 mmHg, compared with 20 or 30 mmHg in the untrained individual (i.e., athletes have a greater ventilation–perfusion inequality with exercise). Accumulation of lung fluid may also occur in the highly trained individual. As a consequence, arterial hypoxemia is observed in these athletes at maximal exercise, when they have an arterial PO_2 of 55 to 75 mmHg.

This fall in arterial oxygen level at maximal exercise in the endurance athlete may have a significant negative impact on maximal arterial-venous oxygen difference, resulting in a fall in potential $\dot{V}O_2$max. A typical reduction in oxygen saturation to 86% or 92% is reflected in a 6% to 8% lower $\dot{V}O_2$max. At the extreme of this effect, athletes whose saturation falls to 84% or 86% can expect a 12% to 14% reduction in maximal aerobic power (14).

While studies of these factors have been performed largely in males, there is evidence that arterial hypoxemia also occurs at peak exercise in adult female athletes. Harms et al. reported that 40% of healthy young distance runners demonstrated a decrease in oxygen saturation at both submaximal and maximal exercise (22). In a second study with the same subjects, mean arterial oxygen saturation fell to 91.8 ± 0.4 % at maximal exercise (21). By administering oxygen and observing changes in oxygen saturation, they estimated that the resulting decrease in $\dot{V}O_2$max was 6.3%.

Arterial Oxygenation in Children

Limited data are available in children. Eriksson et al. measured arterial blood gases during progressive maximal cycle testing in six boys ages 13 to 14 years (15). Average PO_2 was 98 mmHg at rest, 95 mmHg at 600 km · h^{-1}, and 94 mmHg at exhaustion. Alveolar-arterial oxygen difference rose from 8 mmHg at rest to 14 mmHg at submaximal exercise to 24 mmHg at maximal exercise. These values are similar to those typically seen during exercise testing of adults.

Laursen et al. studied the incidence of exercise-induced arterial hypoxemia in premenarcheal girls (27). Nineteen nontrained girls age 11.1 ± 1.6 years performed a maximal cycle test, and percentage arterial oxyhemoglobin saturation was estimated using an ear oximeter. Peak $\dot{V}O_2$ for the group was 43.7 ± 7.1 ml · kg^{-1} · min^{-1}, and peak minute ventilation was 66.3 ± 12.5 L · min^{-1}. The average percentage arterial saturation during the final 30 s of exercise was 96.6 ± 1.2 % (95% confidence limits: 95.5% to 97.7%). No relationship was observed between change in oxygen saturation and peak $\dot{V}O_2$ (figure 7.9).

These findings suggest that exercise-induced arterial hypoxemia does not occur in untrained boys and girls. How well normoxemia is maintained during

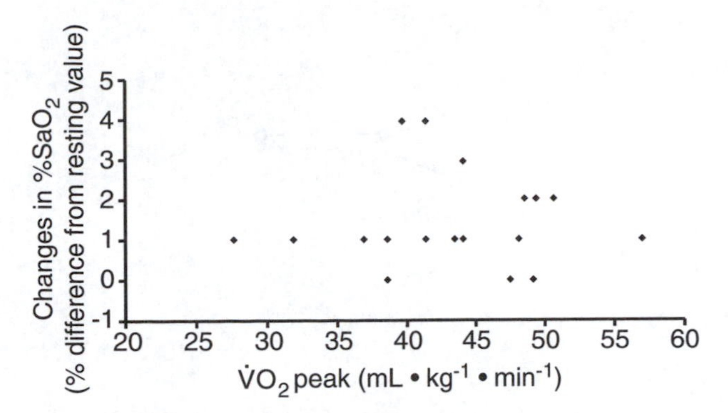

FIGURE 7.9 Changes in percentage oxygen saturation ($\%SaO_2$) from rest to maximal exercise in relation to peak $\dot{V}O_2$ in adolescent girls (from reference 27).

From P.B. Laursen et al., "Incidence of exercise-induced arterial hypoxemia in prepubescent females," Pediatric Pulmonology 34:37-41 copyright © (2002, John Wiley and Sons, Inc.). Reprinted by permission of Wiley-Liss, Inc., a subsidiary of John Wiley & Sons, Inc.

exercise in child athletes—and the effect of any changes on $\dot{V}O_2$max—remains to be investigated.

Lung Diffusion Capacity

The ability of gases to cross the alveolar-capillary interface is usually estimated by measurement of the diffusion capacity of carbon monoxide (D_LCO). This value increases with exercise, and some data suggest that lung diffusion capacity might be greater in children than adults. Shephard et al. found that the slope of the linear $D_LCO{:}\dot{V}O_2$ ratio during exercise was approximately twice that observed in studies of adults (45). However, Johnson et al. did not observe any differences in membrane diffusing capacity at maximal exercise in a comparison of four children 8 to 12 years old and six subjects ages 15 to 28 years (24).

Prolonged Steady-State Exercise

When an adult performs prolonged, steady-load exercise (45–60 minutes) at a moderate exercise intensity, a set of changes termed *ventilatory drift* is observed (13). Breathing frequency increases by 15% to 40%, while tidal volume declines 10% to 15%. A small rise is observed in $\dot{V}_E$, which is out of proportion to $\dot{V}O_2$ and $\dot{V}CO_2$. Arterial blood gases in this situation typically reflect a mild respiratory alkalosis (arterial pH is typically 7.40 to 7.50) with a small depression in arterial PCO_2.

Traditionally, the patterns observed in ventilatory drift have been ascribed to the effects of increased core temperature. Dempsey et al. described it like this:

> During prolonged heavy work a significant ventilatory stimulus, linked to metabolic heat production, is implicated by the magnitude of the observed increases in core and blood temperatures and by the predominant tachypneic breathing pattern. This almost "panting-like" tachypnea, with increasing dead-space ventilation, is a unique response to ventilatory stimuli in humans, in whom the major mode of response to chemoreceptor stimuli and to exercise, per se, is a rising tidal volume. (13, p. 101)

This interpretation implies that increases in breathing frequency in response to elevations in core temperature have a primary role in ventilatory drift. The greater f_R/V_T at a given $\dot{V}_E$ increases minute dead-space ventilation ($\dot{V}_D$), possibly contributing to a compensatory rise in V_E. Other etiologic mechanisms for ventilatory drift have also been suggested, including responses to elevated circulating epinephrine levels, pulmonary venous congestion, respiratory muscle fatigue, and psychologic factors.

Rowland and Rimany compared ventilatory drift in 9- to 13-year-old premenarcheal girls and young women ages 20 to 31 years (43). Subjects cycled for 40 minutes at an average intensity of 63% $\dot{V}O_2$max. No significant differences in the increase in $\dot{V}_E$ (7.1% for the girls and 11.7% for the women) were observed between the two groups (figure 7.10). Breathing frequency rose 15% in the girls and 14% in the women, while V_T fell by 6.0% and 2.0%, respectively. No differences in rise of tympanic membrane temperature between the two groups were recorded.

These findings are similar to those reported by Asano and Hirakoba in a comparison of boys and men (4). These limited data that show no maturational differences in ventilatory drift therefore suggest that respiratory sensitivity to the triggering mechanism for these changes (probably core temperature) is similar in children and adults.

Conclusions

The basic patterns of ventilatory response to exercise are similar in children and adults. However, there are certain anatomic and functional characteristics of prepubertal children that distinguish their ventilatory changes during exercise from those of mature individuals. The explanation for these differences is

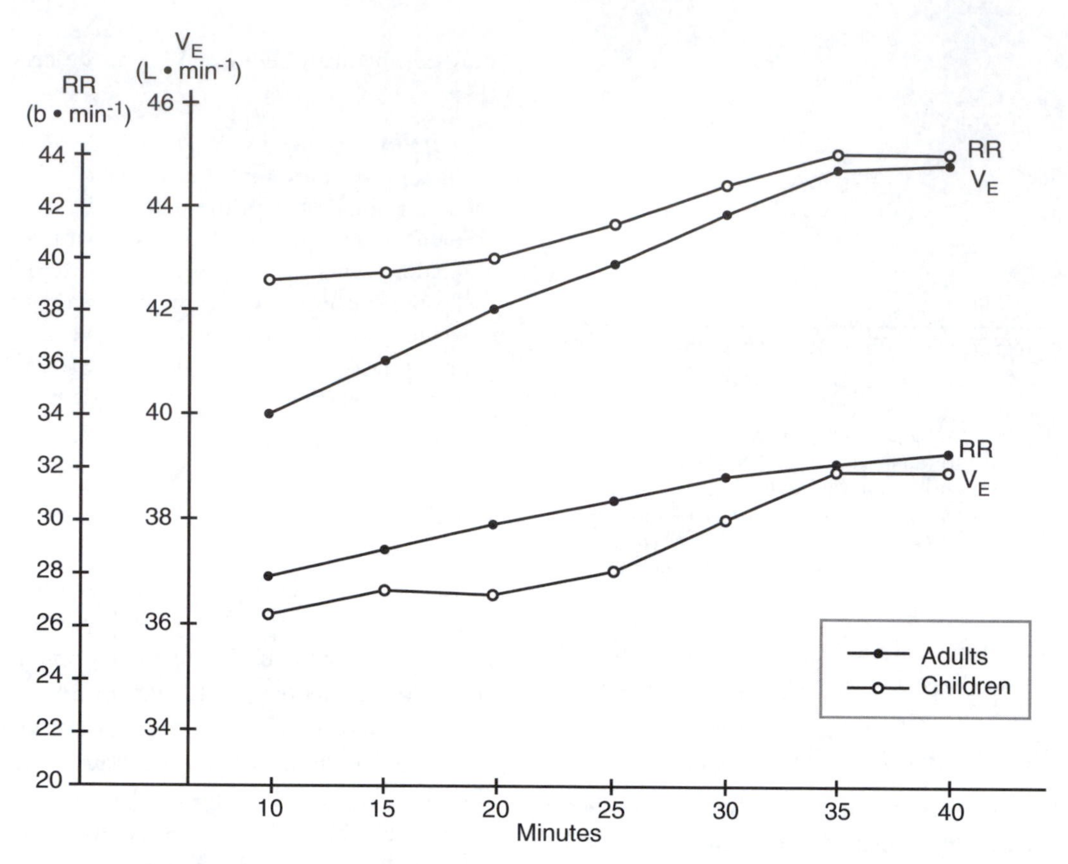

FIGURE 7.10 Changes in respiratory variables during sustained steady-load cycling at 63% $\dot{V}O_2$max in girls and women. $\dot{V}_E$ = minute ventilation; RR = respiratory rate (from reference 43).

Reprinted by permission from T.W. Rowland and T.A. Rimany 1995.

not straightforward, and understanding the quantitative and qualitative variations in children's responses remains a challenge for pediatric exercise physiologists. Likewise, the implications for athletic performance or, at the other end of the spectrum, chronic lung disease of maturational differences in ventilatory responses to exercise remain speculative.

Children hyperventilate during exercise compared with adults. There is abundant evidence to support this conclusion: Their $\dot{V}_E$ for a given metabolic rate ($\dot{V}_E/\dot{V}O_2$) is higher at all intensities of exercise. Arterial PCO_2 is lower and pH greater at maximal exercise in young people, and the slope of the relationship between $\dot{V}_E$ and $\dot{V}CO_2$ decreases with age. Although children develop less metabolic acidosis during exercise (as a result of lower lactate production), their ventilatory compensatory response is exaggerated. Ventilation kinetics at the onset of exercise are more rapid in children than in adults. This hyperventilatory response becomes less obvious as children grow, suggesting that factors related to biologic maturation that do not include the humoral influences of puberty are involved.

Children breathe more rapidly than adults to achieve a particular minute ventilation. This might be assumed simply to reflect that children have smaller lungs (and therefore smaller tidal volumes). However, a greater f_R/V_T has been observed even when $\dot{V}_E$ is adjusted for age and body size. A higher f_R/V_T might be expected to raise minute dead-space ventilation, possibly contributing to a relatively greater $\dot{V}_E$ as a compensatory mechanism.

Lung compliance is less and airway resistance greater in children, but these approach adult values in the later childhood years. The influence of these factors on ventilatory muscle work as children grow is unknown.

Normal arterial oxygenation is maintained in children at maximal exercise. It is not known if trained child endurance athletes experience exercise-induced hypoxemia as adults do. This issue is important, since a fall in oxygen saturation at peak exercise can have a significant negative effect on maximal aerobic power.

Children demonstrate characteristics of ventilatory drift during sustained steady-state exercise similar to those of adults. Assuming that these changes reflect a response to rises in core temperature, this finding implies that (a) temperature increases are independent of maturational status and (b) the link between temperature and increase in f_R is similar in children and adults.

Discussion Questions and Research Directions

1. What is the explanation for children's relative hyperventilation with exercise compared to adults? What does this observation tell us about the mechanics responsible for the hyperpnea of exercise?
2. Do trained child athletes experience exercise-induced hypoxemia? If so, what is the mechanism? And how much is aerobic fitness compromised?
3. Are there maturational differences in lung diffusion capacity during exercise?
4. How much do changes in the mechanical properties of the lungs (compliance, resistance) as children grow affect the metabolic demands of ventilatory musculature during intense exercise?
5. How important are changes in minute dead-space ventilation that might result from the greater f_R/V_T in children?
6. How can $\dot{V}_E$ and V_T values during exercise best be related to differences in body size?

8

Energy Demands of Weight-Bearing Locomotion

The energy cost of travel increases with size, but the unit cost is cheaper by the truckload (cheaper yet by the trainload).

—William Calder (1984)

In this chapter we discuss

- how the energy cost of locomotion relative to body size decreases with age,
- mechanisms that might account for age-related improvements in exercise economy in weight-bearing activities, and
- the influences of sex, physical fitness, and running grade on exercise economy.

Children's greater energy need to move body mass at a certain running or walking speed is perhaps the most consistently documented aspect of pediatric exercise physiology. Robinson was the first to describe this maturational difference in economy of locomotion in his prescient work at the Harvard Fatigue Laboratory in 1938 (39). Groups of male subjects ranging in age from 6 to 75 years walked on a treadmill at 5.6 km · h^{-1}, 8.6% grade, for 15 minutes. $\dot{V}O_2$ fell from an average of 33.0 ml · kg^{-1} · min^{-1} in the youngest group to 23.9 ml · kg^{-1} · min^{-1} in the oldest. During the childhood years, mean mass-relative $\dot{V}O_2$ declined 16% by the late teen years. This investigation indicated, then, not only that children are less economical than adults at a given treadmill setting but also that improvement in submaximal energy economy occurs continuously through the growing years.

Since that time this trend during weight-bearing exercise has been documented by multiple authors. In his review published in 2000, Morgan outlined information from 17 studies that compared walking or running economy either between children and adults or between younger and older children (30). As experimental conditions and age differential varied widely in these reports, so did reported differences in $\dot{V}O_2$ per kilogram between study groups. However, the magnitude of differences between children and adults in mass-specific energy cost for treadmill locomotion was similar to that described by Robinson (39) almost 70 years ago (approximately 15% to 20%).

While the great majority of these studies were cross-sectional, longitudinal investigations have revealed similar changes in economy over time. In the Amsterdam Growth, Health, and Fitness Study, $\dot{V}O_2$ during treadmill running in males decreased from 37.6 to 30.3 ml · kg^{-1} · min^{-1} between the ages of 13 and 27 years (54). Respective values in females were 36.5 to 29.8 ml · kg^{-1} · min^{-1}. In their longitudinal study of treadmill walking, Rowland et al. reported a decline in energy demand from 31.0 ml · kg^{-1} · min^{-1} at age 9 years to 26.5 ml · kg^{-1} · min^{-1} at age 13 (44). Based on these longitudinal and cross-sectional data, it is possible to predict that $\dot{V}O_2$ per kilogram during a given level of treadmill exercise will decrease during childhood by an average of about 1.0 ml · kg^{-1} · min^{-1} each year.

The Meaning of Economy: Is Allometry Appropriate?

There is some confusion and conflict of opinion regarding how these developmental changes in metabolic requirements relative to body mass with submaximal locomotion should be interpreted. Many view this simply as a change in energetics to move body mass. That is, exercise economy is defined as the amount of energy expended per kilogram at a given treadmill setting, and body mass serves as the exercise load. That is, "per kilogram" is interpreted in the context of the load that needs to be moved. The observation, then, that economy improves with age is taken as an indication that at a given level of work, younger children need to use a greater amount of energy to move a given load of body mass. From this viewpoint, it seems that analysis of economy differences between individuals by allometry might not be appropriate.

An alternative perspective, however, suggests that differences in $\dot{V}O_2$ per kilogram during locomotion in childhood might reflect a broader relationship between energy utilization and body size. This is suggested by observations of exercise economy in animals, particularly the similar allometric relationships seen between $\dot{V}O_2$ and body dimensions at rest and during submaximal work. We have iterated several times in these pages that resting or basal metabolic rate per kilogram of body mass is inversely related to body size, with a mass scaling exponent of 0.75 (47). While the explanation is unknown, a popular concept holds that energy production is related to body surface area to match heat loss and thus maintain thermal homeostasis (the surface rule). Thus, the metabolic rate at rest is geared to body surface area rather than body mass, and since smaller animals' ratio of body surface area to mass is greater, they have a higher resting metabolic rate per kilogram of mass than larger ones. The more intense "metabolic fire" of small animals is considered a reflection of greater metabolic rate per cell as well as a relatively larger contribution by organs with higher metabolic rates (see chapters 4 and 5).

When an animal performs submaximal exercise, these relationships do not change. The mass scaling exponent for $\dot{V}O_2$ for submaximal locomotion is approximately 0.67. This observation suggests that the factors responsible for the relationship of $\dot{V}O_2$ and body mass at rest and during exercise might be similar. That is, metabolic rate might be regulated in respect to a separate factor—such as body surface area—rather than the mass being transported. This possibility is suggested by the observation that child–adult differences in mass-specific submaximal energy cost are eliminated when $\dot{V}O_2$ is expressed relative to body surface area (14, 40, 41). And, as in animals, the mass scaling exponent of 0.58 described by Holliday et al. (18) for $\dot{V}O_2$ at rest in children is similar to the value of 0.65 empirically observed in children during submaximal exercise (40). It might be, then, that metabolic arrangements defined at rest in respect to body dimensions simply persist when metabolic rates increase in response to exercise.

These two viewpoints are not dissimilar from the alternative interpretation of maximal aerobic power. Does $\dot{V}O_2$max per kilogram of body mass indicate (a) the oxygen requirement to move a kilogram of body mass at exhaustive exercise or (b) the highest oxygen utilization, normalized to body size?

This issue becomes important when we wish to compare running economy between two groups of subjects. If we accept that economy reflects strictly an energy need to move mass, it could be argued that allometric analysis is not appropriate. On the other hand, the alternative interpretation of economy differences during exercise—as a metabolic response keyed to body surface area that reflects mechanisms at rest—seems to require an allometric approach for valid group comparisons.

It could be, too, that the interpretation we use doesn't make any difference. When Davies et al. compared the use of the ratio standard, allometric scaling, and ANCOVA to compare running economy in adult male and female distance runners, the results were the same (no sex difference), regardless of the means of statistical adjustment (13). In children, Walker et al. (55), Armstrong et al. (2), and Welsman and Armstrong (58) all found that group comparisons of running economy were not affected by use of either the ratio standard or allometric analysis.

Stride Frequency

Biologists have long recognized that smaller animals run less economically than larger ones. Figure 8.1 shows how oxygen consumption is related to body mass with increasing running speed among animals of greatly different sizes (47). Quite obviously, the smaller the animal, the more energetically expensive locomotion is for a particular level of work. Not only is the $\dot{V}O_2$ per kilogram at a given speed inversely proportional to body mass, but the rate of rise in mass-specific energy demands with greater running velocity increases as animal size diminishes.

Biologists have chosen to express running economy in animals by the slope of the relationship between $\dot{V}O_2$ per kilogram and running speed (in milliliters per kilogram per kilometer). This value is independent of velocity, and its use allows comparisons of energy economy between animals that habitually move at different speeds. A plot of this slope versus body mass is depicted in figure 8.2. Again, we see clear evidence that the energy expended to move body mass is indirectly related to animal size.

This relationship between body size and economy is precisely that seen between children and adults (figure 8.3). Moreover, treadmill running or walking at a given speed becomes progressively cheaper from an energy standpoint during the childhood years. From the parallel phylogenetic observations in adult animals, we can infer that (a) the improvement in running economy as children grow reflects a

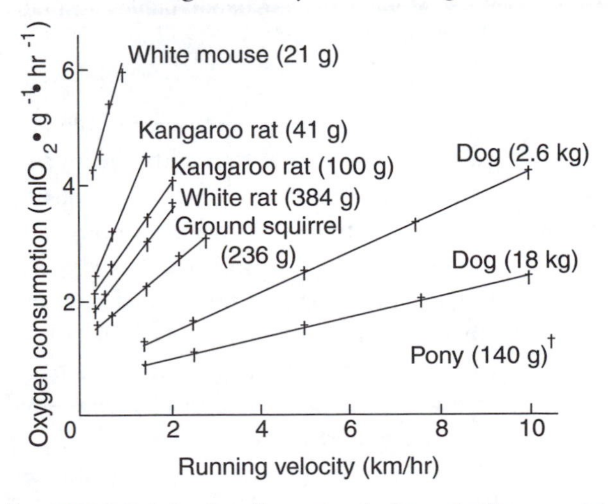

▶ FIGURE 8.1 Smaller animals have both a greater mass-specific metabolic requirement at a given running velocity and a steeper increase in $\dot{V}O_2$ than larger animals (from reference 47).

Reprinted by permission from C.R. Taylor, N.C. Heglund, and G.M.O. Maloiy 1982.

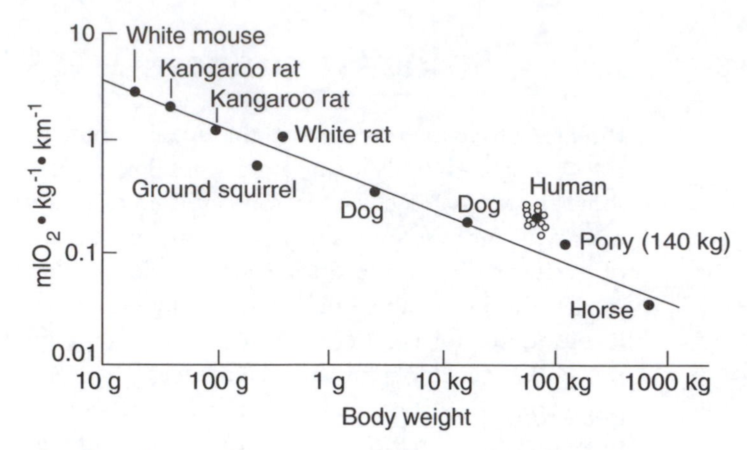

FIGURE 8.2 The energy cost of running (oxygen demand for transporting 1 kg over 1 km) falls with increasing body size in animals (from reference 47).

Reprinted by permission from C.R. Taylor, N.C. Heglund, and C.M.O. Maloiy 1982.

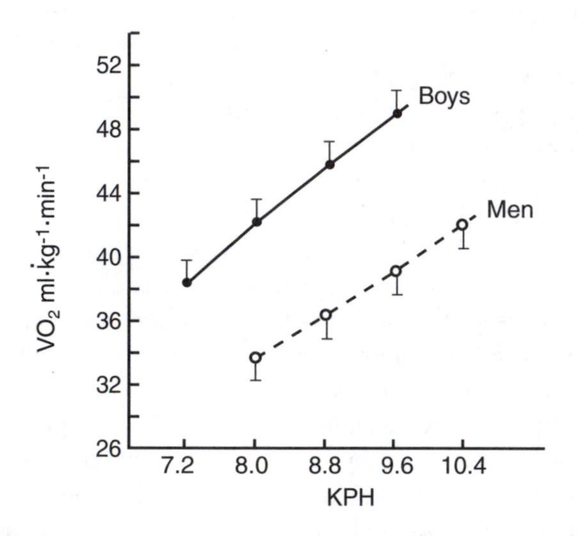

FIGURE 8.3 Submaximal treadmill running economy in boys and men (from reference 41).

Reprinted by permission from T.W. Rowland 1987.

broad biologic principle, not simply a unique change occurring during human growth, and (b) the mechanism responsible for this trend in economy during childhood somehow reflects increases in body size.

By the traditional explanation, the changes in animals' running economy in respect to body size are due to the fact that, at a given velocity, smaller animals have a greater stride frequency than larger ones. This is based on the assumption—backed by empiric evidence—that the energy needed to perform a single stride should be independent of animal size. Taylor et al. documented this, calculating that the metabolic energy expended (5 J) per stride per kilogram of body mass in mammals is the same regardless of the size of the animal (49). Thus, as Schmidt-Nielsen concluded, "The small animal must take more steps to move a given distance, and this costs more. If we related the cost of locomotion to one step (or to any other linear dimension), we find animals to be equally economical, whether large or small" (47, p. 176).

The reasoning for this is straightforward: The work of a single contraction is the product of force and the distance over which the force acts. All mammalian muscles produce the same maximal force per cross-sectional area. Most animals shorten their muscles by a similar fraction of the muscles' resting length. Body mass is generally proportional to muscle mass. Consequently, the work performed in a single contraction should relate to both muscle mass and total body mass (47).

By this explanation, then, the steady improvement in running economy as children age is a reflection of the progressively declining number of strides necessary to run at a given velocity. If this is true, we would expect to find that $\dot{V}O_2$ expressed per kilogram per stride is similar in adults and children and independent of biologic maturation. Studies in children have generally supported this concept.

Rowland et al. (41) found that stride frequency was approximately 17% greater in boys than in men across three treadmill running speeds (8.0, 8.8, and 9.6 $km \cdot h^{-1}$; figure 8.4). Compared to the boys, running economy was superior in the men at all speeds (at 9.6 $km \cdot h^{-1}$, $\dot{V}O_2$ values for the adults and children were 40.0 ± 5.0 and 49.5 ± 4.4 $ml \cdot kg^{-1} \cdot min^{-1}$, respectively), and the $\dot{V}O_2$ cost of increasing treadmill

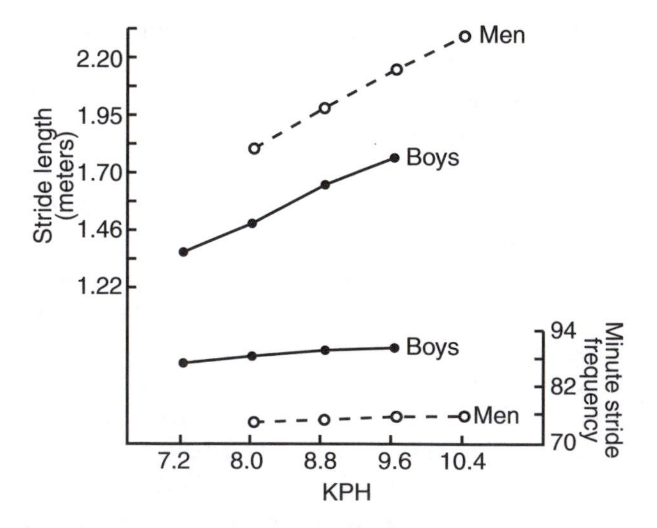

FIGURE 8.4 Stride length and frequency with increasing treadmill speed in prepubertal boys and men (from reference 41).

Reprinted by permission from T.W. Rowland et al. 1987.

speed by 1.6 km · h^{-1}, was less in the men than in the boys (5.6 versus 7.2 ml · kg^{-1} · min^{-1}). However, when $\dot{V}O_2$ per kilogram was expressed relative to a single stride, no differences were observed between the two groups at any speed. The ratio of $\dot{V}O_2$ per kilogram to stride frequency was 0.46, 0.51, and 0.53 for the adults and 0.48, 0.51, and 0.54 for the children at the three respective speeds. Similar findings have been reported by others (14, 26, 51, 52).

In a five-year longitudinal study, Rowland et al. measured stride frequency and submaximal energy cost during treadmill walking in boys and girls ages 9 to 13 years (44). As expected, walking economy improved over the course of the study. However, no significant change in $\dot{V}O_2$ per kilogram per stride was observed except for a decline in the girls in the final two years.

We can examine the validity of the hypothesis that running economy improves with age by (a) comparing economy in children and adults exercising at separate speeds adjusted for body size or (b) studying economy in large children and small adults who have similar sizes and stride lengths. Maliszewski and Freedson observed that when economy is compared between children and adults at the same treadmill speed, the child is exercising at a higher relative work intensity (26). "Such differences in relative intensity," they noted, "draw attention to the lack of proportionality of the task being evaluated" (p. 352). This might alter not only stride frequency but also the relationship between stride frequency and stride length, biomechanical factors, and substrate utilization.

Maliszewski and Freedson (26) compared the energy cost of treadmill running at the same speed (9.6 km · h^{-1}) and at a relative speed adjusted for body size (3.71 leg lengths per second) in 25 adults (18–34 years old) and 21 children (ages 9–11 years). (The speed relative to leg length was derived from pilot work that identified a preferred pace for both groups.) At the same speed, $\dot{V}O_2$ was 34.9 ± 3.2 ml · kg^{-1} · min^{-1} in the adults and 40.6 ± +2.6 ml · kg^{-1} · min^{-1} in the children ($p < .001$). At a speed adjusted for body size, however, no significant difference was observed in economy between the two groups (39.0 ± 3.5 and 37.7 ± 2.8 ml · kg^{-1} · min^{-1} for the adults and children, respectively; figure 8.5). This study indicated, then, that when the exercise task is adjusted for body dimensions, children and adults are equally economical.

Studies in adults indicate that absolute oxygen consumption per stride in a given individual changes dramatically with walking speed, creating a U-shaped curve whose nadir identifies the most economical walking velocity. Workman and Armstrong reported that in adults this optimal velocity decreases with body size and that the absolute $\dot{V}O_2$ cost per stride at the optimal velocity is also directly related to body size (figure 8.6; 60). When $\dot{V}O_2$ cost per stride was expressed per kilogram, the data from all subjects, regardless of size, followed the same curve (figure 8.7).

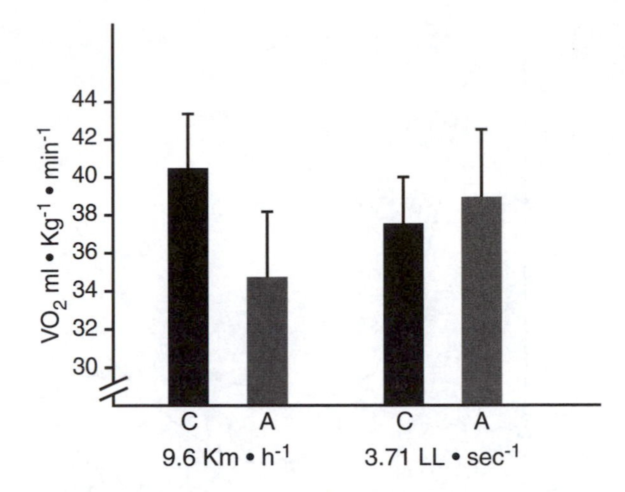

FIGURE 8.5 Treadmill running economy at a speed of 9.6 km · h^{-1} is better in adults than children, but when speed is adjusted for body size [3.71 leg lengths per second (LL · s^{-1})], the difference in economy between the groups disappears (26).

This type of analysis has not been performed in children. Whether children and adults are characterized by similar or different speed–size curves and economically optimal speeds remains to be investigated. The importance of this relationship, however, is suggested by the findings of Bowen et al., who used a portable telemetric oxygen analyzer to assess the metabolic cost of walking at a self-selected, normal speed around an outdoor track in 94 children ages 5 to 15 years (7). No correlation was observed between $\dot{V}O_2$ cost and body mass. The authors postulated that "the self-chosen walking speeds of the subjects are close to their most economical velocity" (p. 592). In accord with the findings of Maliszewski and Freedson (26), then, these results imply that energy economy at a self-selected speed (i.e., related to body size) is independent of age during childhood.

Two studies have examined economy in youth and adults with the same body dimensions, with conflicting results. Both involved young adolescent girls rather than prepubertal subjects.

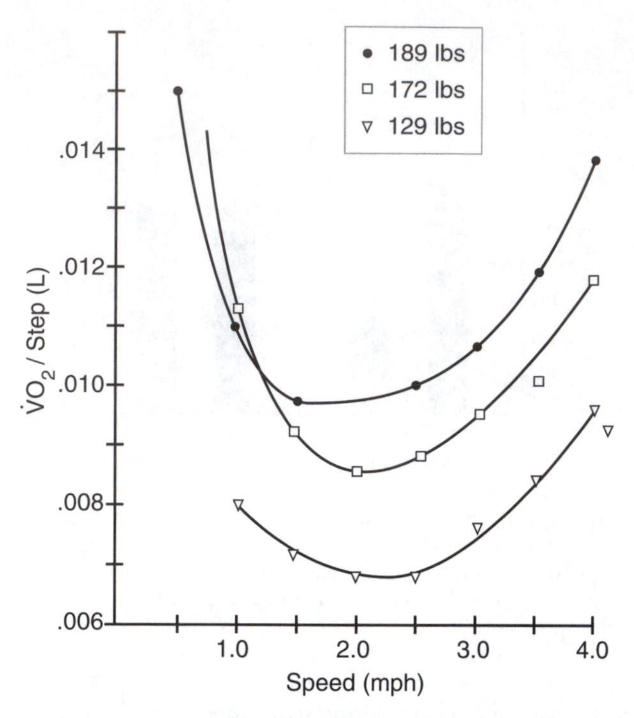

▶ FIGURE 8.6 Oxygen uptake per stride at different speeds of level walking in adults of different weights (from reference 60).

Reprinted by permission from J.M. Workman and B.W. Armstrong 1963.

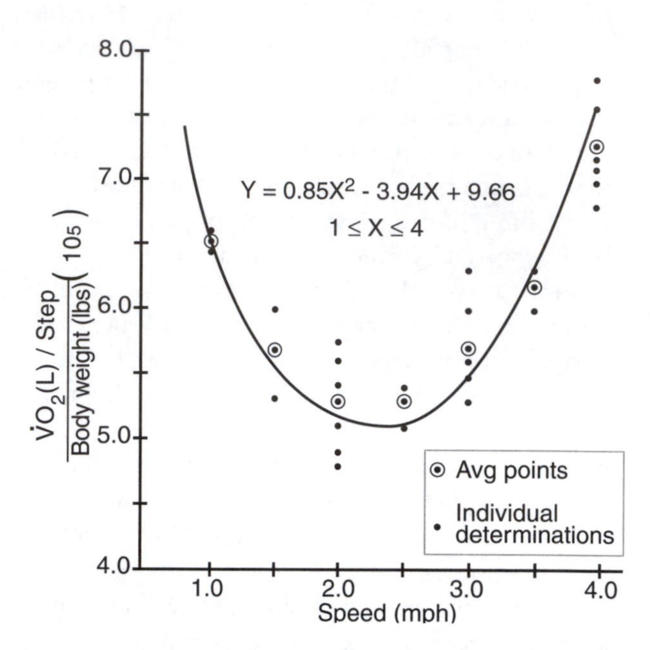

▶ FIGURE 8.7 Oxygen uptake per stride adjusted for body weight in adults during different speeds of level walking (from reference 60).

Reprinted by permission from J.M. Workman and B.W. Armstrong 1963.

Grossner et al. compared the metabolic demands of 10 girls (age 13 ± 0.6 years) and 10 young women (age 22.9 ± 3.0 years) during treadmill running and walking (16). The two groups were matched by height and weight. No significant differences between the girls and women were observed in submaximal $\dot{V}O_2$ at 3.0 mi · h^{-1} (12.3 ± 1.7 and 10.9 ± 1.4 ml · kg^{-1} · min^{-1}, respectively) or at 5.5 mi · h^{-1} (30.5 ± 3.5 and 29.0 ± 2.0 ml · kg^{-1} · min^{-1}, respectively).

In a similar study, however, Perkins et al. found that differences in economy between adolescent girls and young women could not be attributed to variations in stride frequency (34). Thirteen adolescent girls (mean age 13 ± 0.9 years) and 23 women (mean age 21 ± 1.5 years) performed treadmill walking at 3.0 mi · h^{-1} and running at 5.5 mi · h^{-1}. No significant differences in height, weight, leg length, or submaximal stride frequency were evident between the two groups. Treadmill economy was better in the adults than the girls. Submaximal $\dot{V}O_2$ was significantly greater in the girls during both walking (16.4 ± 1.7 vs. 14.4 ± 1.1 ml · kg^{-1} · min^{-1}) and running (38.1 ± 3.7 vs. 33.9 ± 2.4 ml · kg^{-1} · min^{-1}).

Investigations in adults indicate that an individual unconsciously selects the optimal combination of stride frequency and length to run at a given speed. Morgan and his colleagues demonstrated that this is true in children as well (31). The freely chosen stride length at a treadmill speed of 1.34 m · s^{-1} was initially determined for 28 six-year-old boys and girls. The subjects then performed 5-minute level treadmill walks at the same speed with stride lengths that were −10%, −5%, +5%, and +10% shorter or longer than that freely chosen stride length. Subjects displayed the lowest $\dot{V}O_2$ at the freely chosen stride length, and a curvilinear rise in oxygen uptake was observed as stride length deviated from the freely chosen one. These differences were 3.2 and 4.0 ml · kg^{-1} · min^{-1} at the shortest and longest stride length, respectively. The authors concluded that, like adults, children adopt a stride length that minimizes energy cost.

The Cost of Generating Force Hypothesis

The cost of generating force hypothesis holds that muscle force rather than work determines the metabolic cost of running. More specifically, according to this concept, differences in the *rate* of force generation can explain both the linear relationship between running velocity and metabolic consumption and the greater metabolic cost per kilogram of running in small animals (e.g., children) than in large ones (adults).

The essential idea is that when subjects run at a faster pace, the time that the feet contact the ground—during which muscles are required to support or propel body mass against gravity—becomes shorter. The rate of force generation must therefore increase to accomplish this in a more limited period of time. That is, a smaller subject running at the same speed as a larger one needs a quicker rate of force production per stride, and the same is true of a single subject running at a faster speed. In both cases, this accelerated rate of force production results in increased metabolic energy consumption.

Kram (23) developed this concept as follows: A peak vertical ground force of two to three times body weight is expected in subjects of all sizes. Moreover, the amount of muscle force needed to generate vertical ground force does not change with running speed. That is, the vertical force exerted when an animal runs is proportional to body mass at any given velocity. The peak vertical ground reaction force does increase somewhat as running speed increases, but the amount of time spent with the foot in contact with the ground (i.e., the time the force is applied) diminishes. Consequently, "averaged over a complete [period of strides], a running animal exerts a vertical force on the ground that is equal to body weight regardless of running speed [and] the muscle force averaged over a stride does not change with running speed" (p. 139).

The metabolic energy demands of running (E) are dictated by the product of the average vertical force applied to the ground (F) and the rate at which force is applied. This rate, in turn, is inversely related to the length of time that the force is applied, or the time of foot contact ($1/t_c$):

$$E \sim F(1/t_c)$$

During a period of strides, the mean vertical ground reaction force (F) is equivalent to body mass (M), so

$$E/M = c(1/t_c)$$

where c is a cost coefficient. Kram provided data indicating that across both different running velocities and animal sizes, c is indeed relatively constant, which supports the hypothesis.

As indicated by this equation, the metabolic cost of running expressed per kilogram of body mass is inversely proportional to the time of foot contact, and this directly to the rate of generating muscle force. Roberts et al. examined this conclusion by studying the relationship between the metabolic demands of running and foot–ground contact time for five species of running birds and adult human runners (38). The metabolic energy consumption was found to be directly related to the rate of force generation ($1/t_c$) with increasing run speed in all these animals.

Why should rate of force production increase metabolic work? Kram and Taylor explained this as an effect of the recruitment of faster and less energy-economical muscle fibers when an animal runs at a faster velocity with a shorter t_c (24). In vitro studies have demonstrated that the rate of energy consumption of an isometrically contracting muscle is directly related to its maximum shortening velocity (37).

A number of assumptions are inherent in this concept, most of which have been supported by experimental evidence: (a) The rate of crossbridge cycling in active muscles is associated with the speed of leg movements; (b) the major expenditure of metabolic energy during running is to produce force to support body weight (i.e., other demands, such as energy to swing the limbs, are small); (c) muscle fibers operate at a similar force–velocity relationship at different running velocities; (d) horizontal ground forces do not contribute significantly to force production; and (e) the reciprocal of the time of foot contact with the grounds ($1/t_c$) indicates the maximal shortening velocity of the leg extensor muscles.

The cost of generating force hypothesis was developed from studies in which animals and humans did level running. It has been suggested that this concept does not hold for running on a gradient, either uphill or downhill. Minetti et al. described the metabolic expenditure with time of ground contact for adult subjects running at a fixed speed at different slopes (29). The time of foot ground contact was not appreciably altered in running at various gradients, while energy demands changed significantly (e.g., metabolic energy expenditure for running up a 15% gradient nearly doubled that of level running). These authors concluded that Kram and Taylor's [24] approach for level running does not apply to running on a slope.

It seems, however, that the more basic question of whether changes in the metabolic cost of gradient running *relative to body size* were not addressed in this study. Indeed, the issue is complicated: Recall that the metabolic cost per kilogram of performing vertical work against gravity is independent of body size (and presumably of stride frequency and ground contact time). When running is performed on an uphill gradient, a size-independent component of energy expenditure contributes to the overall metabolic demands.

The extent to which changes in foot contact time influence either the metabolic cost of increased running speed or differences in metabolic costs between children and adults running at the same speed should be influenced by the relative proportions that changes in stride length and frequency contribute to greater running velocities. In adults, stride length increases up to moderately high speeds and then plateaus. Stride frequency is then responsible for greater increases in velocity (and shorter ground contact time).

These stride length–frequency relationships have not been well investigated in children. The few studies have generally investigated the normal neuromuscular development of small children, with few subjects and limited study periods (19, 59, pp. 56-57). Rowland et al. reported that both boys and young adult men responded to increased treadmill running speeds (7.2, 8.0, 8.8, and 9.6 km · h^{-1}) with greater stride length and virtually constant stride frequency (41). The ratio of stride length to leg length was identical in the two groups. The children, as expected, used a shorter stride length and greater stride frequency at any speed, but the pattern of change in stride length and frequency with increasing speed was the same as in the adults. Comparisons between older children and adults during field running or walking have not yet been performed.

When Maliszewski and Freedson had children and adults run at 9.6 km · h^{-1}, stride frequency averaged 92 strides per minute in the children and 78 strides per minute in the adults (26). The ratio of stride frequency to stride length was significantly greater in the children (53 vs. 38 in the adults).

Schepens et al. assessed stride frequency–length relationships in adults and children of different ages as they ran across a force platform at varying speeds (46). Maximal running speed increased from 9 km · h^{-1} at age 2 years to 26 km · h^{-1} in adults. At peak velocity, these authors considered it "peculiar" that the stride frequency (4 Hz) was independent of age, meaning that the increase in maximal speed resulted entirely from developmental changes in maximal stride length (figure 8.8).

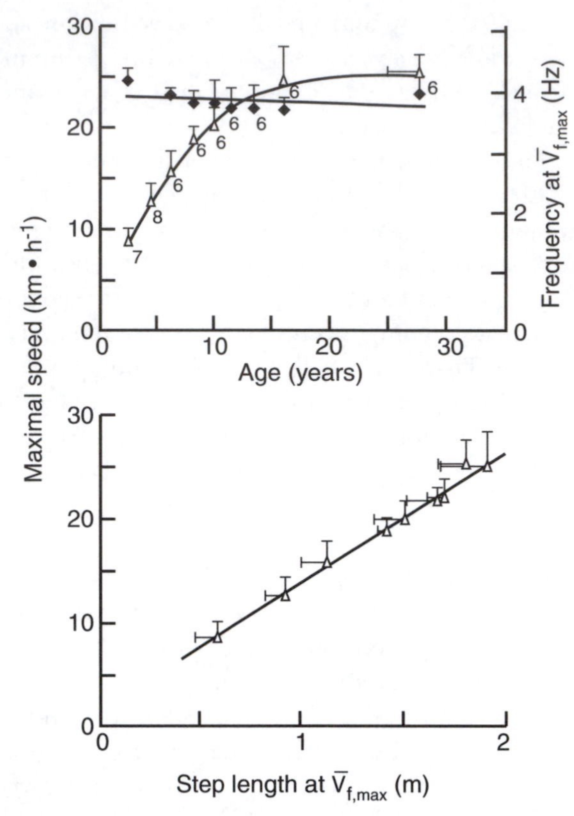

▶ FIGURE 8.8 *Top:* Maximal running speed *(open triangles)* increases during growth up to 16 years of age, but stride frequency at maximal velocity *(solid diamonds)* remains constant. *Bottom:* Changes in step length therefore account for increases in maximal velocity (from reference 46).

In summary, the basic premise of the cost of generating force hypothesis is consistent with observations regarding mass-specific metabolic cost and variations of stride frequency during running as children grow. More information is needed about developmental patterns in ground contact time and ground force velocity vectors as children grow to better assess the validity of this concept.

Stride Efficiency and Elastic Recoil

The concept of stride efficiency and how it changes with increases in body size was introduced in chapter 1. As described Calder, during overground locomotion, excursion distance (L_{exc}) can be defined as the distance from the full forward position of the leg to full rearward extension (8). This excursion distance is a factor of the greatest angle between these leg positions and the length of the leg. In mature animals of various sizes, this angle decreases with body size ($\sim M^{-0.10}$), while the excursion distance (because of the increase in leg length) relates to $M^{0.15}$.

Stride length (L_{st}) is the distance between two successful footfalls, and in animals at a trot-to-gallop transition, L_{st} scales to mass by the allometric exponent 0.38. During walking, stride length and excursion distance are the same, but with running, $L_{st} > L_{exc}$ because of the airborne, nonsupport phase. Stride efficiency in running is defined as

$$L_{st}/L_{exc} \sim M^{0.38}/M^{0.15} = M^{0.23}$$

This means that as animals get larger, their stride efficiency increases, which is an outcome of a greater airborne segment during a given stride. It may be that the greater distance covered for a given leg excursion during running translates into a lower energy demand to cover a given distance in larger animals.

Limited information is available in children. Schepens et al. reported that both leg angle and relative excursion distance were slightly larger in children than adults, and these values decrease with age, attaining the adult value after the age of about 12 years (46). Therefore, the higher step frequency in children is due to a shorter leg length, partly compensated for by an increased amplitude of movement.

Thorstensson (51) found that the ratio between stride length and leg length, consistent with data in animals, was greater in boys (1.10, 1.36, and 1.47 at 8, 10, and 11 km · h^{-1}, respectively) than in men (0.98, 1.21, and 1.31 at the same respective speeds). That is, children have a greater excursion distance and longer stride in respect to their leg length than adults. How these two factors relate to each other in terms of stride efficiency, however, is unclear.

These observations raise questions as to whether changes in elastic recoil account for differences in stride efficiency and, consequently, in energy demands for locomotion. Running can be considered a series of jumps, or one-legged hops, in which the elastic recoil of leg connective tissue and muscle may contribute a spring effect. The degree to which elastic recoil might contribute to the kinetic energy of running and how children might differ from adults in leg "stiffness" has not been well studied. Early studies suggested that young subjects might have lower levels of elastic recoil in their legs (6, 33).

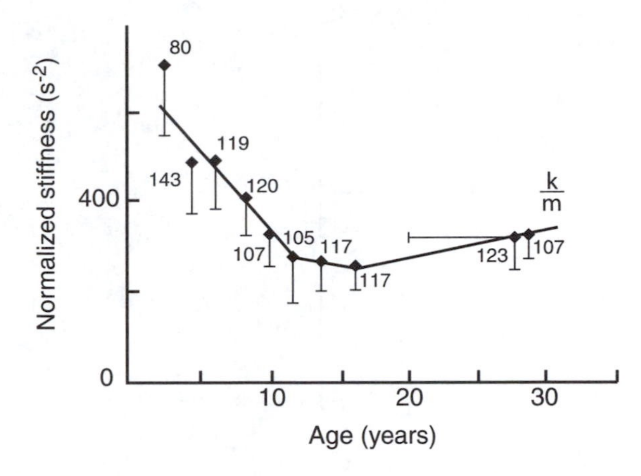

FIGURE 8.9 Vertical stiffness (*k/m*) relative to age (from reference 46).

Schepens et al. studied the mechanics of running in 58 children ages 2 to 16 years and six healthy adults ages 23 to 31 years, with particular attention to elastic recoil (46). Subjects ran across a force platform at varying speeds up to their maximum. The mass-specific vertical stiffness, calculated from the slope of the plot of vertical acceleration and displacement of the center of body mass, decreased with age (figure 8.9).

Muscular Efficiency

Muscular efficiency is the work accomplished during exercise relative to the energy expended. Values are most readily determined during cycle exercise, when work is accurately established by the load applied and energy expended is defined by oxygen uptake converted to its caloric equivalent, considering the concurrently measured respiratory exchange ratio.

Mechanical efficiency during exercise is affected by variables that contribute to resting metabolism, external work production (efficiency of contraction coupling and fraction of energy released from ATP in the muscle cells), and internal work (i.e., that needed to overcome elastic forces). To measure mechanical efficiency in terms of external work alone, efficiency can be expressed as net efficiency (work accomplished divided by energy expenditure above that at rest), work efficiency (work accomplished divided by energy expended above that in unloaded cycling), or delta efficiency (the change in work accomplished between two loads divided by the change in energy expended between the same two loads).

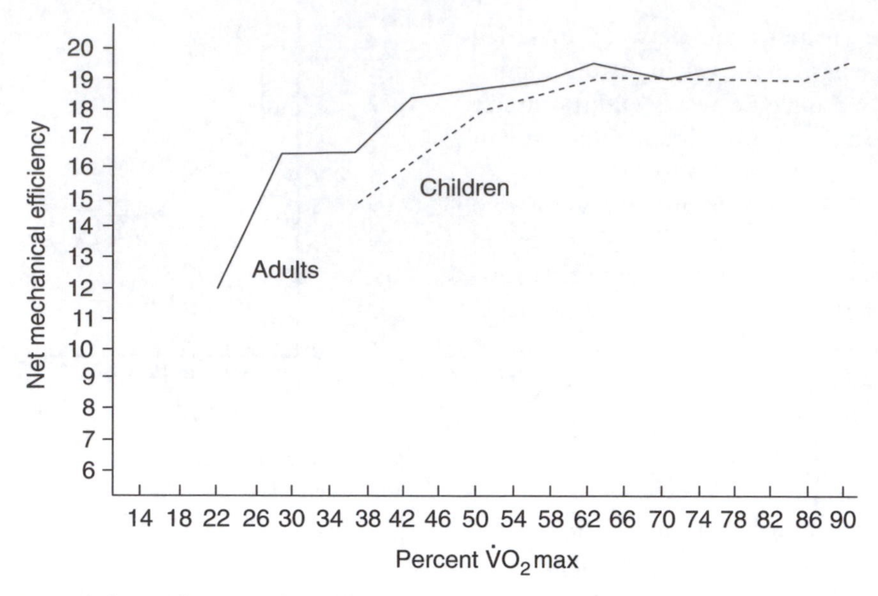

FIGURE 8.10 Estimated net efficiency during cycle exercise relative to work intensity in boys and men (from reference 45).
Reprinted by permission from T.W. Rowland 1990.

It has generally been assumed that mechanical efficiency during exercise is similar in children and adults. That is, the energetics of muscle contraction appear to be independent of biologic maturation. Consequently, differences between children and adults in running economy cannot be explained by variations in muscle mechanical efficiency.

This conclusion is based on a limited set of research data, however, and not all of it is supportive. Cooper et al. reported mean values for delta efficiency of 28% and 29% during cycle exercise in 6- and 18-year-old males, respectively (9). However, in that study a progressive increase in efficiency was observed in females across the same age span. Average efficiency was 29% in the youngest and 33% in the oldest ($p < .05$). On the other hand, Thompson reported net efficiency of 19.2% in six women ages 21 to 28 years and 19.3% in six girls ages 9 to 11 years (50).

Rowland and colleagues (45) studied 19 prepubertal boys (mean age 10.5 years) and 21 male college students (average age 21.3 years). Estimated net efficiency was similar during exercise at an intensity above 50% $\dot{V}O_2$max, but below that level the efficiency tended to be greater in the men (figure 8.10). Mean delta efficiency between workloads of similar relative intensities was 23.2% for the prepubertal boys and 22.5% for the college men. Between equal absolute workloads the delta values were 23.2% and 26.5%, respectively.

Using a different experimental model, Zanconato et al. reported that the $\dot{V}O_2$ cost of work done in high-intensity exercise was greater in children than adults (61). $\dot{V}O_2$ was measured during one-minute bouts of constant-rate cycle exercise at 80% of anaerobic threshold, $\dot{V}O_2$max, and 125% $\dot{V}O_2$max. The cumulative oxygen cost (in milliliters per joule) was consistently about 25% higher in children than in adult subjects. (At these high intensities, however, differences between children and adults could reflect varying contributions of anaerobic metabolism rather than dissimilar efficiency in energy utilization.)

More information is needed before possible differences in muscle mechanical efficiency can be discounted as contributing to maturational differences in running and walking economy. Particularly, more data need to be obtained regarding possible sex differences and the influence of exercise intensity on mechanical efficiency.

Muscle Co-Contraction

During most forms of motor activity, agonist and antagonist muscle groups contract simultaneously to create limb motion, a process termed *co-contraction*. When the antagonist muscle is activated to a greater extent than necessary to maintain joint stability, however, the energy cost of limb movement increases. That is, unnecessary activation of antagonist muscles wastes energy and contributes to overall metabolic expenditure. Frost and her colleagues thought that a greater amount of co-contraction might contribute to children's higher mass-specific metabolic expenditure during treadmill walking and running (15). This idea follows from recognition that decreases in co-contraction may occur during the process of normal motor development in children (5).

Frost et al. (15) used electromyography (EMG) to assess muscle activity during treadmill walking in three groups of subjects: 7- to 8-year-olds, 10- to 12-year-olds, and 15- to 16-year-olds (five boys and five girls in each group). EMG activity was compared between vastus lateralis and hamstring muscles and between tibialis anterior and soleus muscles for a three-second interval during the final minute of each of six 4-minute bouts of running and walking at different speeds. EMG values were indexed to the largest activation recorded during previous measurement of maximal voluntary contractions.

The co-contraction index was highest in the youngest subjects at both the same absolute and relative treadmill speeds (figure 8.11). In general, values in thigh muscles were approximately 30% greater in the 7- to 8-year-olds than in the 10- to 12-year-olds. At the same time, $\dot{V}O_2$ per kilogram was about 20% higher in the youngest than in the middle age group. The authors concluded that "although we cannot determine the exact cause or function of the higher levels of cocontraction in the younger children, there can be little doubt that they are associated with an increased metabolic cost of locomotion" (p. 186).

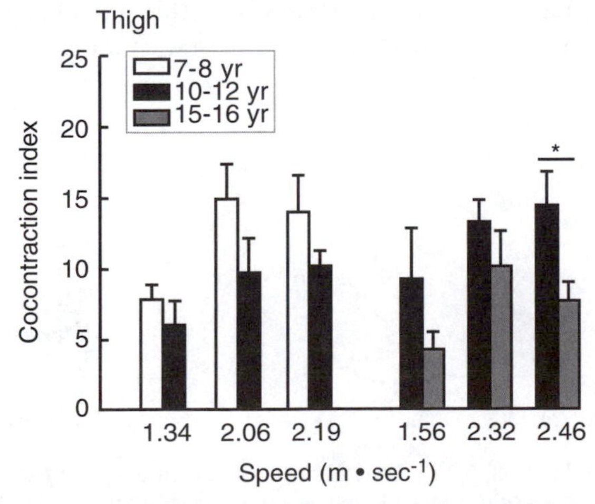

FIGURE 8.11 Co-contraction index for thigh muscles during treadmill exercise in three age groups (from reference 15). * = $p < 0.05$.

Reprinted by permission from Frost et al. 1997.

Sex Differences in Economy

One of the most perplexing questions about exercise economy in children is whether boys and girls differ in their submaximal energy demands. Early studies were more or less equally divided between those showing a greater economy in girls (i.e., lower $\dot{V}O_2$ per kilogram at a given treadmill setting; 1, 3, 11, 56) and those indicating no sex difference (12, 25, 27, 40). Of the approximately 14 investigations addressing this issue to date, however, none has reported superior running or walking economy in boys.

If girls do actually possess greater economy during weight-bearing exercise, why should this be so? Several suggestions have been offered: Girls have a lower basal metabolic rate than boys, carry a greater percentage of body fat and lower lean body mass, and may be characterized by different biomechanical features.

More recent studies have shed no more light on this issue. Walker et al. created generalized equations for predicting $\dot{V}O_2$ during horizontal running in 47 males and 35 females ages 12-18 years (55). After speed was considered, sex failed to account for any significant additional variance in energy expenditure.

Rowland and colleagues could also find no sex differences in treadmill-walking economy in a five-year longitudinal study of 10 girls and 11 boys, beginning at an average age of 9.2 years (44). Economy was determined during the final minute of a four-minute walk at 3.25 mi · h^{-1} and 8% grade. Over the five years of the study, submaximal $\dot{V}O_2$ fell 16% in the girls (from an average of 30.9 to 25.9 ml · kg^{-1} · min^{-1}) and 13% in the boys (from a mean of 31.1 to 26.9 ml · kg^{-1} · min^{-1}).

Other studies have supported the idea that girls are more economical than boys during treadmill exercise, but the explanation remains elusive. Armstrong and his associates (2) studied the effects of sex and body size on treadmill economy in 97 boys and 97 girls (mean age 12.2 years). Subjects ran at 8, 9, and 10 km · h^{-1} in three-minute stages. During submaximal running, no significant difference was observed between the sexes in absolute $\dot{V}O_2$, but $\dot{V}O_2$ per kilogram or adjusted for mass using allometry was greater in the boys than in the girls.

Two investigations have indicated that sex-related variations in body composition might explain differences in economy between boys and girls. McMurray et al. assessed factors contributing to running economy in 8- to 18-year-old males and females (28). No sex differences were seen in mass-relative $\dot{V}O_2$ at rest, but the males had a significantly greater value while running at 8 km · h^{-1}. However, when $\dot{V}O_2$ was expressed relative to fat-free mass, no sex differences in economy were observed. Multiple regression analysis indicated that absolute $\dot{V}O_2$ during running was most closely related to fat-free mass. This suggested, then, that any sex differences in submaximal running economy reflect the variations in body composition of boys and girls.

Similar findings were described by Morgan et al. (32) in their study of 35 six-year-olds (15 boys and 20 girls). Resting energy expenditure was not significantly different between the boys and the girls. Economy during a five-minute level run at 5 mi · h^{-1} was better in the girls ($\dot{V}O_2$ 36.5 ± 2.9 ml · kg^{-1} · min^{-1}) than in the boys (39.1 ± 2.8 ml · kg^{-1} · min^{-1}). $\dot{V}O_2$ values expressed relative to estimated lean body mass were similar in the two groups (46.9 ± 3.4 and 44.8 ± 4.0 ml · kg^{-1} · min^{-1} for the boys and girls, respectively). The authors concluded that "the higher submaximal aerobic demand of running displayed by the boys appears to be related to the presence of a greater muscle mass" (p. 127).

Others have failed to document such an effect of body composition. Welsman and Armstrong looked at this question in a longitudinal three-year study of 118 boys and 118 girls who were 11.2 ± 0.4 years old at the beginning of the study (58). Submaximal $\dot{V}O_2$ responses were recorded during horizontal treadmill running at a speed of 8 km · h^{-1}. Data were analyzed by sex, age, and level of sexual maturity using an allometric approach with multilevel regression modeling. Changes in submaximal $\dot{V}O_2$ over the three years were explained mainly by changes in body mass and skinfold thicknesses. No independent influence of sexual maturity was observed. Girls were more economical than boys, even when differences in skinfold thicknesses were considered. This sex difference became more exaggerated with increasing age. The explanation for the greater running economy in girls was not readily apparent.

Ayub and Bar-Or specifically addressed the question of the contribution of adiposity to energy cost of walking in children (4). They studied 23 children ages 8.5 to 13.5 years who varied in percentage body fat from 11% to 43% and in mass from 27.1 to 82.5 kg. The subjects performed treadmill walking for four minutes at 3, 4, and 5 km · h^{-1} at 50% and 70% $\dot{V}O_2$max. Correlations between absolute $\dot{V}O_2$ and percentage body fat were high (*r* = .32 to .98). However, the partial correlation of $\dot{V}O_2$max and percentage body fat dropped dramatically when the effect of body mass was adjusted for. The authors concluded that after body mass was considered, percentage body fat accounts for only a small variance in the $\dot{V}O_2$ demands of walking in children.

Uphill and Downhill Running

It comes as no surprise that the energy required for running uphill is greater than that needed for running on a flat surface or downhill. Anyone who has run a road race in hilly terrain is aware of the strain during the uphill climb and the relief coming downhill. There are theoretical reasons to expect that the stress of uphill running might be less, at least to some degree, in children than in adults. This hypothesis comes from information in animals indicating that the work of moving mass against gravity is performed at the same energy expense, independent of animal size (see chapter 1). That is, we know that it takes a greater $\dot{V}O_2$ per kilogram to increase speed by 1.0 km · h^{-1} during level running in small animals (such as children) than in larger ones (such as adults). But the energy required to lift 1 kg of body mass a given vertical distance against gravity is identical in large and small animals. This means that the *relative increase* in energy expenditure when an animal moves upward against gravity is less in the small animal than in the large.

Figure 8.12 demonstrates how this would apply to a child A and an adult B who are running at the same speed on the flat and then encounter a very steep hill to climb. Let's say that while running hori-

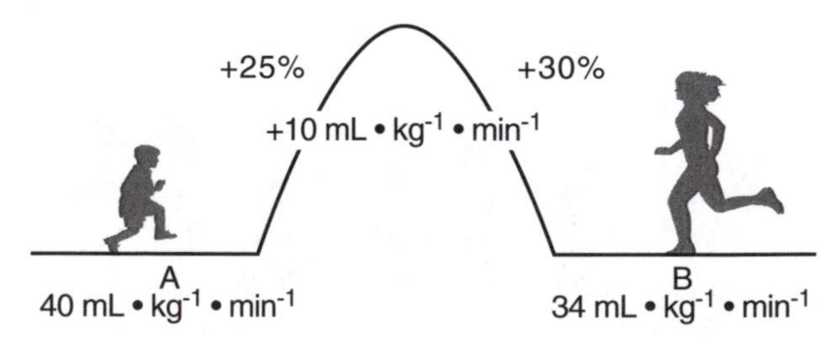

FIGURE 8.12 During horizontal running at the same speed, child A has a larger $\dot{V}O_2$ per kilogram than adult B. The mass-specific energy cost climbing a hill, however, is independent of mass. As a result, the relative increase in work intensity is greater in the adult (30%) than in the child (25%).

zontally, the child expends 40 ml · kg^{-1} · min^{-1}, while the adult, being more economical, has a $\dot{V}O_2$ of 34 ml · kg^{-1} · min^{-1}. The mass-specific metabolic cost of climbing the hill, however, is the same in both, perhaps an additional 10 ml · kg^{-1} · min^{-1}. That means a relative increase in energy expenditure that is greater in the adult (+30%) than in the child (+25%). (Note that this is not true if the child and adult are running on the flat at a speed adjusted for body size, at which the metabolic demands per kilogram and the percentage increase with climbing the hill are the same.)

No adequate research data are available to allow us to evaluate whether these theoretical differences truly occur or, if they do, whether their magnitude carries any practical significance. Hamar et al. (17) compared energy expenditure during running at 16 km · h^{-1} and walking at 6 km · h^{-1} horizontally and at a 5% incline in 22 adolescent distance runners ($\dot{V}O_2$max 68.2 ± 4.8 ml · kg^{-1} · min^{-1}). Oxygen cost rose by 37% during uphill running and 26% during uphill walking. No adult comparisons were conducted in this study. Pivarnik and Sherman (36) reported a 20% increase in $\dot{V}O_2$ per kilogram between horizontal running at 9.6 km · h^{-1} and running at a 5% grade in aerobically trained men (mean age 27.1 ± 4.8 years, $\dot{V}O_2$max 61 ml · kg^{-1} · min^{-1}).

An analysis of maturational effects on the energy demands of downhill running would be more complex, given the contribution of factors such as recovery of kinetic energy. $\dot{V}O_2$ decreases as treadmill slope becomes more negative, but the rate of this decline is less than that of the increases observed with uphill running (36).

Webber et al. (57) compared $\dot{V}O_2$ changes during downhill treadmill running in 16 children (10.4 ± 0.3 years) and 15 adults (27.1 ± 0.8 years). Subjects ran flat for 5 minutes and then at a −10% grade for 30 minutes at a speed that elicited a heart rate of 80% to 85% of the age-predicted maximum. As expected (since running speed was adjusted for relative work intensity), $\dot{V}O_2$ per kilogram values during level running were not different between the two groups. Similarly, the decline in mass-specific energy expenditure was the same when the children and adults began running downhill (about −25%; figure 8.13). For reasons that are unclear, values of $\dot{V}O_2$ per kilogram in the two groups diverged during the 30 minutes of constant-rate exercise; the children's became greater than the adults'.

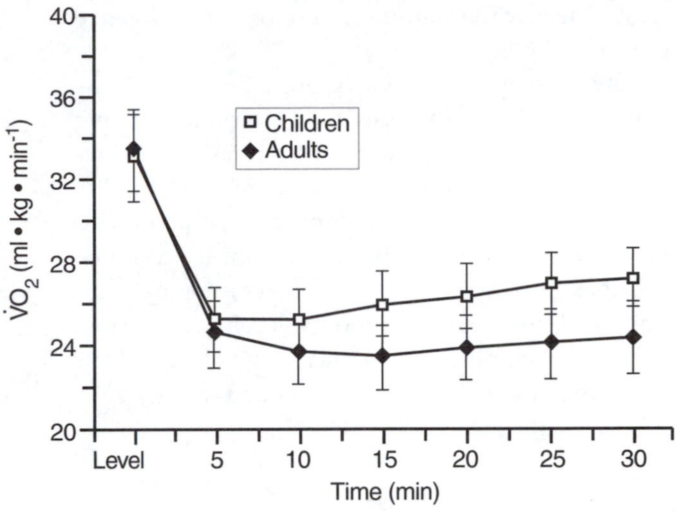

FIGURE 8.13 Oxygen uptake in children and adults while running on a treadmill on the level and then downhill at a −10% grade at a speed adjusted for relative intensity for 30 minutes (from reference 57).

Reprinted by permission from L.M. Webber et al. 1989.

Implications for Aerobic Fitness

We might intuitively expect that improvements in running or walking economy as children grow should contribute to developmental improvements in endurance exercise performance. This concept was already addressed in chapter 5. If $\dot{V}O_2$ per kilogram remains stable (in boys), a lower $\dot{V}O_2$ per kilogram at a given running speed during childhood means that the relative intensity (percentage of $\dot{V}O_2$max) at that speed progressively declines. This should translate into a child's ability to run faster over a given distance or longer at a given speed with advancing age. (Such an analysis is more complicated in girls, in whom both maximal and submaximal $\dot{V}O_2$ per kilogram decrease with age.)

For instance, in a five-year longitudinal study of children, decreases in submaximal $\dot{V}O_2$ caused percentage $\dot{V}O_2$max to decline from 66% to 60% in the girls and from 65% to 51% in the boys (44). Krahenbuhl et al. compared running economy and performance on a nine-minute run test in children when they were an average age of 9.9 years and again 7 years later (20). Submaximal energy demand per kilogram fell by 13%, while $\dot{V}O_2$max remained stable. Meanwhile, endurance performance distance improved by 29%.

Interestingly, in children, having a better submaximal exercise economy does not appear to confer any advantages in ability to perform endurance activities or superior physiologic fitness ($\dot{V}O_2max$; 21, 22, 42). This is true even for child endurance athletes who have similar $\dot{V}O_2max$ values, in whom no relationship has been observed between competitive run times and submaximal economy (10, 53). Moreover, endurance training that results in small to modest improvements in $\dot{V}O_2max$ does not typically alter submaximal economy (35, 43, 48).

Conclusions

The efficiency of converting chemical energy to muscle work is not appreciably different between children and adults. A greater increase in energy expenditure (relative to body size) during exercise in children becomes apparent (a) only during weight-bearing exercise such as running and walking and (b) only when children perform that exercise at the same workload as adults. Exercise economy defined in this manner improves (i.e., submaximal $\dot{V}O_2$ per kilogram decreases) continuously over the course of childhood.

A number of mechanisms have been suggested to explain the lower submaximal running and walking economy in small children, a trend observed in phylogenetic studies of mature animals as well. It may be a key observation, however, that at a given level of treadmill work, the child is running at both a greater relative exercise intensity and a faster rate relative to body size than the adult. Both of these deviations relate to a number of anatomic and physiologic variations, including stride frequency, stride frequency–length relationship, rate of force production, substrate utilization, and biomechanical factors.

Moreover, it is not clear what the biologic relevance is of the observation that at a running speed of 8 km · h^{-1} an 8-year-old boy has a $\dot{V}O_2$ of 39 ml · kg^{-1} · min^{-1} compared with 32 ml · kg^{-1} · min^{-1} in a 20-year-old adult. Eight-year-olds and 20-year-olds do not run at the same velocities either recreationally or in sports competition. Instead, their stride length and frequency are controlled by anatomic features and body dimensions. A key study by Maliszewski and Freedson demonstrated that when children and adults run at a velocity adjusted for body size (in this case, leg length), no maturational differences are observed in running economy (26). The question posed by the title of their study—Is running economy different between adults and children?—was answered by their findings: It depends on the test design. If children and adults run at the same speed and incline on a treadmill, the adults are more economical. But if the speed is adjusted for body size, no differences in economy are observed. The authors suggested that "a test model that adjusts speed to body size may be preferable for the determination of the physiological and biomechanical differences during running due to human development" (p. 359). From this perspective, the energy costs of weight-bearing locomotion for children and adults are similar.

Discussion Questions and Research Directions

1. What is the meaning of submaximal $\dot{V}O_2$ in relationship to body mass in children? Does mass serve as a load that incurs a certain metabolic cost? Or is energy expenditure as work intensity increases linked to other factors, such as body surface area (as is observed at rest)?
2. Do girls really have superior running economy to that of boys? If so, what mechanisms explain this sex difference? Would sex differences in submaximal $\dot{V}O_2$ have any practical significance?
3. How does the stride frequency–stride length relationship change during maturation? How might this affect the metabolic demands of exercise?
4. Do maturational differences in elastic recoil of the legs influence variations in running economy?

5. What role do changes in running economy play in developmental improvements in endurance running?
6. Does uphill running reveal biologically important differences in metabolic expenditure during exercise between children and adults?

9

Short-Burst Activities and Anaerobic Fitness

In evolution, the anaerobic capacity has been an essential component for survival, maybe even more so for our ancestors as hunters than a high aerobic capacity. Today, a high maximal anaerobic capacity has no practical significance other than in certain sport disciplines. What should be remembered, however, is the paramount role anaerobic energy release has in allowing for very quick alteration in muscle power output.

—Bengt Saltin (1990))

In this chapter we discuss

- the metabolic basis of anaerobic fitness,
- the development of markers of anaerobic fitness as children grow, and
- the relationships between biochemical processes, laboratory testing, and field performance of short-burst activities in youth.

When a subject's results on a series of exercise tests at various workloads is plotted, external power is related to exercise duration by a hyperbolic curve (i.e., the lighter the load, the longer the exercise can be sustained). This idea was addressed in a discussion of the critical power concept in chapter 5. Figure 9.1 compares this power–duration curve between children and adults and indicates no obvious maturational differences (66).

At any point along this curve, the process of muscle contraction (the sliding of actin-myosin filaments) and the energy source for accomplishing this (ATP) are the same. What differ as work intensity and duration change are the metabolic processes—and resulting by-products—by which this energy requirement is satisfied. In this chapter we examine those activities that involve the steeper portion of the power–duration curve, those that take place in the first one or two minutes of exercise, when small changes in external power production are associated with more dramatic changes in ability to sustain exercise. This portion of the curve is traditionally associated with the energy contributions of anaerobic glycolysis, and the ability to perform short-burst activities has therefore often been assumed to reflect "anaerobic fitness," or the limits of anaerobic metabolic capacity.

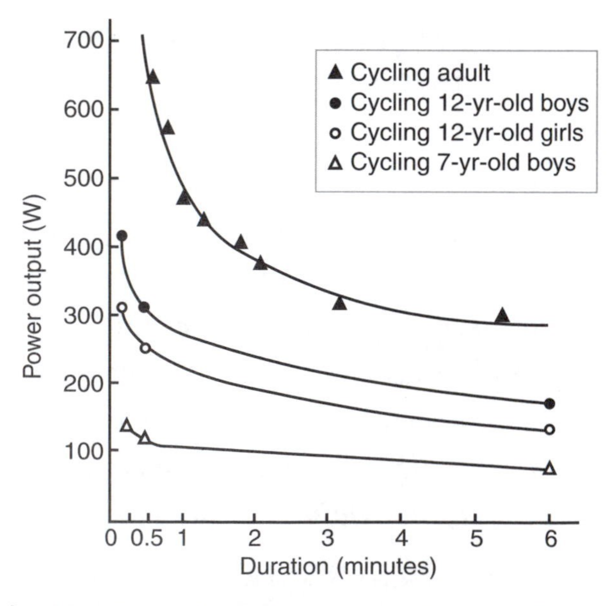

FIGURE 9.1 Power–duration curves for adults (76) and children (66), compiled by Van Praagh (66).

In the past, short-burst activities were felt to be fueled exclusively by anaerobic metabolism, longer ones by aerobic processes. It is now recognized that while this trend is valid, there is considerable overlap in the anaerobic and aerobic contributions to the power–duration curve. In fact, both forms of metabolism contribute to exercise of almost all intensities and durations.

Anaerobic metabolism responds rapidly to the energy requirements of exercise and is capable of satisfying the needs of very high intensity work. Aerobic metabolic processes react more slowly to exercise but provide greater energy for sustained work. Gastin indicated that in a short, 15-s burst of maximal exercise, 88% of the energy contribution is anaerobic and 12% aerobic in adults (32). In a 60- to 75-s all-out test, the anaerobic and aerobic energy input are about equal. At four minutes, 80% of the energy needs for maximal exercise are met by aerobic metabolism.

Limited data are available in children. Chia et al. reported that, depending on a child's mechanical efficiency, from 19% to 44% of energy used in a 30-s Wingate test comes from aerobic sources (18). Counil et al. found that the aerobic contribution to the energy demands of Wingate cycle testing was 26% in children (20). In a child-to-adult comparison study, Hebestreit et al. (36) reported that the estimated aerobic inputs to Wingate testing in boys (9–12 years old) and in men (ages 19–23 years) were 34% and 23%, respectively ($p < .01$). Van Praagh et al. found that $\dot{V}O_2$ reached 60% to 70% of maximal values in children during a Wingate test (67). The importance of these data is that tests interpreted as indicating "anaerobic fitness" may include a substantial aerobic component.

As the parents of every three-year-old know, a high level of short-burst activities is characteristic of children. Using observational techniques, Bailey et al. (6) found that over a 12-hour period the average duration of intense exercise in a group of 6- to 10-year-old children was 3 s (figure 9.2). They recorded not a single episode of vigorous exercise lasting 10 consecutive minutes, and 95% of all intense activity was shorter than 15 s. The average duration of activities of all intensities was 6 s. Similar findings have been documented by others (7, 34). Clearly, "these

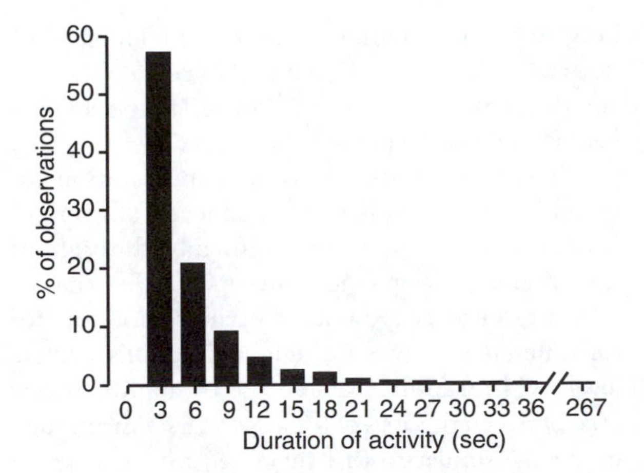

FIGURE 9.2 Duration of children's high-intensity activity determined by observation in natural conditions (from reference 6).

Reprinted by permission from R.C. Bailey 1995.

results indicate that children engage in very short bursts of intense physical activity interspersed with varying intervals of activity of low and moderate intensity" (6, p. 1033).

During the course of childhood this short-burst activity progressively declines, whether measured as activity level or caloric expenditure (55, pp. 34-36). Between the ages of 6 and 16 years, total daily energy expenditure (in kilocalories per kilogram) declines by almost 50%. From this observation it might be expected that mass-relative utilization of anaerobic metabolism during routine daily activities would decline with age. As already discussed in chapter 4, anaerobic glycolytic capacity paradoxically appears to *increase* as children grow.

As indicated by the quotation that opened this chapter, short-burst activities have not traditionally been considered important to physical health. However, it can be argued that short-burst activities bear critical significance to health outcomes in the pediatric population. In the current paradigm of the exercise–health connection in adults, habits of regular, moderate exercise appear to be the best predictor of positive health outcomes. Since short-burst activities are what children do all day and what they can be expected to do if encouraged to exercise, it is this form of exercise that should be the focus of efforts to improve pediatric habits of physical activity. Short-burst activity can be expected to increase caloric expenditure in the treatment and prevention of obesity, maintain muscle development, and promote bone health. Thus, the study of anaerobic fitness and determinants of short-burst activities in children is highly relevant from a preventive health care standpoint.

In addition, short-burst activities are characteristic of youth sports. Basketball, soccer, hockey, and swimming involve frequent, high-intensity, short bouts of exercise that can be expected to rely on anaerobic metabolic capacity. Understanding how children respond physiologically to exercise of this nature is important in optimizing performance and providing for participant safety.

Anaerobic fitness is a somewhat elusive concept. At the onset, assessment of glycolytic capacity is problematic since it is not feasible to measure anaerobic enzymatic function and substrate utilization directly during high-intensity exercise. As an alternative, levels of biochemical markers of anaerobic glycolysis, particularly blood lactate, can be measured. However, given the influence of variables such as elimination rate, metabolism, and cell membrane transfer, the extent to which these markers provide an accurate view of the intracellular metabolic processes is not certain.

Anaerobic metabolic input can be estimated by subtracting oxygen uptake from the total energy demand of a motor task (i.e., by calculating the oxygen deficit, or oxygen debt). Power production during short-duration, high-intensity cycle or treadmill exercise has been used as a laboratory measure of anaerobic fitness, assuming a high reliance of these activities on anaerobic glycolysis. Such an interpretation is limited by the contribution of aerobic metabolism to these tests and by the influence of factors such as muscle size, strength, and subject motivation. Field tests of short duration, such as sprints, have also been considered an expression of anaerobic fitness, but other variables, particularly neuromuscular and anthropometric variables, play important roles in such performances. Consequently, short-term exercise fitness is not necessarily synonymous with anaerobic metabolic capacity.

In this chapter we discuss alternative methods of defining and indirectly measuring anaerobic fitness in the context of biologic maturation. The focus is on understanding how these separate constructs change with growth, how development in these aspects can be related to body size, and how they are related to one another. For a more comprehensive overview of anaerobic performance in children, the reader is referred to excellent reviews by Van Praagh (66) and Van Praagh and Doré (68).

Metabolic Anaerobic Fitness

Evidence of change in the anaerobic glycolytic capacity of muscle as children grow was reviewed in chapter 4. That information is therefore only summarized here. In general, evidence from lactate production during exercise and from changes in glycolytic enzyme activity in both human and animal studies indicates that the anaerobic glycolytic machinery becomes increasingly more effective with age.

Limited biopsy data have indicated that activity of glycolytic enzymes increases as children grow. Berg et al. (12) examined samples of vastus lateralis muscle in children from 4 to 18 years old and observed positive correlations of both pyruvate kinase activity and aldolase activity with age (r = .45 and .35, respectively). Between these ages, glycolytic enzyme activity rose by almost 50%. Likewise, Haralambie described increases in the glycolytic enzymes 3-phosphoglycerate kinase, aldolase, and pyruvate kinase until the pubertal years (35; figure 9.3). Eriksson and Saltin described levels of phosphofructokinase in 11-year-old boys that were approximately 30% of those in untrained adults (27). As Van Praagh pointed out, however, these studies all reflect values in the resting state and say nothing about capacity during exercise (66).

The findings in growing children mimic those in adult animals of various sizes, suggesting that the mechanism for maturational changes in anaerobic enzyme activity in humans is somehow related to body size. For example, Emmet and Hochachka described a direct relationship between animal size and the activities of pyruvate kinase, lactic dehydrogenase, and glycogen phosphorylase (26).

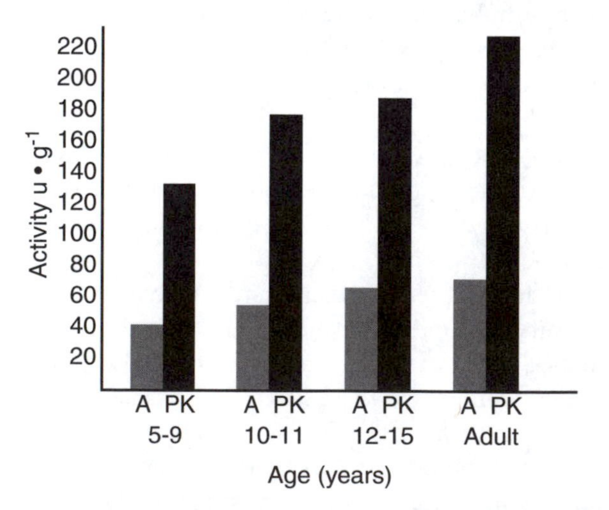

FIGURE 9.3 Activity of the glycolytic enzymes aldolase (A) and pyruvate kinase (PK) in vastus lateralis biopsies from children of different ages and from adults (from reference 35).

Children demonstrate lower concentrations of blood lactate than adults do at all levels of exercise. Maximal lactate increases uniformly throughout the childhood years, indicating that the hormonal effects of puberty are not primarily responsible for the differences between children and adults. Lactate levels with maximal exercise rise by approximately 50% between the ages of 6 and 14 years, a magnitude of change similar to that suggested for increases in resting glycolytic enzyme activity.

Laboratory Anaerobic Fitness: Wingate Cycle Testing

A number of laboratory exercise tests that demand a high energy output over a brief time have been used as indicators of anaerobic metabolic capacity. These include bouts of cycle exercise, stair running, and motorized and nonmotorized treadmill tests (see references 8 and 72 for review). The most popular is the Wingate anaerobic cycling test, devised by Cumming (21) and further developed by Bar-Or (10). In this 30-s test, two indicators of anaerobic power can be obtained: peak power (the greatest value, usually obtained within the first 5 s) and mean power, or the average power output during the entire test. These are markers of anaerobic metabolic function, but the 30-s duration is not sufficient to measure total anaerobic capacity and is long enough to allow a significant contribution by aerobic metabolism. The duration of the Wingate test, then, represents a compromise between these two disadvantages.

Findings from Wingate testing can be used to predict performance in short-burst field tests such as sprinting. As reviewed later in this chapter, however, the associations reported between peak and mean power and sprint velocity have been moderate, causing Bar-Or to conclude that "correlation between the Wingate anaerobic test power indices and 'anaerobic performance' is quite high, but not high enough for using the Wingate anaerobic test as a predictor of success in these specific tasks" (10, p. 388).

Absolute values of anaerobic power increase progressively during childhood in boys and girls (figure 9.4), with an acceleration at puberty in boys. Between

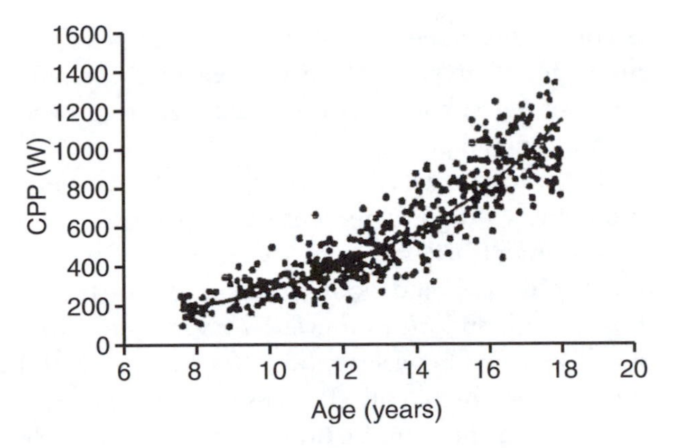

FIGURE 9.4 Relationship of peak anaerobic power and age during childhood (from reference 24).
Reprinted by permission from E. Doré 2000.

the ages of 12 and 17 years, peak anaerobic power increases by 121% in males and 66% in females (3). Most data also indicate that anaerobic power adjusted for body mass also rises during the pediatric years. This is the basis for inferring that anaerobic fitness improves during childhood at a greater rate than can be explained by changes in body size alone. This conclusion is tempered by concerns that body mass may not be an appropriate means to normalize measures of anaerobic power between individuals (discussed later). Early studies examining sex differences in anaerobic performance on Wingate tests were conflicting, but Santos et al. concluded that, overall, both absolute and mass-specific peak power differences are minimal until the time of puberty (56). A concern has been raised that some results of Wingate testing, particularly as they relate to sex differences, may be influenced by the braking force applied (1). More-recent studies have added both to our understanding and to the confusion about these issues.

Recent Studies

We begin with an older report whose results provide a good overview of maturational differences in anaerobic power. In 1995, Gaul and her associates compared performance by 11- to 12-year-old prepubescent boys and adult men in an all-out 90-s cycle test (33). They divided findings into short-term anaerobic power (work at 10 s), intermediate-term anaerobic capacity (work at 30 s), and long-term anaerobic capacity (work at 90 s). Mean blood lactate concentrations at the end of exercise rose from a mean of 1.0 ± 0.4 mmol · L^{-1} at rest to 8.2 ± 1.4 mmol · L^{-1} in the boys and from 1.5 ± 0.4 to 12.5 ± 2.8 mmol · L^{-1} in the men.

Absolute power output was greater in the adults than in the children at all stages of measurement in this study (33). The difference was approximately halved by controlling for body mass, thigh volume, or height cubed. In general, after these size adjustments, anaerobic performance in the boys was approximately two-thirds that of the men.

When the short-, intermediate-, and long-term components of anaerobic performance were compared between children and adults, the children's performance expressed as a percentage of the adults increased with longer duration. This trend was observed regardless of the means of size normalization. For instance, when performance was related to body mass, the boys' short-, intermediate-, and long-term capacities expressed as percentages of the men's were 65%, 68%, and 79%, respectively.

This study, then, illustrates several basic premises regarding maturational differences in anaerobic fitness: (a) boys produce less lactate than men in a 90-s test, supporting the theory of diminished anaerobic glycolytic capacity in prepubertal subjects, (b) factors other than body size contribute to the differences in anaerobic performance with biologic maturation, and (c) children demonstrate anaerobic performance greatly inferior to that adults in tests of shorter duration, which presumably is an expression of the increasing contribution of aerobic fitness in tests of longer duration.

The question of how power indices on laboratory exercise tests should best be related to body size or composition remains controversial. Van Praagh concluded that it makes no sense to relate peak and mean power to body size on a non-weight-bearing test and argued that power should be instead related to the size of the muscles performing the work (66). The group at Exeter University has provided a compelling argument that allometric scaling techniques are important in valid analyses of anaerobic fitness (3, 22).

However, not all are in agreement. Doré et al. examined the relationship between short-term power and body dimensions in prepubescent, adolescent, and adult females (23). Subjects performed three all-out sprints on a friction-loaded cycle ergometer. When peak power was adjusted for body mass or lean body mass, values were significantly greater in the adults than in the two younger groups. The differences decreased but were not eliminated when lean leg volume was used to adjust power measurements.

In their discussion of these findings, the authors made several observations regarding size normalization for anaerobic power (23). Measurement techniques for assessing leg volume in mature individuals may not be equally appropriate for growing children. Moreover, leg muscle volume does not include all the muscles involved in a high-output cycle test (i.e., arm, trunk, and gluteus maximus muscles are excluded). The authors concluded that "because of its practicality and accuracy compared to estimation of lean leg volume and percent body fat, body mass remains a suitable scaling variable in prospective growth studies" (p. 479). This conclusion was supported by their finding that correlation coefficients between peak power and body mass, fat-free mass, and lean leg volume were similar (r = .57 to .64), regardless of age group.

This study was similar in design to one involving boys ages 11 to 19 years by Mercier et al. (45), who assessed the effects of age and anthropometric features on maximal anaerobic power (using a force–velocity test). Lean body mass and total muscle mass were calculated from skinfold measurements, and leg volume was estimated anthropometrically. Absolute maximal power increased with age and was closely related to leg volume (r = .84) and total muscle mass (r = .88). The highest correlation, however, was with lean body mass (r = .94), which accounted for 88% of the variance in maximal anaerobic power.

In both this study and that of Doré et al. (23), maximal aerobic power expressed relative to any anthropometric variable increased significantly with age. That is, in both males and females, improvements in anaerobic power with age could not be explained by increases in body mass, lean body mass, or leg muscle volume alone.

De Ste Croix et al. (22) did not concur with the conclusions of Doré et al. (23). They used a multilevel modeling approach to examine the development of anaerobic power during Wingate testing of prepubertal children. The subjects were age 10.0 ± 0.3 years at first testing, and testing was repeated at age 11.8 ± 0.3 years. With mass and body composition adjusted for, no sex-related effects on anaerobic performance were observed. In their first analytical model, body mass and height were significant explanatory variables, with scaling exponents of 0.76 and 2.55, respectively, for peak power and 0.75 and 2.92 for mean power. The introduction of a sum of skinfolds measure improved the fit of the model, and the mass exponent increased to 1.49 and 1.48 for peak and mean power, respectively. Estimated thigh volume (but not leg strength) contributed significantly to both mean and peak power.

These authors felt that "changes in body composition affect the development of short-term power output and [this] indicates the need to consider both body mass and body composition as predictors of peak and mean power. The variation of the mass exponent between models for both peak and mean power highlights the sensitivity of the mass exponent to the introduction of other explanatory variable and highlights the limitations of the use of the conventional ratio ($W \cdot kg^{-1}$)" (22, p. 146). This study also indicated that increases in thigh muscle volume contribute significantly to changes in anaerobic power, beyond the influences of body mass and skinfold thickness. And, consistent with other studies, these authors found that even when all of these anthropometric factors were taken into account, anaerobic power still improved with age.

Using a similar analytical approach, Armstrong and his colleagues studied the development of anaerobic power in the same subjects at ages 12, 13, and 17 years (3). They found that mean and peak power were greater in boys than girls, but the differences were reduced (but persisted) when body composition was considered. For mean power the sex difference increased with age. In the initial model, the mass exponent for peak power was 0.88 and for mean power, 0.61. When the statistical model included skinfold thickness, the relative exponents changed to 1.23 and 1.12, indicating that peak and mean anaerobic power increase with age during adolescence independently of body size and composition.

The findings of Welsman et al. suggest the importance of considering thigh muscle volume in interindividual comparisons of anaerobic power (73). These researchers examined the relationship between thigh muscle volume determined by magnetic resonance imaging and anaerobic power during Wingate cycle exercise in 9- to 10-year-old boys and girls. Mean and peak anaerobic power were greater in boys than girls, whether expressed relative to $M^{1.0}$ or to allometrically scaled mass. However, there were no sex differences in peak or mean anaerobic power when expressed relative to thigh muscle volume. This was considered to indicate that boys and girls possess no qualitative differences in anaerobic function.

However, Van Praagh et al. were not able to confirm these findings in a group of 12-year-old boys and girls (71). Peak power was determined by the force–velocity test and mean power by Wingate testing. The boys demonstrated an average absolute peak power 33% greater than girls'; this difference was reduced to 31% and 15% when corrected for body mass and fat-free mass, respectively. When adjusted for lean thigh volume (by anthropometric measurement), the boys had a 21% greater value for peak power. Similar findings were obtained for mean anaerobic power. When analysis of covariance was used to remove the effects of body composition, the same sex differences persisted.

These conflicting data make it difficult to draw conclusions regarding sex differences in anaerobic power and the optimal means of adjusting power values for body dimensions. All these recent studies, however, consistently support the concepts that (a) anaerobic power improves during the growing years and (b) something beyond body or muscle size accounts for these improvements.

Some aspects of short-burst power testing are underexplored. For instance, Van Praagh and his colleagues found age differences in the *rate* that power was developed during the 30 s of Wingate testing (70). They compared power outputs in 18 boys age 7.4 ± 0.2 years and 14 boys age 12.9 ± 0.5 years. Time analysis indicated that the younger group did not reach either maximal work or maximal velocity before 15 s, while the older group was capable of achieving both by 10 s. Consequently, cumulated supramaximal work was significantly lower in the younger than in the older boys during the first 5 s. The authors suggested that "developmental, metabolic, and neuromuscular factors may explain these differences in 'anaerobic fitness' in pre- and circumpubertal boys" (p. 92). Doré et al. described similar findings in a group of 506 males ages 7 to 18 years (24; figure 9.5).

Mechanisms

As defined by short-duration power production and indicators of glycolytic capacity, anaerobic fitness relative to body size increases as children grow. The explanation for this trend is not readily apparent. Why should children become relatively more metabolically anaerobic with increasing age? And what is the functional significance of this pattern of change?

Whatever the mechanism, it is not a trivial one. In the growth curves for absolute and mass-relative anaerobic power provided by Bar-Or (9), the mass-independent influence is observed to account for approximately 30% to 40% of the rise in peak and mean values between the ages of 8 and 14 years. By necessity, then, any proposed mechanism for the relative development of anaerobic capacity and power in children needs to account for an effect of this magnitude.

Changes in the ratio of body mass to volume of actively contracting muscle may account for some of this trend (44). But, as noted earlier, considerable interindividual differences in anaerobic power are observed even when values are related to lean leg volume or lean body mass. Developmental changes in neuromuscular activation could affect anaerobic performance, but this has not yet been clearly demonstrated (see chapter 10). Differences in muscle architecture (15) or muscle–joint relationships (44) between children and adults might contribute.

It appears unlikely that any changes in the proportion of fast-twitch muscle fibers during the pediatric years can account for the extent of size-independent improvement in anaerobic power production. On the other hand, the apparent increase in the glycolytic capacity of type II fibers as children age, as suggested by changes in enzyme activity and lactate production, is a plausible explanation. According to the study of Berg et al., glycolytic enzyme activity (at rest) increases by almost 50% between the ages of 4 and 18 years, consistent with the magnitude of rise in blood lactate levels at maximal exercise over

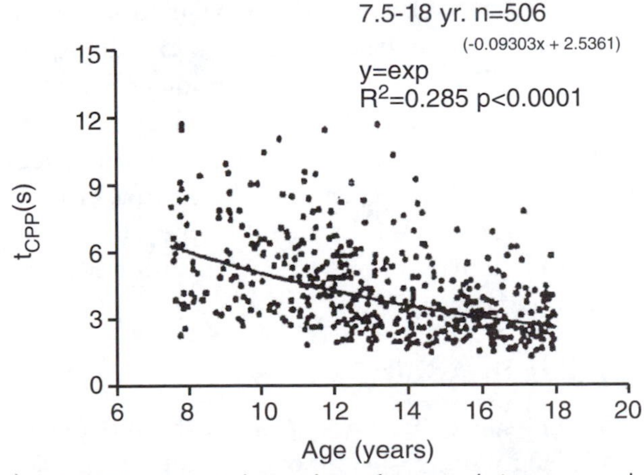

FIGURE 9.5 Relationship of age and time to reach maximal power (t_{CPP}) during a cycle sprint (from reference 24).

the same period (12). In turn, the extent of these changes is similar to those observed for the mass-independent rises in peak and mean anaerobic power during Wingate testing as children grow (9).

Equally intriguing is the mystery of the biologic meaning of improvements in anaerobic metabolic fitness and power production. As noted before, the parallel increases in glycolytic activity witnessed in phylogenetic studies of adult animals suggest that this trend is linked to body size rather than biologic maturation. In addition, the functional implications of the changes in glycolytic capacity and anaerobic power apparent in laboratory testing are not clear. As we shall see in the next section, these changes are not necessarily translated into improved field performance of short-burst activities.

Short-Burst Fitness in the Field: Sprinting

Sprinting is a particularly appropriate model for studying the development of short-burst fitness in children. The duration of a 40- to 50-m sprint is generally no more than 12 s, well within the time domain expected for anaerobic glycolysis. Sprint running is a natural exercise for children and is easy to test. Moreover, short-burst running is a typical feature of many youth sport activities.

Sprint speed over a given distance progressively improves during childhood. The factors that contribute to this improved performance are not clear. Particularly, the relationship of sprint capacity to changes in body size and anaerobic metabolic capacity remains to be clarified. We can begin to gain some perspective on this issue, however, by examining the larger body of experimental data on adult subjects. In this discussion, however, it is important to recognize that the factors that define interindividual differences in sprint performance, either among adults or children, are not necessarily the same ones that determine ontogenetic improvements in sprint capacity as children grow.

Studies in Adults

Running speed during a sprint is the product of stride length times the number of strides taken. Factors influencing sprint speed, then, can affect either the frequency at which the legs move or the force generated in a single stride to propel the body mass against gravity. The interaction of these two variables is highly complex. For instance, greater stride frequency can be achieved not only by more rapid repositioning of the legs but also by decreasing the time that the foot is in contact with the ground in each stride. Elastic recoil might contribute in a variable amount to stride length beyond the force applied. The relative horizontal and vertical vectors of force applied to the ground affect the amount of forward propulsion. In addition, external factors such as wind resistance, shoe construction, terrain, and ambient temperature all affect sprint performance.

The relationship between velocity and time during a sprint has two distinct phases: an initial acceleration during the first 30 m of a 100-m race, followed by a constant (or maximal) speed. (In some individuals a later third phase of deceleration is observed.) Sprint performance, then, depends on the ability to accelerate rapidly at the onset and then maintain a high velocity to the finish line. Because these phases involve differences in force production and biomechanics, the determinants of performance in each phase may need to be considered separately (39, 65).

Stride Length

The individual with a greater stride length is generally thought to have an advantage in sprint performance by being able to produce greater forward propulsion. Weyand et al. studied the relative contributions of stride frequency and length to treadmill sprint velocity in adults 18 to 36 years old (75). Subjects responded to increasing speed predominantly by increasing stride length at lower speeds and stride frequency at higher velocities. The stride length at the top speed was 1.69 times greater in fast than in slow runners, while stride frequency was 1.16 times greater. These authors concluded that stride frequency was not a determinant of sprint performance: "Certainly, top sprinters have faster muscle fibers and greater muscular power available to reposition their limbs, yet do so little or no faster than average or slow human runners do." (p. 1998). Others, however, have suggested that stride rate may be more important than stride length in determining maximal sprint velocity (49).

Force Production

Greater stride lengths are an outcome of more force applied to the running surface to propel the body against gravity. In a study by Weyand et al., the average mass-specific force during level treadmill running was 1.26 greater in faster than in slower runners (74). Chelly and Denis reported that forward-leg power measured on a treadmill was related to both the initial acceleration phase (r = .80) and maximal running velocity (r = .73) during sprinting by 16-year-old runners (17).

Young et al. (77) studied the relationship of explosive strength of the leg extensor muscles during a vertical jump with 50-m sprint performance in junior elite athletes (ages 16–18 years). The best predictor of starting performance (time for the first 2.5 m) was the peak force (relative to body mass) produced during a vertical jump from a 120° knee angle. Maximal sprint speed correlated closely with force applied at 100 ms from the start of a jump. These findings indicate that explosive power as defined by vertical jump performance is a critical determinant of sprint speed.

Muscle Strength

Young et al. also examined the relationship between sprint performance and maximal muscle strength as determined from an isometric squat (77). Absolute strength was closely related to maximal sprint velocity (r = .79) and starting ability (r = .72). Similarly, Mero et al. described a correlation between maximal absolute isometric strength and maximal sprint speed of r = .62 (50).

Dowson et al. found that isokinetic strength was related to sprint running performance in elite runners (25). However, this relationship was largely explained by differences in leg length and body mass.

Anaerobic Metabolic Capacity

Anaerobic glycolysis contributes approximately 55% to 75% of the energy demands of a sprint lasting 10 s, while aerobic metabolism is responsible for 13% (54). Consistent with this, maximal blood lactate production in adult runners has been linked to performance on a 30-m sprint (51).

In an important study, Weyand et al. found, however, that the power output during maximal sprints was not limited by the rate of anaerobic metabolism (74). They demonstrated no significant difference in maximal running speeds in normoxic (20% O_2) and hypoxic (13% O_2) conditions. The subjects' anaerobic metabolic input (as estimated by oxygen deficit) increased to compensate for the reductions in aerobic power. This finding suggests that sprint performance is not normally limited by anaerobic metabolic capacity, and maximal velocity during sprinting consequently cannot be used as a marker of an individual's anaerobic metabolic fitness.

Muscle Fiber Types

Individuals who excel at sprinting possess a greater percentage of type II (fast-twitch) fibers (19). The fast repetitive motions necessary for optimal sprint performance and the greater rate of force development observed in sprint athletes than in nonathletes are consistent with this finding (54). The presumed genetic basis of fiber-type proportion suggests a high heritability of sprint capability.

Neural Influences

Muscle force is controlled by not only the number of recruited motor units but also the firing rate of individual motoneurons. The influence of these neural factors on sprint performance has not yet been determined. Lehnert and Weber demonstrated that motoneuron conduction velocity is faster in adult athletes trained for speed (cited in reference 49).

Sprint Performance Correlates in Children

Numerous studies have described improvements in sprint performance as children grow (60). In some studies the improvement in sprint speed with growth was found to be similar in boys and girls. In others, particularly those involving a longer age span, boys perform better. For instance, in the AAHPERD testing results, average 50-yd sprint velocity improved by about 50% in boys and 23% in girls between the ages of 7 and 17 years (2; figure 9.6).

As Sargeant noted, "At first glance therefore it may not seem surprising that as the muscle size increases so does performance in short duration events. However, while muscle size and hence power will be increasing with age, so too will the size of the whole body which has to be moved" (57, p. 144). Thus, determinants of improvements in sprint velocity during childhood need to be examined in the context of changes in body dimensions. The variables that contribute to running performance in adults, outlined earlier, can be considered. Establishing cause-and-effect relationships in this analysis is difficult, and it must be remembered that cross-sectional differences among adults may not reflect longitudinal, ontogenetic changes in children.

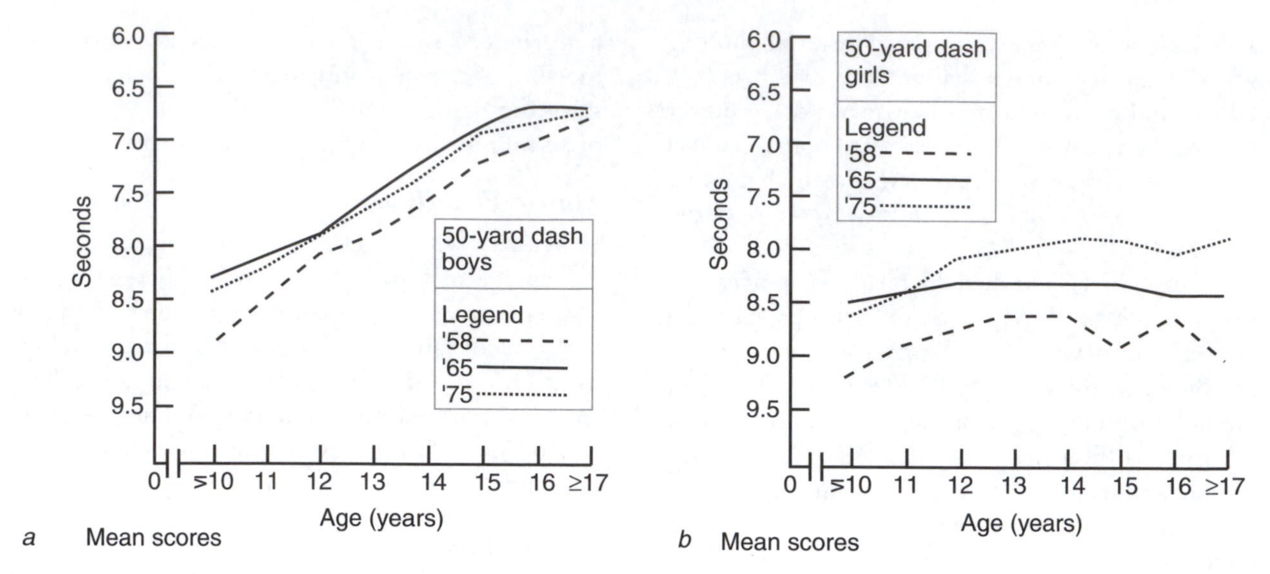

▶ FIGURE 9.6 Sprint performance in *(a)* boys and *(b)* girls with increasing age (from reference 2).
Reprinted by permission from AAHPERD 1976.

Stride Length

The development of 50-yd sprint performance was assessed in the longitudinal Trois Rivieres Study (61). Run times between the ages of 6 and 12 years were related to height by the exponent –1.04 in boys and –0.87 in girls. Assuming that height and leg length are closely related during this time period, these findings suggest that improvements in sprint performance during childhood are associated with progressive increases in stride length. Dimensionality theory also indicates that sprint performance in this study should be related to mass by the exponent 0.33; that is, running velocity per kilogram of body mass should decrease with increasing body mass.

Mero studied athletic boys from ages 11 to 15 years and found that improvements in maximal 20-m sprint speed were achieved by increases in stride length (46). Rowland examined the relationship of 50-yd sprint performance and morphologic measures in a five-year longitudinal study of children who were initially nine years old (unpublished data). Velocity relative to body height remained essentially stable over the study period, while velocity in respect to body mass declined. At age nine, sprint times related to mass by the exponent 0.22. These data, then, support the idea that when running performance is considered in respect to body mass, it decreases as children age.

This mass-specific decline contrasts with the observed trend for improved glycolytic capacity as children grow. This observation suggests that while the energy for sprinting is derived from anaerobic sources, anaerobic capacity does not serve as a limiting factor for sprint performance during childhood. Moreover, it appears that sprint performance should not be interpreted as a marker of anaerobic glycolytic capacity. This finding and interpretation are consistent with the study by Weyand et al. in adults, which indicated that sprints do not exhaust anaerobic metabolic capacity (74).

Asmussen and Heeboll-Nielsen described a height scaling exponent of 0.68 for sprint performance in boys 7 to 16 years old who ran barefoot down a long corridor (5). The scaling exponent was 1.86 for the velocity during the first 4 m of the sprint, consistent with the need to generate more force to accelerate body mass against gravity during the initial phase of the sprint.

Force Production

Although speed during sprinting is related to body height, the force applied per stride should be expected to correlate with body mass (since force is applied to move body mass against gravity). Shephard et al. found that leg extension force was related to height by the exponent 2.80 in boys and 2.96 in girls, indicating that force increased in direct proportion to mass over the six years of the study (61). Rowland found similar values for height of vertical jump per kilogram in children between the ages of 9 and 13 (unpublished data).

Muscle Strength

Increases in muscle strength generally follow body mass as children grow, although considerable variability is observed in this relationship, depending on the muscle group in question. Cross-sectional studies of children have found muscle strength to be related to sprint ability. Thorland et al. compared isokinetic peak torque in national junior-level sprinters age 11.1 ± 0.2 years and middle-distance runners age 10.9 ± 0.2 years (64). The sprinters exhibited significantly greater peak torque values at all contraction velocities. Similar findings were apparent in adolescent sprinters and distance runners. On the other hand, Mero et al. could find no differences in maximal isometric force between sprinters, distance runners, and nonathletic control boys ages 10 to 13 years (48).

Muscle Fiber Types

Little information is available regarding the relationship of muscle fiber types and sprint performance during growth in children. In general, distribution of muscle fiber types is thought to genetically based and fixed early in life (52). It is not expected, then, that an alteration in the population of type I or type II fibers would contribute to changes in short-burst performance during childhood. The matter is not entirely settled, however. The autopsy study of Lexell et al., for instance, indicated a relative increase in type II (fast-twitch) fibers from 35% at age 5 years to about 50% at age 20 (41).

A cross-sectional study of athletic youth revealed similar fiber type differences in relationship to sport specialization as observed in adults. Mero and colleagues divided 11- to 13-year-old athletic boys into two groups according to muscle fiber distribution (47). The "fast" group, with more than 50% fast-twitch fibers, included sprinters, weightlifters, and tennis players. The "slow" group, with more than 50% slow-twitch fibers, included endurance runners and tennis players. The fast group was superior to the slow group in reaction time, rate of force production, and rise of the body's center of gravity in the squatting jump.

Neural Influences

There appears to be no change in nerve conduction speed during biologic maturation once the school years are reached. Motoneuron conduction velocity in infants is approximately one-third that of mature individuals. However, by age five or six years, values have generally reached adult levels (30, 31). In a study of 184 subjects ages 10 to 75 years, no age-related changes were seen in the amplitude or duration of the action potentials of sensory nerves (40).

Blimkie evaluated age differences in the degree of motor unit activation in a group of 10- to 16-year-old children (14). No age-related differences were seen in average percentage of motor activation for the elbow flexors (89.4% for the 10-year-olds and 89.9% for the 16-year olds). However, activation for knee extensors was significantly higher in the older subjects (95.3% vs. 77.7% for the younger subjects). The question of possible maturational differences in patterns of motor unit recruitment therefore remains unanswered.

Possible Conclusions

This information regarding the factors involved in the development of sprint performance in children is highly fragmentary and precludes any definite conclusions. A number of concepts, however, are suggested. First, these data reveal no improvements in sprint performance with growth beyond those that can be explained by increases in body size. Longer stride length, in association with greater force applied per stride (a consequence of increasing muscle mass), appears to account for the improvements in sprint velocity during childhood.

The role of anaerobic metabolism in the development of sprint performance in youth remains problematic. While sprinting depends largely on energy derived from glycolytic pathways, information from adults suggests that the limits of anaerobic capacity are not taxed during sprint performance. That anaerobic metabolic capacity is not linked to improvements in sprint speed with growth is further indicated by the observation that glycolytic capacity (as indicated by lactate production during exercise and activity of glycolytic enzymes) improves during childhood, while mass-specific sprint velocity decreases.

It is reasonable to suggest, then, that the ontogenetic development of sprint fitness in children is related to anthropometric rather than metabolic factors. This should be even more true for those short-burst activities whose performance relies highly on developmental changes in neuromuscular coordination, balance, and motor skill (e.g., skipping, jumping rope, hopping). In general, it seems that improvements in these activities during childhood cannot be interpreted as changes in anaerobic metabolic fitness. (The same may not be true of cross-sectional comparisons of sprint performance in youth, in which interindividual differences may be associated with percentage of fast-twitch fibers and anaerobic capacity.)

In some sports, such as sprint cycling, however, anaerobic metabolic capacity might be expected to have a dominant effect. In swimming, on the other hand, the influence of anaerobic glycolysis is uncertain, given the potential influences of skill in swim stroke, the effects of body habitus on economy, and so on. Bar-Or et al. conducted an analysis that suggested that anaerobic capacity might not play a critical role in competitive performance by age-grouped swimmers (11). These authors reasoned that because anaerobic capacity is considered relatively inferior in children, their performance in short-sprint swim events should be expected to be relatively poorer than adults' in long-distance events. A review of performance times by Canadian swimmers failed to support this idea. These authors concluded, then, that "during the years of growth, performance [in swimming] depends primarily on mechanical factors and less so on the relative anaerobic-aerobic performance" (p. 204).

Van Praagh and França compiled similar data in child and adult runners (72). They compared performance times for short- and middle-distance runs by 10-year-old U.S. champion girls and boys with the respective world records by adults. A similar difference of about 25% was observed in 100-m, 200-m, and 400-m event times for both boys and girls. These data support the idea that maturational differences in anaerobic glycolytic capacity do not contribute significantly to the improvement in the competitive performance of young runners and swimmers.

Sprint–Power Correlations

The relationship between sprint performance and anaerobic power has been examined in several cross-sectional studies of children. It is not altogether clear, however, how these measures should be expressed relative to body size when making such comparisons. Absolute values of both sprint velocity and anaerobic power in Wingate or other laboratory testing increase with age. But sprint velocity appears to be linked to increases in body height (i.e., leg and stride length) and therefore scales allometrically to increasing body mass by an exponent less than 1.00. Peak and mean anaerobic power, on the other hand, rise with age in greater proportion to body mass, so the mass scaling exponent is greater than 1.00.

Almuzaini reported correlation coefficients between absolute 50-m sprint time and measures of peak and mean anaerobic power in 12-year-old children at various braking resistances (0.065–0.080 kp per kilogram of body mass) expressed in absolute terms and relative to body weight and height (1). For peak power, a significant relationship between power and sprint time was observed only for the highest braking load, and coefficients were similar regardless of how power was expressed ($r = -.47$ to $-.53$). For mean power, correlations were similar but were seen at all braking forces.

Van Praagh et al. (71) found a correlation of $r = .71$ for peak power per kilogram of body mass and 30-m dash time in 12-year-old boys, but the relationship was insignificant in girls ($r = .23$). In a study of seven-year-old boys (69), the same authors reported that 30-m indoor run times correlated moderately well with both maximal peak power measured by force–velocity test ($r = -.45$) and mean power measured by Wingate testing ($r = -.60$).

Ventilatory Anaerobic Threshold

The anaerobic threshold is defined as the point in a progressive exercise test when blood lactate levels begin to rise. This threshold was initially interpreted as an anaerobic glycolytic response to insufficient oxygen for aerobic metabolic support of contracting muscle. According to this model, the increased rate of glycolysis and lactate production are part of a default mechanism for providing energy for locomotion. This "oxygen starvation" construct has been replaced by the idea that lactate accumulation occurs when the rise in pyruvate from accelerated glycolytic metabolism—which occurs independently of oxygen supply—exceeds that which can be utilized by aerobic metabolic pathways.

In either model, however, the anaerobic threshold can potentially be influenced by either aerobic or anaerobic capacity. The anaerobic threshold has therefore been promoted as a useful noninvasive means of assessing fitness that does not require maximal effort by the subject, a particular advantage for people with cardiopulmonary disease.

The rising lactate at the anaerobic threshold is buffered by bicarbonate, with a subsequent rise in CO_2 production beyond that caused by metabolic activity. This excess CO_2 stimulates an acceleration in minute ventilation out of proportion to the increase in oxygen uptake. The ventilatory anaerobic threshold (VAT), the point when the increases in minute ventilation and $\dot{V}O_2$ diverge, has therefore been used as a noninvasive means of estimating the anaerobic threshold.

Several approaches have been used for estimating VAT, including (a) the v-slope method, in which a computerized linear regression analysis determines the point at which $\dot{V}CO_2$ begins to increase more rapidly than $\dot{V}O_2$, and (b) visual methods, most commonly the point at which the ventilatory equivalent for oxygen ($\dot{V}_E/\dot{V}O_2$) is observed to increase without a concomitant rise in the ventilatory equivalent for carbon dioxide ($\dot{V}_E/\dot{V}CO_2$). Fawkner et al. found that the two methods produced similar values for VAT expressed both as absolute $\dot{V}O_2$ and as a percentage of $\dot{V}O_2$max in 22 children (28). These authors felt that the main advantage of the v-slope over the visual method was the ability of the former to detect VAT more frequently (all of the cases in their study, compared with a 16% failure by the visual method). On the other hand, Singh et al. (63) found that VAT values determined by the v-slope method were significantly lower than values determined by observational methods (18.9 ± 2.7 vs. 22.3 ± 3.7 ml · kg^{-1} · min^{-1}, respectively, $p < .001$).

Evidence supports the use of VAT as an accurate marker of lactate anaerobic threshold in children. Ohuchi et al. assessed its validity as an indicator of anaerobic threshold in a group of patients ages 7 to 21 years that included 8 with a history of Kawasaki disease and 17 with complex cyanotic congenital heart disease (53). VAT was defined as the oxygen uptake at which an increase was observed in the ventilatory equivalent for oxygen without a rise in the ventilatory equivalent for carbon dioxide. No significant differences were seen between mean values for lactate anaerobic threshold and VAT either in absolute terms or as mass-relative $\dot{V}O_2$. The correlation between VAT and lactate anaerobic threshold was r = .91. The researchers found similar results when VAT was determined by the v-slope method. In this study, VAT was also observed to be a reliable marker of aerobic fitness, with a close relationship to $\dot{V}O_2$max (r = .75).

Hebestreit and his colleagues also found VAT to serve as a valid marker of aerobic fitness in children (37). In a study of 6- to 12-year-old children, these authors reported a close relationship between VAT (determined by visual methods) and peak $\dot{V}O_2$ (r = .92; figure 9.7).

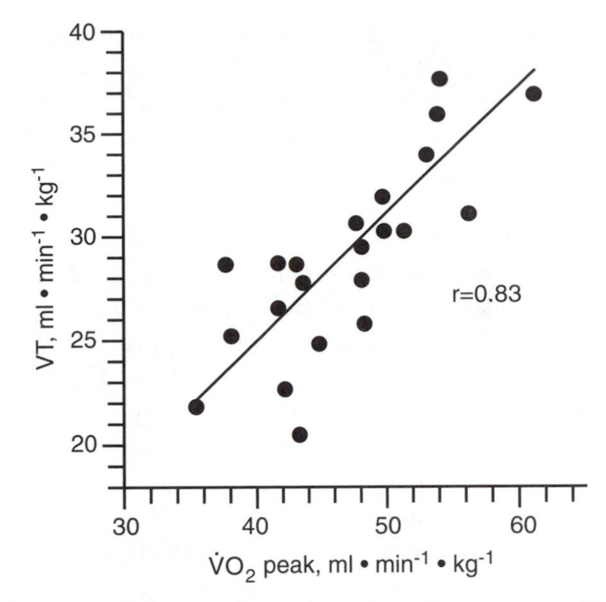

FIGURE 9.7 Relationship of ventilatory anaerobic threshold (VT) and peak $\dot{V}O_2$ in children 6 to 12 years old (modified from reference 37).

Adapted by permission from H. Hebestreit, B. Staschen, and A. Hebestreit 2000.

VAT as a percentage of $\dot{V}O_2$max has been demonstrated to decrease during childhood in most studies. This trend has been interpreted to indicate an increase in anaerobic capacity as children grow. However, as Mahon and Cheatham pointed out, "The data from these studies are far from unanimous with regard to the effect of growth and development on VAT during childhood. The primary limitation within these studies is that the cross-sectional design precludes the establishment of cause and effect between growth and development and VAT" (42, p. 23).

Explosive Power: Vertical Jump

The vertical height achieved on a single, maximal jump has been interpreted to indicate explosive muscular power. Although a number of protocols have been used, most mimic the one designed by Sargent in 1921 (58), in which subjects jump after a countermovement (i.e., from a crouching position) and jump height is recorded as the height achieved minus the vertical distance of the upstretched arm at rest.

This seemingly simple maneuver is, in fact, highly complex, involving vertical acceleration of the center of mass by rotation of body segments in a sequential fashion (16). The height of a vertical jump (a) is effectively determined by the velocity (v) of the body at takeoff by the equation $a = v^2/2g$, where g is the acceleration due to gravity (4). The work of the muscles to propel the body at this velocity is expressed by $fd = mv^2/2$, where f is force, d is the distance over which the force works (from the crouch position to toe extension), and m is body mass. Combining and rearranging these equations, given that mg is equal to body weight (w), gives $a = fd/w$. Changes in vertical jump height during growth in children, then, depend on improvements in muscle strength after the influence of the ratio of distance and mass is accounted for.

In addition, the muscle force applied at takeoff contributes not only to effective energy (i.e., that utilized to raise the center of mass) but also to the energy "wasted" on variables such as horizontal kinetic energy and rotation of the body segments. A coordinated jump, then, optimizes the *efficacy ratio,* or the ratio of effective energy to total mechanical muscle work. Therefore, many potential variables could influence developmental changes in vertical jump performance during childhood.

Before describing the empiric findings in jump performance in children, it is of interest to examine the observations in animals briefly. Zoologists delight in identifying biologic phenomena that are counterintuitive, and jump performance relative to animal size is one of them. By theoretical considerations, geometrically similar animals should all be expected to jump to the same height, *regardless of body mass.* The reasoning, as explained by Schmidt-Nielsen, goes as follows:

> "Muscle force is proportional to its cross-sectional area, and the shortening is proportional to the initial length of the muscle. The cross section times the length is the volume of the muscle, and the energy output of a single contraction is the product of force and distance. Energy output is therefore proportional to the muscle mass and, in turn, to body mass. In a jump, an animal uses a single contraction of the jumping muscles, and the energy available for takeoff and used for acceleration is therefore the same relative to the body mass. The same amount of energy (work) per unit mass can lift the bodies to the same height and no more. The conclusion is that similar animals of different masses should jump to the same height, provided that their muscles contract with the same force" (59, pp. 177–179).

Although observation tells us that this is not entirely true (the jumping height of a flea, locust, and adult human are clearly not identical), the basic concept seems valid. As Schmidt-Nielsen pointed out, the jumping heights of animals varying in body mass by a factor of over 100 million are found to vary no more than about threefold (59).

Violating theoretical expectations, then, vertical jump performance is observed to improve progressively as children become older. Malina and Bouchard combined the findings of several studies to describe age-related changes in vertical jump (43). Between ages 5 and 13 years, jump heights doubled and were similar in boys and girls (figure 9.8). At puberty, height increases for boys exceeded those for girls.

Klausen et al. described longitudinal changes in vertical jump height in two groups of boys and girls studied between the ages of 10 and 12 years and between 13 and 15 years (38). Mean jump height increased in the first group from 15 cm to 19 cm, with no sex difference. Between the ages of 13 and 15 years, mean vertical jump increased from 18 cm to 24 cm in the males, but no significant change was observed in the females over the three years.

When these jump height values were expressed relative to body mass, no significant change was seen over time in the younger group (0.45 cm · kg^{-1} initially and 0.43 cm · kg^{-1} at follow-up). In the older group, jump height related to body mass did not change in the boys over the three years of testing (from 0.43

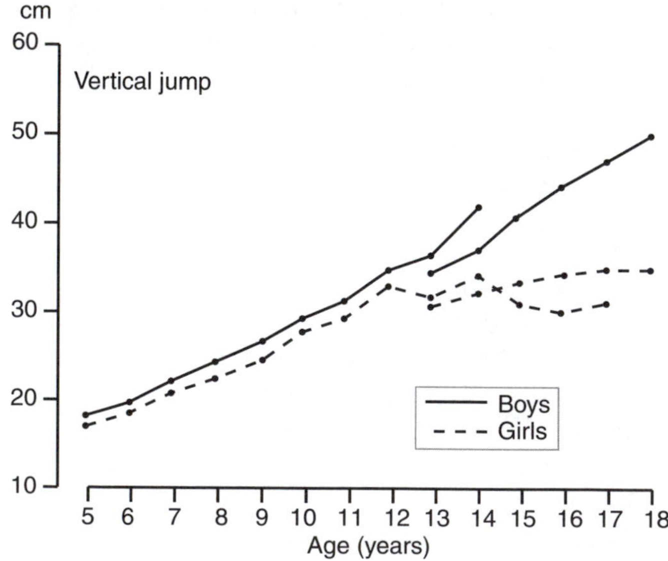

FIGURE 9.8 Increases in vertical jump height between ages 5 and 18 years. Data from multiple sources compiled by Malina and Bouchard (43).

Reprinted by permission from R.M. Malina and C. Bouchard 1991.

to 0.40 cm · kg^{-1}), but values declined in the females (from 0.42 to 0.34 cm · kg^{-1} at final testing).

Rowland compared the vertical jump height of nine boys and six girls measured when they were 9.2 ± 0.5 years old and then again at age 17.0 ± 0.6 years (unpublished data). Vertical jump height increased from 23.8 ± 3.0 cm to 44.1 ± 11.4 cm over the eight years. Values relative to body mass, however, were similar: 0.72 cm · kg^{-1} at the first study and 0.65 cm · kg^{-1} at follow-up. However, the ratio of jump height to body height was 0.172 at age 9 and 0.252 at age 17.

These data indicate that over an eight-year period, the ability of the muscles of a child's lower extremities to propel body mass vertically in a single contraction increases by almost 100%. Moreover, improvements in explosive power as defined by vertical jump height appear to be closely associated with increases in body mass (i.e., vertical jump height is directly related to mass by the exponent 1.00).

However, when components of force and velocity are considered in jump performance, the findings may be different. Ferretti et al. compared vertical jump performance in 13 children ages 8 to 13 years and young adults ages 20 to 35 years (29). The maximal instantaneous muscle power (W_{peak}) was determined from force (measured by force platform) and velocity (calculated by time integration of the instantaneous acceleration, considered equal to the ratio of force to the mass of the subject). Absolute W_{peak} was 3,151 ± 528 W in the adults and 1,103 ± 393 W in the children. Values continued to be significantly greater in the adults when expressed relative to body mass (43.1 ± 5.8 vs. 31.6 ± 8.4 W · kg^{-1} in the children) or muscle cross-sectional area estimated by anthropometric methods (5.15 ± .07 vs. 3.37 ± .07 W · cm^{-2} in the children). The W_{peak} was therefore about 65% lower in children than would be expected from differences in muscle cross-sectional area between the groups.

The matter is clearly complex, and the numerous confounding factors cloud our attempts to understand how jump performance improves in children with respect to body size and the mechanisms that drive this change. By dimensionality theory, muscle strength or force should be related to the cross-sectional area of muscle, which in turn should be related to height by the exponent 2.0 or to mass by the exponent 0.67. The analyses of Klausen et al. (38) and Rowland (unpublished data) indicate that vertical jump height in children increases at the same rate as mass (i.e., it is related to mass by the exponent 1.0), while the findings of Ferretti et al. (29) indicate that the mass scaling exponent is greater than 1.00.

Vertical jump performance, then, joins the list of other markers of anaerobic performance that indicate that developmental improvements during childhood occur at a greater rate than can be accounted for by increases in muscle size alone. The size-independent variable responsible for this trend has not been identified. The energy needs of such instantaneous exercise should be met by stored ATP and phosphocreatine rather than glycolytic metabolic pathways. In the jump study by Ferretti et al., the relative resting concentrations of ATP and phosphocreatine estimated by nuclear magnetic resonance imaging were not significantly different between children and adults (29). Despite the attractiveness of a neural explanation for developmental changes in muscle power production, there are no studies that convincingly document this. Possible neurologic contributions are discussed further in chapter 10.

Vertical jump capacity appears to be under considerable genetic control in children and adolescents. Beunen et al. quantified the genetic and environmental contributions to explosive strength (vertical jump) in 105 twin pairs studied longitudinally from ages 10 to 18 years (13). Genetic factors reportedly explained 48% to 92% of the interindividual variation in performance. This magnitude of genetic influence is similar to that observed for other forms of anaerobic performance (62).

Conclusions

Most research attention in developmental exercise physiology has been directed toward aerobic rather

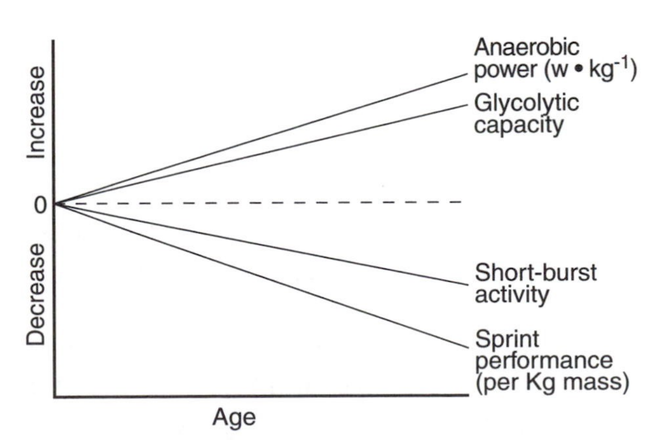

FIGURE 9.9 Developmental trends with age in different aspects of anaerobic fitness.

than anaerobic fitness because (a) aerobic capacity is better defined and easier to study and (b) aerobic fitness has been linked to health outcomes. As Van Praagh and França stated, however, this is surprising, since the exercise habits of children are almost entirely short-burst activities (72). Moreover, salutary effects of exercise on health appear for the most part to be related to habitual moderate activity rather than to physical fitness. Therefore, we should be interested in understanding and promoting short-burst activities in children from the standpoint of preventive health care.

The principal challenge facing researchers in this area is defining the relationships between developmental changes in anaerobic glycolytic activity, laboratory measures of power production, and practical outcomes in the performance of short-burst activities. In particular, it is not clear how each of these measures is best expressed in relation to changes in body dimensions, a critical point for understanding the biologic development of anaerobic fitness.

There is, in fact, a peculiar divergence of trends in these different aspects of anaerobic capacity (figure 9.9). Amount of habitual short-burst activity declines in respect to body size as children grow. At the same time, glycolytic activity, as indicated by lactate production and enzyme function, appears to increase, and this factor may explain why power production in brief, all-out laboratory tests improves at a rate greater than can be explained by increasing body size as children grow. Ability to perform short-burst field activities seems to be less related to anaerobic energy capacity than to mechanical, neuromuscular, and anthropometric factors. In fact, the relationship of such activities to body dimensions as children grow is different from that of anaerobic power or glycolytic capacity to body dimensions (e.g., sprint running velocity in respect to body mass *decreases* through the pediatric years).

These observations confuse the meaning of the expression "anaerobic fitness" in children. As Sargeant summarized,

> "Anaerobic performance is a convenient but misleading shorthand expression used to refer to the performance of short duration exercise lasting for seconds rather than minutes. . . . The rate at which work is done—that is, the power delivered—ultimately depends upon the rate of energy turnover in the muscle. . . . There is very little direct evidence, although much speculation, on the development of the major energy pathways and their contribution to total energy turnover in short-term exercise performed by children. . . . [Therefore] it might be considered preferable to refer to 'short-term power output' or 'short-duration exercise performance' for example rather than 'anaerobic power or performance'" (57, p. 143).

Discussion Questions and Research Directions

1. How can laboratory tests of anaerobic power be best designed to measure glycolytic metabolic capacity? What methods can be devised to more directly but noninvasively assess glycolytic rate during exercise in children?
2. How are changes in glycolytic capacity with growth reflected in performance of short-burst activities?
3. What mechanism drives changes in anaerobic metabolic capacity with growth?
4. How are markers of anaerobic power or short-burst performance best related to changes in body size as children grow?
5. Do sex differences in anaerobic fitness exist independently of the effects of body composition?
6. How much do factors such as motoneuron activation pattern, muscle architecture, and changes in muscle fiber type populations contribute to changes in anaerobic fitness (by any definition) during the growing years?

10

Muscle Strength

It appears that the maximum force, or
stress, that can be exerted
by any muscle is inherent
in the structure of the muscle filaments.
This force is body-size-independent
and is the same for mouse
and elephant muscle. In fact,
the structure of mouse muscle
and elephant muscle is so similar
that a microscopist would have difficulty
identifying them except for
a larger number of mitochondria
in the muscles of smaller animals.

—Knut Schmidt-Nielsen (1984)

In this chapter we discuss

- the development of muscular strength during childhood,
- qualitative and quantitative factors responsible for improvements in strength as children grow, and
- effects of exercise on muscle damage in youth.

We have looked at fitness in children in terms of the power–duration curve: The greater the force output, the shorter its sustainable duration before fatigue sets in. The basic muscle contractile mechanism is identical at any point on the curve, but the factors that limit that contractile process depend on the time course of the exercise (i.e., the power output). That is, performance during low-intensity, sustained exercise (aerobic fitness) at one end of the curve is defined by energy derived from oxidative metabolic processes and substrate availability, while short-burst activities at the other end of the curve are constrained by factors such as anaerobic glycolytic capacity, muscle strength, neuromuscular development, and anthropometric features. The developmental patterns of aerobic and anaerobic fitness during biologic matration are thus under the control of different influences.

In this chapter we focus on the very beginning of the power–duration curve, the maximal force that can be generated in a single contraction, or muscular strength. That contraction can occur against a fixed object with no shortening of muscle (static, or isometric, strength) or against a moving resistance through an arc of motion (isokinetic strength). We examine the factors that are responsible for improvements in muscle strength as children grow and that distinguish maximal force production from other forms of fitness along the power–duration curve.

The principal determinant of strength is muscle size, and it therefore comes as no surprise that absolute measures of muscle strength improve as children grow. What is less clear is whether the strength of the prepubertal child is different from that of the adult when expressed per unit muscle mass. This is another way of asking if increases in muscle mass can fully account for improvements in strength as children grow. There is, in fact, compelling evidence that size-independent factors beyond muscle size contribute to the development of strength in children. Supporting this hypothesis is the observation that children can experience significant gains in strength from resistance training without concomitant increases in muscle bulk. Changes in muscle architecture, innervation, and muscle contractile properties are the primary candidates to explain this phenomenon, but at present there is insufficient evidence to build a convincing case for any particular influence.

In this chapter we examine data indicating that factors other than muscle size are involved in the development of strength in the growing child. The evidence for determinants that might be responsible is then reviewed. We finish by addressing possible maturational differences in muscle damage during exercise. A discussion of muscle responses to resistance training in children is postponed until the next chapter.

Dimensionality Theory and Allometric Scaling

Before beginning this discussion, the role of dimensionality theory and allometric scaling in constructing the picture of normal strength development in youth is pertinent. As will be evident throughout this chapter, these techniques have commonly been employed to relate changes in strength to body dimensions. This adjustment is a critical necessity, for without an appropriate means of relating such changes to somatic growth, we are left with a rather meaningless set of data. That is, we cannot define sex differences, effects of qualitative factors on muscle function, or the influence of strength training without knowing how strength development normally relates to muscle size.

The key word here is *appropriate.* How can we select normalizing factors for muscle or body dimensions that accurately reflect size-to-strength relationships? Many authors believe that the principles of geometry and physics should provide useful, predictable means of understanding these relationships. Not all, however, are in agreement.

Jaric emphasized the importance of applying the correct allometric exponent derived from dimensionality theory to specific forms of muscle strength (28). Given geometric similarity between subjects (discussed later), linear dimensions should relate to height (H) by the exponent 1.0, areas (such as muscle cross-section) to $H^{2.0}$, and volumes (in essence, body mass) to $H^{3.0}$. Since height relates to mass by the exponent 0.33, the corresponding exponents for mass (M) are 0.33, 0.67, and 1.00.

Muscle force, such as that recorded with a handgrip dynamometer, is expected to relate to the cross-sectional area of the contracting muscle. This means that strength could be expressed as $H^{2.0}$ or $M^{0.67}$ if muscle size were the only factor responsible for strength. If this is true, the handgrip force (F) of subjects of different sizes and ages could be compared by $F/H^{2.00}$ or $F/M^{0.67}$. We shall see in the course of this chapter that some empiric data indicate that the scaling factors for muscle strength are actually higher than these predicted values. These data, then, lead us to suspect that size-independent factors beyond muscle cross-sectional area contribute to increases in muscle strength as children grow.

Jaric contended that muscle torque measured during isokinetic testing should be related to $M^{1.00}$ (28). Torque is determined not only by muscle cross-sectional area but also by the lever arm of the muscle, which, being a linear construct, relates to $M^{0.33}$. Thus, he argued, muscle torque should be expressed in relation to $M^{0.67} \times M^{0.33} = M^{1.00}$.

It has further been suggested that measures of motor performance should be scaled to exponents appropriate for the type of activity involved. For instance, in events that require strength to move body mass (e.g., sit-ups, chin-ups), the capacity to exert force ($M^{0.67}$) increases at lesser rate than body mass ($M^{1.00}$). Therefore, performance of such events should be negatively related to body mass ($M^{0.67}/M^{1.00} = M^{-0.33}$). These considerations might be critical in studies examining relationships between strength and motor performance. Indeed, failure to recognize these relationships might be responsible for the surprisingly poor correlations reported between laboratory measures of muscle strength in children and their performance of motor tasks (56).

These principles are conceptually sound, but some researchers consider their applicability to understanding developmental changes in strength to be limited. Blimkie and Sale reasoned, "While the dimensionality theory provides an interesting theoretical means by which to account for the effects of size on strength development, it is too simplistic and problematic, and it detracts from the search for more basic explanations for the age changes and gender differences in strength in childhood" (9).

As shall be seen in the discussion that follows, considerable variation is observed in children's allometric scaling exponents for strength relative to muscle group, athleticism, sex, and body composition. Muscles are peculiarly shaped tissues that may not conform to the simple geometric expectations used in dimensionality theory. Moreover, the architecture of different muscle groups can be quite disparate. Individual muscles vary in fiber length, type of bony insertion, division by fibrous septa, and angle of pennation. It is not surprising, then, to find that contractile features and the relationship of force production to muscle cross-sectional area, height, and weight vary among different muscle groups.

Development of Muscle Strength

Increases in muscle size as children grow occur as muscle fibers become larger (hypertrophy) from increased protein content rather than by cellular multiplication (hyperplasia; 30). Muscle fiber number is fixed at or soon after birth, but between the ages of one year and adolescence, fiber diameter increases almost threefold (42). This muscle hypertrophy is reflected in an increase in total-body muscle mass during the growing years (see figure I.3 in the introduction). Estimated muscle mass rises linearly with age in the prepubertal years, with mean values slightly higher in boys. At puberty, androgenic hormones cause a rise in the rate of muscle growth in boys, while minimal changes are seen in girls. These curves for muscle bulk by sex mimic those for muscle strength.

It may be pertinent that muscle mass as a proportion of body mass increases during growth (34, pp. 126-128). In males the average muscle mass as a percentage of body mass rises from 42% at age 5 years to 53% at age 17. This change is not observed in females, who have values of 41% and 42% at these ages.

The development of muscle mass in the growing child is reflected in a progressive increase in muscle strength. Assessment of developmental changes in strength has been performed in cross-sectional and longitudinal studies by testing isometric and isokinetic strength using different testing modalities (handgrip dynamometers, isokinetic dynamometers) in multiple muscle groups. While certain variations have been reported, the overall pattern of strength development observed by using these different approaches is remarkably similar. These developmental changes in strength in children have been well reviewed by Jones and Round (30), Blimkie and Sale (9), and Froberg and Lammert (19).

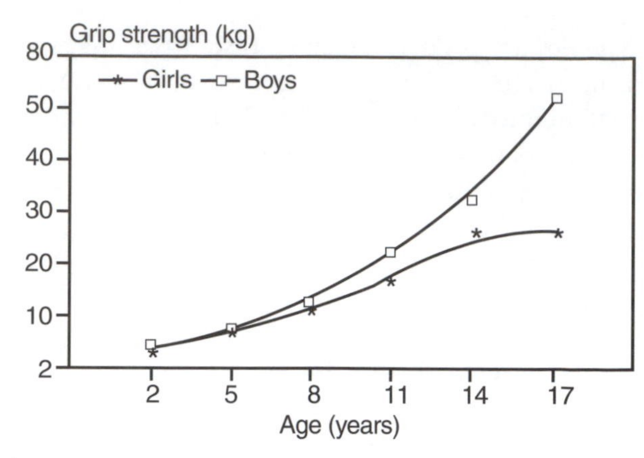

FIGURE 10.1 Developmental improvements in single-hand grip strength with age in boys and girls (from reference 9).

Reprinted by permission from C.J.R. Blimkie and D.G. Sale 1998.

These reviews indicate that measures of strength in boys improve more or less linearly in the prepubertal years. As puberty approaches, muscle mass grows in response to increased levels of circulating testosterone, and strength improvement accelerates. The prepubertal trend is similar in girls, but mean values are somewhat less than for the boys. This small sex difference is observed in handgrip strength as early as three years of age. At puberty the strength curve in girls either continues to rise slowly or, in some studies, even plateaus with increasing age. Consequently, a significant sex-related gap in muscle strength occurs in the adolescent years. By age 17 years, grip strength is almost twice as great in males and females (figure 10.1).

Longitudinal studies have provided insight into the timing of static strength development in relationship to the adolescent growth spurt. For most muscle groups, the adolescent strength spurt in males occurs approximately one year after age of peak height velocity and corresponds instead with the time of peak weight velocity. The more limited data in females are less clear and show considerable interindividual variability (figure 10.2).

Performance on measures of strength appears to be under at least moderate genetic control (7). When static strength is compared between siblings, heritability estimates have ranged from 0.44 to 0.58, and a review of 14 twin studies, most with a small number of subjects, indicated heritability ranging from 0.24 to 0.83. Beunen et al. measured static strength by arm pull in 105 twin pairs annually between ages 10 and 18 years (7). A model including additive genetic and specific environmental factors indicated an additive genetic contribution varying between 0.44 and 0.83.

The early large-scale studies of strength development in children involved measurement of absolute isometric strength (typically by handgrip) or relative strength (flexed arm hang). More recent investigations have added assessment of maturational differences in dynamic strength measured with isokinetic dynamometers (see reference 5 for a review). This testing allows assessment of muscle force in the rotation of limb segments around joints. Thus, the maximal moment exertion of muscles is recorded through the range of motion.

Gaul provided an overview of the different measurements obtained during isokinetic testing (22). The most commonly used indicator of muscle function is the peak torque produced over a range of angular velocities. As more information regarding isokinetic strength has become available, increases with age and sex-related differences appear similar to those seen in static strength (14, 15, 45, 48).

Determinants of Muscle Strength Development

Clearly, the increase in muscle size is primarily responsible for developmental improvements in

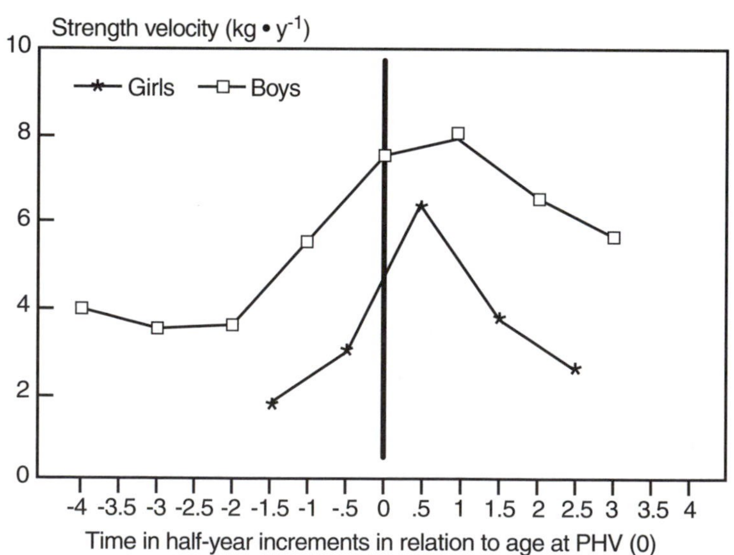

FIGURE 10.2 Development of strength relative to age at peak height velocity (PHV; from reference 9).

Reprinted by permission from C.J.R. Blimkie and D.G. Sale 1998.

strength as children grow. A child becomes progressively stronger as a result of the endocrine and paracrine effects of agents such as growth hormone and IGF-I, whose anabolic actions promote muscle protein synthesis and stimulate fiber hypertrophy (see chapter 2). The nearly identical growth curves for muscle mass and body strength in relationship to both age and sex attest to this.

The first clue that other, size-independent factors might contribute to the development of strength in children came from early studies indicating that improvements in strength with age in relationship to body size do not fit theoretical expectations. We have seen that muscle strength should reflect muscle cross-sectional area, which by dimensionality theory should relate to height by the allometric scaling exponent 2.0 and to weight by 0.67. Fifty years ago, Asmussen and Heeboll-Nielsen found in their study of 7- to 16-year-old Danish schoolboys that the exponents for height relative to strength of the leg extensors, elbow flexors, and finger flexors were 2.89, 3.89, and 3.27, respectively (3). That is, strength improved at a greater rate with age than would be expected by height (2.0) or, presumably, the cross-sectional area of muscle alone. The authors felt that "it is obvious that the deviations must be due to qualitative changes in the motor system," a result they considered most reasonably "due to an increased ability to exert maximal strength caused by maturation of the nervous system" (p. 603).

Shephard et al. reported similar exponents when subjects were studied longitudinally between the ages of 6 and 12 years (50). Height exponents for boys for handgrip force, back extension force, and leg extension force were 3.29, 2.69, and 2.80, respectively. Values in girls were similar.

More recent studies have demonstrated findings consistent with these earlier reports. Rauch et al. described the relationships of height and weight to maximal isometric grip force measured using a handheld dynamometer in 315 children ages 6 to 19 years (46). In the girls, maximal force was related to $H^{3.92}$ and $M^{1.16}$. In the boys, the exponents for height and mass were 3.87 and 1.29, respectively. At all body heights, values of strength were greater in boys than in girls. Moreover, the pattern of change in maximal force with age was different between the boys and girls. In boys, age and maximal force rose in a linear relationship, while in the girls a plateau was observed at approximately 15 years.

For analyses by dimensional constructs to be valid, subjects of different size must be *geometrically similar.* That is, their different body parts and segments need to be equally proportional to total body size. Geometrically similar individuals have a different body size but the same shape. In fact, children and adults, or younger children and older children, are not exactly geometrically similar. Specifically, younger and smaller subjects have relatively larger heads and shorter legs than older ones.

Within the range of body size generally of interest to pediatric exercise physiologists, however, these variations are minimal. In the study of Asmussen and Heeboll-Nielsen, for example, the ratio of trunk length to total body height was 39% in subjects with a height of 126 cm and 38.7% in those 181 cm tall (3). Head length represented 15% of height in the smaller boys and 11% in the tallest. It seems, then, that in a dimensional approach to strength development in children, no substantial error will be created by assuming geometric similarity among subjects of different ages. Not all researchers agree, however. Blimkie and Sale concluded that "despite the assertions [of Asmussen and Heeboll-Nielsen], the assumptions about geometric similarity and constancy of tissue composition during childhood, which underlie this theory, are highly questionable" (9, p. 199).

Allometric studies of muscle strength in children have found wide variability in height scaling exponents for different muscle groups. In the study by Asmussen and Heeboll-Nielsen, for instance, recall that the height exponent for elbow flexor strength was 35% greater than that for leg extensors (3). It was just this variability that caused these authors to conclude that "the fact the [exponents] are different in different situations seems to point to changes in the nervous system as the reason for the great variations" (p. 602).

A recent study by Jaric et al. pointed out just how wide that variability in exponents can be (29). These researchers measured isometric leg strength in groups of adult, pubescent, and prepubescent athletes involved in different sports to determine the optimal mass scaling exponent for strength (S), that is, the exponent that would eliminate the effect of body size on strength in these populations of subjects. In other words, they sought to identify in these various groups the ideal exponent *b* that would cause $S/M^b \sim M^{0.0}$.

The most striking finding in this study was that the optimal mass exponent varied dramatically depending on muscle group, sport, and age of subjects. For hip extensor strength, for example, the ideal exponent in the adults ranged from 0.25 to 1.75, depending on the sport involved. In the prepubescent subjects the ideal mass exponent for hip flexor strength varied from 0.35 to 1.15. Even in the most consistent group—hip extensor strength in prepubertal subjects—the optimal exponent for strength ranged from 0.45 to 0.70 (figure 10.3).

The *average* ideal exponents for all muscle groups and sports were 0.61 and 0.63 for the adult and prepubescent athletes, approximating the expectation by dimensionality theory of $M^{0.67}$. However, the mean value was 1.14 in the prepubescent subjects. These data suggest that scaling exponents for muscle strength can vary markedly depending on muscle group involved, sport played by the subjects, and—perhaps—maturational differences. These findings preclude any quick and uncritical acceptance of mean scaling exponents in allometric analyses of strength.

As Froberg and Lammert commented, allometric analyses of back-muscle strength suggest a relationship with body size that may be different from those of other muscle groups (19). They reported a height scaling exponent for back strength of 2.36 in girls and 2.90 in boys, values lower than those found for muscles of the extremities. Sinaki et al. described height scaling exponents for maximal back strength in 137 boys and girls 5 to 18 years old (51). Strength of back extensors and flexors was measured with an isometric dynamometer. As expected, trunk strength increased with age, and strength in the boys began to increase more rapidly than in the girls by the age of 9 or 10 years. The height exponent for back extensor strength was 0.83 and for back flexor strength 1.69.

The question of whether size-independent qualitative factors contribute to the development of strength during childhood is more appropriately answered by comparing changes in strength to changes in muscle cross-sectional area. Age comparison cross-sectional studies in children have revealed moderately strong correlation coefficients (r = .60 to .90) between both isometric and isokinetic maximal voluntary strength and muscle cross-sectional area (as estimated by anthropometry, ultrasound, or computed tomography; 8). Davies reported that changes

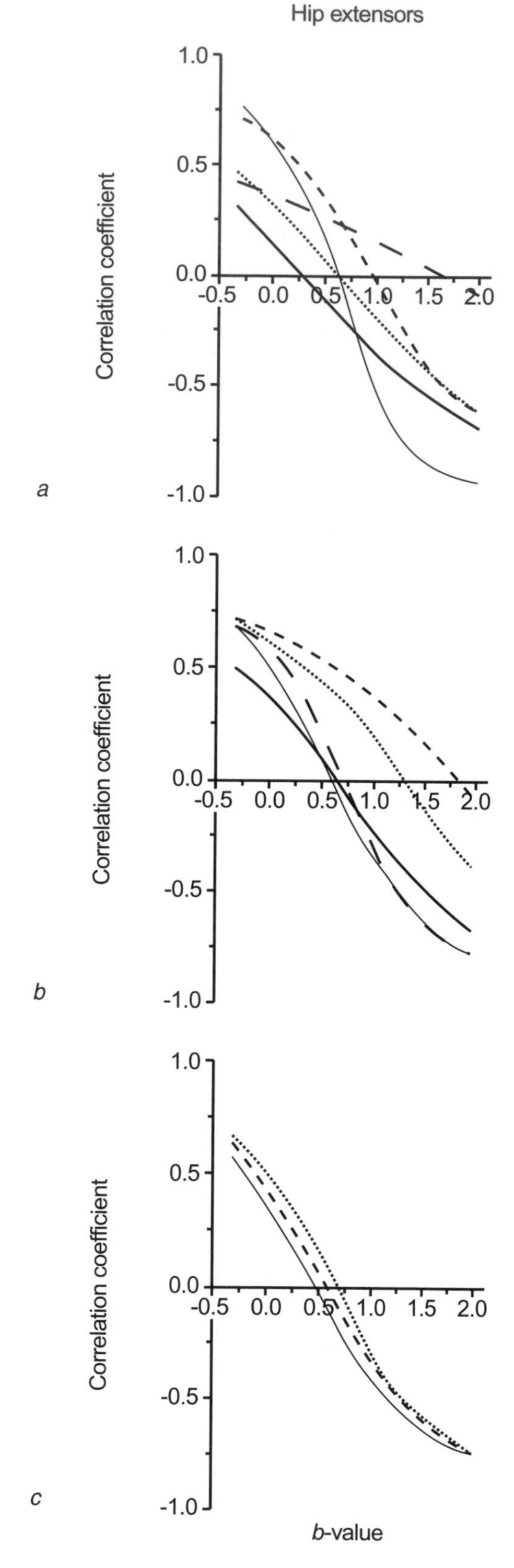

FIGURE 10.3 Correlation between S/M^b versus different values of b in adult *(a)*, pubescent *(b)*, and prepubescent *(c)* athletes. The value of b at a correlation coefficient of 0.0 represents the optimal exponent for each group of subjects (from reference 29). The different curves represent various types of sports.

Reprinted by permission from S. Jaric, D. Ugarkovic, and M. Kukolj 2002.

in strength with age and differences between boys and girls disappeared when strength was related to muscle cross-sectional area (12). These findings, then, fail to support the idea that qualitative changes other than the increase in muscle size affect the development of strength during childhood.

More recently, however, Neu and colleagues provided evidence supporting a role for size-independent factors in ontogenetic improvements in strength (39). They analyzed the relationship between cross-sectional area of forearm muscles determined by computed tomography and maximal isometric grip strength in 366 children and adolescents (6 to 23 years old). Patterns of change in absolute grip force and muscle size relative to age and sex mirrored those of earlier studies. Grip force relative to muscle cross-sectional area nearly doubled between the ages of 6 and 23 years. When grip force was expressed relative to muscle cross-sectional area adjusted for forearm length (explained later under "Muscle Fiber Architecture"), values rose by almost 50% between the ages of 6 and 20 years in both boys and girls (figure 10.4). These findings strongly suggest that size-independent factors act to improve muscle strength with age and also that sex differences in grip strength are due entirely to males' greater muscle size.

In a longitudinal study, De Ste Croix et al. (15) used multilevel regression modeling to investigate the influences of age, sex, body size, sexual maturation, and body fat on development of isokinetic knee extension and flexion strength in 41 boys and girls (initially 10.0 ± 0.3 years old). Subjects were studied on eight occasions over a four-year period. On one test occasion, the relationship of muscle cross-sectional area (determined by magnetic resonance

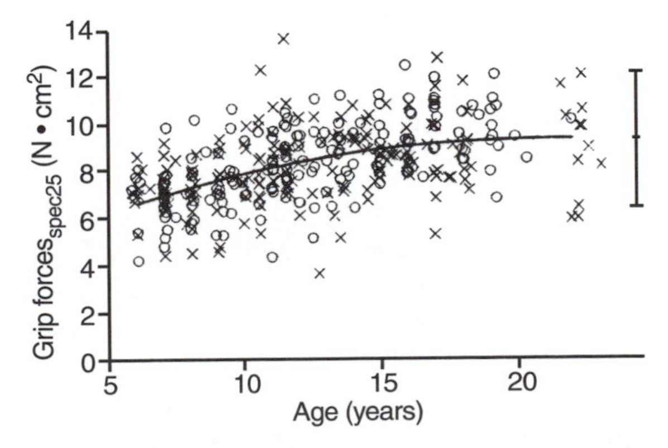

FIGURE 10.4 Change in grip strength related to muscle cross-sectional area and normalized to a forearm length of 25 cm with age (from reference 39).

Reprinted by permission from C.M. Neu 2002.

imaging) and isokinetic leg strength was assessed. Isokinetic strength was moderately related to muscle cross-sectional area (r = .49 to .69). While muscle cross-sectional area, age, and maturity all were related to the development of isokinetic strength, these factors became insignificant, nonexplanatory variables once body size (height and mass) was accounted for.

In this study, cross-sectional area was related to flexion and extensor torque in 14-year-old subjects by the allometric scaling exponents of 1.05 and 0.71, respectively (15). This finding, plus the inability of age to account for increases in strength apart from changes in body size, fails to support the argument for the effect of qualitative muscle changes on strength development, at least in this age span.

However, Nevill et al. found that isometric strength improved with age in a cross-sectional study of 8- to 17-year-old boys, even after accounting for increases in stature and body mass (40). And, as discussed later, Kanehisa et al. found that relating isokinetic strength to muscle cross-sectional area did not eliminate differences between children and adults (31).

Much of these data, then, provide evidence that qualitative changes beyond increases in muscle size are important in the development of strength in children. Not all the information is consistent with this view, however, and a great deal of variability seems to be contributed by factors such as muscle group, sex, type of testing, and so on. De Ste Croix et al. concluded from their study (described earlier), "These data highlight the need to examine explanatory variables concurrently to elucidate the influence they may have on the development of isokinetic leg strength [and] emphasize the insecurity of making general statements based on the study of single muscle groups or actions" (15, p. 60).

Explaining Qualitative Changes

At this point we will make the assumption that qualitative, size-independent changes that improve muscle strength beyond the effect of increases in muscle size alone occur during biologic maturation. (This assumption is supported by data on increases in muscle strength with resistance training in children that are discussed in the next chapter.) What could that factor or factors be? As previously alluded to several times in these chapters, neurologic influences are strongly implicated, but so far with little supportive evidence. Changes in the intrinsic contractile properties of muscle during development could contribute, as might alterations in muscle architecture. Finally, central inhibitory limits to voluntary muscle strength production might evolve during the childhood years. In this section we examine each of these candidates briefly. Brevity is appropriate, since little information is available about these processes in children as a result of ethical constraints on investigative techniques that largely restrict our view. Where pertinent, however, we add data on adult subjects to provide a framework for the developmental changes that *could* occur in children.

Neurologic Factors

A motor unit consists of a single motoneuron, located in the anterior horn of the spinal cord, and its axons, which are attached to the motor end plates (synaptic junctions) of a number of muscle fibers. Each unit can innervate as few as 3 or as many as 2,000 fibers, depending on the precision of motor control required. The excitation of the muscle fibers in a particular motor unit is an all-or-none phenomenon. An increase in force production by a muscle can therefore be accomplished by either increasing the frequency of stimuli by a given motor unit or recruiting additional motor units. For a muscle to generate an increasing amount of tension, these two processes occur in tandem: As the firing rate of a given motor unit reaches its maximum, new units are recruited. Eventually all motor units are stimulated at their maximal frequency, and this defines the maximal voluntary contraction of the muscle (55).

Different types of motor units are characterized by their metabolic capacity, their capacity for force production, and the electromyographic features of the muscle they innervate. Type I motor units with small neurons typically innervate fibers containing a high concentration of mitochondria, capillaries, and aerobic enzymes. These units can produce low tension over an extended period of time. Type II motor units are characterized by large neurons, innervate muscle fibers that rely on anaerobic metabolism, and can produce high forces in a short time. This categorization, then, describes slow-twitch aerobic and fast-twitch anaerobic muscle fibers but reminds us that the contractile features of these fibers are defined not simply by their metabolic features but also by their type of innervation. There is a distinct link between motoneuron properties and both the metabolic and contractile properties of muscle fibers.

In muscle work of increasing intensity, the size of recruited units is directly related to the intensity of the work. At low levels, the smallest units (type I) are recruited, and as intensity increases, the larger units (type II) come into play. The relative contribution of glycolytic anaerobic and aerobic metabolism during exercise is thus directed by neural mechanisms.

Fatigue, or depression of muscle tension, could result from a slower firing rate or a limitation in recruitment of muscle fibers. A reduction in firing frequency has been observed during sustained isometric contractions, but this has generally not been considered a primary factor in muscle fatigue (24). Still, the potential role of neurologic mechanisms in limiting muscle function must be considered. Green commented,

> "Some investigators have assumed reductions in force output are invariably caused by impaired energy availability. However, these conclusions may not be justified. If the cause of the force loss is more central, for example, resulting from inhibition of the motoneuron, the activation of the muscle cell would be reduced. Reductions in activation would result in reductions in actin-myosin turnover, with consequent reductions in energy expenditure. The depressed glycolysis may well be a *response* to a reduced energy *demand* [italics mine] rather than an inability of glycolysis to respond to an increase challenge" (24).

From this brief overview, it is obvious that developmental changes in innervation of muscle could easily contribute to the maturational development of metabolic function, muscle strength, and motor performance in children. Whether this actually occurs, however, is problematic, as research information is scant. We can nevertheless examine the potential roles of several neurologic features.

Nerve Conduction Velocity

Most research in children has involved motor and sensory nerve conduction velocity, as this neurologic variable is most readily measured noninvasively. Garcia et al. demonstrated that motor conduction velocity, sensory conduction velocity, and the amplitude and morphology of action potentials increase during the first year of life (21). In that study, maximal conduction velocities were twice as fast in adults as in neonates. However, values of all these measures approached adult levels by age four or five years. Others have observed similar findings (17, 18).

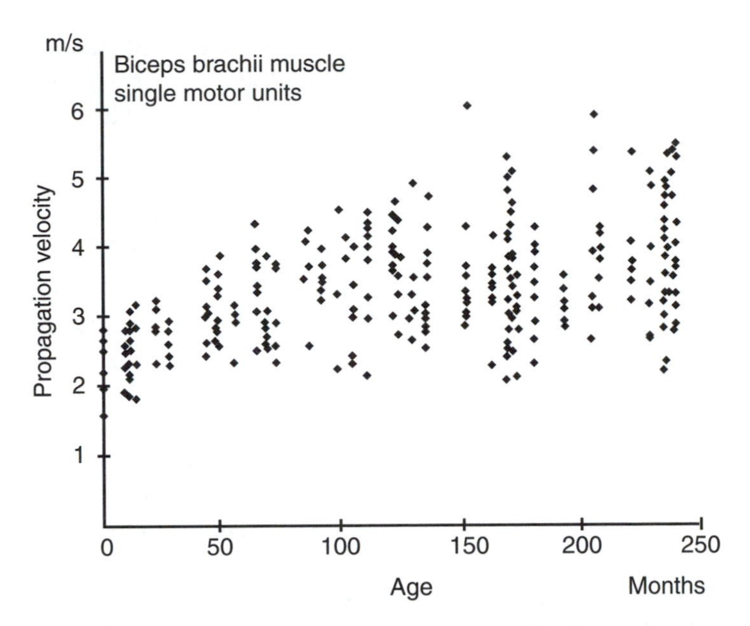

FIGURE 10.5 Nerve conduction velocity of the biceps brachii muscle with increasing age in healthy children (from reference 35).

From C. Patten, G. Kamen, and D.M. Rowland, "Adaptations in maximal motor unit discharge rate to strength training in young and older adults," Muscle and Nerve 24: 542-550, copyright © (2001, Jown Wiley and Sons, Inc.). Reprinted by permission of John Wiley & Sons, Inc.

However, continued increases in peripheral nerve velocity in later childhood were seen by Malmstrom and Lindstrom in their study of 63 children and adolescents ages 0 to 20 years using surface electromyography (figure 10.5; 35). Propagation velocities were found to relate not only to age but also to height and muscle diameter. Between the heights of 100 and 200 cm, conduction velocity rose by almost 20%. The authors concluded that "all of these studies [showing an increase in nerve conduction velocity only up to age five years] rely on small numbers of observations in each child, and should therefore not be taken as final evidence on the development of the muscle potential velocity during childhood" (p. 408).

Lang et al. also reported that peripheral nerve conduction velocities change in later childhood and adolescence (33). They found small but significant increases in propagation speed in the upper extremities but a decrease in the lower extremities in both males and females in this age group. They concluded that the evolution of nerve conduction velocity is a complex phenomenon, which may relate to the fact that as limbs grow in length, the distal portions of the axons may become thinner than more proximal portions.

The cerebral cortex can be stimulated to trigger motor discharges by a rapidly changing magnetic field evoked via a coil on the scalp (20). This technique has been used to compare corticospinal conduction velocities in children and adults. Heinen et al. measured conduction times in the corticospinal tract of 6- to 9-year-old children and adults ages 22 to 26 years (26). In a facilitated condition (i.e., during muscle contraction), no difference in central conduction time was seen between the two groups. However, in the relaxed state, conduction was significantly longer in the children (10.5 ± 2.2 versus 8.6 ± 0.7 ms in the adults).

Motoneuron Firing Rate

Studies in adults using needle electrodes have indicated that the rate of motor unit discharge may limit force production. Maximal motor unit discharge rate declines with age in the adult years, and this corresponds to declines in muscle strength. It has been suggested that this fall in firing frequency relates to muscle contractile characteristics that become slower with age. The fall in motor unit firing frequency with age in adults, then, may represent a matching of neural properties to muscle contractile function (43, p. 194). Because use of intramuscular needle electrodes is considered inappropriate in healthy children, we have no information on changes in motoneuron firing rates that might occur during biologic maturation.

Number and Recruitment of Motor Units

It has generally been thought that the number of neurons that will be present by early adulthood already exist at birth or early infancy. But populations of motor units can change; McComas et al. witnessed progressive decreases throughout the adult years (37). For example, they noted that the estimated number of thenar muscle units steadily declined between ages 20 and 80 years, such that number in the oldest subjects was one half that in the youngest.

Torque acceleration energy (TAE) indicates the amount of work in the initial portion of an isokinetic contraction and has been used as a marker of motor unit recruitment (22). Studies comparing TAE between children and adults might be useful in assessing maturational differences in motor unit recruitment patterns, but no such studies have yet been published. Brodie et al. measured TAE at various velocities ranging from 0.52 to 3.15 rad · s^{-1} in

24 prepubertal boys (10). Values varied from 0.16 to 1.74, indicating that TAE is strongly dependent on velocity. The authors took this to mean that only those motor units capable of high contraction velocity are recruited at high limb velocities.

Whether motor unit recruitment patterns during exercise can change with age is unknown, but any such alterations would be expected to affect metabolic and physiologic responses to physical activity. There has been a good deal of recent interest in the plasticity of neuromuscular function (44). Animal studies have suggested, in fact, that alterations in neurologic firing can modify the metabolic and functional characteristics of skeletal muscle. How such phenomena might relate to development of muscle fitness in children remains to be seen.

In summary, on the basis of the preceding evidence, a courtroom lawyer would have a difficult time convincing a jury that neural mechanisms are responsible for qualitative improvements in strength during the course of childhood. On the other hand, the idea is conceptually attractive, given the several plausible mechanisms by which this could occur. The jury would probably find that the limited information—due to the lack of investigative techniques appropriate for children—precludes any final verdict on the matter. We return to this issue in the next chapter when we address the improvements in muscle strength that occur without increases in muscle size following resistance training in children.

Muscle Contraction

The several studies that have examined muscle function in children have not made it clear whether contractile processes change during the course of biologic development. Going et al. characterized force-versus-time relationships during a single maximal contraction in 8- to 11-year-old children (23). Normally, a plot of force versus time describes a sigmoid curve, with a slow initial rise, followed by an extended plateau, and then a more gradual rise is observed to the point of maximal force. In this study involving multiple trials of finger flexors, forearm flexors, and forearm extensors, maximal rate of force increase proved to be the most reproducible measure. There were no adults in this study, but the maximal rate of force and the pattern of force production in the children were different from those previously reported in adult subjects.

In children the average maximal rates of force for finger flexors, forearm flexors, and forearm extensors were 103, 67, and 60 $kg \cdot s^{-1}$, respectively (23). In studies of adults, rates of 144 and 138 $kg \cdot s^{-1}$ have been described for forearm flexors (cited in 23). The same adult studies indicated that maximal force is reached faster in adults than the children in this study. Going et al. reported times to reach maximal force of 1.87 and 1.55 s for forearm flexors and finger flexors in children they studied, respectively, compared with previous reports of 0.42 s (men) and 0.75 s (women) for forearm flexors.

Backman and Henriksson evaluated muscle contractile response to electrical stimulation of the adductor pollicis muscle in children 9 to 15 years old (4). Relaxation rate, which has been used as a marker of energy turnover, was independent of age and sex. Similarly, no differences in force produced relative to stimulation frequency were seen with increasing age or between boys and girls. The extent to which this testing protocol mimics short-burst activities in children, however, is debatable.

Kanehisa et al. examined the relationship between force-generating capacity and muscle size in children and adults from the perspective of the decline in force output with repeated maximal contractions (32). Fifty repeated maximal knee extensions at 3.14 $rad \cdot s^{-1}$ over one minute were performed by 26 boys age 14 years and 27 young adult men ages 18 to 25 years. Muscle cross-sectional area was estimated by ultrasound. Force production was greater in the adults regardless of whether it was expresses as an absolute value or relative to muscle cross-sectional area or to area multiplied by length. The average percentage decline of force with 50 contractions was greater in the adults than the boys (48% vs. 36%; figure 10.6).

While the observations of this study might reflect maturational differences in intrinsic muscle properties, the slower decline in strength with repeated contractions in the children might reflect a reduced accumulation of cellular lactate and less metabolic acidosis. It is well recognized that these metabolic by-products can interfere with the effectiveness of excitation-contraction coupling in the ensuing contraction. Hebestreit and his colleagues reported that boys recover faster metabolically from brief intense exercise than men do (25). These authors found that 10 minutes after a 30-s, all-out cycle sprint, blood $[H^+]$ in men was 66.1 ± 5.9 $nmol \cdot L^{-1}$, compared with 47.5 ± 1.2 $nmol \cdot L^{-1}$ in boys, while blood lactate levels were 14.2 ± 1.8 and 5.7 ± 0.7 $mmol \cdot L^{-1}$, respectively (figure 10.7).

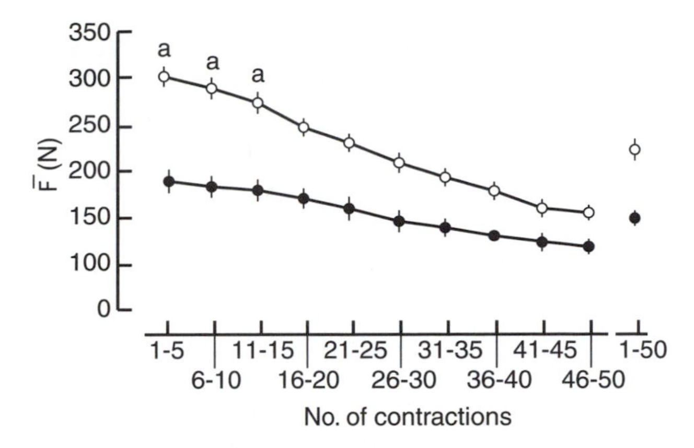

FIGURE 10.6 Force production with 50 repeated maximal contractions in children and adults, indicating a greater level of fatigue in the older subjects (from reference 32).● = 14-year-old boys; ○ = young adults; a = $P < 0.05$ boys versus adults.

Reprinted by permission from H. Kanehisa 1994.

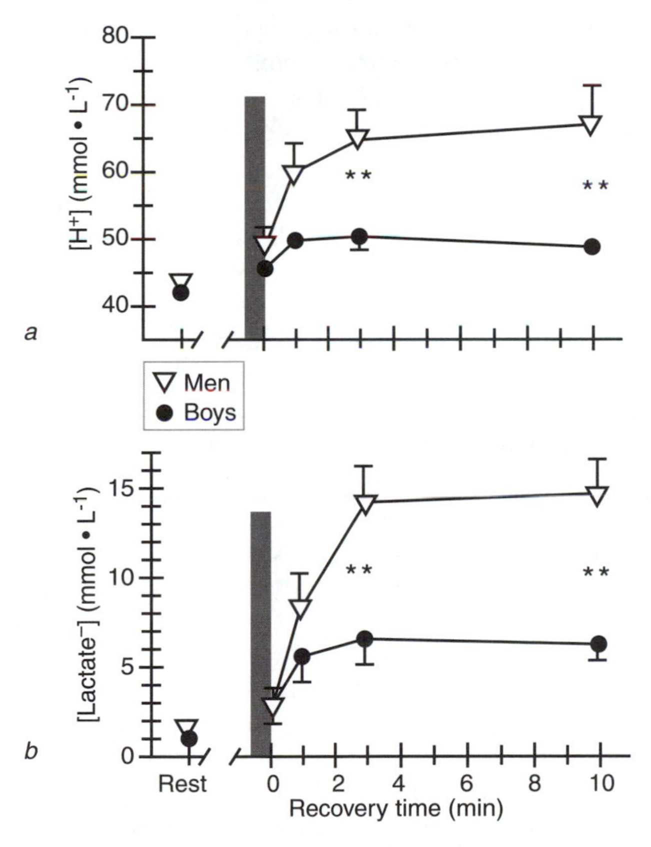

FIGURE 10.7 Plasma concentrations of *(a)* hydrogen ions and *(b)* lactate in five boys and five men at rest and following a 30-s, all-out cycle sprint (from reference 25).

Reprinted by permission from H. Hebestreit et al. 1995.

Other investigations have not supported the idea that muscular function is qualitatively different between children and adults. When Belanger and McComas compared prepubertal and postpubertal subjects, no overall differences were observed in muscle fatigue or relaxation times in tibialis anterior and plantar flexor muscles (6). McComas et al. found no age-related differences in maximal twitch contraction time of the extensor hallucis brevis muscle in males ranging from 3 to 22 years old (38).

Davies and coworkers studied electrically stimulated and voluntary maximal force of the triceps surae in girls and boys 11 to 14 years old and compared these findings to their earlier findings in adults (13). Values of force-generating capacity relative to muscle cross-sectional area, muscle fatigability, and contraction and relaxation times were independent of age.

Muscle Fiber Architecture

The amount of tension that a muscle can produce is a function of the cross-sectional area of the number its sarcomeres in parallel. Rather than a simple slice across the muscle, however, this area is more correctly described by the angle of pennation of the muscle fibers in relationship to the direction of the insertion of the muscle's tendon. The more the fibers diverge from the axis of action of the muscle, the less tension is produced. The angle of pennation, then, describes the *physiologic* cross-sectional area of the muscle (as distinct from the *anatomic* area).

Wickiewicz et al. reported the difference between physiologic and anatomic cross-sectional area (i.e., pennation angle) in 27 different muscles from 300 adult human cadaver specimens (54). The pennation angle relative to the direction of pull of the muscle varied among muscles from 0° to 30°.

There is some evidence that the angle of pennation in muscle may change as children grow. If so, this change might cause the force production of muscle to differ from expectations derived from dimensionality theory. Fukunaga and Kawakami used ultrasound measurements to assess muscle fascicle angle by age and sex in the vastus lateralis and gastrocnemius muscles (cited by reference 9). As indicated in figure 10.8, a trend for increased angulation with age was observed in both sexes and both muscles. These data indicate that the pennation angle in the vastus lateralis increases by 35% in males between childhood and young adulthood. As Blimkie and Sale concluded, the importance of these changes in pennation for strength development during the childhood years remains to be clarified (9).

The question of the role of physiologic cross-sectional area is a difficult one, since it cannot easily be measured. However, the physiologic cross-sectional

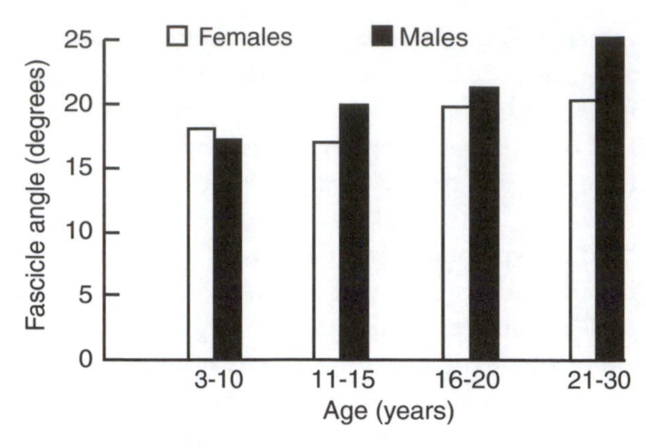

FIGURE 10.8 Change in muscle pennation angle during childhood as estimated by ultrasonography (from reference 9, data from Fukunaga and Kawakami).
Reprinted by permission from C.J.R. Blimkie and D.G. Sale 1998.

area of a muscle is linearly related to its volume. Considering this, Kanehisa et al. examined differences in maximal voluntary isokinetic torque of the right knee between children and adults to determine if maturational variations in strength could be accounted for by muscle volume and, by inference, by muscle pennation angle and physiologic cross-sectional area (31). Sixty boys and girls six to nine years old and 71 young adults performed leg exercises at velocities ranging from 1.05 to 5.24 rad $\cdot$ s^{-1}, and torque was measured with an isokinetic dynamometer. Anatomic cross-sectional area of the quadriceps femoris muscle estimated by ultrasound measurements was multiplied by muscle length to estimate muscle volume.

At the different velocities, high correlations were observed between strength and estimated muscle volume (boys, r = .71 to .76; girls, r = .72 to .83; men, r = .71 to .76; women, r = .53 to .70). The absolute strength of the children was only 18% to 33% that of the adults. These differences decreased when values were expressed relative to estimated muscle volume or anatomic cross-sectional area, but differences between children and adults still persisted. These limited findings suggest that differences in anatomic and physiologic cross-sectional area of muscles between children and adults do not play a major role in maturational trends in muscle strength.

Changes in Central Inhibition

Electrically evoked peak muscle contractions have been measured to determine the extent to which maximal voluntary contractions reflect the true peak contractile capacity of muscle (22). *Why* observed gaps might exist between maximal voluntary and electrically stimulated peak contractions has long been debated; that is, do differences indicate that neural rather than muscular factors limit muscle contractile performance (20)? Asmussen concluded, "It is ordinarily not possible to produce the highest tensions of which muscle should be capable [and] there must be a reserve of force that will only be called forth under extraordinary conditions" (2, page 75).

It has been postulated that if "maximal" voluntary contractions are not truly maximal, there must be some inhibitory factor that limits exertion during tests of strength, presumably to prevent muscle fiber damage. Negative feedback from peripheral receptors might be responsible, or inhibitory signals might arise primarily in the cortex or reticular formation of the brain (2). Any maturational change in the threshold of this inhibitory factor as children grow could account for size-independent increases in muscle strength. This idea would be supported by findings of maturational changes in percentage of motor unit activation in studies using electrically evoked contraction in children and adults. Limited research data are available to examine this premise.

Blimkie described the degree of motor unit activation (MUA) using the interpolated twitch technique in 10- to 16-year-old children (8). The amount of force output from a supramaximal electrical stimulation beyond that produced by a maximal voluntary contraction defined the percentage MUA. In the MUA of elbow flexors, no age-related differences were seen (89.4% and 89.9% for 10- and 16-year-olds, respectively). However, for the knee extensors, MUA was significantly higher in the older subjects (95.3% versus 77.7%).

Belanger and McComas examined the extent of MUA during peak voluntary contractions relative to age in two groups of males 6 to 13 years old and 15 to 18 years old (6). They found that complete motor activation was achieved by voluntary contraction of the tibialis muscle at all ages. Two of the 10 prepubertal subjects (but none of the postpubertal boys) did not fully activate the ankle plantar flexors (MUA < 95%). About this finding the authors commented, "There was a suspicion in this study, not verified statistically, that younger children were less able to activate plantarflexor motoneurons than older ones and adolescents. Indeed, it would seem logical to anticipate a critical age during development before which full motor unit activations cannot be achieved, and this might differ from one muscle to another, depending on the strength of the descending voluntary pathways" (p. 566).

Maximal voluntary torque with concentric contractions decreases with greater angular velocity, while electromyographic (EMG) values are greatest at the fastest velocity. However, eccentric torque remains unchanged with increasing angular velocity. This has been attributed to a neural tension-regulating mechanism that inhibits voluntary contractile force, thereby protecting muscles and joints from excessive loads (48). Seger and Thorstensson suggested that "it is conceivable that such a mechanism would be even more pronounced in children than adults" (48, p. 55). They were, however, unable to demonstrate differences in these factors in a cross-sectional study of prepubertal boys and girls and adult men and women (49).

In a later longitudinal study they studied isokinetic strength in association with EMG measures in subjects from age 11 to age 16 years (48). Eccentric and concentric muscle strength was measured at angular velocities of 45°, 90°, and 180° $\cdot$ s^{-1}. The pattern of the torque-to-velocity relationship remained essentially unchanged with age in both sexes. In general, the ratio of eccentric to concentric torque (relative to EMG changes) did not change with age, but a significant change was observed in this ratio at the highest speeds postpuberty. The authors concluded that a neural inhibitory mechanism that protects against excessive force was partly supported by these findings.

Muscle Damage

Intense exercise, particularly that involving eccentric contractions, causes microscopic muscle damage, with disruption of fibers and infiltration of inflammatory cells. This process is associated with release of creatine kinase and other biochemical markers into the bloodstream and characterized clinically by delayed-onset muscle soreness, diminished strength, and reduced range of motion (11, 41).

Creatine kinase (CK) exists in three isoenzyme forms: CK-MM, CK-MB, and CK-BB. Mature skeletal muscle in nonathletic adults contains almost exclusively CK-MM. During early fetal development CK-BB predominates, with increasing CK-MB during gestation (47). At delivery and in early infancy the CK isoenzyme pattern becomes similar to the adult CK-MM pattern. This suggests that CK-BB and CK-MB are markers of muscle growth and development. Similarly, adult distance runners have increased CK-MB in skeletal muscle, suggesting that the stresses of training cause a reactive shift from the usual CK-MM pattern (1).

Increased CK-MB might also reflect the higher aerobic capacity of adult distance runners. The percentage of CK-MB has been linked directly to both percentage of slow-twitch fibers and concentrations of aerobic enzymes in skeletal muscle (27). That is, CK-MB in skeletal muscle might serve as a marker of aerobic fitness, and the release of CK-MB with exercise could then serve as a noninvasive indicator of the aerobic biochemical profile of muscle.

This concept has not been tested in children, but several studies have compared markers of muscle damage with intense exercise in prepubertal subjects and adults. These studies found a number of differences between children and adults, suggesting that children experience less muscle damage than adults.

Soares et al. measured markers of muscle damage after five series of bench presses at 80% maximal strength in 10 children and 10 adults (52). Maximal isometric strength, subjective muscle discomfort, and serum CK activity were measured before exercise and again 48 hours, 72 hours, and one week after exercise. The adults demonstrated the typical findings of reduced strength, muscle pain, and elevated total CK levels in response to exercise. Mean CK activity rose from 48 $\pm$ 17 μmol $\cdot$ L^{-1} before exercise to 1,190 $\pm$ 1,523 μmol $\cdot$ L^{-1} at 72 hours. The children, on the other hand, reported minimal muscle discomfort, and no significant change was seen in CK levels postexercise. The authors noted that "given that the exercise protocol was performed with the same relative intensity for all subjects, the lack of decrease in the child group suggests less muscle injury induced by the exercise [and] the alterations in CK support these assumptions" (p. 365).

Webber et al. compared muscle soreness and CK levels in children (mean age 10.4 $\pm$ 1.2 years) and adults (27 $\pm$ 3.6 years) following a single bout of downhill running (53). Subjects ran for 30 minutes at a –10% grade at an intensity of 83% of age-predicted maximal heart rate. Preexercise CK values were not significantly different between the children

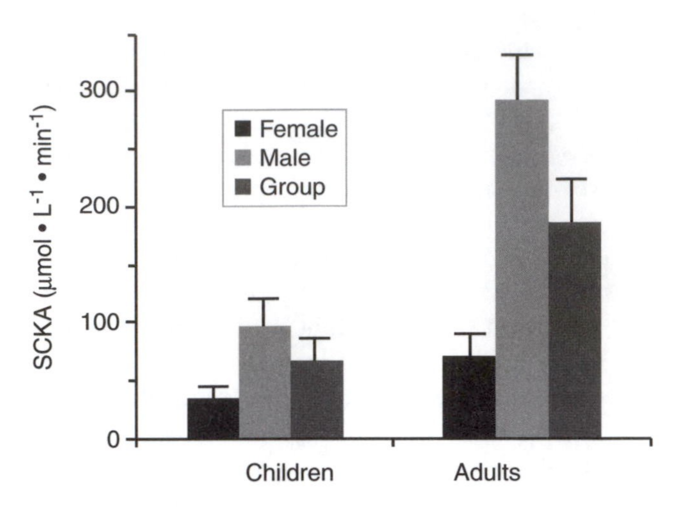

FIGURE 10.9 Change in serum creatine kinase activity (SCKA) 24 hours after a bout of downhill running in children and adults (from reference 53).

Reprinted by permission from L.M. Webber et al. 1989.

and adults (91.7 ± 34.2 and 79.1 ± 27.2 µmol · L^{-1}, respectively). Although CK levels were elevated 24 hours postexercise in both groups, the increase was significantly greater in the adults (178 ± 149 µmol · L^{-1}) than in the children (69 ± 65 µmol · L^{-1}; figure 10.9). However, when the CK changes were adjusted for body weight, no significant differences were seen between the adults and children. Muscle soreness ratings 24 hours after the downhill run were similar in both groups.

Duarte et al. drew similar conclusions in their study of markers of muscle overuse in 13 boys, although they made no direct comparison with adults (16). Twenty 13-year-old boys performed two different protocols of one-leg stepping exercise to exhaustion. Half the subjects performed more eccentric exercise than the other group. All markers of damage were more exaggerated in the eccentric-contraction group. Decrease in muscle strength appeared to be less than that described in adults, and recovery from muscle soreness was more rapid. Rises in CK levels were small and blunted in comparison to the increases seen in adult studies.

Marginson et al. compared the relationship between delayed-onset muscle soreness and isometric strength loss in boys and men after high-intensity eccentric exercise (hopping) (36). The strength losses in the two groups were similar, but the time course was different, with an earlier decrement in the children.

The general picture from these studies, then, is that children experience less muscle damage from intense exercise than adults. Webber et al. suggested that "adults, because of their greater body weight, generate more force per unit fiber area during eccentric contractions compared to children, thus resulting in greater damage and greater release of intracellular creatine kinase into the serum" (53, p. 357). These authors also pointed out that there may be other potential explanations for the lower CK response to exercise in children. Children might have lower CK concentrations in muscle than adults, or the time course of CK release into the bloodstream or its clearance could be different.

Conclusions

Children gain isometric and isokinetic strength as they get older largely because their muscles get bigger. Muscular enlargement, in turn, is an expression of fiber hypertrophy in response to the actions of anabolic hormones and growth factors such as growth hormone and IGF-I. Before adolescence this process is similar in boys and girls. At the time of puberty, however, increases in muscle size and strength accelerate in males from the superimposed effect of circulating testosterone, creating a significant advantage in muscular strength over females.

While this scenario is straightforward, the extent to which qualitative changes in muscle function influence the development of strength during growth remains unresolved. The fact that significant improvements in strength with resistance training are observed in children in the absence of hypertrophy indicates that size-independent changes *can* occur. Comparisons of empirically derived allometric exponents relating anthropometric measures to strength with those predicted by dimensionality theory also suggest such qualitative improvements. A strict interpretation of dimensionality constructs in this analysis, however, may not necessarily be sound; considerable variability is observed in height- or mass-related scaling exponents for different muscles and groups of individuals. Studies in which strength measures were directly related to muscle size have not provided a consistent picture of the role of qualitative effects on strength development. The extent to which size-independent factors stimulate improvements in strength during childhood therefore remains debatable.

If such qualitative changes occur, a neurologic mechanism is likely. Developmental changes in motor unit firing rate, recruitment, or conduction velocity could all possibly contribute. Alternatively, changes in muscle fiber architecture, particularly developmental alterations in pennation angle, or an increase in intrinsic muscle contractile force itself could also contribute. Particularly intriguing is the possibility that a progressive diminution of central inhibitory influences on maximal voluntary contraction might occur in the course of childhood.

A dearth of research data—in some areas a complete lack—allows little more than speculation on these matters, however. Insight into the role of these influences on strength development in children awaits improved measurement techniques that are ethically appropriate for this age group.

Discussion Questions and Research Directions

1. Does dimensionality theory provide a valid basis for separating the influences of muscle size from qualitative influences on the development of strength in children?
2. How are laboratory measures of muscle strength and performance of motor tasks related? By which size-normalizing methods can they most appropriately be compared?
3. How much do architectural and contractile features of muscle contribute to improvements in strength as children grow?
4. How can neural influences on muscular performance accurately be assessed in the pediatric age group? Do motor unit discharge frequency and recruitment change during biologic maturation?
5. Does a central inhibitor of maximal voluntary muscle strength exist? If so, does it change during childhood? If so, why?
6. Do children demonstrate less muscle damage during high-intensity exercise than adults? If so, why should this difference exist? What is its functional significance?
7. Can the extent of release of muscle creatine kinase into the bloodstream with intense exercise serve as an accurate marker of cellular aerobic capacity in children?

11

Responses to Physical Training

It is a grave mistake to submit children to the training programs of adults. After all, children are not simply little adults.

—Tudor Bompa (2000)

In this chapter we discuss

- morphologic and qualitative responses to resistance training in children,
- evidence for a dampened response to aerobic training in youth, and
- trainability of children in short-burst activities.

In the healthy individual, biologic systems respond to repeated stress with an increase in functional capacity. This is the essence of the general adaptation syndrome proposed by Hans Selye, which describes neurohormonal alterations with stress that, within certain limits, produce desirable, positive alterations in performance (113). Today such adaptations are considered in the context of sport performance, physical rehabilitation, and preventive health care, but they date back to a much earlier time. For our prehistoric ancestors, fitness had survival value for escaping human and animal predators and permitting greater mobility to obtain food and avoid harsh climates—a true Darwinian survival of the fittest.

Malina makes the interesting point, however, that physical fitness and Darwinian, or genetic, fitness are not necessarily the same (70). From one perspective, in fact, they may be in conflict. Darwinian fitness refers to reproductive efficiency, or fertility. While physical fitness might achieve this through improved survival from hunger and enemies, the highly fit state is, in fact, characterized by a decrease in reproductive capacity (at least in females). As reviewed in chapter 3, a high level of exercise training is associated with menstrual dysfunction, often with sustained amenorrhea and infertility.

In the continuum of the power–duration curve, the muscle contractile mechanism is the same, regardless if a maximal contraction occurs in 0.5 s or repeated contractions of lower tension that lead to fatigue 60 minutes later. We have seen, however, that the innervation, muscle fiber population, and metabolic support of muscle contraction is highly dependent on the intensity and duration of exercise. This is nowhere more obvious than in the response of improved function to exercise training, in which the nature of the stimulation is highly specific to its outcome. To build strength, one performs high-intensity resistance training, while repeated low muscle forces are critical to improving physical endurance. This stimulus-to-outcome specificity is sometimes striking, as even small variations in contractile velocity, angle of contraction, or use of muscle groups during training can diminish functional improvements.

There is no particular reason to expect a priori that children should respond to physical training differently, either quantitatively or qualitatively, than adults. Yet the idea that children's ability to respond to training is in some way inferior to adults' has persisted throughout the history of exercise science. In at least one case this appears to be true. When children are placed in an endurance training program, improvements in $\dot{V}O_2max$ are less than would be expected in adult subjects. Whether this translates into a reduced response in performance to endurance training (e.g., long-distance run or swim times) remains to be seen.

In other cases, such as in strength training, earlier preconceptions that children have inferior responses to adults' have proven incorrect. Contemporary evidence indicates that relative strength improvements with resistance training are similar in prepubertal and postpubertal subjects. In respect to trainability in short-burst, so-called anaerobic activities, the verdict is not in, pending further research. We address maturational changes in trainability of these specific forms of fitness in the sections that follow.

Information regarding children's training adaptations is obviously important for sport coaches and physical educators, as well as for physicians and physiatrists who plan exercise rehabilitation programs for children with chronic cardiac, musculoskeletal, or pulmonary disease. In addition, insights into maturational differences in trainability may help us understand the basic mechanisms responsible for adaptations to exercise stress. Many questions, in fact, need to be answered. Why should maturational differences be evident in aerobic but not strength trainability? (The implication is that the mechanisms for these two forms of fitness are not only different but are also dissimilarly influenced by biologic development during childhood.) How are the factors that improve maximal aerobic power with training impaired in children? How do maturational differences affect the ability to improve physical performance with training?

As we address these questions, it will be useful to keep in mind how adaptability to exercise occurs and how developmental changes might occur in the context of this design. As depicted in figure 11.1, the key factor in the training response is the turning on of genetic actions that create a phenotypic expression,

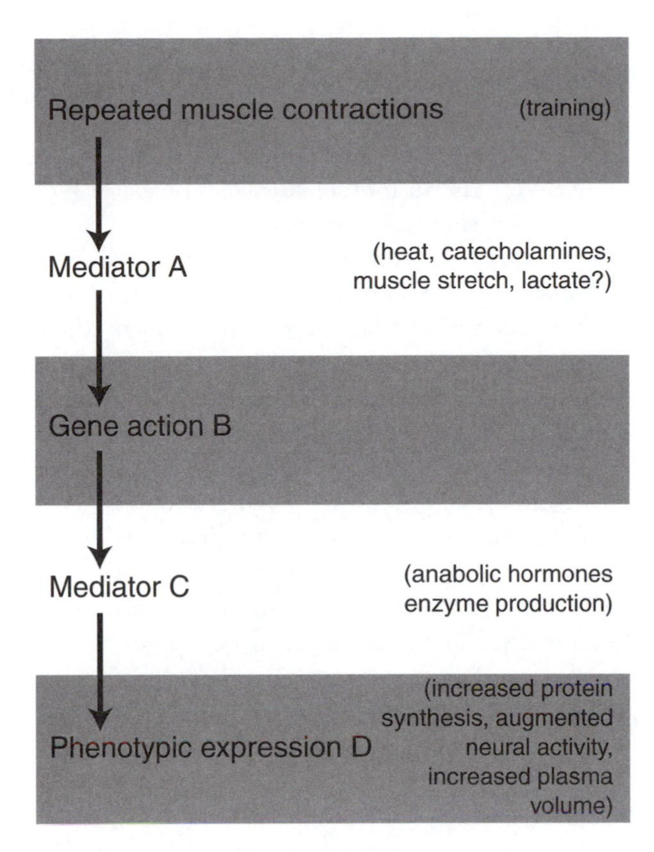

FIGURE 11.1 Pathways for the training response.

the end result of which is improved physiologic function. The first step in understanding this process is identifying the mediator A by which repeated muscle contractions signal an increase in gene action B. That is, what is it about running five miles a day (increases in core temperature? muscle stretch?) that tells the gene to trigger the adaptation process? Then there needs to be a mediator C (anabolic hormones? increased enzyme activity?), a direct product of gene action, that stimulates phenotypic expression D (increased cellular protein synthesis, augmented neural activity, increased heart size).

In fact, if we look at a multifaceted physiologic variable such as $\dot{V}O_2max$, endurance training produces a number of physiologic outcomes D (increased plasma volume, greater cellular aerobic enzyme content, angiogenesis), which have no obvious connection with one another. That is, *multiple* levels of phenotypic expression must occur independently (yet somehow in concert) during endurance training. Adding to the complexity, some of these outcomes are physical (enlarged ventricular diastolic size), others biochemical (cellular enzyme activity), and still others functional (motoneuron firing rate). The goal, then, is to understand this alphabet scenario and how it differs between mature and immature individuals. Simple answers are not anticipated.

Resistance Training

"As to the general effect of strength training in children, the findings of all authors are in general concordant. It seems that strength development is closely related to sexual maturation. Therefore, specific strength training can only be effective in the postpubertal age " (127, p. 152). When Vrijens wrote these words in 1978 he was expressing popular opinion at the time: In the absence of circulating testosterone, children were incapable of strength gains with resistance training. He would probably now regret these words. The accumulated research since that time indicates quite convincingly that both prepubertal girls and boys are, in fact, capable of improving strength with a period of resistance training. Moreover, the relative increases in strength observed in these studies are similar to those observed in adults. Vrijens' conclusion, often quoted in these recent reports, has consequently gained notoriety as a case of misguided perceptions.

How did earlier authors come to this mistaken conclusion? Faigenbaum suggested that methodological limitations, such as the short duration of the training program, low training volume, or inadequate control data, may have been responsible (29). However, in Vrijens' review of the literature as well as portions of his own study, some information indicated that muscle strength could, in fact, be developed with training during childhood.

Vrijens was correct from one standpoint. Children do not generally develop muscle hypertrophy while gaining strength during resistance training, and this appears to be due to a lack of testosterone. Improvements in strength with training during the prepubertal years appear to result exclusively from size-independent factors, presumed to be neural in origin. From this standpoint, then, the influences that drive ontogenetic strength gains as children grow are different from those that generate improvements in strength with resistance training. In the former case, an increase in muscle size is largely responsible, probably with some superimposed qualitative changes related to neurologic maturation. In the latter case, neurologic alterations triggered by resistance training appear to be entirely responsible for strength gains.

The recognition that strength can be effectively increased with training by pubertal children has given rise to a host of interesting issues. What health benefits can children accrue from resistance training? Will improvements in strength at this age enhance sport performance? If children are stronger, will they be protected against musculoskeletal injuries during physical activities? What training regimens are most effective? Should there be concerns regarding the safety of resistance training for young children?

These and other aspects of strength training have been comprehensively reviewed by Faigenbaum (29), Sale (107), Kraemer et al. (65), and Blimkie and Bar-Or (6). In this section, discussion is largely limited to the mechanisms responsible for strength gains with resistance training by children. We start with information available about adults to address questions regarding maturational differences in children.

Studies in Adults

When an adult undertakes a resistance training program, substantial gains in strength occur in the first three to five weeks in the absence of muscle hypertrophy. After this period, increased muscle fiber size contributes predominantly to strength improvements. Overall, as pointed out in a review by Fleck and Kraemer (34), the relationship between changes in muscle size and increased muscle strength in such programs is surprisingly low. Ikai and Fukunaga, for example, reported a 92% increase in maximal isometric strength following training that resulted in only a 23% increase in muscle cross-sectional area (53). This information has been interpreted as indicating that neural adaptations play a critical role in responses to resistance training in adults.

Since adult women have much lower blood testosterone levels than men, it might be expected that their responses to resistance training would be similar to those of prepubertal children. To some extent this seems to be true. Absolute increases in muscle strength and size with training are smaller in women, but when improvements are expressed relative to initial pretraining findings, they are similar to men's (108). Since men demonstrate not only higher resting testosterone levels but also greater increases in those levels with exercise, other anabolic factors (growth hormone, IGF-I) may play a role in the limited hypertrophic response in women.

Hakkinen and Pakarinen reported training responses in 10 women after three weeks of intense, heavy-resistance strength training (44). Maximal peak force of leg extensor muscles increased by 9.7%, with only a 4.6% rise in muscle cross-sectional area (i.e., there was a significant increase in maximal force relative to muscle size). Maximal neural activation, as indicated by EMG recordings, rose by 15.8%. The authors concluded that improvements in maximal strength from short-term resistance training in women are mainly a result of increased activation of muscles, with limited changes in muscle fiber size.

Resistance training in adults causes an increase in myofibrillar protein with hypertrophy of fast-twitch fibers. As a result, mitochondrial density and capillary density decrease. This trend is opposite that seen with endurance training and may explain why aerobic endurance capacity can be diminished by a program of resistance training (121). That is, the specific metabolic and anatomic changes that occur with strength and endurance training are so dissimilar that skeletal muscle cannot respond to both simultaneously (67).

While genetics moderately influence interindividual differences in muscle strength, little is known regarding the genetic effect on an individual's ability to improve muscle strength with resistance training. Based on findings in a study of adult male monozygotic and dizygotic twin pairs, Thomis et al. concluded that 20% of the strength responses to a 10-week resistance training program for elbow flexors could be explained by training-specific genetic factors (122).

Neural Factors

As noted earlier, there is good evidence in both adult men and women that neural adaptations play a critical role in strength adaptations to resistance training. This conclusion is supported by findings of increased electromyographic activity after both isometric and isokinetic training (106). This should not be surprising, since, as Bawa emphasized, "the muscle, no matter how sophisticated in design and composition, acts in response to neural commands to produce the required range of motor outputs" (4, p. 59). The mechanisms by which this occurs, however, remain uncertain. Little is known, in fact, regarding the plasticity of neural factors in response to training nor the triggers for these changes. There are, however, several sites in the pathway of muscle innervation where changes might occur. Neural alterations that could occur with training include changes in central input to motoneurons, alterations in the firing and transmission of the motoneuron itself, and adaptations at the level of synaptic connections with muscle fibers (21).

Maximal Firing Frequency The maximal firing frequency of motoneurons has been reported to rise with a period of resistance training in adults and is higher in trained weightlifters than in nonathletes (1, 58). The implications of this change are unclear. Maximal firing rates at the beginning of a maximal voluntary contraction are considerably higher than those at the point of maximal force generation, which occur 250 to 400 ms later (100–200 Hz versus 15–35 Hz). Aagaard contended that "it is possible, therefore, that supramaximal firing rates in the initial phase of muscle contraction serve to maximize the rate of force development rather than to influence maximal contraction force per se" (1, p. 62). That is, changes in the *rate* of contractile force development could be the central neural influence on muscle strength development with resistance training.

Not all studies have indicated such increased firing rates with training (95). Patten et al. showed an early rise in discharge rates during training, but after two weeks a decline was observed (88). At six weeks of training, values during maximal contractions were similar to those at baseline.

Rate of Force Development In adults, it typically takes at least 300 ms to develop maximal force in contraction of elbow flexors and knee extensors. This is considerably longer than the contraction times required in activities such as sprint running or karate (50 to 250 ms). During such events, then, the limited time for contraction does not permit achievement of maximal muscle force. Consequently, the rate of force development (RFD) is an important measure, since it defines the fraction of maximal strength that can be produced in activities requiring very short contraction times.

RFD is influenced by many factors, including motoneuron firing frequency, recruitment of nonactivated motoneurons, muscle fiber size, and fiber type. RFD has been documented to increase with a period of resistance training in adults. For example, Aagaard et al. studied RFD and rate of EMG rise (as an indicator of neural drive) in 15 young adult men before and after a 14-week period of heavy-resistance strength training (2). Maximal isometric quadriceps strength increased from 291 ± 10 to 339 ± 10 N · m after training. When normalized to maximal voluntary contraction, overall RFD rose by 15%, and mean muscle EMG increased by 41% to 106%. The enhancement of RFD could therefore be explained by a greater neural drive. The authors pointed out that an increase in RFD could translate into the ability to perform the same short-burst activity with greater force and thus, presumably, enhanced performance.

Based on such observations, it has been suggested that the velocity of force production for a specific motor task is important in sport training. That is, training regimens that simulate the velocity and acceleration characteristics of performance tasks such as throwing, sprinting, or jumping may be critical for optimal training results (18).

Co-contraction During a motor task, the action of the muscles primarily responsible (agonists) is countered by simultaneous contraction of antagonistic muscle groups. This co-contraction may be important for joint stability, but it also is thought to serve an inhibitory function, protecting the agonists from excessive stress. The action of antagonistic muscles, though, prevents full activation of the agonists. It has been suggested that training may reduce innervation of antagonistic muscles, lessening the level of co-contraction and thereby permitting greater force production by agonistic muscles (106).

Other Factors In animals, high-intensity training increases the area of the neuromuscular junction (18). In addition, trained rats showed changes in synapse morphology, with a greater total length of dendrite branching.

Control input from motor areas of the brain to motoneurons in the spinal cord may be influenced by a balance of inhibition and excitation to particular motor areas. There is some evidence that this balance can be altered with physical training, by augmented overall stimulation to certain muscle groups (4).

Muscle Fiber Hypertrophy

There are a number of agents that cause skeletal muscle hypertrophy (see chapter 2), but the roles of these various anabolic factors on hypertrophic responses to resistance training are uncertain. The observation that some (such as testosterone, growth hormone, and IGF-I) rise during acute bouts of exercise suggests that repeated "doses" during training might produce chronic alterations in muscle size.

The most obvious candidate for stimulating strength changes with training is testosterone, which clearly induces muscle hypertrophy and is directly correlated with muscle fiber size (118). Circulating levels of this hormone in adult men are 10 times those of women, and a rise with strength exercise is seen only in males (44, 93). Resting levels of testosterone are probably not influenced by resistance training (64).

The importance of both testosterone and muscle hypertrophic changes to resistance training in adults was highlighted by the report of Bhasin et al. (5). These investigators administered supraphysiologic doses of testosterone to healthy men alone or in addition to a 10-week resistance training program. Compared with both a nontraining placebo group and the testosterone-treated nontraining group, those who weight-trained while receiving testosterone demonstrated greater gains in both muscle strength and size. These findings indicate that testosterone facilitates muscle hypertrophic response to resistance training, which in turn is reflected in greater increases in muscle strength.

The influences of other anabolic agents on strength development with resistance training is unknown. There is currently no evidence to suggest estrogen plays a role in muscle tissue growth (64).

Adult strength-training athletes (weightlifters, wrestlers) show increased heart ventricular wall thickness that corresponds to skeletal muscle hypertrophy (27). This suggests that strength training is accompanied by myocardial as well as skeletal muscle hypertrophy, and echocardiographic findings during the course of a strength training program support this idea (33). The parallel changes in heart and skeletal muscle also indicate that similar mechanisms might be responsible for both.

Follard et al. found that the allele of the angiotensin-converting enzyme (ACE) genotype is related to strength gains with resistance training in adults (35). Following nine weeks of isometric or isokinetic training, a significant interaction was observed between ACE genotype and increases in strength, with greater gains in subjects with the D allele. A similar role of the ACE genotype on both cardiac hypertrophy and physical fitness has also been described (104, 135). These findings, then, support the concept that the responses of cardiac and skeletal muscle to functional overload may share a common mechanism.

Goldspink and Yang demonstrated the importance of local autocrine anabolic factors in muscle hypertrophy and provided ideas on how mechanical factors might translate into strength gains (39). They found that four days of passive stretching of arm muscles produced mechanogrowth factor (MGF), which, when injected into mouse tibialis anterior muscle, produced a 25% increase in muscle mass. Electrical stimulation of the arm muscle without stretching failed to reproduce this effect. These findings demonstrated a link between mechanical stimulation and gene expression of muscle hypertrophy and suggested that mechanical strain on the muscle cell rather than contractile activity itself is the trigger for gene action.

Studies in Children

With these concepts from studies in adults in mind, we can now consider the patterns of strength response to resistance training by children. Training is as effective at improving strength in prepubertal subjects as it is in adults if comparisons are made on a relative (i.e., percentage increase) basis. Studies in children, however, have seldom detected evidence of muscular hypertrophy with strength gains. Instead, improvements in strength with resistance training in prepubertal subjects appear to be entirely due to qualitative changes, presumed to be neural in origin. It might be assumed that these neural adaptations to training are the same as those responsible for strength gains in adults in the early phases of training programs.

Gains in Strength

In a meta-analysis published in 1997, Payne et al. identified 28 studies that examined the effects of resistance training in subjects under 18 years of age (90). Training in these studies, which included children as young as six years old, included isometric and isokinetic regimens using weight machines and free weights, calisthenics, and specific sport or activity training (such as wrestling and martial arts). Overall, improvements in strength were in the range of 30% to 40% following programs of 8 to 12 weeks' duration. These relative gains are similar to those observed in training programs in young adults. No differences in training effect with age were observed in this meta-analysis.

Similar findings were observed in another meta-analysis of strength training studies confined to those involving children under age 12 or 13 years (31). Among the nine studies that provided sufficient data for analysis, gains in strength were between 13% and 30%. Most of these studies involved boys, but the limited information regarding strength training in girls revealed no obvious effect of sex. Lack of injuries in these studies attested to the safety of supervised strength training programs in youth.

With only two exceptions, the duration of resistance training in these studies did not exceed 20 weeks. The authors concluded that the effect of training duration on strength production in children was yet to be determined. Likewise, these reports differ in the muscle groups studied, and possible maturational effects on trainability of specific muscles cannot be ruled out.

A number of weaknesses were pointed out in both of these meta-analyses. The learning effect, which can account for early strength gains with training, was not often assessed. Adherence rates have not been regularly reported. Failure to randomly assign participants to training and control groups was common. And, as noted before, few of studies included female subjects.

The effects of resistance training in adolescent boys are of particular interest, since improvements in strength in this group might be expected to reflect the combination of (a) the anabolic actions of testosterone, (b) stimulation by factors such as growth hormone and IGF-I, which are responsible for normal increases in muscle bulk with growth, (c) size-independent adaptations, presumed to be neural, reported in both children and adults, and (d) maybe improvements in intrinsic muscle fiber contractile force.

Studies in high school wrestlers have indicated that age effects on increasing peak torque cannot be fully accounted for by changes in muscle mass (49, 50). Housh et al. commented, "These findings lend indirect support for the hypothesis that neural maturation during adolescence influences the expression of muscular strength and may account for the age effect for increases in peak torque" (50, p. 181).

Neural Mechanisms

Only limited information is available on neural adaptations during resistance training by children. While this information suggests possible changes in neurologic innervation, the mechanisms by which they might occur remain obscure.

Ozmun et al. demonstrated that children's EMG activity increases as they train, supporting the possibility of improved neural activation of muscles (86). Eight prepubertal children performed three sets (of 7–11 repetitions) of dumbbell biceps curls three times a week for eight weeks. Isometric and isokinetic arm strength improved with training by an average of 22.6% and 27.8%, respectively. At the same time, EMG amplitude rose by 16.8%. No significant changes in any of these measures were observed in the control group (figure 11.2).

Ramsay et al. sought to determine whether increases in muscle size or neurologic function were responsible for strength gains with training in

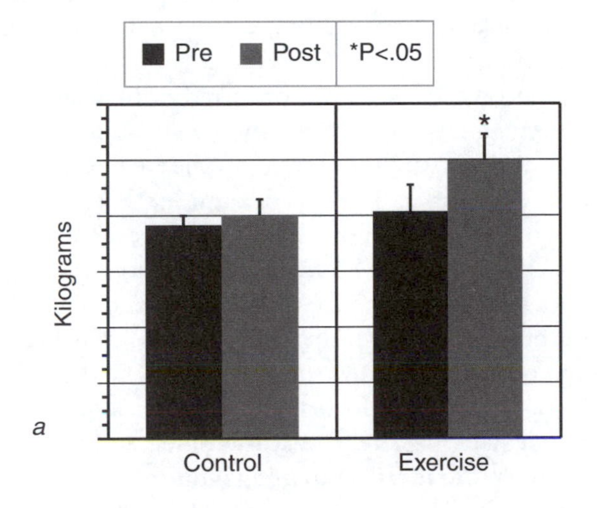

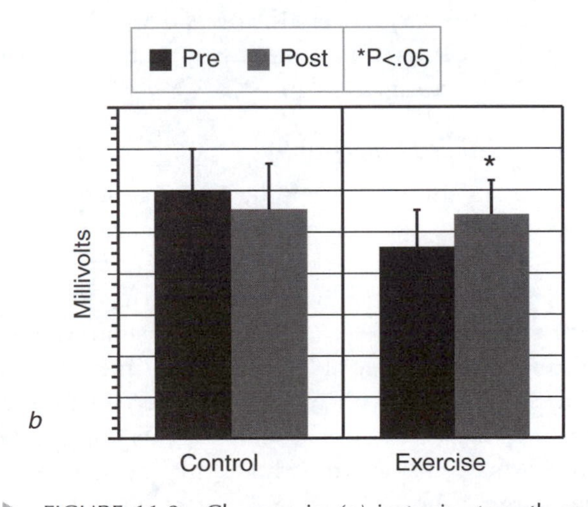

FIGURE 11.2 Changes in *(a)* isotonic strength and *(b)* EMG activity after eight weeks of resistance training in prepubertal children and nontraining controls (from reference 86).

Reprinted by permission from J.C. Ozmun, A.E. Mikesky, and P.R. Surburg 1994.

13 prepubertal boys ages 9 to 11 years (94). After 20 weeks of training three times a week, 1RM bench press and leg press strength increased by 35% and 22%, respectively, with only a 12.3% overall increase in nontraining controls. Training also induced similar increases in isokinetic peak torque of both the elbow flexors and knee extensors at four contraction velocities. At the same time, no training effect was observed on muscle cross-sectional area as measured by computerized tomography. Likewise, no changes were recorded in muscle contractile properties (time to peak torque, half-relaxation time). There was a trend for increased motor unit activation in the elbow flexors and knee extensors, but these changes did not achieve statistical significance. (Other possible neurologic adaptations, such as motor unit coordination, recruitment, and firing frequency, were not measured.)

The only change that occurred with increased strength in this study was an improvement in electrically evoked twitch torque of the elbow flexors and knee extensors (94). This measure has been interpreted as an indicator of the intrinsic force-producing capacity of muscle. The authors pointed out that this is not necessarily so, since tetanic stimulation (which was not performed in this study because of the discomfort it would cause subjects) would be the only way to obtain that information. They concluded, "We cannot state for certain that the observed increases in twitch torque reflect training induced increases in intrinsic force-producing capacity of the elbow flexor or knee extensor muscles. [If this is true], change in excitation-contraction coupling appears to be the most likely explanation" (p. 612).

Muscle Fiber Hypertrophy

Almost all studies have indicated that significant improvements in strength with resistance training by children are not accompanied by increases in muscle size. This has been attributed to the lack of a testosterone effect in prepubertal subjects. Testosterone is not available to facilitate increases in muscle fiber size with training, but certainly other anabolic factors are. In particular, agents responsible for normal gains in strength and hypertrophy with age (growth hormone, IGF-I, paracrine and autocrine factors, insulin) are all acting on muscles when a prepubertal child undergoes strength training. If absence of testosterone is, in fact, the reason for lack of hypertrophy with training in children, we would be forced to conclude that this hormone facilitates strength gains with training, while other anabolic agents—the ones responsible for increased muscle growth in the prepubertal years—do not.

It should be recognized that some studies have shown small increases in muscle bulk with resistance training in children. Mersch and Stoboy used magnetic resonance tomography to assess changes in quadriceps cross-sectional area with training of the left leg in two preadolescent twin boys (79). After 10 weeks of isometric strength training, the trained leg increased in strength by 35% to 40%, while strength in the untrained leg improved by 10%. Muscle cross-sectional area increased in the trained by 4% to 9% and in the untrained leg by 2%.

In a study by Fukunaga et al., 50 boys ages 6.9 to 10.9 years underwent a 12-week strength training program (37). Cross-sectional area of muscle was estimated by ultrasound. Significant improvements in both isometric strength and muscle cross-sectional area compared with control subjects were observed. This was particularly true in the older children, in whom average increase in muscle cross-sectional area was 15.1% for boys and 12.8% for girls. Blimkie and Sale (7) identified some curious findings in the study by Fukunaga et al. (37), such as the observation that a training program based on isometric elbow flexion exercise produced greater strength gains in the elbow extensors than flexors. They concluded that "whether muscle hypertrophy is possible or not, however, it is evident from the results of all these studies . . . that the magnitude of this morphological adaptation is small in comparison to the reported strength gains and in comparison to adults. Therefore, other factors besides changes in muscle size must account predominantly for the strength gains observed in these studies" (7, p. 215).

In contrast to the failure of skeletal muscle to hypertrophy with training in children, it is interesting that cardiac muscle demonstrates a very robust hypertrophy in response to stress in children with congenital heart disease. The requirements to increase ventricular pressure in response to outflow obstructions such as aortic valve stenosis or coarctation of the aorta are met with ventricular hypertrophy that is equal to or greater than that seen in adults with the same anomalies. There is more than one way of explaining this skeletal muscle–myocardial difference. It is possible that children's lack of skeletal muscle hypertrophy with training is a quantitative rather than a qualitative difference from adults. That is, the prepubertal child may be capable of such hypertrophy, but the duration or level of training stimulation may need to be greater than for the adult (who, you will recall, also shows no muscle hypertrophy until well into the course of a training program). Cardiac muscle may show greater hypertrophy because it faces a greater challenge: force produced 80 to 100 times a minute, 24 hours day, 365 days a year, with no rest period—a truly intense workout! Alternatively, however, the mechanisms responsible for myocardial hypertrophy in children with heart disease may differ from those controlling skeletal muscle fiber response to resistance training in healthy youngsters.

The extent to which the myocardium responds to strength training in children in comparison with adults would be of interest, but as yet, there are no experimental data on this point (107). Servidio et al. reported that left ventricular volume, but not wall thickness, increased in six prepubertal boys after an eight-week training program consisting of Olympic-style lifts (114). This single study found that the myocardium does not change with training as it does in adults.

Trainability of Fitness in Short-Burst Activities

We face some difficulty in trying to define anaerobic trainability in children, given the many facets of performance and capacity for short-burst activities. We could focus on training responses to metabolic capacity; power production during brief, high-intensity laboratory exercise tests; or performance in short field activities such as sprints. It may, in fact, be appropriate to view this in a stratified, hierarchical way. We can start with metabolic factors as perhaps the "purest" form of anaerobic capacity. Differences in trainability at this level would reflect variations in the response of determinants such as glycogen storage, neuroendocrine influences, and activity of rate-limiting enzymes. Power production during tests such as the Wingate test are influenced not only by energy supply but also strength. So trainability on a 30-s, all-out cycle test would also include changes in muscle size and neural influence that were previously identified as contributing to training-induced improvements in muscle strength. When assessing trainability of speed (i.e., sprinting) and other short-burst activities, we must also add the plasticity of neuromuscular coordination and skill. Clearly, oversimplistic considerations of anaerobic trainability are to be avoided.

Standards of training intensity and duration that improve different aspects of short-burst fitness are not as clearly defined as for aerobic fitness. Nonetheless, certain characteristics of training have been considered important if improvement in these measures is to be expected (12). High-intensity exercise should be brief and performed until near exhaustion. Rest periods between exercise bouts should not permit complete recovery. Repeat bouts with three to five minutes of recovery may fit these criteria (74). Anaerobic and aerobic training should not be mixed. And work rate must progressively increase to improve performance.

In adult subjects, a period of high-intensity training following these guidelines has been reported to increase anaerobic enzyme activity, fast-twitch muscle size, lactate production, and performance in short-burst activities (74). Jacobs et al., for example, found that young men demonstrated a 16% rise in maximal lactate and a 16% increase in phosphofructokinase (PFK) activity after six weeks of cycle sprint training (54). Despite these metabolic responses, however, no significant improvement was observed in 30-s Wingate cycle performance.

Some training programs for sprint performance in adults consider the acceleration and the maximal speed portions of the sprint as different forms of motor performance (see chapter 9). Various types of training to develop muscle hypertrophy and neural adaptations separately have been used, on the assumption that such training will improve sprint performance. The scientific literature supporting this, however, is limited (see reference 19 for a review). In the studies that do demonstrate improvement, decreases in sprint times have been small.

Jansson et al. described an increase in the proportion of fast-twitch fibers in adult men following sprint training (55). Similar findings have been described in animals (129). However, others have inexplicably described increases in slow-twitch (type I) fibers during sprint training programs in adults (68). Thus, Van Praagh concluded, "With respect to conflicting data in the literature, it seems a) difficult to draw any definite conclusions, and b) unreasonable to extrapolate these findings, derived from adults, to children" (125, p. 168).

Metabolic Trainability in Children

Information regarding anaerobic metabolic changes with training in youth is largely limited to reports after endurance training or following high-intensity training by postpubertal subjects. Eriksson et al. (25) reported that muscle PFK activity rose from 8.4 ± 1.5 to 15.4 ± 1.6 $\mu mol \cdot g^{-1} \cdot min^{-1}$ in five 11-year-old boys after six weeks of cycle endurance training (30 minutes three times a week). A small but significant rise was also seen in maximal blood lactate concentration (from 4.7 ± 0.6 to 5.9 ± 0.7 $mmol \cdot L^{-1}$).

Fournier et al. examined alterations in muscle PFK and fiber area in 16- to 17-year-old boys with sprint training four times a week for three months (36). Training consisted of interval runs varying from 50 m to 250 m. $\dot{V}O_2max$ rose with training by an average of 10%. No change was seen in muscle fiber size or distribution. Pretraining PFK activity was 28 ± 7 $\mu mol \cdot g^{-1} \cdot min^{-1}$, which is less than that expected in adults. With training, PFK activity rose by 21%.

Despite these enzyme findings, children usually fail to reveal any improvements in maximal blood lactate levels after physical training, but again, these regimens have generally involved endurance activities (62, 76, 78, 119).

Prado compared lactate responses to exercise following anaerobic training in 12 boys 10.8 ± 0.7 years old and 12 adults 24.0 ± 5.7 years old (92). Swim training, consisting of an anaerobic series intended to minimize alterations in aerobic capacity, was performed three times a week for six weeks. Testing involved three maximal anaerobic tests (25-m and 100-m sprints and distance swum in 45 s at maximal intensity) performed on different days. Blood lactate levels were determined one and three minutes after exercise. The adults' performance improved in the 100-m sprint and 45-s test, but no effects of the six-week training period were observed in the children. Lactate levels were two to three times greater in the adults than the children at all measurement points. However, no significant changes were observed in maximal lactate with training in either group.

Changes in Anaerobic Power

Several studies have indicated that children's power in short-burst exercise tests can improve after anaerobic training (41, 96, 110). In general, however, these increments in performance have been small. For example, Grodjinovsky et al. discovered a 3.4% and a 3.7% increase in mean anaerobic power during Wingate testing in 11- to 13-year-old boys after high-intensity cycling and sprint training, respectively (41). Sargeant et al. studied 13-year-old boys in an eight-week program of doubled physical education time that included a mixture of short-burst and aerobic activities (110). Maximal power on an isokinetic cycle test improved 8.5%, compared with an average 3.7% in nontraining controls. Rotstein et al. reported 10% and 14% increases in mean and peak anaerobic power on Wingate testing in 28 boys ages 10 to 11 years after a nine-week interval training program (96).

McManus et al. (75) described fitness changes following an eight-week aerobic cycling program (*n* = 12) or a sprint-running program (*n* = 11). Both groups demonstrated a rise in $\dot{V}O_2max$ (10% following cycling and 8% with sprint training), but no changes were seen in control subjects. Cycle training and sprint training both resulted in significant improvements in peak power after 5 s during Wingate testing. At the same time, no significant changes were observed in either group in mean power after 30 s.

Other evidence exists that training in distance running can improve anaerobic power in children. Obert et al. reported changes in maximal power during a short-term exercise test in 10- to 11-year-old girls and boys who participated in a 13-week distance-running program (84). Findings were compared with those of 16 nontraining children. Training involved two weekly one-hour sessions of predominantly aerobic interval running (10 × 300 m, 12 × 250 m, 4 × 600 m) at a heart rate of 75% to 80% maximum. Lower-limb mass was estimated by dual X-ray absorptiometry. The training produced significant increases in maximal power on a force–velocity test even when changes in muscle mass due to normal growth were accounted for. No such changes were seen in the nontraining control subjects.

Gutin et al. compared anaerobic capacity—defined as the number of revolutions completed in a 30-s, all-out cycle sprint—in highly trained 8- to 13-year-old distance runners (who also participated in other sports, such as soccer and basketball) and age-matched nontrained controls (43). Anaerobic capacity was significantly greater in the runners than in the controls (54 ± 7 vs. 44 ± 9 revolutions in 30 s, respectively).

Training Improvements in Short-Burst Activities

Mero reasoned that since (a) strength is an important contributor to the performance of short-burst activities and (b) children improve in strength with resistance training, prepubertal subjects should be trainable in such activities as sprinting and jumping (77). The limited research data, however, are conflicting.

Diallo and colleagues confirmed that anaerobic training can improve short-burst performance in prepubertal male soccer players (22). Thirty 12- to 13-year-old boys were divided into a plyometric training group (jumping, hurdles, skipping), a cycle sprint training group, and a nontraining control group. Training was performed three days per week for 10 weeks. The sprint cycling performance and countermovement jump height of both the jump- and sprint-trained children significantly improved compared with those of the control subjects. Interestingly, when the subjects were tested again after an eight-week detraining period, no decrements in any of these performances were observed (23).

Mosher et al. reported responses to a 12-week period of high-speed activity training in 10- to 11-year-old elite-level soccer players (81). Compared to nontraining soccer players, the experimental subjects' performance on a high-intensity treadmill run ($7 \text{ mi} \cdot h^{-1}$ at 18% grade) improved by 20%, but no changes were seen in 40-yd sprint times. As noted earlier, Prado could find no improvement in swim sprint times of younger swimmers after training (92).

From this collection of information, it is impossible to say whether children are trainable in short-burst activities, either metabolically or functionally, or, if training responses do occur, whether they are similar to those in adults. Much of the data, in fact, conflict on these points. Many studies describe training programs that were not focused on short-burst activities or involved athletes who were already well trained. Moreover, testing modalities and outcome measures varied. A more standard approach to defining anaerobic fitness and its testing methodology will be important for understanding maturational changes in trainability in short-burst activities.

Aerobic Trainability

It became apparent when studies of children were first performed that prepubertal subjects' $\dot{V}O_2$max responses to a period of endurance training were often less than the 15% to 30% increase typically seen in adults. It was not unusual, in fact, for maximal aerobic power in children (usually boys) to improve with training by less than 10% (3). Some viewed this inferior trainability of children as reflecting biologic differences between pre- and postpubertal subjects. That is, "there is one critical time period in a child's life (termed the 'trigger point') which coincides with puberty in most children, but may occur earlier in some, below which the effects of physical conditioning will be minimal, or will not occur at all. . . . This trigger phenomenon is the result of modulating effects of hormones that initiate puberty and influence functional development and subsequent organic adaptations" (59, p. 242).

It is recognized, however, that these early studies often suffered from methodological weaknesses, including absence of controls, too few subjects, or inclusion of athletic children. Some researchers believe that children might train less effectively than adults or that the stimulus intensity needs to be greater for children because of their relatively higher ventilatory threshold (i.e., it takes a higher training intensity to tax their aerobic systems). It has also been pointed out that since children are more active than adults, they are, in effect, in a constant state of self-training, such that further improvements in fitness from structured training programs might be less.

Critical reviews that included only pediatric studies that seemed to conform to criteria in adults for training intensity, duration, and frequency still showed that increases in $\dot{V}O_2$max with training were only about 10% to 14% (87, 89, 97, 99). In the 1993 meta-analysis of Payne and Morrow, the average rise in $\dot{V}O_2$max in 23 training studies of children was only 5% (89).

Between 1995 and 2001 a number of endurance training studies of children were performed that avoided the methodological problems of earlier investigations (table 11.1). Subjects in these studies were previously untrained youth who exercised at an documented, appropriate intensity (heart rate 160–170 bpm), frequency, and duration. Findings were compared with those of either a nontraining control group or the same subjects in a nontraining period. In effect, by all accepted criteria (developed in adults), subjects in these studies would be *expected* to demonstrate a rise in $\dot{V}O_2$max similar to that seen in postpubertal individuals. As is evident from table

TABLE 11.1 Recent endurance training studies assessing $\dot{V}O_2$ response in children

Study	*n*	Age (years)	Sex	Duration (weeks)	% Change in $\dot{V}O_2max$
McManus et al. (75)	12	9.6	F	8	7.8
Rowland and Boyajian (100)	37	10-12	M, F	12	6.7
Welsman et al. (131)	17	10	F	8	NS
Williford et al. (133)	12	12	M	15	10.3
Rowland et al. (102)	31	10-12	M, F	13	5.4
Shore and Shephard (116)	15	10	M, F	12	NS
Tolfrey et al. (123)	12	10	M	12	NS
Tolfrey et al. (123)	14	10	F	12	7.9
Williams et al. (132)	13	10	M	8	NS
Mandigout et al. (71)	28	10-11	M	13	4.6
Mandigout et al. (71)	22	10-11	F	13	9.1
Ignico and Mahon (52)	18	8-11	M, F	10	NS
Eliakim et al. (24)	20	9	F	5	9.5
Yoshizawa et al. (136)	8	4-6	F	72	18.9
Mobert et al. (80)	12	13	M	28	12.2

NS = no significant change.

11.1, even when training studies in children are well designed, improvements in maximal aerobic power are small. Most are in the range of 0% to 10%, and the overall average is 5.8%, closely matching the findings in the earlier meta-analysis of Payne and Morrow (89).

A number of these studies provide specific information regarding aerobic trainability in children. First, overall, there are no obvious sex differences in aerobic trainability in the prepubertal age group (89, 100). However, in two studies that directly compared boys and girls, the girls demonstrated greater changes in $\dot{V}O_2max$. Mandigout et al. (71) reported that 10- to 11-year-old girls had a twice greater rise in $\dot{V}O_2max$ after a 13-week training program than boys (9.1% vs. 4.6%). The authors felt that this could be explained by the lower initial fitness of the girls, since a significant relationship was observed between percentage rise in $\dot{V}O_2max$ and pretraining values.

Tolfrey et al. came to the same conclusion when they examined training responses to a 12-week endurance cycle program in 14 girls and 12 boys (123). Average peak $\dot{V}O_2$ per kilogram increased by 7.9% and 1.3% in the two groups, respectively. ANCOVA analysis revealed, however, that the sex-by-time interaction was statistically insignificant. This indicated to the authors that the sex difference in peak $\dot{V}O_2$ increase was a function of lower pretraining values in the girls.

No clear association between training duration and $\dot{V}O_2max$ response is evident from the studies in table 11.1. However, note that three of the studies with the longest training periods (15 weeks, 28 weeks, 18 months) found the highest rises in $\dot{V}O_2max$ (10.3%, 12.2%, and 18.9%, respectively). This raises the possibility that the dampened response of prepubertal subjects to endurance training could represent a quantitative, volume-related characteristic rather than any qualitative, mechanistic difference from adults.

The remarkable study by Yoshizawa et al. should be noted (136). These investigators had eight young girls perform a 915-m run six days a week for 18 months. Treadmill tests were performed every six months to compare the runners' $\dot{V}O_2max$ with untrained controls. $\dot{V}O_2max$ at the onset of the study was nearly identical in the two groups (42.2 ± 2.2 and 42.4 ± 3.1 $ml \cdot kg^{-1} \cdot min^{-1}$ for exercisers and controls, respectively). The rise in $\dot{V}O_2max$ was greater in the trained girls than in the controls (figure 11.3), a

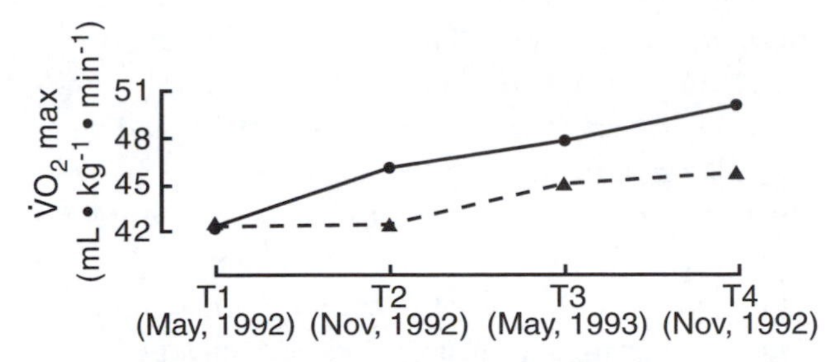

FIGURE 11.3 Changes in mass-relative $\dot{V}O_2$max in eight young girls during an 18-month run training program *(solid line)* compared with untrained children (*dashed line;* from reference 136).

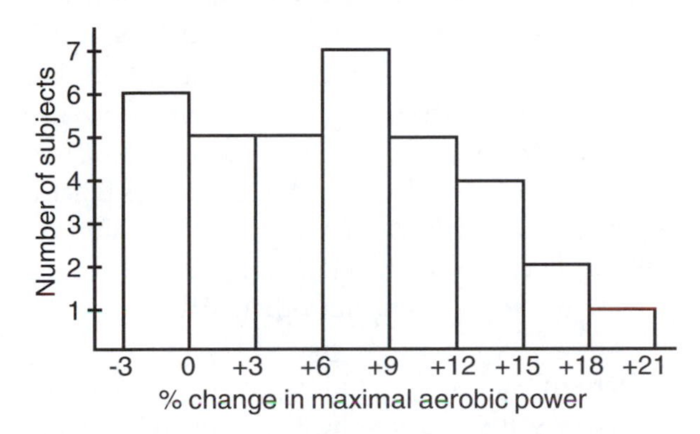

FIGURE 11.4 Distribution of $\dot{V}O_2$max responses to a 12-week training program in children 11 to 13 years old (from reference 100).

difference that reached statistical significance by 12 months.

Considerable interindividual difference in $\dot{V}O_2$ responses to training have been observed in training studies of children, but some evidence suggests that this variability might be less than in adults. For instance, when Rowland and Boyajian trained 35 children for 12 weeks, one third of the subjects demonstrated a rise in $\dot{V}O_2$max of less than 3%, while the single greatest increase was 19.7% (100; figure 11.4). This contrasts with the 20-week training study of Lortie et al. involving sedentary adults, in which increases in $\dot{V}O_2$max ranged from 5% to 88% (69).

Possible Explanations

Recent training studies in children have avoided the methodological pitfalls that beset earlier reports. Consequently, inappropriate study design cannot be used to explain inferior training responses in prepubertal subjects. These most recent studies have been well structured with adequate controls, appropriate subject populations, and documented, high training intensities that should be expected to improve $\dot{V}O_2$max.

Similarly, the idea that the higher habitual physical activity levels of children might "pretrain" young subjects has been largely discounted. First, although children are more active than adults, this activity is short-burst in nature and would not be expected to trigger aerobic responses. Second, there is no strong association between level of physical activity and $\dot{V}O_2$max in the pediatric population (see chapter 5). Finally, during the course of childhood and early adolescence, daily energy expenditure (relative to body size) decreases progressively, while over the same age span, values of $\dot{V}O_2$max per kilogram (at least in males) remain stable.

One of the tenets in adult studies is that an inverse relationship exists between pretraining $\dot{V}O_2$max and magnitude of response of $\dot{V}O_2$max to a period of endurance training. Training studies in adults usually involve subjects who have pretraining values in the range of 30 to 45 ml · kg^{-1} · min^{-1}, while in pediatric training studies, initial $\dot{V}O_2$max values typically are considerably higher. Thus, it could be that the dampened response to aerobic training in children as compared with adults could reflect the higher initial level of physiologic fitness in the former. In fact, in adult studies in which the initial $\dot{V}O_2$max was in the range of 50 to 55 ml · kg^{-1} · min^{-1}, the rise in $\dot{V}O_2$max with training was typically only about 10% (109), similar to that described in pediatric reports.

This line of thinking is complicated by use of the ratio standard rather than a more appropriate allometric approach. Moreover, this explanation for the limited training rise in $\dot{V}O_2$max in children is weakened by the observation that children with lower initial $\dot{V}O_2$max still show a reduced $\dot{V}O_2$max response. For example, Ignico and Mahon reported the effects of a 10-week aerobic fitness program in 8- to 11-year-old children who had failed to meet at least three of four performance standards in routine school fitness testing (52). Initial $\dot{V}O_2$max was 45.6 ± 5.2 ml · kg^{-1} · min^{-1}, and training failed to produce any significant improvement. (Although frequency and intensity requirements were satisfied, the authors noted that the duration of each exercise task varied from a few minutes to 20 or 25 minutes. They therefore thought that the subjects' inability to maintain an elevated heart rate for a sufficient period of time might explain the lack of rise in $\dot{V}O_2$max.)

Biologic Mechanisms

The preceding research supports the suspicion that true biologic differences between children and adults exist that restrict improvements in aerobic fitness with training in immature subjects. The mechanism by which this restriction might occur, however, remains obscure. In this section we review current concepts regarding the biologic factors responsible for improvement in $\dot{V}O_2$max in studies of adults and attempt to predict how developmental immaturity might influence these determinants.

It generally thought that differences in aerobic trainability between children and adults are demarcated by the age of puberty. There are, however, no good data to confirm this impression (61, 130). Nonetheless, it seems appropriate to seek biologic mechanisms that separate trainability between children and adults among the changes that occur at puberty. This means that we should pay particular attention to the potential roles of the sex hormones as we review the candidates for explaining maturational effects on responses to aerobic training in the following sections.

Much of this book has focused on the healthy, nonathletic child, but in examining aerobic trainability, we may gain insights from highly trained child endurance athletes. The aerobic characteristics of these young athletes (greater maximal cardiac output, larger left ventricular dimensions) can be due to either the effect of training or genetic endowment, and we cannot distinguish between these two influences. On the other hand, it can be assumed that any such characteristic *includes* the effect of training. Thus, if we find that a particular feature is similar in child athletes and nonathletes, we can conclude that this variable is not influenced by training in children.

General Principles

Unlike strength training, the adaptations to endurance training involve a multitude of biologic systems. Indeed, we can view the aerobic training effect as a generalized enlargement and enhancement of the entire oxygen delivery system: a greater maximal stroke volume, larger ventricular size, expanded plasma volume, higher maximal minute ventilation, and increased cellular aerobic enzyme activity. It is important to note that these responses involve disparate sets of biochemical, functional, and anatomic characteristics that do not have any obvious etiologic connection. This immediately raises questions about triggers and outcomes: How can repetitive actions of large muscles (i.e., training) be translated into functions as different as Krebs cycle activity and lung ventilatory capacity? Is there a common mechanistic pathway that activates genes to create these different outcomes? Are there separate mechanisms for the different effects of aerobic training, or are certain outcomes primary, causing a reactive change in the others?

Another interesting aspect of the aerobic training effect is the components of the oxygen delivery chain that are *not* enhanced by endurance training. Maximal heart rate is not altered, nor is myocardial contractile capacity. Despite the fact that skeletal muscle endurance is improved, there is no increase in muscle size. There are probably no increases in sympathetic nervous input nor circulating cathecholamines. That is, repetitive muscular activity as occurs in endurance training is clearly connected with certain genetic phenotypic triggers but not others that could contribute to increases in $\dot{V}O_2$max.

Determinants of aerobic training have traditionally been sought using the Fick equation, which indicates that $\dot{V}O_2$max is the product of maximal cardiac output and arterial-venous oxygen difference. In adults, arterial-venous oxygen difference is usually maximized (i.e., venous oxygen content is minimal) in an exhaustive, acute bout of exercise, and therefore little or no change is expected in maximal arterial-venous oxygen difference with training (134). Consequently, research attention has focused on factors that affect maximal cardiac output to explain training increases in $\dot{V}O_2$max. This is surprising since, as outlined in chapter 6, it is well accepted that peripheral rather than central cardiac factors control blood flow with exercise. It is therefore more appropriate to examine training-induced changes in circulatory flow—and $\dot{V}O_2$max—from a broader perspective that includes changes in skeletal muscle pump function, circulatory blood volume, and arteriolar dilatation.

Blood Volume

Blood volume typically expands in adult subjects by about 5% to 10% during a three- to four-month period of endurance training (15). This increased volume aids thermoregulatory control and contributes to increased cardiac filling and greater stroke volume. Consequently, exercise at a given submaximal intensity is performed after training with a lower heart rate and a greater stroke volume. Training-induced increases in blood volume can thereby contribute to improvements in $\dot{V}O_2max$ by augmenting resting and maximal stroke volume. Blood volume is, in fact, closely correlated with $\dot{V}O_2max$, and experimentally increased blood volume in adults results in improvements in maximal stroke volume and cardiac output (66).

As blood volume increases with training, no changes are observed in plasma protein concentration (16). This observation led Convertino to conclude that "increased circulatory protein represents a primary mechanism for expansion of plasma volume during exercise training by increasing oncotic pressure across capillary membranes" (15, p. 213). Other factors may also contribute, including the actions of aldosterone and attenuation of volume reflex control mechanisms.

The magnitude of increase in blood volume in adults, however, does not entirely explain improvements in maximal stroke volume or $\dot{V}O_2max$. Hopper et al. estimated that approximately one half of the difference in stroke volume between trained and untrained men could be accounted for by the trained men's greater blood volume (48).

Very little information is available regarding plasma volume changes with training in children. Eriksson and Koch reported a 12% increase in estimated blood volume and a 16.8% rise in $\dot{V}O_2max$ in nine boys ages 11 to 13 years after a four-month training period (26). Blood volume in that study was calculated from an estimate of total hemoglobin content and hemoglobin concentration in the blood.

Whether children might respond to a period of endurance training by producing less serum protein—and consequently experiencing a smaller rise in plasma volume—is not known. When $\dot{V}O_2max$ elevations are equal in children and adults, no differences in serum protein concentration or plasma volume have been identified. Koch and Rocker (63) found that plasma volume and intravascular protein mass (adjusted for body size) were not different among eight well-trained 13- to 15-year-old boys and six young athletic men who had similar $\dot{V}O_2max$ levels (59.6 ± 6.5 and 63.3 ± 4.1 ml · kg^{-1} · min^{-1}, respectively). In that study the boys were examined again at 14 and 15 years of age, and on each occasion, absolute values of total blood hemoglobin and blood volume were considerably higher than previously reported levels for untrained boys (figure 11.5).

There is evidence in animals that testosterone acts to increase blood volume. Broulik et al. reported that castration in mice caused a decrease in blood volume from 90.3 ± 3 ml per kilogram of body weight to 82 ± 2 ml · kg^{-1} after 21 days (10). With administered testosterone propionate, blood volume returned to normal within seven days. This at least suggests that increases in aerobic trainability in children

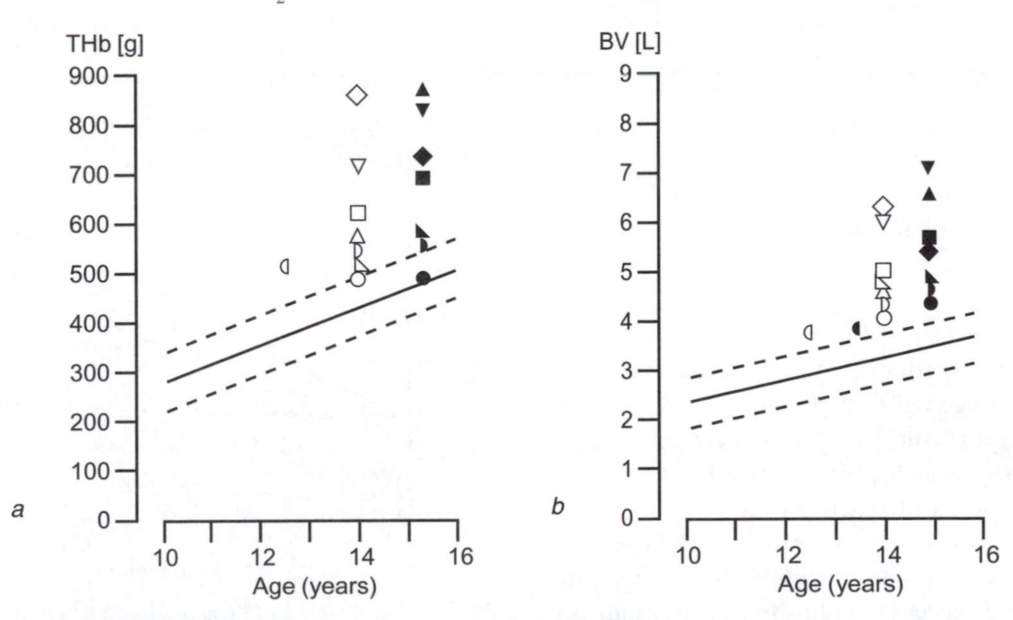

FIGURE 11.5 Total hemoglobin (THb) and blood volume (BV) in highly trained youth at 14 and 15 years of age *(open and solid symbols)* compared with expected values (*continuous* and *dashed lines* indicate mean ±1 standard error of the estimate) in nontrained children (from reference 63).

(particularly boys) at puberty might reflect a greater capacity to augment plasma volume as a result of circulating testosterone.

Estrogen also acts to increase blood volume. Stachenfeld et al. produced data indicating that estrogen decreases the transcapillary escape rate of serum proteins, thereby increasing intravascular osmotic pressure (120).

Maturational differences in other hormones that might affect blood volume with training—aldosterone, vasopressin, and atrial natriuretic peptide—have not been investigated (99). Falk et al. provided evidence that plasma aldosterone and plasma volume responses to acute exercise are not affected by level of sexual maturation (30).

Oxidative Enzymes

In adults, the activity of skeletal muscle enzymes and concentration of mitochondria increase in response to endurance training (46). For some enzymes the augmented activity is quite dramatic, with a rise to two or three times pretraining values. These changes are associated with improved endurance, lower blood lactate concentration, and greater reliance on fat oxidation at a given exercise intensity.

The increased metabolic capacity of skeletal muscle has generally not been considered to contribute to the rise in $\dot{V}O_2$max with training, which has instead been attributed to improved oxygen delivery. However, increased cellular oxidative capacity may have a role in enhancing skeletal muscle pump function, thereby improving circulatory responses and oxygen delivery following training (discussed later). Since $\dot{V}O_2$ at a given intensity is not changed by training, the increased cellular metabolic capacity has been linked to the ability to sustain submaximal exercise (which, of course, is the model of performance in distance sport events).

The stimulus for genetic expression of accelerated aerobic enzyme synthesis with training is unknown. It has been suggested that plasma catecholamines might serve this role by way of beta-adrenergic receptors (56). Booth pointed out, however, that most data suggest that increased mitochondrial density is limited to the muscle undergoing training, indicating a local rather than a systemic stimulus (8). A recently published symposium on mitochondrial function reviews recent investigations of cellular signals that stimulate adaptations during training (124).

A single study, by Eriksson et al. in five 11- to 13-year-old boys, has provided the only information regarding the responses of oxidative enzymes to aerobic training in children (25). The subjects underwent cycle training for 30 minutes, three times a week for six weeks. The activity of succinate dehydrogenase (SDH; an oxidative enzyme in the Krebs cycle present only in mitochondria) determined by vastus lateralis biopsies increased by 30%. The authors considered the magnitude of this rise to be similar to that reported for adult men (126). In a cross-sectional study, Gollnick et al. reported SDH levels to be at least 50% greater in athletically trained men than sedentary individuals (40).

These data are far too sparse to reveal whether maturational differences exist in enzyme or mitochondrial responses to endurance exercise. One particularly intriguing study in animals, however, suggests that such differences could exist (117). Simoneau et al. found that the extent of enzyme response to chronically stimulated muscle was directly related to the size of the animal. The left hindlimb tibialis anterior muscle of adult animals of different species was stimulated for 10 hours a day by means of implanted electrodes. Increases in the activity of the oxidative enzyme citrate synthase were consistently observed and were greater in the larger animals, ranging from 1.2-fold increase in mice to a 3-fold rise in rabbits (figure 11.6). Similar patterns of responses were seen in other aerobic enzymes, including malate dehydrogenase and 3-hydroxyacyl CoA dehydrogenase. Moreover, relative increases in mitochondrial volume and density for the different species paralleled the magnitude of rise in aerobic enzymes.

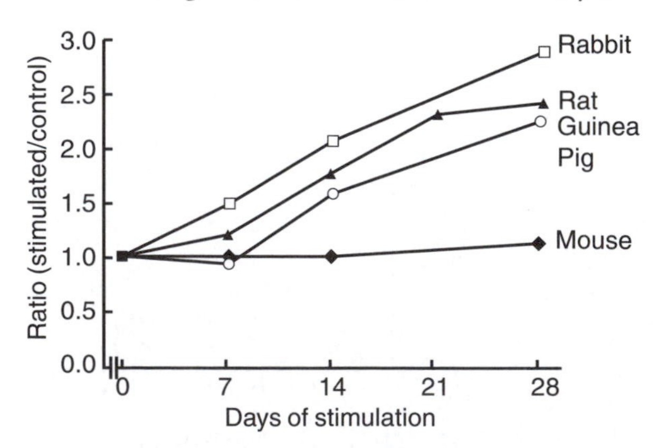

FIGURE 11.6 Changes in citrate synthase activity in chronically stimulated tibialis anterior muscles of different animals. Values are indicated as the ratio of stimulated to unstimulated muscles (from reference 117).

The authors noted that "the induced increases in enzymes of aerobic-oxidative metabolism seemed to follow a pattern such that the increments observed were inversely related to the basal enzyme activity levels" (p. 100). That is, the smaller the animal, the higher the baseline level of aerobic enzyme activity and the lower the increase with repetitive muscle contractions.

As indicated in chapter 4, children demonstrate higher activity of cellular aerobic enzymes than do adults. If prepubertal humans can be assumed to be similar to rodents, this higher baseline enzyme activity might translate into a smaller response to endurance training than in large people (i.e., adults) with lower baseline enzyme activity. Whether this idea is fact or fantasy remains to be seen.

Skeletal Muscle Pump

Several of the many anatomic and physiologic responses to endurance exercise training might serve to improve the functional capacity of the skeletal muscle pump. However, no specific information regarding muscle pump function in response to training is available in either adults or children. That this aspect of aerobic training response has been largely ignored is perplexing, given the importance of this peripheral pump to circulatory responses to exercise (see chapter 6). Indeed, it is not unreasonable to propose that the endurance capacity of the skeletal muscle pump may define the limits of circulatory responses to exercise and thus of oxygen delivery.

By limiting lactate accumulation and the fall in cellular pH, increased oxidative metabolic capacity of muscle cells could extend the endurance capacity of the skeletal muscle pump with increasing exercise intensity. In combination with increased plasma volume, this effect would increase maximal circulatory flow rate (erroneously interpreted as a primary rise in maximal cardiac output) and $\dot{V}O_2max$ (47).

Both endurance training and electrical stimulation increase capillary density in skeletal muscle. A number of angiogenetic factors may be involved in this response, particularly vascular endothelial growth factor and fibroblast growth factor 2. Some evidence implicates cellular hypoxia as a stimulus (128). But an increase in these agents has also been observed in response to a rise in muscle blood flow as well as to muscle stretching (42).

Information regarding changes in muscle capillary density with training in children is not available. From observations in certain disease states, however, it is apparent that children are capable of capillary proliferation (e.g., the "clubbing" of the digits in patients with cyanotic heart disease), particularly in response to hypoxia.

Increases in capillarization result in greater area for oxygen flux and also decreases in diffusion distance and blood flow velocity, thereby increasing the transit time for gas exchange. This greater capillary density in combination with increased blood volume also serves to increase the "stroke volume" of the skeletal muscle pump.

Several adaptive changes following a period of endurance training, then, contribute to augmented skeletal muscle pump function. Increases in blood volume and capillary density enhance pump stroke volume through a greater preload. The ability of the muscle pump to sustain contractions and pumping capacity at high work intensities may be augmented via greater cellular oxidative enzyme activity. At puberty in males, rising testosterone concentration increases the size of the skeletal muscle pump, but how this influence might affect training responses is not evident. Further research is needed to investigate these adaptations to training and possible influences of biologic maturation.

Cardiac Responses

During an acute bout of progressive exercise, the heart of a highly fit individual can generate a greater stroke volume than an unfit person's, and this is responsible for higher maximal cardiac output and oxygen uptake. The superior stroke volume at maximal exercise is in turn related a greater stroke volume at rest. Thus, factors that influence variations in resting stroke volume have been considered critical in differentiating $\dot{V}O_2max$ among individuals. Among the determinants of stroke volume, variations in ventricular preload (or diastolic filling) appear to be most important in defining these differences in aerobic fitness (see chapter 6).

The same cardiac characteristics that determine interindividual differences in $\dot{V}O_2max$ appear to be at play in responses to a period of endurance training. That is, the features that differentiate the athlete from the nonathlete, that define variations in $\dot{V}O_2max$ in the nonathletic population, or that change with endurance training appear to be the same. It is likely, then, that each of these situations involves identical mechanisms of phenotypic expression.

From this Fickian perspective, the principal determinants of individual differences in $\dot{V}O_2max$ and improvements in $\dot{V}O_2max$ with training should be sought in the means by which left ventricular diastolic volume is increased. We have already seen that this perspective is not necessarily correct, given that peripheral rather than central cardiac factors govern the circulatory response to acute exercise. To begin with, we have seen that a noncardiac factor—expansion of blood volume—already explains about half of the increase in diastolic filling, maximal stroke volume, and $\dot{V}O_2max$ with endurance training.

In this section we review the cardiac responses to endurance training in children and adults and then examine candidates within the heart itself that could account for increases in maximal stroke volume (and thereby in $\dot{V}O_2max$) with aerobic training. We then look at potential means by which these responses might be related to level of biologic maturation.

Previously sedentary adults respond to endurance training by several well-documented cardiovascular adaptations (134): (a) Resting heart rate falls by approximately 15%, an effect of increased vagal tone. Maximal heart rate, however, does not change. (b) Resting and maximal stroke volume rise by 20%. An increase of similar magnitude is therefore seen in maximal cardiac output and, since maximal arterial-venous oxygen difference changes little, in $\dot{V}O_2max$ as well. (c) Left ventricular end-diastolic dimension increases, with a 10% expansion in left ventricular resting volume.

Increases in intrinsic myocardial contractility with training have been observed in small mammals (mice and rats). However, most evidence indicates that the systolic contractile function of humans, unlike that of rodents, does not improve with endurance training (14).

Three studies have assessed cardiac performance before and after endurance training in children (26, 80, 83). Using the dye dilution technique, Eriksson and Koch measured the cardiac responses of nine 11- to 13-year-old boys to 16 weeks of training (26). Mean resting heart rate fell with training from 82 ± 8 to 71 ± 8 bpm. Mass-relative $\dot{V}O_2max$ rose by 17.9% (from 38.6 to 45.5 ml · kg^{-1} · min^{-1}), resulting entirely from a 13-ml rise in maximal stroke volume. Resting stroke volume was greater after training by a similar amount. Heart volume (determined by X-ray) increased with training from 499 ± 113 ml to 548 ± 137 ml. No changes were observed in either maximal arterial-venous oxygen difference or rate of decline in peripheral vascular resistance at maximal exercise.

Mobert et al. studied 12 boys (13.7 ± 0.3 years old) who underwent a seven-month training program (80). Subjects performed aerobic exercise for 90 minutes twice a week. Only six of the subjects participated "regularly and efficiently" in the training sessions, and their results were compared with those of the other six, who were considered "untrained." Average $\dot{V}O_2max$ improved from 49 to 55 ml · kg^{-1} · min^{-1} (12.2%) in the trained group, while values actually fell in the untrained group. Training resulted in a 5.7% decline in resting heart rate (from 88 to 83 bpm), while stroke volume (measured by impedance cardiography) rose by 20% (from 55 to 66 ml). At a submaximal (but nearly exhaustive) exercise work level of 150 W, stroke volume increased by 9% (77 to 84 ml).

Obert et al. studied the cardiac responses of 10 girls and 9 boys (10.5 ± 0.3 years old) to 13 weeks of aerobic training using echocardiographic techniques (83). $\dot{V}O_2max$ increased by 15% in the boys and 9% in the girls, which reflected a 15% rise in maximal stroke index in the former and 11% in the latter. Changes in resting and maximal stroke volume with training were closely correlated (r = .73). Resting heart rate fell from 86 ± 10 bpm to 76 ± 11 bpm in the boys (–12%) and from 95 ± 8 bpm to 86 ± 6 bpm (–9%) in the girls. No changes in maximal arterial-venous oxygen difference were observed in either sex.

In this study, a greater fall in systemic vascular resistance during exercise was seen after the training period compared to pretraining levels. Values at maximal exercise declined from 7.1 ± 1.5 IU to 5.8 ± 0.9 IU in the boys and from 7.5 ± 1.2 IU to 6.6 ± 0.8 IU in the girls. No changes in any of these measures were seen in nontraining control subjects. The change in systemic vascular resistance with training was highly associated with the increase in maximal cardiac output (r = .86).

Resting left ventricular end-diastolic dimension increased with training in both boys and girls, but increases (although smaller) were also seen in the control subjects. Resting left ventricular shortening fraction did not change with training.

Findings in these training studies are consistent with those of cross-sectional investigations of athletically trained versus nonathletic children. Young athletes generally show larger left ventricular volumes

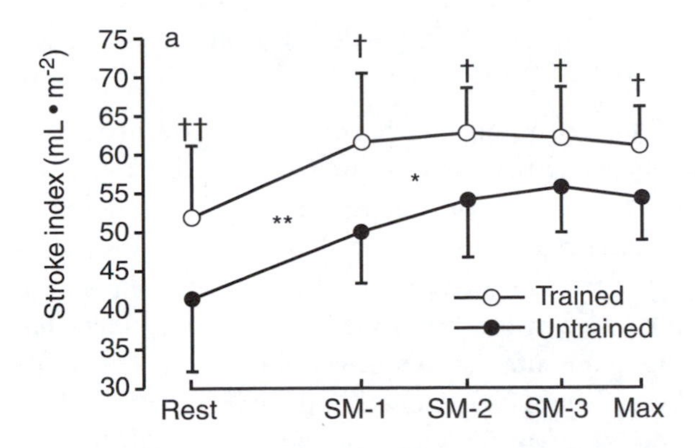

FIGURE 11.7 Changes in stroke index during a progressive exercise test in trained child cyclists *(open circles)* and untrained children (*close circles;* from reference 82).†= $p < 0.01$ between child cyclists and untrained children; †† = $p < 0.001$ between child cyclists and untrained children. Difference between workloads for both groups: * = $p < 0.05$; ** = $p < 0.001$.

and lower heart rates at rest than nontrained children (see reference 97 for a review). Using Doppler echocardiography, Nottin et al. compared cardiovascular responses of 10 trained child cyclists (age 11.2 ± 1.0 years) with those of age-matched untrained controls during a maximal cycle test (82). $\dot{V}O_2max$ values were 58.5 ± 4.4 ml · kg^{-1} · min^{-1} in the cyclists and 45.9 ± 6.7 ml · kg^{-1} · min^{-1} in the nonathletes. Maximal arterial-venous oxygen difference was similar in the two groups. Maximal stroke index was significantly greater in the cyclists (63 ± 5 ml · m^{-2} vs. 56 ± 5 ml · m^{-2}), which reflected a similar difference at rest (52 ± 2 ml · m^{-2} vs. 41 ± 9 ml · m^{-2}, respectively; figure 11.7). Left ventricular end-diastolic size was greater in the athletes both at rest and at maximal exercise. The magnitude and pattern of the fall in systemic vascular resistance were similar in the two groups. Other studies have indicated similar findings (101, 104).

In Rowland's laboratory, cardiac responses to exercise were compared between highly trained child cyclists and nonathletic children (104) and then between adult male cyclists and untrained men using the same measurement techniques and testing protocol (103). The child and adult cyclists had 28% and 55% higher mass-relative $\dot{V}O_2max$, respectively, than their nontrained peers. In both comparisons, maximal arterial-venous oxygen difference was similar in the athletes and nonathletes. The child cyclists had 27% greater stroke indexes than the nontrained children, while the adult cyclists' stroke indexes were 39% higher than their nontrained peers.

In summary, the cardiac responses to endurance training are the same in children and adults, at least qualitatively. Whether quantitative differences relative to level of biologic maturation exist in these adaptations is not yet possible to decipher. No direct comparisons have been made between children and adults during training, and no conclusions are possible from cross-sectional data, given differences in pattern, duration, and frequency of training among studies.

This leaves us with three questions: Do the cardiac adaptations to training represent a primary or secondary response? If the cardiac changes are primary, what are the mechanisms responsible? And why might the magnitude of these responses in children differ from adults'? In the following discussion we seek to identify mechanisms that might be primary determinants of both cardiac responses and improvements in $\dot{V}O_2max$ with training. If cardiac changes with endurance training are primary, we should be able to identify some potential basis by which these determinants might be influenced by biologic maturation.

Decrease in Resting Heart Rate The heart rate at rest reflects autonomic vagal tone superimposed on the intrinsic firing rate of the sinus node. With endurance training, augmented vagal influence lowers heart rate. As a result, diastolic filling of the ventricle is prolonged, and this increases end-diastolic filling volume and stroke volume. If this is a primary effect, enhanced parasympathetic activity following endurance training could at least partially explain the increased left ventricular size and the enhanced stroke volume at rest and during exercise. (It is equally possible, too, that the vagal-induced bradycardia is a secondary response to increased left ventricular size from another influence.)

Assuming the training-induced fall in heart rate to be primary, De Maria et al. (20) calculated the increase in left ventricular size that would occur from the usual fall in heart rate seen with endurance training in adults (about 10 bpm). This turned out to be about 1.0 mm in left ventricular end-diastolic dimension, which is less than the average reported increase of 2.1 mm (91). From this analysis, Perrault and Turcotte concluded that in adults, the fall in heart rate with endurance training could account for a portion, but not all, of the increase in resting ventricular size and stroke volume (91).

The resting heart rate in a child rises approximately 40 bpm after autonomic blockade, indicating that the influence of vagal tone at rest is at least equal to that of adults (72). The relative decline in heart rate with training is also probably similar between children and young adults. When Perrault and Turcotte reviewed the literature on over 1,200 adult endurance athletes and 800 age-matched controls, the average resting heart rates were 55 ± 5 bpm and 66 ± 6 bpm for the two groups, respectively (91). In a similar analysis of five reports in child endurance athletes (swimmers, runners, cyclists), trained subjects had an average resting heart rate of 69 bpm and controls 81 bpm (98). Thus, adult and child athletes had lower resting heart rates by 16.7% and 17.4%, respectively, than nonathletic peers. This suggests that (a) the magnitude of increased parasympathetic activity induced by training is similar in children and adults and (b) the effect of training bradycardia on the increase in left ventricular diastolic size is not related to level of biologic maturation.

The Stretch Hypothesis Enlargement of the left ventricle with aerobic training has often been attributed to the repetitive volume overload placed on the ventricle by the increased cardiac output that accompanies endurance exercise. Shapiro stated, "The heart responds to most athletic disciplines in a manner similar to volume overloading, and the increases in ventricular cavity size and wall thickness are appropriate physiologic compensation for the chronic volume loading" (115, p. 374).

This model of ventricular enlargement is based on observations of people with forms of heart disease that create ventricular volume overload. In patients with aortic valve regurgitation, for example, the left ventricle is faced with the added work of pumping blood that it has already expelled. The consequent increase in ventricular end-diastolic volume results in ventricular enlargement, and the extent of chamber dilatation indicates the degree of valve leakage.

Colan, echoing popular opinion, thought that the same phenomenon occurs with endurance-trained athletes:

> "Athletes who experience a large rise in cardiac output without pressure loads (endurance athletes) have a diastolic load and manifest a larger ventricular volume with normal mass-to-volume ratio. . . . Increased preload causes a rise in diastolic stress, which, in turn, induces the addition of fibers in series, resulting in a larger diastolic volume, thereby reducing diastolic stress towards normal" (14, p. 361).

The problem with this otherwise attractive concept is that, as discussed in chapter 6, the left ventricle does *not* enlarge during acute bouts of endurance exercise; in fact, diastolic dimension actually decreases. Volume overload of the ventricle is prevented by the increasing heart rate, which matches the augmented systemic venous return. The heart beats faster with exercise, but the volume of each beat does not increase. Thus, there is no stimulus (i.e., no stretch) for myocardial fiber remodeling to increase left ventricular cavity size.

Buttrick pointed out, however, that during endurance training, the left ventricle does, in fact, undergo a stretch stimulus at rest and during submaximal exercise, as the augmented vagal tone and decrease in heart rate create greater diastolic filling (11). Thus, it could be that the stretch hypothesis is correct because of autonomic changes rather than a response of the myocardium to volume overload during exercise. If so, differences between children's and adults' training responses might be expected maturational variations in vagal response to training. As discussed in the previous section, however, changes in resting heart rate with training in children and adults suggest that this is not the case.

Effect of Tachycardia Endurance training is characterized by recurrent bouts of tachycardia. Could recurrent increases in heart rate with training create structural or functional cardiac adaptations? Ianuzzo et al. examined the effects of chronic tachycardia on myocardial energetics and contractile characteristics in pigs (51). Animals trained for 16 to 22 weeks with daily 85-minute sessions during which heart rate rose to 200 or 300 bpm. No changes were seen in glycolytic or mitochondrial capacities of the myocardium. Also, no switch was seen in myosin isoforms, an expected expression of increased cardiac work.

However, in a rat heterotopic heart transplant model, Geenen et al. found that pacing the heart at 420 bpm for one week resulted in a 15% increase in cardiac mass and 30% higher myosin ATPase activity (38). At present, therefore, it is not clear whether repetitive tachycardia of the extent experienced with endurance training can trigger favorable metabolic, functional, or anatomic changes in the heart (or whether children and adults might differ in this respect).

Anabolic Agents A number of hormonal influences can affect cardiac size and function in animals and humans. It is not clear, however, how these agents might cause the typical characteristics observed in the human heart during endurance training: increases in left ventricular end-diastolic dimension, minimal ventricular wall thickening (eccentric hypertrophy to match increased volume and to limit wall stress), and absence of change in intrinsic myocardial contractile function.

Cardiac myocyte number is established shortly after birth. Consequently, myocardial morphologic responses to stress must involve either hypertrophy or remodeling (i.e., realignment) of existing cells. In pathologic states, hypertrophy is typical of pressure (or systolic) overload, as in patients with semilunar valve stenosis, and dilatation is observed with volume (diastolic) overloads, as experienced by those with aortic insufficiency. In a seemingly parallel process, resistance exercise training by adults is typically associated with ventricular hypertrophy, while increased chamber size is observed in endurance athletes and occurs as a result of endurance training. The extent of the similarity between the two types of pathologic and training stress-response patterns, however, is uncertain.

- *Locally acting agents.* A great deal of research attention has focused on the mechanisms responsible for myocardial hypertrophy in response to stress and the means by which heart work triggers genotypic and phenotypic expression (9, 45, 112). Several locally acting anabolic agents (fibroblastic growth factor, IGF-I, angiotensin II, epidermal growth factor) that respond to myocardial stress have been identified (105). It seems unlikely, however, that these factors, which stimulate myocyte hypertrophy and improved contractility, are responsible for the type of ventricular changes observed with endurance training (cardiac chamber enlargement without augmented contractility or significant hypertrophy). Unfortunately, experimental models of diastolic volume overload in animals, such as surgically created arterial-venous fistulas or mitral valve regurgitation, result in myocardial hypertrophic changes that do not mimic the cardiac responses typically seen with endurance training in humans (112).
- *Testosterone.* As reviewed in chapter 3, testosterone is thought to play an important role in improving physical fitness at puberty in males. Sex-related increases in muscle size and strength, $\dot{V}O_2max$, and maximal arterial-venous oxygen difference (via elevated arterial oxygen content due to greater hemoglobin concentration) are all related to pubertal increases in this hormone. In fact, a number of findings suggest a role for testosterone in trainability: (a) Testosterone levels rise with acute bouts of exercise (i.e., training causes repeated "doses" of this agent), but not in prepubertal subjects (28). (b) Testosterone has recognized anabolic effects on heart muscle in animals (111). (c) Strength gains with resistance training in adults have been facilitated by administered testosterone (5). (d) The timing of increases in testosterone at puberty matches the age when trainability improves. Still, it has been difficult to obtain a clear idea of just how testosterone might directly stimulate an increase in ventricular chamber dimensions.
- *Growth hormone.* Growth hormone has characteristics similar to those of testosterone: Levels increase with exercise, are high in pubertal subjects, and have recognized anabolic effects on heart muscle. These features might indicate a role in improved heart size (and $\dot{V}O_2max$) with endurance training. Such an effect was not supported by the study of Cooper et al., who compared cardiorespiratory responses to a four-week training period in young rats with suppressed growth hormone secretion and controls (17). The difference between maximal and resting $\dot{V}O_2max$ after training was 60 $ml \cdot kg^{-1} \cdot min^{-1}$ in the growth hormone–suppressed animals and 48 $ml \cdot kg^{-1} \cdot min^{-1}$ in the normal trained rats. This study indicated that suppression of pituitary growth hormone function in immature rats does not interfere with aerobic responses to exercise training (figure 11.8). This suggests that the anabolic effects of growth hormone are not essential to normal improvements in cardiac stroke volume (and $\dot{V}O_2max$), at least in prepubertal rodents. In addition, growth hormone concentrations are highest at puberty but then decrease in the young adult years, but trainability does not similarly decline. There is, then, no solid information to suggest a link between maturational differences in growth hormone secretion and aerobic trainability.

These data provide no convincing evidence that, as often supposed, cardiac alterations during endurance training are a primary factor in improving $\dot{V}O_2max$. Indeed, the changes in cardiac performance

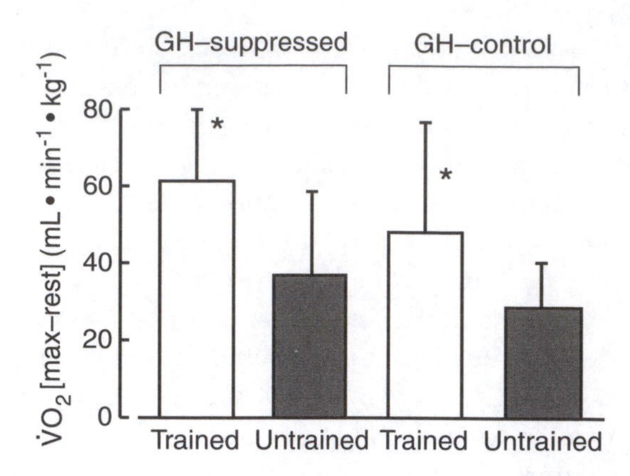

FIGURE 11.8 Changes in $\dot{V}O_2$max with training in growth hormone–suppressed rats and control rats with normal growth hormone compared with untrained animals (from reference 17). * = $p < 0.05$.

Reprinted by permission from D.M. Cooper 1994.

with training can all be explained by increases in diastolic filling due to enhanced plasma volume and vagal-induced bradycardia. The heart enlarges with training, with an associated augmentation of stroke volume, but myocardial mechanics, contractility, and sinus node function remain unchanged. Moreover, little of the preceding information suggests that the adaptations are different between children and adults. Maturity-related differences exist in the secretion of myocardial anabolic agents (testosterone, growth hormone), but at present there is no evidence that these variations influence cardiac responses to endurance training.

Vascular Conductance

Skeletal muscle arterioles dilate during acute exercise, and this increased vascular conductance is considered principally responsible for augmenting circulatory blood flow (see chapter 6). It might be expected, then, that an exaggeration of this phenomenon during endurance training could contribute to improvements in blood flow and oxygen delivery.

There is, in fact, evidence in adults that regular endurance exercise increases both arteriolar conductance (i.e., vasodilatation) and arterial compliance (13). Although statistically significant, however, the differences in systemic vascular resistance between trained and untrained subjects have generally been small (32, 85). For example, in their study of endurance-trained versus active women, Ferguson et al. found a drop in systemic vascular resistance during a progressive exercise test from 21,913 to 3,638 dyn · s^{-1} · cm^{-5} in the former and from 21,655 to 4,569 dyn · s^{-1} · cm^{-5} in the latter (32). Some data suggest that increased action of nitric oxide, a vasodilator produced by vascular endothelium in response to factors such as acetylcholine and shear stress, may be responsible for these changes (57, 60, 73).

In the two longitudinal training studies of children that estimated changes in vascular resistance with acute exercise, one showed a decrease (83) and the other no effect (26). In the report that described greater conductance with training (83), values of peripheral vascular resistance in boys at rest and at maximum exercise were 20.8 ± 5.0 and 7.1 ± 1.5 IU, respectively, before training and 19.4 ± 3.5 and 5.8 ± 0.9 IU after training. Similar changes were seen in girls. As noted earlier, Nottin et al. found no differences in peripheral resistance responses to progressive exercise between child cyclists and nonathletes (82).

The data are insufficient, then, to determine whether changes in systemic vascular resistance with training are different between children and adults. However, considering the relatively small changes described, it seems unlikely that this factor contributes appreciably to maturational differences in aerobic trainability.

To sum up, it is difficult to identify a biologic basis for diminished aerobic trainability in children since the basic physiologic mechanisms underlying such responses themselves are generally unclear. Still, given the preceding collection of data, certain speculations are perhaps not entirely inappropriate. Training-induced plasma volume, skeletal muscle angiogenesis, and enhancement of cellular aerobic enzyme activity serve to improve the pumping capacity of skeletal muscle. This augments systemic venous return, which increases cardiac diastolic filling and stimulates an adaptive enlargement of the ventricular cavity. This, in turn, increases maximal stroke volume, cardiac output, and $\dot{V}O_2$max. This scenario is consistent with the recognized role of peripheral, noncardiac factors that control circulatory responses to acute exercise. It relegates the heart to a reactionary function with training, much like its apparent role during acute bouts of exercise.

Within this model, children's reduced capacity for improving $\dot{V}O_2$max with training might be most logically attributable to impaired responses of plasma volume and cellular enzyme activity. In this chapter we have seen that some mechanism by which each of these might occur is at least possible: a lack of testosterone effect to increase blood volume in prepubertal subjects and a limited rise in aerobic enzyme activity because of higher resting levels. Both of these ideas are based on scant information in animals, but they suggest future lines of research in children.

Conclusions

Investigations of children undergoing exercise training have provided an increasingly clear picture of how biologic maturation influences physiologic adaptations to physical stress. Relative strength gains with resistance training are the same in children and adults. Strength improves with training in prepubertal subjects, however, by mechanisms other than increases in muscle size. Neurologic adaptations are presumed to be key, but the precise means by which they occur is not clear. Adults' strength also improves by size-independent mechanisms during the early phases of a resistance training program. It can be assumed, then, that similar neural factors are involved in both pre- and postpubertal subjects.

The trainability of children, and of adults, in short-burst activities that rely largely on anaerobic metabolism is difficult to clearly define. Improvements have been documented with training in certain biochemical factors (e.g., increased phosphofructokinase activity) but not others (maximal lactate production). Short-burst power production (in Wingate cycle testing) can be improved with training in children, but the extent to which this reflects metabolic, strength, or even aerobic adaptations is not clear. At present, then, it is not possible to judge whether children are more or less capable of improvements in the various forms of anaerobic fitness than adults.

The literature is clear about the diminished level of physiologic aerobic trainability in children. With a period of endurance training, increases in $\dot{V}O_2max$ in pediatric studies are generally no more than one third of those expected in adults. While certain methodological explanations have been proposed, it seems most likely that a biologic mechanism is responsible. Among the factors that contribute to improvements in $\dot{V}O_2max$ with endurance training, increased plasma volume and cellular aerobic capacity are two reasonable candidates to explain maturity-related differences.

Discussion Questions and Research Directions

1. What neural mechanisms are presumed to be responsible for strength gains with resistance training in the prepubertal years?
2. How do different components of metabolic, strength, and neurologic maturation contribute to improvements in short-burst activity as children grow?
3. How do the different forms of fitness training in children alter biochemical factors (i.e., enzyme activity) within the skeletal muscle cell?
4. How do pre- and postpubertal subjects respond to different forms of athletic training in the same training study?
5. What biologic mechanisms might explain differences in aerobic trainability between children and adults? What is the functional (i.e., performance) importance of this maturity-related variability?

12

Thermoregulation

Study Shows New Findings About Effects of Exercise in Tropical Climates

—Headline in the *San Juan Star* the morning before this study (43) was scheduled to begin

In this chapter we discuss

- how children differ from adults in thermoregulatory responses to exercise in the heat,
- factors responsible for exercise intolerance in hot climatic conditions, and
- fluid balance changes in youth during exercise in the heat.

The skeletal muscle "motor" of children and adults alike works at about 20% efficiency. That means that during exercise four times as much energy generates heat as produces locomotion. This accumulated heat must be dissipated, for failure to do so results in a rise in core temperature and inhibition of temperature-sensitive cellular enzymatic processes. The ultimate result is not only a decrement in physical performance but a risk of cardiovascular collapse.

The body dissipates heat by (a) increasing blood flow to the skin for convective heat loss and (b) increasing sweating rate for evaporative cooling. Although these methods act synergistically to provide body cooling, they do so in different patterns. Sweat production responds directly to increases in body heat. The evaporative rate of sweat—and thus its effectiveness at cooling—is related to the water pressure gradient between the skin and the environment. Therefore, the effectiveness of sweating as a cooling mechanism is reduced in humid conditions. Heat loss from sweating, then, is determined by both sweat production (an effect of body temperature) and evaporative rate (a response controlled by ambient humidity).

Cooling by convection, on the other hand, becomes progressively less effective as ambient temperature rises, since heat loss by this mechanism depends on the gradient between the skin and ambient temperature. As Nadel concluded, "When environmental temperature exceeds 36°C, all of the metabolic heat of exercise must be dissipated from the body by the evaporation of sweat, because radiative and convective losses cannot occur when environmental temperature is close to or above mean skin temperature" (34, p. 134). Heat loss by convection also depends on adequate cutaneous blood flow, which may become compromised when dehydration causes a reduction in plasma volume.

Core temperature during exercise is related to metabolic rate: The higher the energy expenditure, the greater the temperature. Certain factors are responsible for accelerating a rise core temperature in warm conditions, particularly dehydration from insufficient fluid intake to match sweat losses (53). Even a mild degree of fluid deficit (i.e., 1% of body weight) can cause an exaggerated rise in core temperature with exercise, and fluid loss as a percentage of body weight is directly related to increase in core temperature. The increased sweating rate with exercise, then, has both positive and negative consequences. Sweating must be profuse enough to cause evaporative cooling at the skin surface, but this fluid loss can result in dehydration and increased core temperature if fluid replacement during exercise is inadequate (34).

Both increased sweating and heat loss by convection are expected to occur at the cost of increased stress on the cardiovascular system. Sweating leads to dehydration and the risk of reductions in central blood volume and cardiac filling. During convective heat loss, the need for increased blood flow to the cutaneous circulation diverts blood away from the exercising muscle. It is therefore generally believed that a reduction in circulatory flow to meet the combined demands of locomotion and thermoregulation may compromise exercise capacity in the heat.

In general, children are thought to tolerate exercise in the heat more poorly than adults. A number of maturational factors might contribute. First, boys demonstrate a significantly lower sweating rate with exercise than adult men. (This does not appear to be true in females.) This leaves prepubertal boys more reliant on skin-convection heat loss to maintain thermal stability during exercise. On the other hand, boys' lower sweating rate might reduce the adverse of effects of dehydration.

Children create more heat relative to their body mass during exercise than adults, but this is compensated for by a relatively larger body surface area (BSA). The greater BSA may be maladaptive at very high ambient temperatures, however, when a reversal in the temperature gradient between skin and environment may occur. Some researchers also think that cardiac functional capacity may be less in children, limiting their circulatory responses to the cutaneous vascular shunt and dehydration that limit central blood volume during exercise in the heat.

This chapter begins with an overview of current knowledge regarding thermoregulatory differences between children and adults. Readers wishing further details may consult comprehensive reviews of this subject by Armstrong and Maresh (2), Bar-Or (4), and

Falk (15). The discussion then addresses two questions about the potential physiologic consequences of these maturational differences: Are differences in cardiovascular functional capacity responsible for children's relative intolerance, compared with adults, to exercise in high ambient temperatures? And are the thermoregulatory characteristics of children expressed in alterations in fluid balance during exercise?

Maturational Changes

Discussion of the differences in thermoregulatory responses to exercise between children and adults begins with two primary observations: (a) Before puberty, sweating rate during exercise in boys is limited in comparison with that of mature individuals (the data is less conclusive in females). Consequently, boys are expected to rely to a greater extent on increased cutaneous blood flow for convective skin heat loss than men do. (b) At a given work rate, the metabolic rate of a child relative to his or her body mass progressively decreases with growth. That is, heat production relative to body mass is greater in a 5-year-old boy running at 5 mi $\cdot$ h^{-1} than in a 14-year-old running at the same speed. However, a correspondingly greater ratio of body surface area to body mass in the young child allows more effective heat loss. As a result, heat production relative to body surface area is independent of age.

Sweating Rate

Whole-body sweating rates (g $\cdot$ m^{-2} $\cdot$ h^{-1}) are about 40% greater in adult men than in prepubertal boys (2). This is demonstrated not only during exercise in warm environments but also when sweating is induced artificially by pilocarpine iontophoresis at rest in thermoneutral conditions. Girls tend to sweat less than boys during childhood, but the sex differences become more accentuated at the time of puberty. The sweating rate of an adult man may be three times that of a woman. Most studies have found no difference in sweating rates between pre- and postpubertal females, but other data are conflicting (5).

This information suggests that maturational changes in thermoregulation during exercise are more likely to be restricted to males. Bar-Or cautioned, however, that more studies directly comparing thermal responses by sex and maturation need to be performed before definite conclusions can be drawn (5).

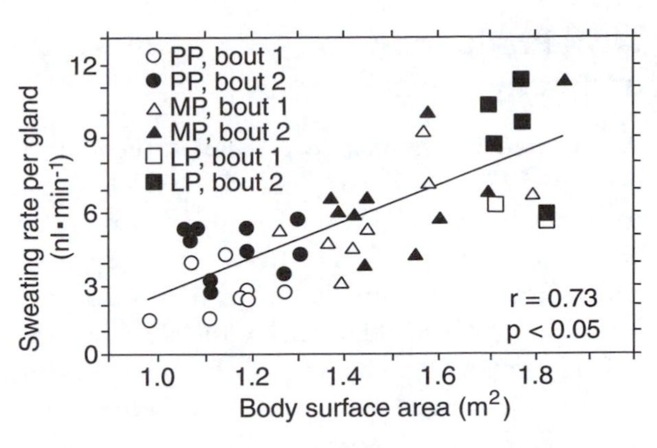

FIGURE 12.1 Relationship between sweating rate and body surface area during exercise bouts in the heat in prepubertal (PP), midpubertal (MP), and late pubertal (LP) boys (from reference 16).

Reprinted by permission from B. Falk 1992.

The number of eccrine sweat glands is fixed early in infancy. Thus, the density of glands on the skin surface actually decreases during body growth. The change in sweating rate that occurs in boys at puberty is therefore an expression of increased output per gland. Falk et al. measured sweating in circumpubertal boys cycling at 50% $\dot{V}O_2$max in a climatic chamber (16). A linear relationship was observed between sweating rate per gland and pubertal status as well as body surface area (r = .73; figure 12.1). Regression analysis indicated that BSA and sexual maturity together accounted for 66% of the variance in sweat production per gland.

An explanation for increased sweating rates in boys at puberty can be sought within the recognized controlling mechanisms for sweat production. Stimulation of the preoptic area of the brain in response to increased body heat causes an outflow of sympathetic activity, which triggers sweat production via cholinergic nerve endings on or near the sweat gland. Sex or maturational influences are not readily obvious in this scenario. Given sex and pubertal effects on sweating, it is reasonable to suspect that a sudorific (sweat-stimulating) action of testosterone might be responsible. Bar-Or reviewed experimental data indicating, however, that administration of testosterone or antiandrogenic agents does not affect sweating rate (4). Nonetheless, the matter is far from clear. Bar-Or cited the conclusion of Rees and Shuster that sweating rate is determined "by androgenic-induced gene expression during puberty and not by androgen modulation in adult life" (42, p. 691).

Heat Production and Body Surface Area

As dictated by geometric principles, the ratio of body surface area to body mass is inversely related to body mass. At age 5 years the ratio of body surface area to mass is approximately $4.0\ m^2 \cdot kg^{-1} \cdot 10^{-2}$, and by age 15 the value has fallen to less than $3.0 \cdot 10^{-2}$. The smaller child, then, has the advantage of a larger "radiator" by which to lose body heat (except in very hot environments, where the larger surface area becomes a disadvantage). This is important to thermal homeostasis, since the smaller child generates more heat per body mass during exercise than a larger one. In the 10-year age span from 5 to 15, energy expenditure (and thus heat production) in respect to body mass during treadmill running at 5 $mi \cdot h^{-1}$ falls by 30% (4).

Except at very high temperatures, a smaller child, because of his or her relatively greater body surface area, has more effective heat loss by convection than a larger child. This offsets the child's greater heat production relative to body mass. The combined effect of a progressively smaller BSA-to-mass ratio and decline in mass-specific energy expenditure as a child grows results in close relationship between heat production and body surface area at all ages. For example, Rowland et al. compared metabolic expenditure during submaximal treadmill running between 20 boys ages 9 to 13 years and young adult men (48). As expected, at a speed of $9.6\ km \cdot h^{-1}$, oxygen uptake relative to body mass was 25% greater in the boys (49.5 ± 4.4 vs. $40.0 \pm 5.0\ ml \cdot kg^{-1} \cdot min^{-1}$). However, no differences were seen when $\dot{V}O_2$ was adjusted for body surface area ($1{,}551 \pm 157$ and $1{,}557 \pm 158\ ml \cdot min^{-1} \cdot m^{-2}$ for the boys and men, respectively).

This, of course, is the biologic explanation offered for the relationship of metabolic rate to mass observed among animals of different sizes (the surface law). Although resting $\dot{V}O_2$ (i.e., heat production) per kilogram is inversely related to body mass ($M^{-0.25}$), core temperatures are similar across all animal sizes. This phylogenetic thermal stability is achieved because the greater BSA-to-mass ratio in smaller animals means a relatively greater area for heat loss. According to the surface law, then, the higher mass-specific metabolic rate of the small child is interpreted as an *adaptation* to the child's larger BSA.

Thermoregulatory Patterns in the Heat

Adults demonstrate two thermoregulatory responses to increases in exercise intensity or ambient temperature. First, sweating rate rises directly with body heat load, augmenting evaporative heat loss on the skin. While promoting cooling, the associated fluid loss leads to dehydration, which accentuates core temperature responses to exercise and can lead to diminished cardiac filling. Second, more blood is shunted to the cutaneous circulation for convective heat loss, which is governed by the temperature gradient between the skin and the ambient air. Consequently, convective heat loss is progressively less effective as a means of dissipating heat as the ambient temperature rises. When the environmental temperature is very high, the gradient may be reversed. At this point convective heat loss is ineffective, and total body heat may be gained instead of lost. When exercise is performed in high ambient heat, then, sweating is the only effective means of heat loss.

Differences that can be expected between children and adults in this scenario can be predicted from the maturational differences in sweating capacity and body surface area outlined earlier. As long as an effective temperature gradient exists between skin and ambient temperature, no thermoregulatory differences should be evident between children and adults. But in conditions of high environmental heat, thermal stability should be more difficult for immature subjects, who sweat less and must rely on diminishingly effective convective loss to prevent a rise in core temperature. The problem is further compounded for children because their greater BSA relative to body mass becomes a liability when the skin-to-air temperature gradient is reversed and heat flux is *toward* the body. The only theoretical advantage for children in this situation is that their lower sweating rate should reduce the impact of dehydration on thermal responses during exercise.

These conceptual thermoregulatory differences between children and adults have been addressed in several comparison studies. The findings have been mixed. Davies found that evaporative rate accounted for 51% of dissipated heat in 13-year-old boys and girls during treadmill running in a thermoneutral environment, compared with 65% in men (13). Similar findings were described by Wagner et al. in boys 11 to 14 years old and adult men (56).

When Drinkwater et al. studied five prepubertal girls and five college women who walked on treadmills at 35°C and 48°C, the percentage of total thermal load lost through sweating was similar (14). This finding confirms the idea that sweating rate is not influenced by puberty in females.

Shibasaki et al. described thermal changes in 7 prepubertal boys and 11 young men who cycled at 40% $\dot{V}O_2max$ for 45 minutes at 30°C and 45% relative humidity (54). The boys exhibited the expected lower sweating rates and higher cutaneous blood flow. Skin temperatures, however, were higher in the men.

Falk et al. (16) compared the responses of prepubertal, midpubertal, and late pubertal boys to 60 minutes of cycling in hot, dry conditions (42°C, 20% humidity). Forearm blood flow was highest in the prepubertal subjects, who also demonstrated the lowest sweating rates. However, no significant differences were seen in the three groups in evaporative heat loss, and skin and rectal temperatures were highest in the late pubertal subjects. The authors concluded that maturational differences in temperature regulation, at least in the environmental and exercise conditions of this study, were not established during puberty.

Dehydration

Meyer and Bar-Or reviewed data from six studies that assessed level of dehydration in children and adults in similar climatic and exercise conditions (31). They concluded from this information that "in general, the magnitude of the potential degree of hypohydration is similar in children and adults. [Therefore] when correcting for body mass, children are generally similar to adults with regard to their water losses during exercise" (p. 6).

The concentration of sodium in sweat tends to be lower in children than adults. Given the lower sweating rates in young subjects (at least among males), total sodium losses through sweating can be expected to be less in children (figure 12.2). No maturational differences are seen in total sweat potassium losses, however (32).

Core Temperature and Heat Tolerance

Variations in sweating and evaporative rates, body surface area, heat production, skin heat convection, level of dehydration, skin temperature, cutaneous blood flow, and circulatory reserve may all contribute to differences in thermoregulation as children mature.

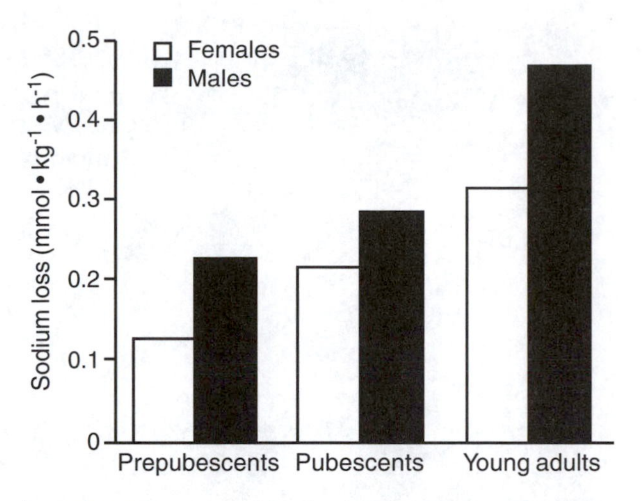

FIGURE 12.2 Sodium losses in sweat by maturation and sex (from reference 31, data from reference 32).
Reprinted by permission from F. Meyer 1992.

The physiologic bottom line of all these adaptations is preservation of core temperature. Given the varying mechanisms for thermoregulation during exercise outlined earlier, we would expect prepubertal children (at least boys) to demonstrate an exaggerated rise in rectal temperature compared with adults, most particularly at high ambient temperatures.

As Armstrong and Maresh emphasized, the experimental data do not bear this out (2). Table 12.1 outlines data from six studies that directly compared rectal temperatures in children and adults while they exercised at different ambient temperatures but similar relative intensities. In these reports, the rise in core temperature in children did generally not exceed that of adults. These findings suggest that children are as able as adults are to prevent a detrimental rise in core temperature with exercise in the heat, even at extremes of ambient temperature.

While an exaggerated rise in rectal temperature during exercise is not seen in children, some studies suggest that prepubertal subjects do not tolerate exercise in a hot climate as well as adults do. Tolerance in these studies has been defined by subjects' symptoms: Dizziness, nausea, abdominal discomfort, headaches, and inability to persist are commonly experienced by children, and these symptoms have been interpreted as evidence of circulatory instability. Bar-Or pointed out that children tolerate exercise as well as adults in thermoneutral environments or when ambient temperature does not exceed skin temperature by more than 5°C to 7°C (about 29°C, or 85°F) and 45% to 65% humidity (4). Symptoms of fatigue are more common in children compared to adults when

TABLE 12.1 Changes in core temperature with exercise in pre- and postpubertal subjects

	T_{amb}(°C)	Humidity (%)	Exercise	Subjects	ΔT_{rec} (°C)
THERMONEUTRAL					
Drinkwater et al. (14)	28	45	Walk	Girls Women	0.7 0.7
Davies (13)	21	67	Run	Boys, girls Men	1.9 1.8
WARM					
Drinkwater et al. (14)	35	65	Walk	Girls Women	1.2 1.2
Shibasaki et al. (54)	30	40	Cycling	Boys Men	0.5 0.5
Rivera Brown et al. (44)	33	55	Cycling	Girls Women	0.9 1.1
VERY HOT					
Drinkwater et al. (14)	48	10	Walk	Girls Women	0.9 1.0
Wagner et al. (56)	49	<10	Walk	Boys Men	1.0 1.5
Falk et al. (15)	42	20	Cycling	Prepubertal boys Midpubertal boys Postpubertal boys	0.7 1.0 1.2

All studies involved direct comparisons of rectal temperatures in pre- and postpubertal subjects while they exercised at different ambient temperatures but similar relative intensities.

T_{amb} = ambient temperature; T_{rec} = rectal temperature.

air temperature is greater than skin temperature by over 10°C. Indeed most of the studies documenting this intolerance to exercise in immature subjects have been performed in very hot environments (47–49°C, or 117–120°F).

It appears that children's symptomatic intolerance to exercise in the heat is not related to any inadequacy of their thermoregulatory responses to limit a rise in core temperature.

Heat and Exercise Intolerance

Increases in ambient temperature dramatically limit endurance exercise performance. Galloway and Maughan, for example, demonstrated that adult men cycling at an intensity of 70% $\dot{V}O_2$max endured an average of 93 minutes at 11°C, 80 minutes at 21°C, and 50 minutes at 31°C (22; figure 12.3). Suzuki found that adults could exercise at 66% $\dot{V}O_2$max for 91 minutes at an ambient temperature of 0°C but only 19 minutes at 40°C (55).

There is evidence from a very small number of subjects that the negative impact of heat on exercise tolerance may be exaggerated in children compared with adults. Drinkwater et al. compared the physiologic effects of two 50-minute walks at 30% $\dot{V}O_2$max at 28°C, 35°C, and 48°C ambient temperature in five prepubertal girls and five college women (14). All child and adult subjects completed the two walks at 28°C. At 35°C, only two of the five girls but all

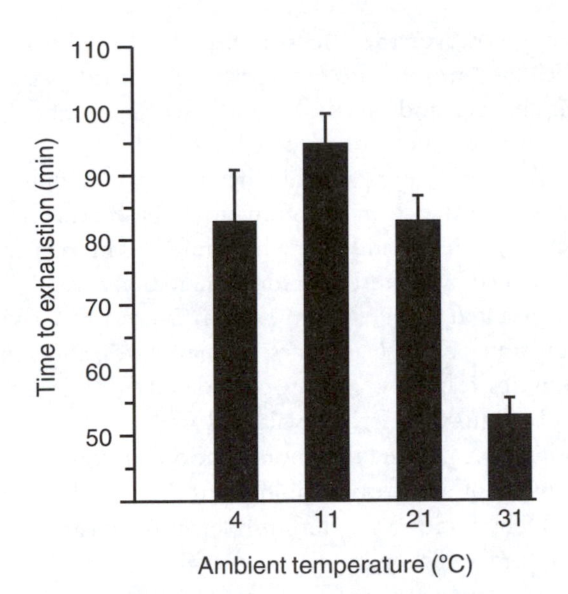

FIGURE 12.3 Endurance capacity in adult men exercising at 70% $\dot{V}O_2$max in different ambient temperatures (data from reference 22).

From S.D.R. Galloway and R.J. Maughan 1995.

of the women could finish the full 100 minutes. In the 48°C environment, four of the five girls were stopped from completing their exercise during the first 50-minute walk, when their heart rates reached 90% maximum (a predetermined safety criterion for removing a subject from the climatic chamber). The authors reported, "There were overt indicators that the girls were experiencing cardiovascular difficulty when they were stopped. Faces were extremely flushed with marked signs of distress, and two of the girls reported feeling dizzy in the 48° condition" (p. 1051). All five of the women, however, completed the first walk.

Haymes et al. (28) reported that 9- to 11-year-old girls could not sustain treadmill work as long as previously reported for adult women, and Wagner et al. (56) found that endurance walking times in hot ambient conditions were greater for adult men than for 11- to 14-year-old boys. However, in both of these studies, the subjects exercised at the same absolute rather than relative intensity, which would be expected to result in superior endurance performance by the older group.

This section reviews the explanations offered for reduced exercise tolerance in the heat and examines how these might be responsible for particular limitations in prepubertal subjects. The much larger volume of research on adults' exercise tolerance in the heat is reviewed first. We then look at how this information might be applied to interpreting the few investigations performed in children.

Studies in Adults

Despite the common experience of exercise fatigue in the heat, it is interesting that no compelling argument exists for the primary role of any single physiologic determinant. Possible factors include

- cardiovascular insufficiency from the shunting of central blood volume to the cutaneous circulation for convective cooling,
- dehydration and depressed blood circulation from fluid losses as sweat,
- metabolic or substrate alterations,
- depression of central nervous system drive, and
- impairment of myocardial contractility.

Data on each of these candidates from studies of adults is reviewed next. How each might be affected by biologic immaturity remains problematic.

Cutaneous Circulation "Theft"

The traditional explanation for exercise intolerance in the heat involves the need for the cardiovascular system to supply blood flow to the cutaneous circulation for convective heat loss. This shunts blood away from the contracting muscle, and in this competition, circulatory support to the muscle becomes limited. (As we shall review later, this idea has been invoked to explain differences between children's and adults' tolerance to exercise in the heat, since prepubertal subjects' cardiac functional capacity is thought to be limited compared with adults'.)

There is, however, no experimental evidence that redistribution of blood to satisfy the cutaneous circulation in the heat occurs at the detriment of muscle blood flow. Studies indicate, in fact, that no decrease is seen in leg muscle blood flow during exercise in the heat (36, 38, 51). Nielsen et al. studied seven young adult males as they walked uphill on a treadmill, initially for 30 minutes in a cool environment (38). Then they moved to an adjacent room with a temperature of 40°C, where they walked for an additional 60 minutes or to exhaustion. Core temperature in the hot environment exceeded 39°C. The flow of blood to the leg stayed the same in the hot and the cool conditions (figure 12.4). Moreover, there were

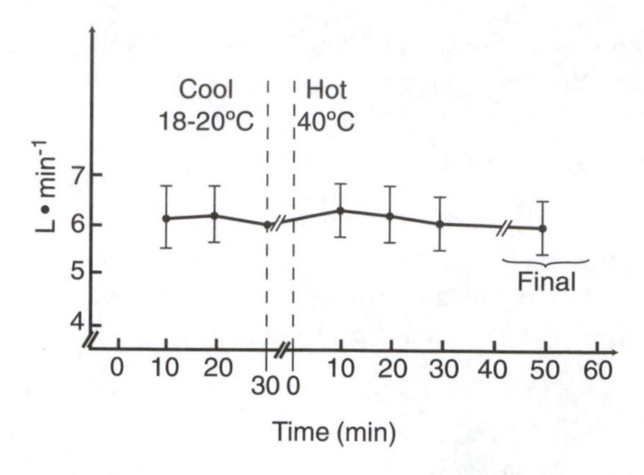

FIGURE 12.4 Leg blood flow in adult men while walking in cool and hot climatic conditions (from reference 38).

Reprinted by permission from B. Nielsen et al. 1990.

no differences in arterial-venous oxygen difference or arterial or venous lactate concentrations between the two temperatures. The authors concluded, then, that (a) no cutaneous circulatory theft occurred and (b) blood flow to muscle was not responsible for limiting exercise performance in the heat.

If increased cutaneous blood flow in the heat is not diverted from muscle blood flow, where does it come from? The answer appears to be that the augmented circulatory needs are satisfied at the expense of further reduction in blood flow to nonexercising tissues. The perfusion of inactive tissues, in fact, has been demonstrated to decrease in direct proportion to the increased need for cutaneous blood flow. Rowell demonstrated that both splanchnic and renal blood flow decreased with increasing work rate in 25°C conditions, but at 43°C these flows declined by an additional 20% at any given workload (46).

Dehydration and Circulatory Insufficiency

Sweating rates in adults can be extremely high during exercise—often 1.0 to 2.0 L · h^{-1}—leading to dehydration if these losses are not matched by fluid consumption (52). Typically, however, ad libitum intake driven by thirst is inadequate, and involuntary dehydration is typical during exercise in the heat. (This issue is discussed in respect to children later in this chapter.)

Subjects who are dehydrated during exercise in hot conditions demonstrate a significant decline in endurance performance. Craig and Cummings showed that exercise capacity decreased 20% with a weight loss of 2%, and the performance decrement was 45% when weight loss was 4% (12). In this study, however, the effects of fluid loss could not be separated from those of hyperthermia, cardiovascular changes, and metabolic responses that might also contribute to exercise intolerance.

Armstrong et al. assessed the effect of dehydration on performance in isolation from these other factors (1). They found that when fluid loss in men was produced by diuretics, a reduction in body weight of 2% resulted in increases in 1.5-km, 5-km, and 10-km run times of 0.16 minutes, 1.31 minutes, and 2.62 minutes, respectively. They considered these findings to be "most logically explained by altered aerobic function, impaired thermoregulation, increased perception of effort or combinations thereof" (p. 459).

Dehydration may diminish left ventricular filling pressure and thereby decrease stroke volume. Some researchers have therefore concluded that "maintenance of cardiac filling pressure is likely to be the most important factor in continuing to exercise in a hot environment" (40, p.48). This conclusion ignores the observation, though, that while stroke volume decreases with exercise in the heat, cardiac output changes little (at least until extremes of core temperature are reached).

For example, Gonzalez-Alonso et al. concluded that "dehydration markedly impairs cardiovascular function in hyperthermic endurance athletes" from their findings in 15 endurance-trained adult cyclists (23, p. 1229). These subjects exercised in the heat for 100 to 120 minutes and then continued to cycle for an additional 30 minutes in either a dehydrated (4% body weight) or euhydrated state. In conditions of hyperthermia (1°C rise in rectal temperature) or dehydration alone, no depression in cardiac output was observed. However, hyperthermia and dehydration together caused "an inability to maintain cardiac output and blood pressure" (23, p. 1229; figure 12.5).

However, these data do not appear to indicate cardiovascular insufficiency (23). In exercising subjects who were both hyperthermic and dehydrated, cardiac output declined only from an average of 21.1 L · min^{-1} to 18.4 L · min^{-1}, and mean arterial pressure dropped from 101 mmHg to 96 mmHg.

Similar results—and conclusions—were reported by Rowell et al. in a study of three men who exercised for 12 minutes at 15% treadmill grade and ambient temperatures of 25.6°C and 43.3°C (47). Average cardiac output was 21.0 L · min^{-1} in the thermoneutral condition and 19.3 L · min^{-1} in the

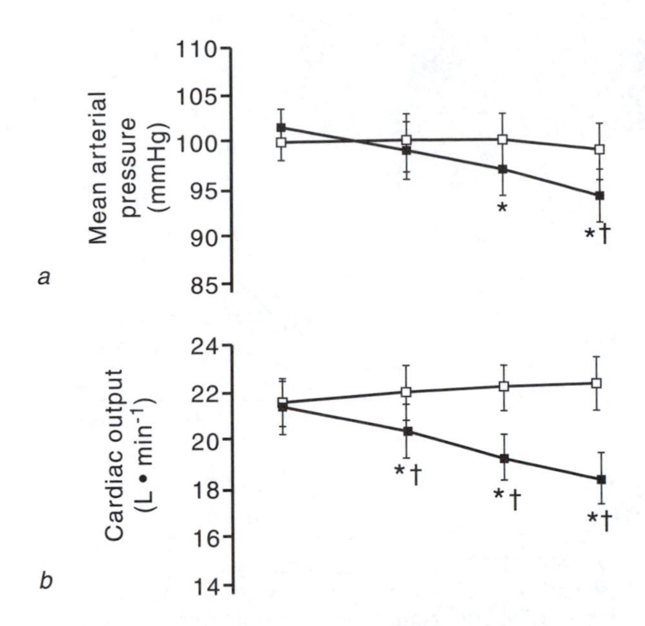

FIGURE 12.5 Changes in *(a)* mean arterial pressure and *(b)* cardiac output during 120 minutes of cycling by men at 63% $\dot{V}O_2$max in 35°C, 48% relative humidity conditions. Open squares, euhydrated; solid squares, dehydrated (from reference 23). * = significantly different from 15-min value ($p < 0.05$). † = significantly different from euhydration ($p < 0.05$).

Reprinted by permission from J. Gonzalez-Alonso et al. 1997.

heat. Surprisingly, these authors concluded, "Failure to maintain the needed level of cardiac output when strenuous exercise is performed in the heat indicates impairment of cardiovascular function" (p. 1814).

Nadel et al. reported that 25 minutes of cycling by men at 70% $\dot{V}O_2$max at an ambient temperature of 36°C caused a 1.8°C mean increase in core temperature (35). Plasma volume fell by 20%, but no appreciable change was seen in cardiac output. In a study by Saltin and Stenberg, subjects exercised for three hours at 75% $\dot{V}O_2$max (50). Cardiac output increased slightly over the 180 minutes, while stroke volume fell from 126 to 107 ml, and mean arterial pressure decreased by 10%.

While dehydration may contribute to performance limits in the heat, then, it appears to do so by elevating core temperature rather than limiting cardiovascular function. As we shall see later, however, this appears to be true only within certain limits, for once high core temperatures are reached (40–41°C), cardiovascular function is clearly compromised.

Metabolic and Substrate Alterations

Force production in mammalian muscle is affected by heat and has an optimal temperature range of 30°C to 37°C (17). It has been proposed that negative effects when muscle temperature exceeds 37°C reflect perturbations in metabolic function (18). Febbraio described in vitro studies showing a number of metabolic consequences of increased temperature that might limit muscle function, including changes in phosphorylative efficiency, calcium uptake by the sarcoplasmic reticulum, and integrity of mitochondrial membranes (18). These disturbances did not occur, however, until temperatures over 40°C were imposed.

Febbraio et al. showed that raising muscle temperature (by a heating blanket) increased muscle glycogenolysis, glycolysis, and high-energy phosphate degradation in men who performed two minutes of cycle work at 115% $\dot{V}O_2$max (19). These changes occurred without alterations in core temperature or catecholamine levels.

Exercise in the heat may influence substrate utilization (18). Carbohydrate utilization increases as fat oxidation declines, with an increase in the respiratory exchange ratio. A more rapid breakdown of glycogen in response to more circulating catecholamines when exercising in the heat is thought to be responsible. This more rapid depletion of glycogen stores has been suggested to limit exercise performance in the heat in animal studies (20). However, studies in humans have consistently indicated that muscle glycogen stores are not depleted at the point of exhaustion (38). Parkin et al. (41) demonstrated that muscle glycogen concentration at fatigue after exercise in the heat was greater (>300 mmol · kg^{-1} dry weight) than after exercise performed in thermoneutral conditions (<150 mmol · kg^{-1} dry weight).

Exercise in the heat does not appear to diminish muscle function as indicated by tests of strength. Nielsen et al. demonstrated that the maximal force of elbow flexor and knee extensor muscle groups was not different before and after 90 minutes of cycle exercise at 40°C (36).

Depression of Central Nervous System Drive

Depressive effects of heat on the motor control centers of the brain may limit exercise in the heat. Certainly, people are more uncomfortable when working in hot ambient conditions. In young adult men who performed two 30-minute bouts of moderate-intensity exercise, Brenner et al. reported average ratings of perceived exertion (RPE) of 10.4 ± 1.6 and 12.4 ± 1.8 at a 40°C ambient temperature and 8.9 ± 1.6 and 9.5 ± 1.6 at 23°C (7).

Nielsen et al. showed that exercise in the heat altered arousal patterns on electroencephalograms (37). Seven cyclists exercised at 60% $\dot{V}O_2max$ in hot (42°C) and cool (19°C) conditions. The ratio of alpha to beta frequencies on the EEG was interpreted as an inverse index of arousal level. In the hot condition, cyclists became fatigued after 34 ± 1 minute, at which time their esophageal temperature was 39.8°C, their RPE was 19.0 ± 0.8, and their alpha-to-beta ratio increased by 188 ± 71 %. In the cool conditions, esophageal temperature with exercise was 38°C, RPE was low, and alpha-to-beta frequency index was not significantly elevated (59 ± 27 %). A close correlation was observed between alpha-to-beta index and increases in esophageal temperature. The relative influences of fatigue versus core temperature on EEG findings are not clear from these results.

Caputa et al. provided evidence that tolerance to exercise in goats is related to brain temperature (8). When these researchers artificially increased hypothalamic temperature of the animals to 43°C, the goats' treadmill running effort was diminished. This occurred despite trunk temperature being maintained at 40°C.

In their study of muscle blood flow and metabolism during exercise in the heat, Nielsen et al. commented,

> "The central nervous system and mental functions are susceptible to high temperatures, as can be observed in the dizziness and confused behavior of heat-stressed participants in long-distance sports events. In the present study, the subjects also reported dizziness and inability to move their legs when they discontinued exercising in the heat, despite the fact that muscle circulation was maintained and substrate availability and turnover were adequate. Therefore, it may be that core temperatures >39°C reduce the function of motor centers and the ability to recruit motor units required for the activity, perhaps via an effect on the 'motivation' for motor performance" (38, p. 1045).

Myocardial Dysfunction

Myocardial contractile function appears to be preserved in the usual range of core temperatures observed with endurance exercise in the heat. However, once core temperature exceeds 40°C to 41°C, several lines of evidence indicate deterioration of heart pumping capacity, which occurs in the clinical presentation of heat stroke.

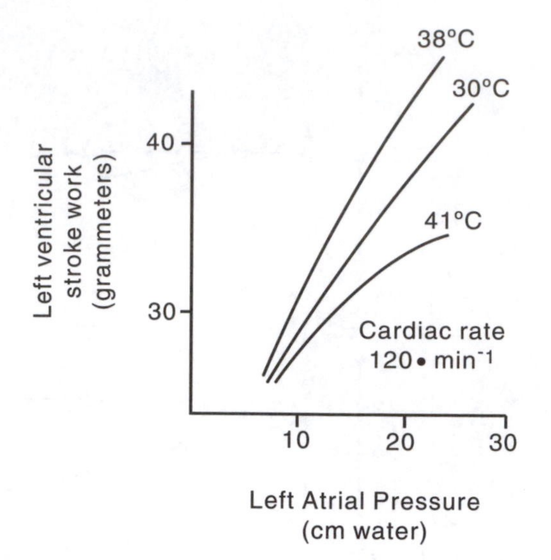

FIGURE 12.6 Ventricular function curves for a single dog at various body temperatures, indicating a decline in myocardial contractility at 41°C (from reference 10).

Cooper et al. studied the effect of hyperthermia on myocardial contractility of dogs (10). Little change was observed in stroke work at any given left atrial filling pressure at internal temperatures between 30°C and 38°C. Above 40°C, however, cardiac output, blood pressure, and stroke work fell at all filling pressures (figure 12.6).

Moore et al. examined force–length relationships in electrically paced, isolated strips of rat ventricle (33). Between 20°C and 37°C no appreciable change was seen in force–length relationships for any given increment of stretch. When bath temperature was elevated to 41.5°C, a decrement was observed in the force of contraction for each increment of stretch.

Yang et al. (57) examined the hemodynamic alterations that occur during heat stroke in 26 dogs placed in an environmental temperature of 54°C to 56°C. Blood pressure and cardiac output were maintained until the rectal temperature reached 41°C to 42°C. Above 42.5°C these variables declined, and central venous pressure rose precipitously, indicative of myocardial failure.

Synthesis of Research Findings

Data in adults fail to identify a single factor that clearly explains limitations of exercise performance in the heat. At the end of exhaustive submaximal exercise in hot compared with cool conditions, cardiac output has fallen little, muscle blood flow has not decreased, muscle glycogen stores are adequate, dehydration is not severe, muscle contractile force is not affected, and myocardial contractility is preserved.

Considering these observations, Nielsen et al. suggested that "the high core temperature per se, and not circulatory failure, is the critical factor for the exhaustion during exercise in heat stress" (36, p. 483). A number of studies support this idea and show that the achievement of a critical core temperature, regardless of fluid or cardiovascular status, limits endurance in the heat. Gonzalez-Alonso et al. (24), for instance, found that seven adult cyclists all fatigued at identical levels of hyperthermia (esophageal temperature 40.1–40.2°C). Time to exhaustion in these subjects was therefore inversely related to initial body temperature. Nielsen et al. (36) reported that adults training at 60% $\dot{V}O_2max$ at 40°C for 9 to 12 days improved time to fatigue from 48 to 80 minutes, but fatigue still occurred at a core temperature of 39.7°C.

By this hypothesis, then, core temperature rises during exercise in the heat because of dehydration and incomplete compensation by heat loss mechanisms. At a certain temperature threshold, symptoms of fatigue and lack of central drive act to limit further motor activity. This protective "governor" provides a safety mechanism against the life-threatening consequences (e.g., cardiovascular collapse) that could occur with longer or higher-intensity exercise. This concept of protective limiting mechanisms that prevent adverse physiologic outcomes is similar to the physiologic governor described by Noakes (39) regarding the limitations of circulatory oxygen delivery that define $\dot{V}O_2max$ (see chapter 6).

Hargreaves and Febbraio concluded that "while such a protective mechanism would act to preserve the organism, more research is required to identify the pathways by which it operates and to understand why it 'fails' in those individuals who overheat during exercise with potentially fatal consequences" (27, p. S116).

Studies in Children

With these concepts in mind, we can now review the more limited data available regarding exercise fatigue in the heat in children. Specifically, we need to ask (a) whether the same cardiovascular, metabolic, and hydration mechanisms are at play in children and (b) whether quantitative or qualitative changes might be responsible for greater heat intolerance in prepubertal children. Studies comparing adults with children have involved responses to heat at rest (in a sauna), extended exercise in normothermic conditions (cardiovascular drift), and exercise in the heat, both in a climatic chamber and in field conditions. We shall see that in these few studies, the relationships between physiologic variables and exercise tolerance in prepubertal subjects mimic those observed in adults.

Sauna Heat

Jokinen et al. studied cardiovascular adjustments and tolerance to sauna heat in four groups of subjects: A, 2 to 5 years old; B, 5 to 10 years old; C, 10 to 15 years old; and D, 15 to 40 years old (29). Subjects sat in a sauna at a temperature of 70°C (158°F), 20% relative humidity, for 10 minutes. Cardiac stroke volume was determined by impedance cardiography. All subjects, regardless of age, tolerated the heat well. The authors stated that only in the 2- to 5-year-old group were symptoms evident: 25% of the very young children complained of "dizziness and feeling too hot." However, when the subjects were moved from the sauna into a 21°C room, "many of the children less than 10 years of age experienced subjective symptoms: 2 children collapsed and 7 of 41 complained of dizziness" (p. 284).

Hyperthermic responses in the sauna tended to be greater in the younger groups (29). The rise in rectal temperature averaged 1.6°C in group A, 1.5°C in groups B and C, and 0.9°C in group D. At the end of the sauna, rectal temperatures were 38.0°C, 38.0°C, 37.6°C, and 37.3°C in groups A, B, C, and D, respectively. Cardiac output did not change in group A but rose by 20% to 30% in the other groups (figure 12.7). This was a reflection of a greater fall in stroke volume in the youngest children (while heart rates rose).

The authors concluded, "The most important finding in this study was the inferiority of children's circulatory adjustment to short heat stress compared with adults" (p. 287). This explanation was offered despite no evidence of depression of cardiac output in any group.

Cardiovascular Drift

Studies of circulatory changes during prolonged submaximal exercise (cardiovascular drift) in children provide no information regarding tolerance limits, and all child-to-adult comparison studies have been performed in thermoneutral conditions. Nonetheless,

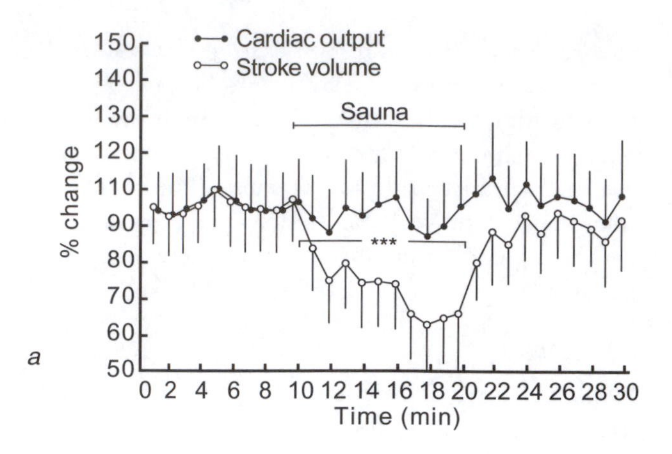

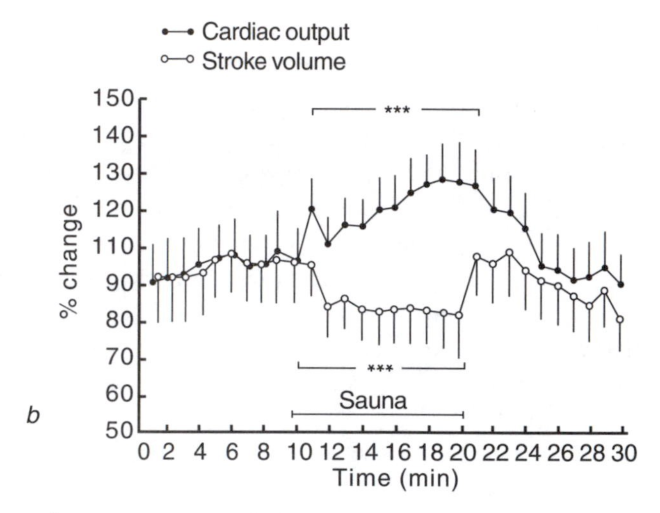

FIGURE 12.7 Percentage changes in cardiac output and stroke volume in *(a)* 2- to 5-year-old children (group A) and *(b)* subjects over 15 years old (group D) in response to sitting in a 70°C sauna (from reference 29). *** = significant difference ($p < .001$) from nonsauna conditions.

Reprinted by permission from Jokinen et al. 1990.

the hemodynamic changes during sustained exercise—rise in heart rate, decline in stroke volume, and stable cardiac output—parallel the steady rise in core temperature. In addition, these cardiovascular changes become magnified when prolonged exercise is performed in increasingly high ambient temperatures. For this reason, most researchers have accepted that cardiovascular drift is a reflection of increases in body temperature. Current evidence suggests, too, that these changes in heart rate and stroke volume are manifestations of dehydration and that a primary increase in heart rate is from augmented sympathetic activity (11, 21).

The studies that have compared cardiovascular drift in children and adults were reviewed previously in chapter 6 (3, 9, 49). These have shown either no or minor differences. In only one of these studies was body temperature measured (49). Premenarcheal girls and young women cycled for 40 minutes at 63% $\dot{V}O_2$max. The patterns of cardiovascular change were similar in the two groups: Heart rate and cardiac output rose, while stroke volume remained unchanged (perhaps because of voluntary fluid intake). The only significant difference between the adults and children was a greater rise in heart rate in the women. The average rise in tympanic temperature was 0.9°F (0.5°C) for the girls and 1.2°F (0.7°C) for the women.

The studies of prolonged exercise suggest that, at least in a thermoneutral environment, cardiovascular responses to increases in body heat are not influenced by level of biologic maturation.

Exercise in the Heat

Two studies have directly compared cardiovascular responses to sustained submaximal exercise in the heat in children and adults. Both of these investigations involved female subjects.

In a study discussed previously, Drinkwater et al. studied five nonacclimated prepubertal girls and five college-age women (14). Subjects walked at a low intensity (30% $\dot{V}O_2$max) for two 50-minute bouts at three ambient temperatures in a climatic chamber: 28°C (83°F), 35°C (95°F), and 48°C (118°F). No fluid replacement was given. Cardiac output was estimated by the acetylene rebreathing technique. Safety criteria for removing a subject from the chamber were (1) a rectal temperature over 39°C, (2) a heart rate over 90% maximum, or (3) subjective signs of distress, such as dizziness, nausea, or headache.

All five subjects in each group finished the first 50-minute walk at 28°C and 35°C. But during the 48°C walk, four of the girls were removed from the treadmill because their heart rates exceeded 90% maximum. At this point, the authors described them like this: "Faces were extremely flushed with marked signs of distress" (14, p. 1051). At the time they stopped walking, however, only one had a rectal temperature over 38.3°C, and rectal temperatures were similar in the girls and women. Although the women had lower heart rates and higher stroke indexes than the girls, no differences were seen in cardiac index or in changes in cardiac index at any of the temperatures (figure 12.8).

When asked to repeated the 50-minute walk, all subjects completed the 28°C condition. However,

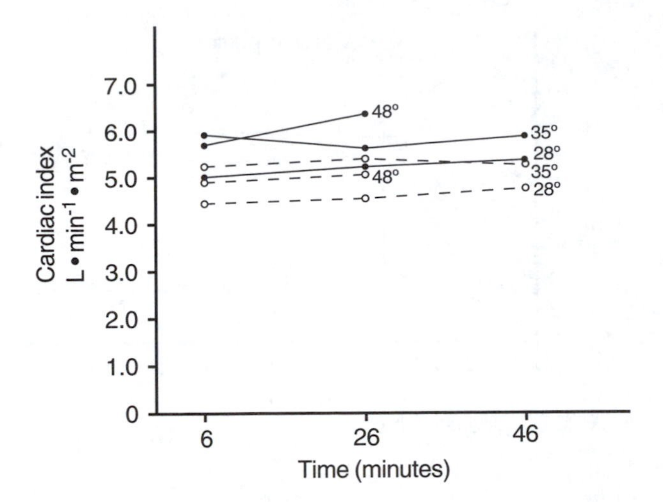

FIGURE 12.8 Changes in cardiac index while walking in the heat in prepubertal girls *(dashed lines)* and adult women *(solid lines)* at three different temperatures (from reference 14).

only two of the girls finished the second 35°C walk (compared to all of the women), and only one girl and three women could start the second exercise at 48°C.

The authors concluded that "the most striking observation" in this study, the low tolerance of prepubertal girls for exercise in the heat, "appears to be primarily a cardiovascular problem" (p. 1051). This was based on the interpretation of their signs and symptoms (facial flushing, dizziness, marked fatigue) as "overt indications that the girls were experiencing cardiovascular difficulty" (p. 1051). From the information provided, however, the development of these findings was not associated with a decline in cardiac output, which was similar in the girls and women.

Rivera Brown et al. evaluated cardiovascular responses to cycle exercise in a different setting (outdoors in the sunshine at an ambient temperature of 33°C) and a different population of girls and women (active or athletic, heat-acclimated females; 44). In this study, as opposed to that of Drinkwater et al. (14), fluid volumes were replaced during exercise relative to body size.

No significant differences were observed in exercise endurance time, sweating rate, increase in rectal temperature, or heat storage between the two groups. At the point of exhaustion, the girls and women demonstrated similar values for rectal temperature, stroke index, cardiac index, forearm skin blood flow, and decrease in average power of the leg extensor muscles during a two-legged jumping test (figure 12.9).

In this study, in which ample fluid was replaced during exercise, no evidence of cardiovascular drift was seen in either group (i.e., stroke volume and heart rate did not change). These findings are similar to those in adults (11) and children (49) that indicate that prevention of dehydration through fluid intake prevents a decline in stroke volume during prolonged exercise.

These limited data suggest that children tolerate exercise in the heat as well as adults except in very high ambient temperatures. There also is no compelling evidence that cardiovascular insufficiency is responsible for possible differences in performance in the heat between children and adults. In accord with the theory for adults, we might speculate that a protective governor that limits exercise in the heat might be set at a lower core temperature in children than adults. Why this would be so is, of course, entirely enigmatic.

Fluid Balance

Preventing dehydration during exercise in the heat is important for limiting the rise in core temperature, maintaining cardiac output, optimizing performance, and preventing heat injury. The only means of accomplishing this is by increasing oral fluid intake to match sweat losses. It has long been recognized that thirst, however, is an inadequate guide to replacing sweat losses with exercise in the heat. A subject drinking ad libitum in such conditions generally consumes no more than 70% of fluid losses. The extent of involuntary dehydration depends on exercise intensity and ambient conditions.

The explanation for this "fluid gap" is unclear. Thirst is driven by multiple mechanisms, including cellular dehydration, serum osmolality, reduced blood volume, levels of angiotensin II, and psychologic factors (25, 30). Any differences in mechanisms for thirst in the pediatric age group are yet to be investigated.

Although comparisons between studies are difficult because of varying experimental conditions, the magnitude of involuntary dehydration appears to be similar in adults and children. Several studies performed in climatic chambers and field conditions have proven informative.

Bar-Or et al. studied the effects of ad libitum fluid intake during exercise on hydration status in children

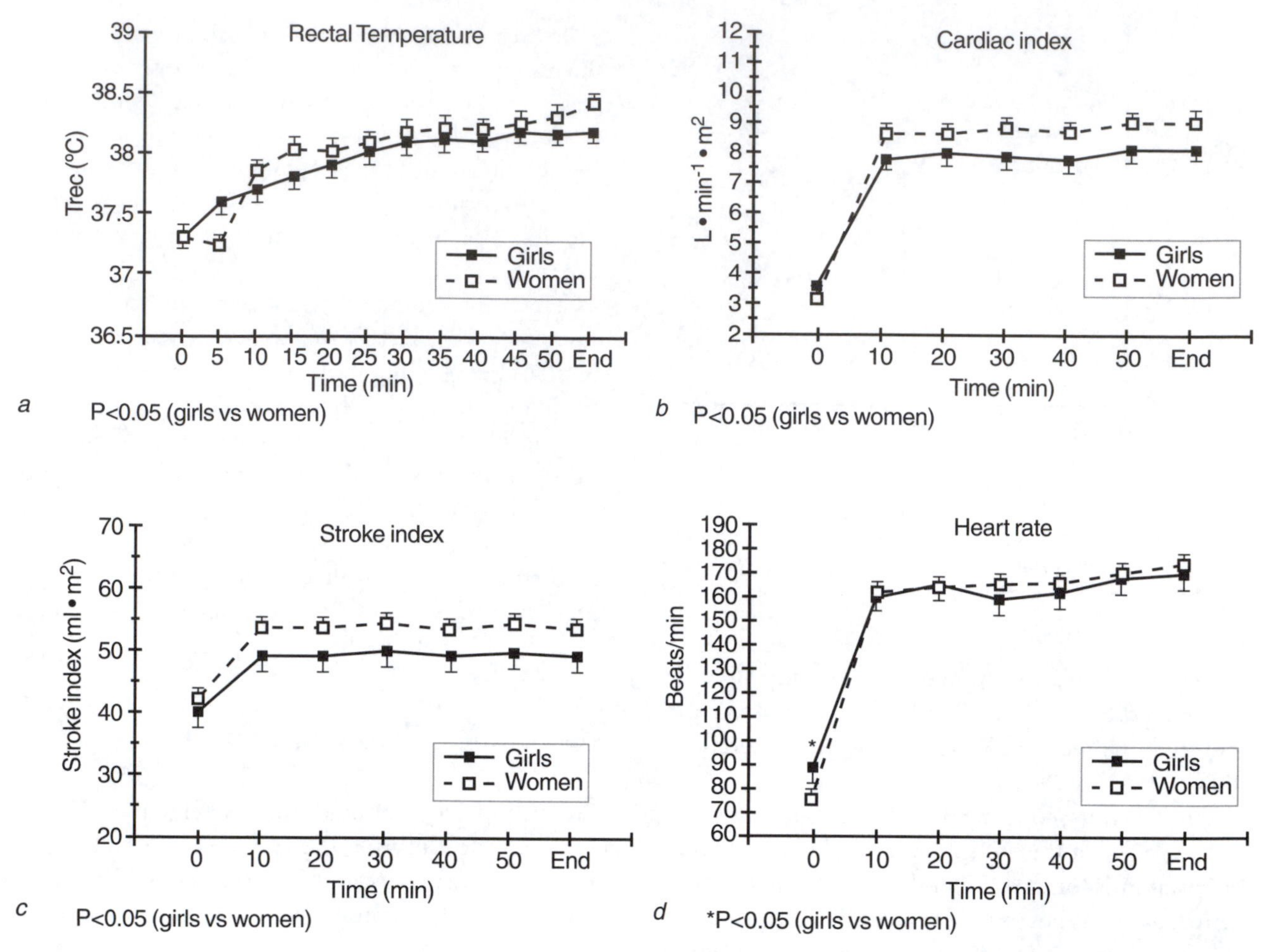

FIGURE 12.9 *(a)* Rectal temperature, *(b)* cardiac index, *(c)* stroke index, and *(d)* heart rate of heat-acclimated athletic premenarcheal girls and adult women cycling in the heat with fluid replacement (from reference 44).

(6). In this investigation, 11 partially heat-acclimated boys ages 10 to 12 years performed two sets of cycle rides at 45%$\dot{V}O_2$max in 39°C ambient temperature, 45% relative humidity. During one exercise session, subjects consumed fluids voluntarily in response to thirst, while in the other, drinking was regulated to replace estimated fluid losses. Voluntary drinking resulted in a fluid intake that was 72% of the regulated consumption. Cumulative fluid loss was approximately three times greater with voluntary drinking than with regulated intake. Final body weight loss with voluntary intake was 1% to 2%. However, no differences were observed in the two drinking conditions in rectal temperature, sweating rate, or skin temperature. The extent to which voluntary drinking fulfilled fluid replacement requirements in this study was similar to the 70% reported by Greenleaf et al. in young men exercising in 39.8°C heat (26).

Rivera Brown et al. (43) studied athletic, heat-acclimated boys during exercise in a hot outdoor environment (33°C, 58% relative humidity). Twelve subjects performed two 3-hour cycling sessions (four 20-minute bouts alternated with 25-minute rests) at 60%$\dot{V}O_2$max. Two fluids were consumed ad libitum: unflavored water during one test and a composite flavored drink with 6% carbohydrate and 18 mmol · L^{-1} of sodium in the other. The voluntary intake of the composite drink was 1,943 ± 190 g versus 1,470 ± 143 g of water. No dehydration was observed with the composite drink, but consumption of water resulted in an average 0.94% loss in body weight (figure 12.10). The gains in rectal temperature and heart rate were similar in the two tests.

Different results were observed when a similar study was performed in young females (45). In this investigation, three fluids were given: unflavored

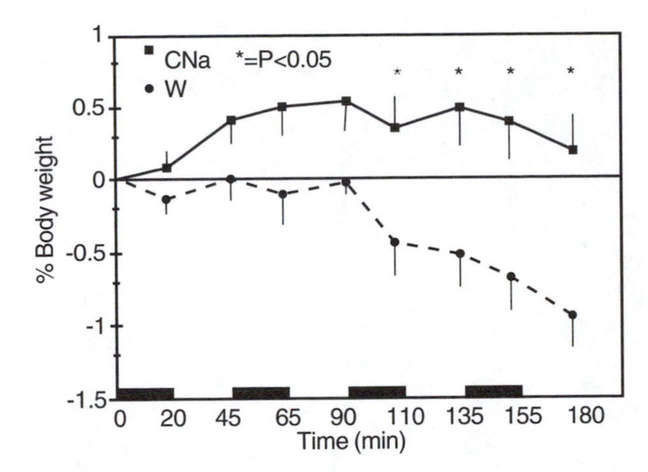

▶ FIGURE 12.10 Cumulative weight loss of heat-acclimated boys while cycling in the heat, drinking either water (W) or a composite flavored drink containing carbohydrate and sodium (CNa; from reference 43).

Reprinted by permission from A.M. Rivera Brown et al. 1999.

water, flavored water, and the flavored composite drink. Total ad libitum fluid intake was similar with all these liquids, as was the mild degree of dehydration (–1.12%, –0.95%, and –0.74% body weight, respectively). Again, no differences were seen in increases in rectal temperature or heart rate among the three fluid-intake sessions.

These studies suggest, then, that progressive dehydration can be expected in prepubertal children during exercise in the heat if thirst serves as the guide for fluid replacement. The limited data indicate that the magnitude of involuntary dehydration in children is similar to that of adults.

Conclusions

The factors that influence thermoregulation and tolerance to exercise in the heat are different in children and adults. Children generate more heat per body mass than adults when performing the same work task, but they have a relatively larger body surface area for dissipating heat. Sweating rate is significantly less in children (at least in boys), causing young subjects to rely more on convective heat loss from increased cutaneous blood flow.

These characteristics of prepubertal subjects should result in inferior thermoregulation and early fatigue at high ambient temperatures. At the same time, children might be expected to be less prone to dehydration because of their smaller fluid losses through sweating. In general, empiric research evidence has not borne out these expectations. There is evidence that children tolerate exercise in very hot climatic conditions more poorly than adults (as defined by symptoms of fatigue). However, there is no clear indication that rise in rectal temperature, level of dehydration, or circulatory function are any more affected by exercise in the heat in children than adults.

Discussion Questions and Research Directions

1. What is the cause of diminished exercise performance in the heat? Are the limits defined by the same controlling mechanisms in children as in adults?
2. How are circulatory variables influenced by exercise in hot climatic conditions? How does biologic maturation affect cutaneous blood flow, central blood volume, and myocardial function?
3. Are the mechanisms of thirst during exercise different between children and adults? How can the drive for voluntary hydration be manipulated to increase fluid intake?
4. Are there differences in thermoregulatory responses with exercise between prepubertal girls and adult women?
5. What is the role of central nervous system factors (i.e., brain function) in limiting exercise in the heat? Is there a protective governor? If so, what defines its threshold?
6. Are children at greater risk for heat injury than adults when playing sports in hot climatic conditions?
7. What is the nature of the symptoms that limit exercise in the heat (dizziness, collapse)? Do these represent circulatory insufficiency or brain perceptive factors?

13

The Central Nervous System and Physiologic Fitness

The fact that one could (or assume to) review all or a major portion of the neural issues related to exercise, activity, and health in a single essay reflects the fact that little is known about the nervous system. On one hand, our lack of understanding is ironic, given the fact that the central nervous system carries most of the responsibility for the regulation of homeostasis of and among most organ systems. On the other hand, the complexity of the nervous system makes it so difficult to study.

—V. Reggi Edgerton and Robert S. Hutton (1990)

In this chapter we discuss

- the role of the central nervous system as a possible limiter of exercise performance,
- maturational differences in perception of exercise stress,
- autonomic influences on exercise physiology, and
- the role of the central nervous system in regulating level of habitual physical activity.

There is an old riddle that asks, "What did George Washington say to his soldiers before they crossed the Delaware River?" The answer is, "Get in the boats, men!" We laugh at this joke because it pokes fun at our failure to grasp the obvious. In the field of exercise physiology, the obvious might be said to be the influence of the central nervous system (CNS), and it is not often considered. While the possibility of central neurologic influences is often acknowledged, their role as critical determinants of physiologic responses to exercise and levels of physical fitness is usually dismissed.

The importance to exercise performance of cognitive, autonomic, and reflex input from the brain is considerable, however. It is, in fact, the central nervous system that determines which muscles contribute during exercise and how vigorously they contract. Signals from the brain determine the contractile force of the skeletal muscle pump. When you want to increase your speed in the second mile of a road race, it is your brain that directs the changes in stride length and frequency that increase velocity. Fatigue, from whatever cause, remains a subjective set of uncomfortable feelings perceived by the brain that define the limits of exercise. Motivation for either peak exercise effort or tolerance of sustained activity is defined by central factors that include not only interpretation of signals from the lungs and muscles but also psychologic traits such as self-perception and confidence.

At an unconscious level the central nervous system also plays an integral role in physiologic response to exercise. Changes in heart rate and blood pressure involve central reflexes. The function of the autonomic system in exercise response is multifold: changes in skin blood flow and sweating rate, control of regional blood circulation, stimulation of glycolytic metabolism, bronchodilatation, myocardial contractility, and so on. Submaximal running economy is affected by running form, controlled by neuromuscular factors that affect gait.

These obvious determinants are often ignored because of the considerable difficulty in measuring them. It is a great deal easier to accurately determine a subject's $\dot{V}O_2max$, lactate response to exercise, or muscle strength than it is to determine the subjective central factors that contribute to the sense of fatigue, which forces a person to stop exercise. Newer techniques such as positron emission tomography (PET) scanning and magnetic induction offer hope that these processes can be more easily determined in the future. Noninvasive techniques are particularly desirable for studying pediatric subjects, in whom use of radioactive tracers or needle techniques is ethically unacceptable.

From the standpoint of the pediatric exercise physiologist, the question is, Can changes in central neurologic factors be responsible for the evolution of physiologic responses to exercise in the growing child? It is clear that neurologic maturation is critical in motor skill development in children, whose abilities to run, catch, and throw improve as they age. But what about physiologic responses to exercise? Could evolution of central neurologic influences contribute to the unique aspects of exercise physiology in children? Do changes in motivation or perception of fatigue discomfort contribute to improvements in Wingate cycle testing as children become older? Could increases in central command account for the size-independent improvements in muscle strength during childhood? Can autonomic changes (i.e., changes in sympathetic stimulation) alter the sweating rates of children at puberty? These are all intriguing questions that deserve to be addressed. In this chapter we examine what little information is available. The principal point, however, is that future research in this area could provide critical insights into the determinants of developmental exercise physiology.

The CNS "Governor"

We might logically consider the CNS to be a stimulant of skeletal muscle performance. Throughout this book, we have several times encountered the concept that CNS factors might inhibit, or limit, different aspects of physiologic function. The underlying theme in each situation has been that this central "governor" exists to prevent injury from excessive exercise stress (figure 13.1). We now briefly review these indications of a central protective mechanism.

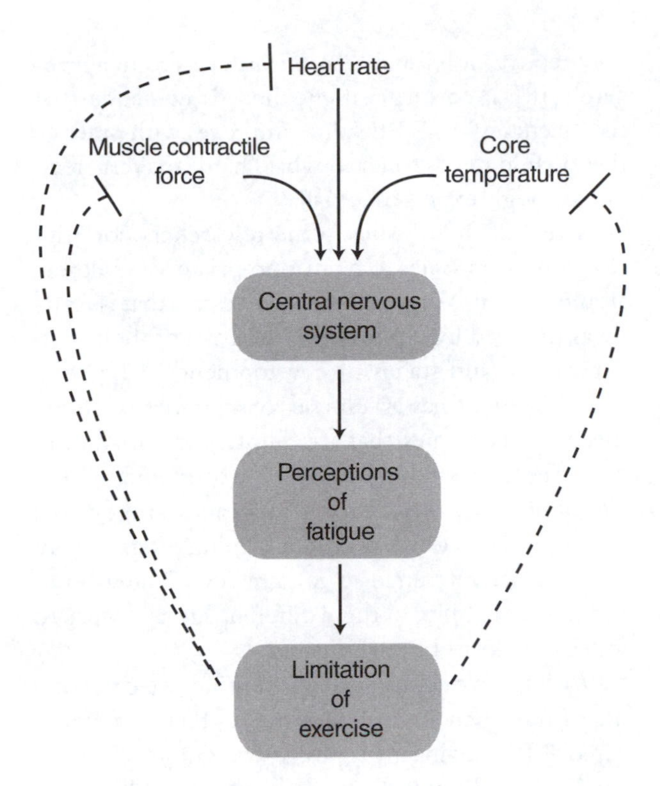

FIGURE 13.1 The possible role of central inhibitory influences on physiologic responses to exercise.

Fatigue in the Heat

It traditionally has been supposed that feelings of dizziness, nausea, weakness, and fatigue that limit exercise in hot climatic conditions are expressions of cardiovascular insufficiency accompanying dehydration and hyperthermia. As outlined in chapter 12, current evidence does not necessarily support this conclusion. Significant declines in cardiac output, blood pressure, and muscle blood flow and metabolic perturbations are not characteristically observed when subjects reach performance limits in the heat.

It has been suggested that these symptoms are instead a reflection of CNS-based perceptive responses to hyperthermia, acting to prevent an excessive rise in temperature that would eventually lead to adverse cardiovascular effects (13, 25). That such effects can occur is evidenced by animal studies that have demonstrated depressed myocardial function and cardiovascular insufficiency once core temperatures reach high levels (>40°C). In humans, the hypothesized central governor can be overcome when extreme exercise is performed in hot climates and high core temperatures and cardiovascular insufficiency (heat stroke) ensue.

Limitations of Oxygen Uptake

A number of candidates among the myriad factors that contribute to oxygen delivery have been considered as the chief limiting factor for $\dot{V}O_2$max. Traditionally, the focus has been on cardiac factors that define maximal stroke volume. Viewed more comprehensively, however, most evidence points toward peripheral factors as the primary limiters of oxygen delivery (see chapters 5 and 6). One problem in this quest is that it is difficult to explain *why* a particular factor should be limiting. That is, some researchers have considered maximal heart rate to be the limiting factor for $\dot{V}O_2$max, since faster rates would compromise diastolic filling time, limiting coronary flow and causing myocardial ischemia. But at maximal exercise, we never see any electrocardiographic or symptomatic markers of myocardial ischemia (26). Or if the contractile reserve of the skeletal muscle pump limits $\dot{V}O_2$max, we might expect muscle damage from excessive mechanical stress and metabolic acidosis. But in a typical maximal exercise test, these outcomes are not observed. This has given rise to the idea of a protective governor that limits exercise before any such adverse events can occur. Since the signal to stop exercise arises from the brain's perceptions (uncomfortable hyperpnea, nausea, light-headedness, feelings of leg fatigue), it has been suggested that a CNS governor might, in fact, be the true limiting factor for $\dot{V}O_2$max.

Voluntary Maximal Muscular Contractions

Many data indicate that skeletal muscle is able to generate force that exceeds the force produced in a voluntary maximal contraction. Inhibition from the CNS has been proposed to explain this difference, again with the assumption that such a protective influence prevents muscle damage from excessive contraction force. Asmussen noted, "It is ordinarily not possible to produce the highest tensions of which muscle should be capable [and] there must be a reserve of force that will be called forth under extraordinary conditions" (3, p. 64).

Proof of a CNS governor to explain these phenomena has not yet been discovered. However, it is interesting that such a mechanism has been suggested for such widely divergent physiologic processes. The idea does raise the possibility that CNS influences are far more important in defining maximal physiologic responses to exercise than has previously been supposed.

The existence of a central governor of these responses might be important to pediatric exercise physiologists. Changes in inhibitory CNS influences as children grow could help explain developmental patterns of factors such as heat tolerance, muscle strength, short-term anaerobic performance, and $\dot{V}O_2$max.

Perception of Exercise Stress

Are perceptions of the stresses of exercise the same in children and adults? If prepubertal children sense the signals of fatigue differently from older people, it might explain a number of maturational changes in physiologic and performance responses to exercise. Obviously, this is a difficult question to address, given the subjective nature of perceptions of fatigue. As a starting point, though, we can examine studies that have compared *ratings of perceived exertion (RPE)* of children and adults.

RPE is a measure in which subjects rate feelings of fatigue and stress as work intensity increases by an objective scale. The subject's report thereby allows subjective sensations of strain to be associated with physiologic changes (e.g., heart rate) during progressive work. Borg developed the popular 15-point RPE category scale in 1970, with subsequent revisions (7). This scale describes levels of stress in terms ranging from "light" to "hard," which the subject selects as the work intensity rises.

Although the validity and reproducibility of these RPE scales have been documented for adults, the question remains of whether they are appropriate for children, particularly young ones (12). Myashita et al. found that correlation coefficients between RPE and exercise intensity (as percentage of maximal heart rate) were greater than .90 in older boys but only .55 to .74 in subjects seven to nine years old (24). Bar-Or and Ward described similar findings and concluded that reasonable ratings cannot often be produced by children younger than eight years old (4, 5).

The association between traditional RPE scales and physiologic demand during exercise in children has varied from tight correlations (16) to virtually none at all (22). Mahon pointed out that the degree of correlation in these studies depends on whether they report individual relationships or group aggregates (18). Not surprisingly, there is evidence that the "accuracy" of RPE values improves with practice (i.e., a child needs to learn what "hard" exercise feels like through experience; 18).

Given all these issues, some researchers consider Borg's written scales to be inappropriate for children, arguing that it is important to use "a scale that is readily assimilated by children on the basis of their own experience and stages of development" (12, p. 88). This sentiment has given rise to a number of child-specific RPE scales that use illustrations, including stick figures, cycling hearts, cartoon characters climbing stairs, ants wearing backpacks, and cyclists pulling wagons of bricks, all becoming progressively fatigued (see reference 12 for a review). The validity and reproducibility of these different age-appropriate scales remain to be investigated.

With the preceding caveats in mind, we can consider findings in a number of studies that have compared RPE during progressive exercise in children and adults. Tolfrey and Mitchell reported RPE values for groups of prepubertal, pubertal, and adult males at 70%, 80%, and 85% peak $\dot{V}O_2$ (36). The prepubertal boys described higher RPE values at each exercise intensity than the other groups.

Bar-Or reported age-group means of RPE relative to percentage maximal heart rate in children and adults (4). As indicated in figure 13.2, values for 10-year-olds and young adolescents were lower than for older subjects, while those for 7- to 9-year-olds were higher. Considering the questionable ability of younger children to report valid RPE data (discussed

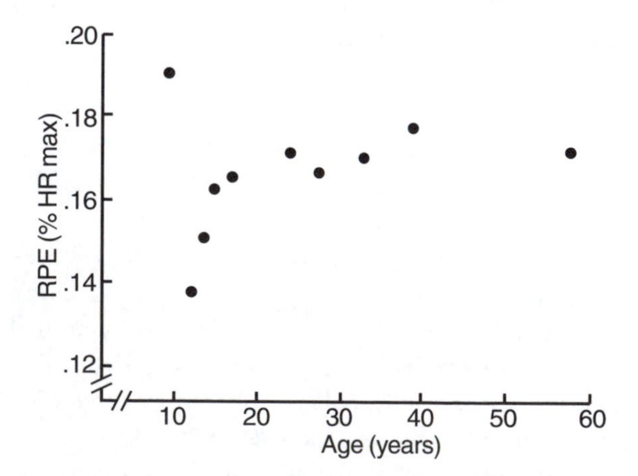

FIGURE 13.2 Age-group means for RPE expressed relative to exercise intensity (percentage of maximal heart rate). (Modified from reference 5.)

Adapted by permission from O. Bar-Or and D.S. Ward 1989.

earlier), the author concluded from these data that "it is apparent that, at any given physiologic strain, children rate their exercise intensity lower than adolescents or young adults. The reason for such low ratings in children is hard to interpret, mostly because one does not yet understand the mechanics for RPE" (4, p. 264).

Mahon et al. compared RPE in children and adults at the ventilatory threshold rather than a fixed exercise intensity (20). Ventilatory threshold values were 62 ± 5 %$\dot{V}O_2$max for the 16 children (mean age 10.9 years) and 65 ± 5 %$\dot{V}O_2$max for the 17 adults (mean age 24.3 years). RPE (using the Borg 6–20 scale) was obtained separately for legs and chest (i.e., breathing). This "differentiated RPE," then, distinguished sensations of cardiorespiratory stress from peripheral leg muscle fatigue.

In this study, all RPE values were higher for the children (20). Leg RPE at the ventilatory threshold was 11.9 ± 2.5 for the adults and 14.1 ± 2.2 for the children. Respective values for chest RPE were 10.5 ± 2.5 and 12.6 ± 2.3. At this exercise intensity, then, the children were feeling more exercise stress than the adults. In attempting to explain these differences, the authors considered factors such as lactate and pH response, mechanical aspects of muscle activation, motor recruitment patterns, heart rate, ventilatory equivalent for oxygen, and respiratory rate. None of these variables appeared likely to account for the higher RPE values in the children, however, leaving CNS perception as a reasonable possibility for the primary explanatory factor.

The findings in this study differ from those of an earlier report from the same laboratory that showed similar RPE values in boys and men (19). Nine boys (10.5 ± 0.7 years old) and nine men (25.3 ± 2.0 years old) exercised for 10 minutes at 80%, 100%, or 120% of $\dot{V}O_2$ at the ventilatory threshold. RPE ranged from 11.2 to 16.2 in the boys and 10.2 to 15.8 in the men.

Finally, Cheatham and colleagues could find no differences in RPE when 8 boys (10–13 years old) and 10 men (18–25 years old) performed a 40-minute sustained bout of steady-state exercise at an intensity equal to the $\dot{V}O_2$ at ventilatory threshold (8). RPE scores for the boys were 12 ± 2, 15 ± 1, 17 ± 1, and 19 ± 1 at 10, 20, 30, and 40 minutes of exercise, respectively, while corresponding values in the men were 13 ± 1, 14 ± 2, 15 ± 2, and 15 ± 2. The increase in RPE during the course of the exercise, however, was significantly greater in the boys than the men.

Whether children perceive exercise stress more than, less than, or the same as adults depends on which study one reads. And when one interprets the findings, the relative abilities of adults and children to report subjective sensations of fatigue during exercise accurately need to be kept in mind. At present, then, it is not possible to decide whether variations in physiologic or performance fitness during biologic maturation are influenced by changes in perceptions of exercise stress. As Eston and Lamb concluded, future studies "should extend beyond the laboratory, take into account the respective cognitive abilities of each age group, and use appropriate methods of assessing the relationships between effort perception and objective markers of effort. This involves scales which are not semantically too advanced for the age group concerned" (12, p. 90).

Autonomic Neurologic Influences

Maturational differences in autonomic neurologic activity, particularly sympathetic tone, might help explain a number of variations in physiologic responses to exercise between children and adults. The rate of sweat production, level of glycolytic metabolism, and regional control of blood flow that could affect variability in lactate clearance are a few examples. Procedures that permit direct measurement of neurologic activity are not ethically acceptable in children, so all estimates of maturational differences in autonomic activity have been based on indirect measures.

Palmer et al. compared heart rate and blood pressure responses to various stimuli of sympathetic activity between 10- to 20-year-old subjects and 42- to 63-year-old subjects (28). Each outcome variable was more prominently affected by cold pressor test, standing upright, and isometric exercise in the older subjects. This suggests that sympathetic neurologic activity, at least at rest, was lower in the younger subjects than in the older ones.

Others, however, have failed to substantiate evidence of maturity-related differences in sympathetic drive. Arnold et al. (2), for example, studied the effect of age on blood flow in one hand when the contralateral hand is immersed in ice water (the cold pressor test, a reflection of sympathetic activity). Among subjects 10 to 48 years old, no relationship was observed between age and percentage decrease in finger blood flow (r = .24) or heart rate (r = .00).

Plasma Norepinephrine Levels

With sympathetic nerve stimulation, norepinephrine generated at postganglionic nerve endings spills into the bloodstream. Levels of circulating norepinephrine are therefore considered an indicator of sympathetic nervous activity. Weise and her colleagues (38) measured catecholamines to assess the effects of age and sexual maturation on sympathetic drive at rest in 80 healthy children ages 5 to 17 years (37 boys and 43 girls). Plasma norepinephrine levels rose significantly with advancing pubertal stage in boys but not in girls (figure 13.3). In the boys, plasma norepinephrine levels were positively correlated with testosterone level (r = .35, p = .03). These findings suggest that in the resting state, *(a)* sympathetic activity is greater in postpubertal than in prepubertal males, reflecting increased levels of circulating testosterone in the former, and *(b)* maturation does not affect sympathetic drive in girls.

Two studies have compared the magnitude of rise in plasma norepinephrine during a maximal exercise test between adult men and boys as a means of estimating maturational differences in sympathetic drive. Rowland and associates (32) measured plasma norepinephrine levels at rest, during two submaximal intensities of cycle exercise, and at maximal exercise in 11 boys (10–12 years old) and 11 men (24–35 years old). Mass-relative $\dot{V}O_2max$ was similar in the two groups. No statistically significant differences in concentrations were observed at any of the measurement points. Values at maximal exercise were 1,385 ± 326 pg · ml^{-1} in the men and 1,196 ± 612 pg · ml^{-1} in the boys (figure 13.4).

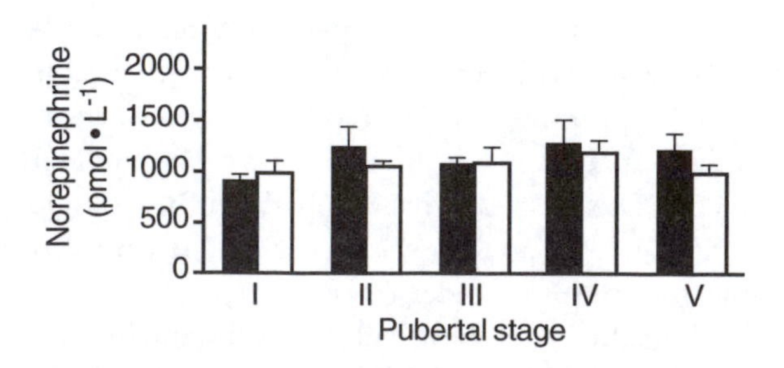

FIGURE 13.3 Resting plasma concentrations of norepinephrine by level of sexual maturation in healthy boys *(filled bars)* and girls (*open bars;* from reference 38).
Reprinted by permission from Weise et al. 2002.

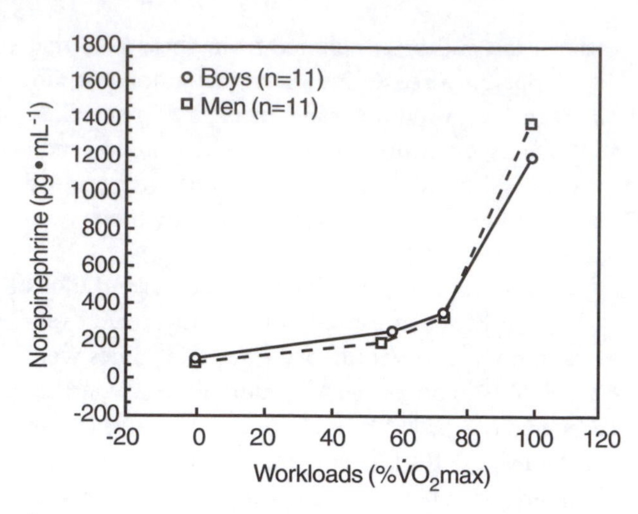

FIGURE 13.4 Average levels of plasma norepinephrine during exercise in boys and men (from reference 32).
Reprinted by permission from T.W. Rowland et al. 1996.

While these findings suggest a lack of maturity-related variability in sympathetic drive during exercise, the authors expressed caution before making this conclusion. First, although the difference was not statistically significant, the men had a mean level of norepinephrine at maximal exercise that was 16% greater than the boys'. Also, mean norepinephrine concentration rose in the men by a factor of 14.7 above that at rest, while in the boys the magnitude of increase was 9.9. The authors concluded, "True maturity-related differences in norepinephrine responses to exercise may have been obscured in this study by the large degree of interindividual variability. [Also] maturity-related differences in change in plasma volume with exercise, which were not measured in this study, might also influence plasma norepinephrine levels" (32, p. 24).

Lehmann et al. reported a similar high variability in plasma norepinephrine levels with exercise (17). They compared responses to maximal treadmill exercise in eight boys (mean age 12.8 years) and seven adults (mean age 27.8 years). Average $\dot{V}O_2max$ values for the boys and men were 45.3 and 51.8 ml · kg^{-1} · min^{-1}, respectively. Norepinephrine levels at maximal exercise were 30% lower in the children than in the adults (3.6 ± 0.8 versus 5.2 ± 1.6 ng · ml^{-1}, respectively; $p < .05$).

Heart Rate Variability

The rate of firing of the sinus node, seemingly steady, actually varies over time. In fact, the beat-to-beat interval changes in a regular phasic manner, such that oscillations in intervals occur with a certain frequency. Spectral analysis by Fourier transformation (known as analysis in the frequency domain) indicates that more than one such frequency can be identified, with a high-frequency (HF) pattern fluctuating every 2 to 7 s and a low-frequency (LF) pattern with peaks at 7- to 25-s intervals. Autonomic influences on the sinus node affect these patterns, and pharmacologic-blocking studies indicate that, at least in adults, changes in the HF band reflect variations in parasympathetic influence, while the LF band is affected by both sympathetic and parasympathetic drive. Consequently, the LF-to-HF ratio has been used to assess the balance between sympathetic and parasympathetic activity on heart rate (39).

Autonomic activity can also be assessed by examining changes in the beat-to-beat interval over time—the so-called time domain. Some measures are considered to reflect parasympathetic activity, such as the square root of the mean of the sum of the squares of the differences between adjacent R-R intervals (RMSSD) and the proportion of pairs of adjacent intervals differing by more than 50 ms (pNN50; 39). Others are felt to indicate sympathetic modulation, such as the standard deviation of all the R-R intervals (SDNN) and the mean of the standard deviations of five-minute cycles over the entire recording period (SDNNINDX).

Measurement of heart rate variability, then, may offer a safe, noninvasive means of assessing sympathetic and parasympathetic activity. As such, its use in children is attractive. However, the biologic significance of, measurement approaches to, and various influences on heart rate variability are just beginning to be explored, and it has so far offered few insights in prepubertal subjects. A number of issues have proven troublesome in adults: failure of heart rate variability findings to always conform to expectations based on known autonomic influences, evidence that nonautonomic factors can affect rate variability, and significant individual variations in day-to-day measurements (39). Although it is possible to measure heart rate variability during submaximal, steady-state exercise, measurements at maximal exercise are not possible because of the marked reduction in rate variance.

Data from subjects at rest suggest that high-frequency power is relatively higher in children than adults. Melanson compared measurements of heart rate variability divided by level of physical activity in 40 adult men and 12 boys 6 to 10 years old (23). The LF-to-HF ratio (the index of sympathetic and parasympathetic balance) was 0.14 ± 0.11 in the boys, while values in the men ranged from 0.28 ± 0.25 to 0.42 ± 0.39, depending on activity level. The author cautioned that the higher HF in the boys might not necessarily indicate greater vagal activity, since pharmacologic-blocking studies that would allow this interpretation have not been performed in prepubertal subjects. It is possible, he noted, that such differences could also reflect mechanical changes in cardiac dynamics (i.e., variation in responsiveness of the right atrium to stretch). Other studies have indicated a higher relative HF power in children than in adults (14, 41).

A comparison of heart rate variability during exercise between children and adults has not been done. The rate of heart recovery after a bout of acute exercise has been thought to be vagally mediated (1). In adults, level of physical fitness is directly related to the rate of heart rate recovery, supporting the concept that trained or fit individuals are characterized by greater parasympathetic activity (10). Ohuchi et al. assessed the role of maturity-related changes in autonomic activity in differences in heart rate recovery between children and adults (27). Seven boys and two girls ages 9 to 12 years and six young adult men and two women ages 17 to 21 years performed a maximal ramp treadmill test and a four-minute constant-load test. Heart rate was measured at one-minute intervals during recovery. High-frequency heart rate variability at rest was significantly greater in the children, and LF-to-HF ratio was higher in the adults. There was a significant inverse correlation between HF variability at rest and the rate of decline in heart rate after exercise in both protocols (figure 13.5). This study supports the idea that parasympathetic activity is higher in children than adults and that greater parasympathetic modulation of heart rate in children is responsible for their more rapid decrease in heart rate after exercise.

Endurance training is accompanied by a fall in resting heart rate in both children and adults. This bradycardic response, then, can be expected to be accompanied by greater parasympathetic activity in measures of heart rate variability. Yamamoto et al. showed this increase in parasympathetic activity in adults (40). A rise in resting HF (and fall in LF) was observed in seven adult subjects who trained for six

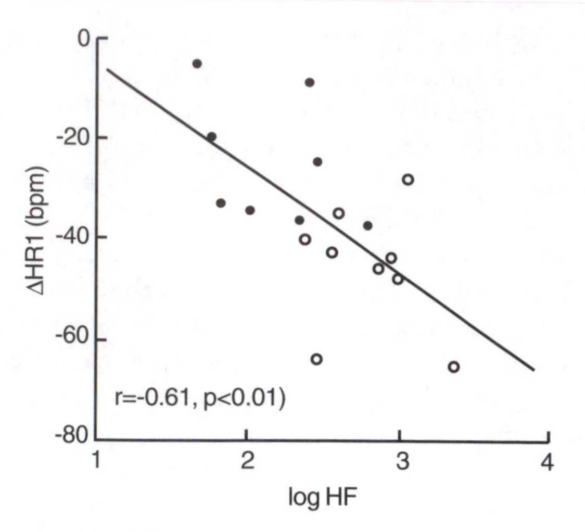

FIGURE 13.5 Relationship between the logarithm of high-frequency (HF) power at rest and rate of fall in heart rate after one minute of peak exercise in children *(open circles)* and young adults (*solid circles;* from reference 27).

Reprinted by permission from H. Ohuchi et al. 2000.

weeks. $\dot{V}O_2max$ rose from 48.2 ± 8.5 to 53.8 ± 9.1 ml · kg^{-1} · min^{-1}, and resting heart rate declined from 68 ± 4 to 53 ± 3 bpm. The decrease in LF-to-HF ratio barely escaped statistical significance (p = .09). In this study a significant correlation was observed between the change in HF with training and the decline in resting heart rate (r = –.68) but not with change in LF-to-HF ratio (r = .33).

Results were not so clear-cut in the only training study performed in healthy children. Mandigout et al. had 10- to 11-year-old prepubertal children participate in a 13-week training program (three sessions per week at 80% maximal heart rate), which improved estimated $\dot{V}O_2max$ by 16% (21). All of the frequency-domain components (measured during sleep) increased with training, and there was no significant change in the LF-to-HF ratio (14.9 ± 0.9 to 16.9 ± 1.0). An increase in the RMSSD in the time domain, however, suggested an increase in parasympathetic activity with training.

In summary, the preceding data are fragmentary, indirect, and not entirely consistent. Overall, however, the evidence at least suggests that vagal parasympathetic activity may be greater in children than adults, while sympathetic drive is less. The magnitude and physiologic implications of such a difference, if it exists, remain speculative. Moreover, maturity-related alterations in autonomic activity that occur with exercise training remain to be discovered.

CNS Control of Physical Activity

There is reason to expect that a center exists within the brain that controls one's level of regular daily physical activity. The biologic role of such an "activity-stat" would be the regulation of daily energy expenditure to maintain energy homeostasis. Two recent essays have addressed this concept (30, 35), and the following discussion is derived mainly from these reviews.

Public health initiatives to increase the physical activity of both children and adults have been based on sound evidence that regular energy expenditure through motor activity has salutary health outcomes. The success of such efforts depends on understanding the factors that determine physical activity, and most investigations have focused on psychologic, social, and environmental variables.

Largely ignored has been the potential influence of biologic controls in affecting levels of physical activity. In fact, abundant evidence indicates the presence of an intrinsic controlling mechanism within the brain that regulates physical activity in both animals and humans. Like other brain centers that control body temperature, hunger, and sexual arousal, it is expected that such an involuntary activity center (a) regulates amount of activity to a specific set level but (b) can at least temporarily be overridden by extrinsic influences. Just as a person does not always eat according to the dictates of CNS hunger centers, central stimulation of activity can be bypassed by personal desires, peer influences, and environmental conditions.

The Rationale

The biologic need for a controller of caloric expenditure via physical activity is the drive for energy homeostasis. The importance of maintaining caloric balance and stable body weight is presumably rooted in prehistoric times, when a means for conserving energy when food supply was low was critical to survival. In contemporary times this mechanism appears to have been heavily influenced by nonbiologic behaviors (overeating, sedentary lifestyle) and deviations in energy balance (perhaps related to deranged function of the controlling center itself) that result in pathologic states such as obesity and anorexia nervosa. However, there is evidence, presented later, that such a biologic drive to maintain

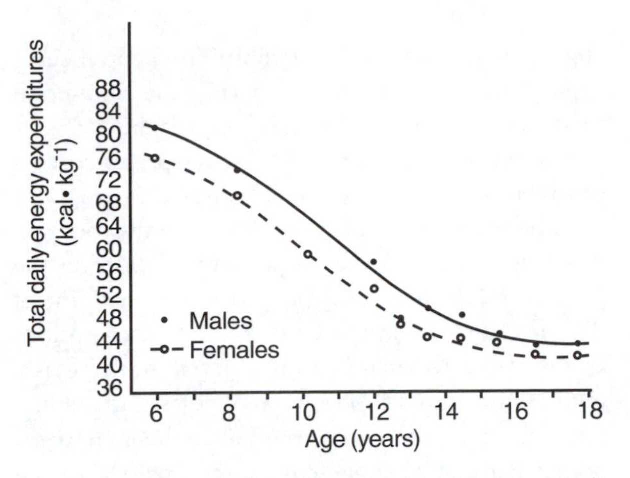

FIGURE 13.6 Fall in daily activity with age as indicated by changes in energy expenditure in children and adolescents (from reference 31).

Reprinted by permission from T.W. Rowland 1990.

energy homeostasis still exists in humans. A hypothalamic appetite center regulates energy intake, and an activity-controlling mechanism affects most of 40% of total energy expenditure not accounted for by resting metabolic rate.

As observed in the changes that occur in physical activity with age, daily energy expenditure (relative to body size) parallels changes in resting metabolic rate. Amount of spontaneous physical activity steadily declines throughout life, with the most exaggerated fall during childhood (31, pp. 34-36; figure 13.6). That this decrease is largely biologic is indicated by a similar fall in spontaneous activity with age in animals (the ambulatory activity of young rats is almost three times that of elderly animals). The rate of this decline in habitual activity closely approximates the normal fall in resting metabolic rate during childhood, suggesting a common mechanism linking the two.

The Messenger

Thorburn and Proietto reasoned that if CNS control of physical activity is important in maintaining caloric homeostasis and body weight, there should be an identifiable messenger that provides the brain with information regarding the size of the adipose mass (35). Leptin, the hormone produced by fat cells that signals brain receptors to modify appetite, is an obvious candidate. However, most data indicate that leptin does not directly influence motor behavior in humans. Its action in energy dynamics appears to be mainly that of altering food intake rather than changing physical activity. In their review, Thorburn and Proietto (35) identified 16 other agents with a recognized ability to alter spontaneous activity in animals or humans, particularly nitric oxide and pancreatic polypeptide (table 13.1).

The Evidence

Several lines of evidence support the existence of biologic control of habitual physical activity. Biorhythmicity, a regular temporal pattern of spontaneous activity, has been observed in children. Cooper et al. found that high-intensity activity in children occurred with significant frequencies of 0.04 to 0.125 per minute (9). Wade et al. found that free-playing children oscillate between levels of high and low activity, with frequencies of 15 minutes' duration superimposed on a greater cycle of 40 minutes (37). This periodicity is typical of other CNS regulatory functions.

Some, but not all, data indicate that a compensatory fall in physical activity can occur following a period of imposed exercise (11, 33). Such studies are confounded by a number of variables, including variations in energy intake and failure to allow sufficient time for possible compensatory changes to occur.

TABLE 13.1 Biologic agents that can influence levels of physical activity

Decrease spontaneous activity	Increase spontaneous activity
Reduced nitric oxide production	Pancreatic polypeptide
Deficient histamine H1 receptors	Melatonin
Deficient cholecystokinin-A receptors	Acetyl-L-carnitine
Low-dose dopamine	D-glucosamine
Low-dose noradrenaline	High-dose dopamine
Serotonin (short-term)	High-dose noradrenaline
Fibroblast growth factor	Serotonin (long-term)
Adenosine	Nerve growth factor
Opioid receptor blockade	
Iron deficiency	

From reference 35.

The ubiquity of spontaneous physical activity expressed as play throughout the animal kingdom strongly implies a biologic basis. Despite a plethora of opinions, biologists and behavioral scientists have to come to no common agreement on the purpose of play. Most recent theories center on the idea that motor activity through play is a means of optimal arousal of the CNS. By this concept, then, play is a reflection of biologically determined motor activity.

Ambulatory activity can be dramatically altered by experimentally induced lesions in the CNS of animals (29). Different pharmacologic interventions (amphetamines, morphine, chlordiazepoxide) can also affect play and motor behavior in animals (34). Abnormalities in blood mineral levels (iron, lead) can also alter physical activity in both animals and humans (15).

Children who have attention-deficit hyperactivity disorder (ADHD) are characterized by high levels of physical activity. The intuitive idea that these children have excessive CNS arousal is not consistent with the observation that CNS stimulants (e.g., methylphenidate) cause a reduction in hyperactive behavior. It has been suggested that children with ADHD instead suffer from *depressed* CNS arousal and that their increased activity is a stimulus-seeking maneuver. There is some evidence, too, that children with ADHD may exhibit depletion of CNS neurotransmitters. At present, there is no unified theory for a neurophysiologic mechanism for ADHD, but the data clearly implicate a biologic basis for the increased motor activity.

The significant genetic influence on physical activity provides further evidence of an intrinsic control mechanism for motor behavior. In a review by Beunen and Thomis, twin and family studies examining the genetic influence on daily physical activity have shown heritability coefficients ranging from .29 to .62 (6). Thorburn and Proietto pointed out that significant differences are observed in physical activity levels betweem different racial groups (35). While these might potentially be explained by sociodemographic factors, "it is conceivable that these effects are, to some extent, attributable to genetic differences between these populations" (p. 90).

The Implications

The aggregate evidence, derived from diverse sources, implies that an intrinsic mechanism in the CNS governs, to some extent, how much an individual engages in regular physical activity. This central effect appears to be most evident during the childhood years, diminishing as adulthood approaches.

Several questions arise. Does the presence of a central intrinsic activity control imply that programs to enhance activity might be less effective because of compensatory decreases in activity outside the program? Not necessarily, because the set point of the "activity-stat" might be reset, and other options are available for maintaining energy homeostasis, such as changes in food intake or resting metabolic rate. Are there forces that could alter the activity set point? Perhaps, as suggested by the analogy to the effects of acclimatization in modulating the core temperature response to exercise in the heat. How might a disturbance in the normal function of the activity control in children account for conditions of chronic energy imbalance (e.g., obesity)? This is an intriguing question, for it opens the door to possible pharmacologic interventions that could modify control center function as a means of treating such conditions.

Conclusions

Maturation of central neurologic factors plays a key role in improvements in motor performance as children grow. The influence of the CNS in the development of physiologic fitness, however, is much less clear. CNS input potentially could have a prominent effect, both cognitive and involuntary, on muscle contractile strength, cardiovascular function, and metabolic capacity. Investigation of central neurologic input on the development of physiologic fitness is hampered, however, by the difficulty in studying these processes, particularly by noninvasive means.

There is evidence from diverse biologic functions that perception of fatigue by the brain may limit physiologic capacity during exercise. In this role as a "governor," the CNS may act to prevent the risks of excessive motor activity (muscle damage, myocardial ischemia, cardiogenic shock). Whether changes in such an inhibitory effect during childhood influence development of muscle strength, $\dot{V}O_2max$, or ability to exercise in the heat is unknown.

Studies addressing the possibility that children might perceive exercise stress differently than adults have provided conflicting results. The use of more age-appropriate RPE scoring scales may help clarify this issue.

Resting and exercise norepinephrine levels and heart rate variability findings suggest that children may be characterized by less sympathetic and greater vagal parasympathetic activity than adults. The magnitude and importance of such differences to physiologic fitness remain to be studied.

Compelling evidence indicates that an intrinsic CNS center influences level of regular physical activity, most particularly in children. This controlling center, which acts to maintain energy homeostasis, may be deranged in people with chronic energy imbalances (e.g., obesity). Identification of ways to modify this "activity-stat" may provide an approach for managing these disorders.

Discussion Questions and Research Directions

1. What is the nature of the inhibitory role of the CNS in defining the limits of voluntary muscle strength, $\dot{V}O_2max$, and exercise in the heat? Does this inhibitory factor change during childhood?
2. Do children perceive exercise stress differently than adults? What measurement scales are most valid for assessing perceived exertion in prepubertal subjects? What physiologic variables do these scales reflect?
3. Do autonomic influences during exercise differ between children and adults? If so, are the maturational variations of sufficient magnitude to significantly alter physiologic capacity?
4. Are the different measures of heart rate variability associated with sympathetic and parasympathetic input in children the same as for adults? Which measures are the most valid indicators of autonomic activity?
5. Can pharmacologic interventions alter the central control of habitual activity and contribute to the management of obesity in children?

GLOSSARY

acromegaly—Abnormal body growth due to excessive action of growth hormone.

adenohypophysis—Anterior lobe of the pituitary gland.

ADP—Adenosine diphosphate.

aerobic scope—Difference between resting and maximal oxygen uptake, expressed as a multiple of the resting uptake.

aldosterone—Steroid secreted by the adrenal cortex that causes sodium retention and potassium loss.

allometric analysis, allometry—The nonlinear relationship of measurement variable *Y* to body size *X*: $Y = a(X^b)$.

anaerobic capacity—Total ability to generate energy by anaerobic metabolic pathways.

anaerobic fitness—Ability to perform short-burst motor activities that depend on anaerobic metabolism for energy.

anaerobic glycolysis—Energy pathway resulting in conversion of glucose to lactic acid without consumption of oxygen.

angiogenesis—Development of new blood vessels.

angiogenetic factor—Agent that triggers formation of new blood vessels.

arterial PCO_2—Partial pressure of carbon dioxide in arterial blood.

ATP—Adenosine triphosphate.

autocrine—Relating to an agent that acts on the same cell that produced it.

Beckwith-Wiedemann syndrome—Genetic condition characterized by growth abnormalities and hypoglycemia.

biologic maturation—Developmental progression to the mature state.

bradycardia—Abnormally slow heart rate.

co-contraction index—Relative proportion of antagonist muscle contraction in response to that of agonists.

corticotrophin-releasing factor—Agent produced in the hypothalamus that causes production of corticotrophin by the pituitary gland.

cytokine—Agent that serves as a mediator for immune responses.

dead-space ventilation (V_D)—Portion of inspired air that does not interface with the circulatory system.

dopamine—Catecholamine that acts as a neurotransmitter in the central nervous system.

electromyography (EMG)—Measurement of electrical activity of striated muscle.

epiphyseal—Relating to the growth region of long bones.

estradiol—Steroid hormone responsible for female traits.

Fick equation—Equation that defines oxygen as the product of cardiac output and arterial-venous oxygen difference.

follicle-stimulating hormone (FSH)—Hormone secreted by the pituitary gland that is responsible for development of ova in the ovary.

GH/IGF-I axis—Chain of agents (growth hormone–releasing hormone, growth hormone, insulin-like growth factor I) that stimulate somatic growth.

gonadotropin—Agent that stimulates gonadal development.

gonadotropin-releasing hormone (GnRH)—Agent produced in the hypothalamus that stimulates release of gonadotropins from the pituitary gland.

growth hormone (GH)—Agent secreted by the pituitary gland that stimulates somatic growth either directly or indirectly via insulin-like growth factor I.

growth hormone–releasing hormone (GHRH)—Agent produced by the hypothalamus that causes the release of growth hormone from the pituitary gland.

hemodynamic overload—Excessive strain on the cardiovascular system.

hypercapnia—Excessive accumulation of carbon dioxide in the blood.

hyperinsulinemia—Increased circulating levels of insulin.

hyperpnea—Excessively high respiration.

hypertrophic cardiomyopathy—Abnormal myocardial thickening.

hypertrophy—Increase in the size of a muscle through the enlargement of muscle fibers

hypoleptinemia—Low circulating levels of leptin.

IGF-I—*See* insulin-like growth factor.

inherent physical fitness—Genetically endowed ability to perform motor tasks.

insulin-like growth factor I (IGF-I)—Anabolic agent released by the liver and other organs after stimulation by growth hormone.

isokinetic strength—Muscle force applied over a range of angular velocities.

isometric strength—Muscle force applied to an immovable object.

Krebs cycle—Series of chemical reactions that results in the production of high-energy phosphate bonds.

kwashiorkor—Disease characterized by edema and abdominal distention, caused by malnutrition.

leptin—Hormone secreted by adipocytes that is thought to suppress appetite.

lipolysis—Breakdown of stored fats into circulating fatty acids.

marasmus—Extreme emaciation.

maximal AVO_2diff—Difference in arterial and venous oxygen content at maximal exercise; that is, the greatest extraction of oxygen from the blood by tissues.

maximal lactate steady state (MLSS)—Highest exercise intensity that does not result in a progressive rise in blood lactate concentration.

melatonin—Hormone secreted by the pineal gland that may regulate sleep, mood, and pubertal changes.

menarche—Onset of menses.

microcephaly—Abnormally small head circumference.

minute ventilation (V_E)—Total ventilation during one minute.

muscle mechanical efficiency—Ratio of work performed by muscle to energy expenditure.

myocardial contractility—Force of heart muscle contraction.

nesidioblastosis—Hyperplasia of the cells of the islets of Langerhans in the pancreas, resulting in hyperinsulinemia and hypoglycemia.

noradrenaline—Agent secreted by the adrenal medulla and at adrenergic nerve endings that mediates sympathetic functions.

normoxemia—Normal arterial oxygen content.

ontogeny, ontogenetic—Biological development of an individual.

opioids—Group of compounds with analgesic and hypnotic actions.

osteoblastic—Promoting bone growth.

oxidative phosphorylation—Process of ATP formation through chemical reactions in the Krebs cycle.

oxygen uptake—Utilization of oxygen by cells.

paracrine—Relating to an agent that acts on cells adjacent to the cells that produce it.

parasympathetic tone—Activity of the parasympathetic nervous system.

peak $\dot{V}O_2$—Highest oxygen uptake value obtained during a progressive exercise test.

phenotypic expression—Observable biologic characteristics.

phylogeny, phylogenetic—Evolutionary change in a group of organisms; comparison of characteristics between mature organisms.

physical activity—Amount of body motion or level of caloric expenditure.

physical fitness—Ability to perform a motor task.

positron emission tomography (PET)—Means of tissue visualization by recording the release of positrons.

power–duration curve—Graphic relationship between work performed and duration until fatigue.

pulsatile pattern—A pattern of regular variation over time.

rating of perceived exertion (RPE)—Subjective description of fatigue based on a numerical scale.

running economy—Energy required for running, measured as oxygen uptake at a given running speed or elevation.

secondary sexual characteristics—Nongamete phenotypic expression of sex (breasts, genitalia).

serotonin—A monoamine agent with broad physiologic activities, including central nervous system transmission.

somatic growth—Development of body size.

somatostatin—Agent secreted by the hypothalamus that inhibits release of growth hormone, thyroid hormone, and corticotrophin from the pituitary gland.

somatotype—Categorization of body build as ectomorph, endomorph, mesomorph.

stroke volume—Amount of blood expelled by the ventricles of the heart in a single contraction.

symmorphosis—Principle that the functional capacity of no single component of a biologic system should exceed that of the other parts of the system.

Tanner staging—Description of level of sexual maturation by characterization of secondary sexual characteristics.

thelarche—Beginning of breast development.

thyrotoxicosis—Excessive activity of the thyroid gland.

tidal volume (V_T)—Amount of air in a single inhalation or exhalation.

ventilatory anaerobic threshold (VAT)—Point in a progressive exercise test when the rate of rise in minute ventilation exceeds the increase in oxygen uptake.

ventilatory drift—Increase in ventilation during a sustained period of constant-level work.

visceromegaly—Enlargement of the internal, particularly abdominal, organs (liver, spleen).

vital capacity (VC)—Greatest amount of air that can be exhaled in a single effort.

$\dot{V}O_2$max—Greatest amount of oxygen that can be utilized during a progressive exercise test, usually defined by a plateau of values as workload increases.

Wingate cycle testing—A 30-s, all-out cycling exercise test designed to assess anaerobic fitness.

REFERENCES

Introduction

1. Taylor, C.R., and E.R. Weibel. Design of the mammalian respiratory system: I. Problem and strategy. *Resp. Physiol.* 44: 1–10, 1981.

Chapter 1

1. Albrecht, G.H., B.R. Gelvin, and S.E. Hartman. Rations as a size adjustment in morphometrics. *Am. J. Phys. Anthropol.* 91:441–468, 1993.
2. American Alliance for Health, Physical Education, Recreation and Dance. *Youth fitness testing manual.* Washington, DC: Author, 1980.
3. American College of Sports Medicine. *Guidelines for exercise testing and prescription.* 4th edition. Philadelphia: Lea & Febiger, 1991.
4. Beunen, G., A.D.G. Baxter-Jones, R.L. Mirwald, M. Thomis, J. Lefevre, R.M. Malina, and D.A. Bailey. Intraindividual allometric development of aerobic power in 8- to 16-year-old boys. *Med. Sci. Sports Exerc.* 34:503–510, 2002.
5. Blaxter, K. *Energy metabolism in animals and man.* Cambridge: Cambridge University Press, 1984.
6. Brett, J.R. The relation of size to rate of oxygen consumption and sustained speed of sockeye salmon *(Oncorhynchus nerka). J. Fish. Res. Bd. Canada* 22:1491–1497, 1965.
7. Brody, S. *Bioenergetics and growth.* New York: Reinhold, 1945.
8. Calder, W.A. *Size, function, and life history.* Cambridge, MA: Harvard University Press, 1984.
9. Cooper, D.M., and N. Berman. Ratios and regressions in body size and function: A commentary. *J. Appl. Physiol.* 77: 2015–2017, 1994.
10. Daniels, S.R., T.R. Kimball, J.A. Morrison, P. Khoury, S. Witt, and R.A. Meyer. Effect of lean body mass, fat mass, blood pressure, and sexual maturation on left ventricular mass in children and adolescents. *Circulation* 92:3249–3254, 1995.
11. DuBois, D., and E.F. DuBois. Clinical calorimetry, tenth paper. A formula to estimate approximate surface area in height and weight be known. *Arch. Intern. Med.* 17:863–866, 1916.
12. Eston, R.G., S. Robson, and E. Winter. A comparison of oxygen uptake during running in children and adults. In: *Kinanthropometry IV.* W. Duquet and J. Day (eds.). London: Spon, 1993 pp. 236–241.
13. Gitin, E.L., J.E. Olerud, and H.W. Carroll. Maximal oxygen uptake based on lean body mass: A meaningful measure of physical fitness? *J. Appl. Physiol.* 36:757–760, 1974.
14. Holliday, M.A., D. Potter, A. Jarrah, and S. Bearg. The relation of metabolic size to body weight and organ size. *Pediatr. Res.* 1:185–195, 1997.
15. Huxley, J.S. On the relation between egg weight and body weight in birds. *J. Linn. Soc. Zool.* 36:457–466, 1927.
16. Janz, K.F., T.L. Burns, J.D. Witt, and L.T. Mahoney. Longitudinal analysis of scaling $\dot{V}O_2$ for differences in body size during puberty: The Muscatine Study. *Med. Sci. Sports Exerc.* 30:1436–1444, 1998.
17. Malina, R.M., and C. Bouchard. *Growth, maturation, and physical activity.* Champaign, IL: Human Kinetics, 1993.
18. Miller, A.T., and C.S. Blyth. Lean body mass as a metabolic reference standard. *J. Appl. Physiol.* 5:311–316, 1953.
19. Nevill, A.M. The appropriate use of scaling techniques in exercise physiology. *Pediatr. Exerc. Sci.* 9:295–298, 1997.
20. Nevill, A.M., R.L. Holder, A. Baxter-Jones, J.M. Round, and D.A. Jones. Modeling developmental changes in strength and aerobic power in children. *J. Appl. Physiol.* 84:963–970, 1998.
21. Nevill, A.M., R. Ramsbottom, and C. Williams. Scaling physiological measurements for individuals of different body size. *Eur. J. Appl. Physiol.* 65:110–117, 1992.
22. Porter, K.R., and M.D. Brand. Cellular oxygen consumption depends on body mass. *Am. J. Physiol.* 269:R226–R228, 1995.
23. Robinson, S. Experimental studies of physical fitness in relation to age. *Arbeitsphysiologie* 10:251–323, 1938.
24. Rowland, T.W., J.A. Auchinachie, T.J. Keenan, and G.M. Green. Physiological responses to treadmill running in adult and prepubertal males. *Int. J. Sports Med.* 8:292–297, 1987.
25. Rowland, T., D. Goff, L. Martel, L. Ferrone, and G. Kline. Normalization of maximum cardiovascular variables for body size in premenarcheal girls. *Pediatr. Cardiol.* 21: 429–432, 2000.
26. Rowland, T., P. Vanderburgh, and L. Cunningham. Body size and growth of maximal aerobic power in children: A longitudinal analysis. *Pediatr. Exerc. Sci.* 9:262–274, 1997.
27. Schmidt-Nielsen, K. *Scaling: Why is animal size so important?* Cambridge: Cambridge University Press, 1984.

28. Schulz, D.M., and D.A. Giordano. Hearts of infants and children: Weights and measurements. *AMA Arch. Pathol.* 74: 464–475, 1962.
29. Shephard, R.J., H. Lavallée, R. LaBarre, J.-C. Jequier, M. Volle, and M. Rajic. The basis of data standardization in prepubertal children. In: *Proceedings of the 2nd International Seminar in Kinanthropometry.* M. Ostyn, G. Bennen, and J. Simons (eds.). Basel: Karger, 1979, pp. 360–369.
30. Taylor, C.R., S.L. Caldwell, and V.J. Rowntree. Running up and down hills: Some consequences of size. *Science* 178: 1096–1097, 1972.
31. Welsman, J.R., and N. Armstrong. Statistical techniques for interpreting body size–related exercise performance during growth. *Pediatr. Exerc. Sci.* 12:112–127, 2000.

Chapter 2

1. Adams, G.R. Role of insulin-like growth factor-I in the regulation of skeletal muscle adaptation to increased loading. *Exerc. Sport Sci. Rev.* 26:31–60, 1998.
2. Attia, N., W.V. Tamborlane, R. Heptulla, D. Maggs, A. Grozman, R.S. Sherwin, and S. Caprio. The metabolic syndrome and insulin-like growth factor I regulation in adolescent obesity. *J. Clin. Endocrinol. Metab.* 83:1467–1475, 1998.
3. Bailey, D.A., R.M. Malina, and R.L. Rasmussen. The influence of exercise, physical activity, and athletic performance on the dynamics of human growth. In: *Human growth 2: Postnatal growth.* F. Faulkner (ed.). New York: Plenum Press, 1978, pp. 475–505.
4. Barton, J.S., S. Cullen, P.C. Hindmarch, C.D.G. Brook, and M.A. Preece. Growth hormone treatment in idiopathic short stature: A preliminary analysis of cardiovascular effects. *Acta Paediatr. Suppl.* 383:35–38, 1992.
5. Beunen, G. Biological age in pediatric exercise research. In: *Advances in pediatric sport sciences: Vol. 3. Biological issues.* O. Bar-Or. (ed.). Champaign, IL: Human Kinetics, 1–40, 1989.
6. Borer, K.T. The effects of exercise on growth. *Sports Med.* 20:375–397, 1995.
7. Brand, T., and M.D. Schneider. Peptide growth factors as determinants of myocardial development and hypertrophy. In: *Cardiovascular response to exercise.* G.F. Fletcher (ed.). Mount Kisco, NY: Futura, 1994, pp. 59–99.
8. Brat, O., I. Ziv, B. Klinger, M. Avraham, and Z. Laron. Muscle force and endurance in untreated and human growth hormone or insulin-like growth factor-I treated patients with growth hormone deficiency or Laron syndrome. *Horm. Res.* 47:45–48, 1997.
9. Brennen, I.K., J. Zamecnik, P.N. Sek, and R.J. Shephard. The impact of heat exposure and repeated exercise on circulating stress hormones. *Eur. J. Appl. Physiol. Occup. Physiol.* 76: 445–454, 1997.
10. Buckler, J.M. The relationship between exercise, body temperature, and plasma growth hormone levels in a human subject. *J. Physiol.* 214:25P–26P, 1971.
11. Caine, D., R. Lewis, P. O'Connor, W. Howe, and S. Bass. Does gymnastics training inhibit growth of females? *Clin. J. Sport Med.* 11:260–270, 2001.
12. Cappon, J.P., E. Ipp, J.A. Brasel, and D.M. Cooper. Acute effects of high-fat and high-glucose meals on the growth hormone response to exercise. *J. Clin. Endocrinol. Metab.* 76: 1418–1422, 1993.
13. Chernausek, S.D. Insulin-like growth factor control of growth. In: *Molecular and cellular pediatric endocrinology.* S. Handwerger (ed.). Totowa, NJ: Humana Press, 1999, pp. 11–21.
14. Cittadini, A., A. Cuocolo, B. Merola, S. Fazio, D. Sabatini, E. Nicolai, A. Colao, et al. Impaired cardiac performance in GH-deficient adults and its improvement after GH replacement. *Am. J. Physiol.* 267:E219–E225, 1994.
15. Cittadini, A., and P.S. Douglas. The cellular and molecular basis for growth hormone action on the heart. In: *Growth hormone and the heart.* A. Giustina (ed.). Boston: Kluwer Academic, 2001, pp. 1–11.
16. Cooper, D.M. New horizons in pediatric exercise research. In: *New horizons in pediatric exercise science.* C.J.R. Blimkie and O. Bar-Or (eds.). Champaign, IL: Human Kinetics, 1995, pp. 1–24.
17. Costin, G., F.R.K. Kaufman, and J. Brasel. Growth hormone secretory dynamics in subjects with normal stature. *J. Pediatr.* 115:537–544, 1989.
18. Courteix, D., C. Jaffre, P. Obert, L. Lespessailles, and L. Benhamou. Skeletal maturation and somatic growth in elite girl gymnasts: A 3-year longitudinal study [abstract]. *Pediatr. Exerc. Sci.* 1:264–265, 1999.
19. Crepaz, R., W. Pitscheider, G. Radetti, C. Paganini, L. Gentili, G. Morini, E. Braito, et al. Cardiovascular effects of high-dose growth hormone treatment in growth-hormone deficient children. *Pediatr. Cardiol.* 16:223–227, 1995.
20. Cuttler, L. The regulation of growth hormone secretion. *Endocrinol. Metab. Clin. North Am.* 25:541–572, 1996.
21. Daly, R.M., P.A. Rich, and R. Klein. Hormonal responses to physical training in high level prepubertal male gymnasts. *Eur. J. Appl. Physiol.* 79:74–81, 1998.
22. Daubeney, P.E.F., E.S. McCaughey, C. Chase, J.M. Walker, Z. Slavik, P.R. Betts, and S.A. Webber. Cardiac effects of growth hormone in short normal children: Results after four years of treatment. *Arch. Dis. Child.* 72:337–339, 1995.
23. Donath, M.Y., R. Jenni, H.-P. Brunner, M. Anrig, S. Kohli, Y. Glatz, and E.R. Froesch. Cardiovascular and metabolic effects of insulin-like growth factor I at rest and during exercise in humans. *J. Clin. Endocrinol. Metab.* 81:4089–4094, 1996.
24. Eliakim, A., J.A. Brasel, T.J. Barstow, S. Mohan, and D.M. Cooper. Peak oxygen uptake, muscle volume, and the growth hormone–insulin-like growth factor-I axis in adolescent males. *Med. Sci. Sports Exerc.* 30:512–517, 1998.
25. Eliakim, A., J. Brasel, S. Mohan, T.J. Barstow, N. Berman, and D.M. Cooper. Physical fitness, endurance training, and the growth hormone–insulin like growth factor I system in adolescent females. *J. Clin. Endocrinol. Metab.* 81:3986–3992, 1996.

26. Eliakim, A., T.P. Scheet, R. Newcomb, S. Mohan, and D.M. Cooper. Fitness, training, and the growth hormone–insulin-like growth factor I axis in prepubertal girls. *J. Clin. Endocrinol. Metab.* 86:2797–2802, 2001.

27. Eriksson, B.O., B. Persson, and J.I. Thorell. The effects of repeated prolonged exercise on plasma growth hormone, insulin, glucose, free fatty acids, glycerol, and beta-hydroxybutyric acid in 13-year-old boys and in adults. *Acta Paediatr. Scand. Suppl.* 217:142–146, 1971.

28. Falk, B., and O. Bar-Or. Longitudinal changes in peak aerobic and anaerobic mechanical power of circumpubertal boys. *Pediatr. Exerc. Sci.* 5:318–331, 1993.

29. Fazio, S., D. Sabatini, B. Capaldo, C. Vigorito, A. Giordano, R. Guida, F. Pardo, et al. A preliminary study of growth hormone in the treatment of dilated cardiomyopathy. *N. Engl. J. Med.* 334:809–814, 1996.

30. Geffner, M.E. The growth without growth hormone syndrome. *Endocrinol. Metab. Clin. North Am.* 25:649–664, 1996.

31. Georgopoulos, N., M. Markou, A. Theodoropoulou, P. Paraskevopoulou, L. Vakow, Z. Kazantzi, M. Leglise, et al. Growth and pubertal development in elite female rhythmic gymnasts. *J. Clin. Endocrinol. Metab.* 84:4525–4530, 1999.

32. Gianotti, L., S. Fassino, G.A. Daga, F. Lanfranco, C. DeBacco, J. Ramunni, E. Arvat, et al. Effects of free fatty acids and acipimox, a lipolysis inhibitor, on the somatotroph responsiveness to GHRH in anorexia nervosa. *Clin. Endocrinol.* 52:713–720, 2000.

33. Golden, N.H., P. Kreitzer, M.S. Jacobson, F.I. Chasalow, J. Schebendach, S.M. Freedman, and I.R. Shenker. Disturbances of growth hormone secretion and action in adolescents with anorexia nervosa. *J. Pediatr.* 125:655–660, 1994.

34. Goldspink, G., and S.Y. Yang. Effects of activity on growth factor expression. *Int. J. Sport Nutr. Metab.* 11:521–527, 2001.

35. Hauspie, R.C., M. Vercanteren, and C. Susanne. Secular changes in growth and maturation: An update. *Acta Paediatr. Suppl.* 423:20–27, 1997.

36. Hefti, M.A., B.A. Harder, H.M. Eppenberger, and M.C. Schaub. Signaling pathways in cardiac myocyte hypertrophy. *J. Mol. Cell Cardiol.* 29:2873–2892, 1997.

37. Hopkins, N.J., P.M. Jakeman, S.C. Hughes, and J.M.P. Holly. Changes in circulating insulin-like growth factor-binding protein-1 (IGBP-1) during prolonged exercise: Effect of carbohydrate feeding. *J. Clin. Endocrinol. Metab.* 79: 1887–1890, 1994.

38. Horswill, C.A., W.B. Zipf, C.L. Kien, and E.B. Kahle. Insulin's contribution to growth in children and potential for exercise to mediate insulin's action. *Pediatr. Exerc. Sci.* 9:18–32, 1997.

39. Hutler, M., D. Schnabel, D. Staab, A. Tacke, U. Wahn, D. Bonin, and R. Beneke. Effect of growth hormone on exercise tolerance in children with cystic fibrosis. *Med. Sci. Sports Exerc.* 34:567–572, 2002.

40. Jahreis, G., E. Kauf, G. Frohner, and H.E. Schmidt. Influence of intensive exercise on insulin-like growth factor I, thyroid and steroid hormones in female gymnasts. *Growth Regulation* 1:95–99, 1991.

41. Jonkman, F.A.M., G. De Jong, and P.M. Fioretti. Growth hormone in the treatment of heart failure: A new tool for the future? *Eur. Heart J.* 18:181–184, 1997.

42. Katzmarzyk, P.T., R.M. Malina, and G.P. Beunen. The contribution of biological maturation to the strength and motor fitness of children. *Ann. Hum. Biol.* 24:493–505, 1997.

43. Kibler, W.B., and T.J. Chandler. Musculoskeletal adaptations and injuries associated with intense participation in youth sports. In: *Intensive participation in children's sports.* B.R. Cahill and A.J. Pearl (eds.). Champaign, IL: Human Kinetics, 1993, pp. 203–216.

44. Lacey, K.A., K. Hewison, and J.M. Parkin. Exercise as a screening test for growth hormone deficiency in children. *Arch. Dis. Child.* 48:508–512, 1973.

45. Leger, J., C. Garel, A. Fjellestad-Paulsen, M. Hassan, and P. Czernichow. Human growth hormone treatment of short-stature children born small for gestational age: Effect on muscle and adipose tissue mass during a 3-year treatment period and after 1 year's withdrawal. *J. Clin. Endocrinol. Metab.* 83:3512–3516, 1998.

46. Leger, J., C. Garel, I. Legrand, A. Paulsen, M. Hassan, and P. Czernichow. Magnetic resonance imaging evaluation of adipose tissue and muscle tissue mass in children with growth hormone deficiency, Turner's syndrome, and intrauterine growth retardation during the first year of treatment with GH. *J. Clin. Endocrinol. Metab.* 78:904–909, 1994.

47. Leigh, S.R., and P.B. Park. Evolution of human growth prolongation. *Am. J. Phys. Anthropol.* 107:331–350, 1998.

48. Lombardi, G., and A. Colao. Physiologic effect of growth hormone on the heart. In: *Growth hormone and the heart.* A. Giustina (ed.). Boston: Kluwer Academic, 2002, pp. 113–122.

49. Malina, R.M. Growth and maturation: Do regular physical activity and training for sport have a significant influence? In: *Paediatric exercise science and medicine.* N. Armstrong and W. van Mechelen (eds.). Oxford: Oxford University Press, 2000, pp. 95–106.

50. Malina, R.M. Growth and maturation of young athletes—Is training for sport a factor? In: *Sports and children.* K.-M. Chan and L.J. Micheli (eds.). Hong Kong: Williams & Wilkins, 1998, pp. 133–161.

51. Malina, R.M. Research on current trends in auxology. *Anthropol. Anz.* 48:209–227, 1990.

52. Malina, R.M. Tracking of physical activity and physical fitness across the life span. *Res. Q. Exerc. Sport* 67 (Suppl. 3): 48–57, 1996.

53. Malina, R.M., and G. Beunen. Matching opponents in youth sports. In: *The child and adolescent athlete.* O. Bar-Or (ed.). Oxford: Blackwell Science, 1996, pp. 202–213.

54. Malina, R.M., and C. Bouchard. *Growth, maturation, and physical activity.* Champaign, IL: Human Kinetics, 1991.

55. Mansfield, M.J., and S.J. Emans. Growth in female gymnasts: Should training decrease during puberty? *J. Pediatr.* 122: 237–240, 1993.

56. Marin, G., H.M. Domene, K.M. Barnes, B.J. Blackwell, F.G. Cassorla, and G.B. Cutler. The effect of estrogen priming and puberty on the growth hormone response to standardized treadmill exercise and arginine-insulin in normal boys and girls. *J. Clin. Endocrinol. Metab.* 79:537–541, 1994.

57. Matsudo, V.K.R. Prediction of future athletic excellence. In: *The child and adolescent athlete.* O. Bar-Or (ed.). Oxford: Blackwell Science, 1996, pp. 92–109.

58. Menon, R.K., and M.A. Sperling. Insulin as a growth factor. *Endocrinol. Metab. Clin. North Am.* 25:633–648, 1996.

59. Mirwald, R.L., D.A. Bailey, N. Cameron, and R.L. Rasmussen. Longitudinal comparison of aerobic power in active and inactive boys age 7.0 to 17.0 years. *Ann. Hum. Biol.* 8: 404–414, 1981.

60. Muller, E.E., and V. Locatelli. Undernutrition and pituitary function: Relevance to the pathophysiology of some neuroendocrine alterations of anorexia nervosa. *J. Endocrinol.* 132:327–329, 1992.

61. Nemet, D., Y. Oh, H.-S. Kim, M.A. Hill, and D.M. Cooper. Effect of intense exercise on inflammatory cytokines and growth mediators in adolescent boys. *Pediatrics* 110:681–689, 2002.

62. Nieman, D.C., D.A. Henson, L.L. Smith, A.C. Utter, D.M. Vinci, J.M. Davis, D.E. Kaminsky, et al. Cytokine changes after a marathon race. *J. Appl. Physiol.* 91: 109–114, 2001.

63. Parks, J.S., H. Abdul-Latif, E. Kinoshita, L.R. Meacham, R.W. Pfaffle, and M.R. Brown. Genetics of growth hormone gene expression. *Horm. Res.* 40:54–61, 1993.

64. Poehlman, E.T., and K.C. Copeland. Influences of physical activity on insulin-like growth factor-I in healthy younger and older men. *J. Clin. Endocrinol. Metab.* 71:1468–1473, 1990.

65. Rankinen, T., L. Perusse, R. Rauramaa, M.A. Rivera, O. Wolfarth, and C. Bouchard. The human gene map for performance and health-related fitness phenotypes: The 2001 update. *Med. Sci. Sports Exerc.* 34:1219–1233, 2002.

66. Roelen, C.A., W.R. de Vries, H.P. Koppeschaar, C. Vervoorn, J.H. Thijssen, and M.A. Blankenstein. Plasma insulin-like growth factor-I and high affinity growth hormone–binding protein levels increase after two weeks of strenuous physical training. *Int. J. Sports Med.* 18:238–241, 1997.

67. Roemmich, J.N., and W.E. Sinning. Weight loss and wrestling training: Effects on growth-related hormones. *J. Appl. Physiol.* 82:1760–1764, 1997.

68. Rosenfeld, R.G., and P. Cohen. Disorders of growth hormone/insulin-like growth factor secretion and action. In: *Pediatric endocrinology.* 2nd edition. M.A. Sperling (ed.). Philadelphia: Saunders, 2002, pp. 289–322.

69. Rousseau, G.G. Growth hormone gene regulation by trans-acting factor. *Horm. Res.* 37 (Suppl. 3):88–92, 1992.

70. Rowland, T.W., A.H. Morris, D.E. Biggs, and E.O. Reiter. Cardiac effects of growth hormone treatment for short stature in children. *J. Pediatr. Endocrinol.* 4:19–23, 1991.

71. Roy, M.-A., D. Bernard, B. Roy, and G. Marcotte. Body checking in PeeWee hockey. *Phys. Sportsmed.* 17:119–126, 1989.

72. Rubin, S.A., P. Buttrick, A. Malhotra, S. Melmed, and M.C. Fishbein. Cardiac physiology, biochemistry, and morphology in response to excess growth hormone in the rat. *J. Mol. Cell Cardiol.* 22:429–438, 1990.

73. Sacca, L., A. Cittadini, and S. Fazio. Growth hormone and the heart. *Endocrinol. Rev.* 15:555–573, 1994.

74. Sargeant, A. Short-term muscle power in children and adolescents. In: *Advances in pediatric sport sciences: Vol. 3. Biological issues.* O. Bar-Or (ed.). Champaign, IL: Human Kinetics, 1989, pp. 41–66.

75. Sartorio, A., E. Palmieri, V. Vangeli, G. Conte, M. Narici, and G. Faglia. Plasma and urinary GH following a standardized exercise protocol to assess GH production in short children. *J. Endocrinol. Invest.* 24:515–521, 2002.

76. Scheet, T.P., P.J. Mills, M.G. Ziegler, J. Stoppani, and D.M. Cooper. Effect of exercise on cytokines and growth mediators in prepubertal children. *Pediatr. Res.* 46:429–434, 1999.

77. Seip, R.L., A. Weltman, D. Goodman, and A.D. Rogol. Clinical utility of cycle exercise for the physiologic assessment of growth hormone release in children. *Am. J. Dis. Child.* 144:998–1000, 1990.

78. Selye, H. *The physiology and pathology of exposure to stress.* Montreal: Acta Medica, 1950.

79. Slack, J.M.W. The mysterious mechanism of growth. *Curr. Biol.* 6:348, 1996.

80. Smith, A.T., D.R. Clemmons, L.E. Underwood, V. Ben-Ezra, and R. McMurray. The effect of exercise on plasma somatomedin-C/insulin like growth factor I concentration. *Metabolism* 36:533–537, 1987.

81. Spagnoli, A., and R.G. Rosenfeld. The mechanisms by which growth hormone brings about growth. *Endocrinol. Metab. Clin. North Am.* 25:615–629, 1996.

82. Suei, K., L. McGillis, R. Calvert, and O. Bar-Or. Relationship among muscle endurance, explosiveness, and strength in circumpubertal boys. *Pediatr. Exerc. Sci.* 10:48–56, 1998.

83. Theintz, G.E., H. Howald, U. Weiss, and P.C. Sizonenko. Evidence for a reduction of growth potential in adolescent female gymnasts. *J. Pediatr.* 122:306–313, 1993.

84. Uberti, E.C.D., P. Franceschetti, and M.R. Ambrosio. Growth hormone and skeletal muscle function. In: *Growth hormone and the heart.* A. Giustina (ed.). Boston: Klewer Academic, 2002, pp. 125–149.

85. Vercauteren, M., and C. Susanne. The secular trend of height and menarche in Belgium: Are there any signs of a future stop? *Eur. J. Pediatr.* 144:306–309, 1985.

86. Wales, J.K.H. A brief history of the study of human growth dynamics. *Ann. Hum. Biol.* 25:175–184, 1998.

87. Weeke, J., and H.J.G. Gundersen. The effect of heating and central cooling on serum TSH, GH, and norepinephrine in resting man. *Acta Physiol. Scand.* 117:33–39, 1983.

88. Weiss, R.E., and S. Refetoff. Effect of thyroid hormone on growth. *Endocrinol. Metab. Clin. North Am.* 25:719–730, 1996.

89. Wells, H.G. *The food of the gods.* London: Unwin, 1925.

90. Weltman, A., J.Y. Weltman, R. Schurrer, W.S. Evans, J.D. Veldhuis, and A.D. Rogol. Endurance training amplifies the pulsatile release of growth hormone effects of training intensity. *J. Appl. Physiol.* 72:2188–2196, 1992.

91. Wirth, A., E. Trager, K. Scheele, D. Mayer, K. Diehm, K. Reischle, and H. Weicker. Cardiopulmonary adjustment and metabolic response to maximal and submaximal physical exercise of boys and girls at different stages of maturity. *Eur. J. Appl. Physiol.* 39:229–240, 1978.

Chapter 3

1. Ahima, R.S., D. Prabakaran, and C. Mantzoros. Role of leptin in the neuroendocrine response to fasting. *Nature* 382:250–252, 1996.

2. Ahmed, M.L., K.K. Ong, D.J. Morrell, L. Cox, N. Drayer, L. Perry, et al. Longitudinal study of leptin concentrations during puberty: Sex differences and relationships to changes in body composition. *J. Clin. Endocrinol. Metab.* 84:899–905, 1999.

3. Allen, D.B. Effects of fitness training on endocrine systems in children and adolescents. *Adv. Pediatr.* 46:41-66, 1999.

4. Amiel, S.A., R.S. Sherwin, D.C. Simonson, A.A. Lauritano, and W.V. Tamborlane. Impaired insulin action in puberty: A contributing factor to poor glycemic control in adolescents with diabetes. *N. Engl. J. Med.* 315:215–219, 1986.

5. Apter, D., and E. Hermanson. Update on female pubertal development. *Curr. Opin. Obstet. Gynecol.* 14:475–481, 2002.

6. Armstrong, N., and J.R. Welsman. Peak oxygen uptake in relation to growth and maturation in 11- to 17-year-old humans. *Eur. J. Appl. Physiol.* 85:546–551, 2001.

7. Armstrong, N., J.R. Welsman, and B.J. Kirby. Peak oxygen uptake and maturation in 12-yr-olds. *Med. Sci. Sports Exerc.* 30:165–169, 1998.

8. Armstrong, N., J.R. Welsman, and B.J. Kirby. Performance on the Wingate anaerobic test and maturation. *Pediatr. Exerc. Sci.* 9:253–261, 1997.

9. Arslanian, S., and C. Suprasongsin. Testosterone treatment in adolescents with delayed puberty: Changes in body composition, protein, fat, and glucose metabolism. *J. Clin. Endocrinol. Metab.* 82:3213–3220, 1997.

10. Ashley, C.D., M.L. Kramer, and P. Bishop. Estrogen and substrate metabolism: A review of contradictory research. *Sports Med.* 29:221–227, 2000.

11. Baer, J.T. Endocrine parameters in amenorrheic and eumenorrheic adolescent female runners. *Int. J. Sports Med.* 14:191–195, 1993.

12. Bar, P.R., G.J. Amelink, and B. Oldenberg. Prevention of exercise-induced muscle membrane damage by oestradiol. *Life Sci.* 42:2677–2681, 1988.

13. Baxter-Jones, A.D.G., N. Helms, N. Maffuli, and M. Preece. Growth and development of males gymnasts, swimmers, soccer, and tennis players: A longitudinal study. *Ann. Hum. Biol.* 22:381–394, 1995.

14. Bethea, C.L., M. Pecins-Thompson, and W.E. Schutzer. Ovarian steroids and serotonin neural function. *Mol. Neurobiol.* 18:87–123, 1998.

15. Beunen, G., and R.M. Malina. Growth and physical performance relative to the timing of the adolescent spurt. *Exerc. Sport Sci. Rev.* 16:503–540, 1988.

16. Beunen, G., and T. Martine. Muscular strength development in children and adolescents. *Pediatr. Exerc. Sci.* 12:174–197, 2000.

17. Bhasin, S., L. Woodhouse, and T.W. Storer. Proof of the effect of testosterone on skeletal muscle. *J. Endocrinol.* 170:27–38, 2001.

18. Booth, A., G. Shelly, A. Mazur, G. Tharp, and R. Kittock. Testosterone and winning and losing in human competition. *Horm. Behav.* 23:556–571, 1989.

19. Boyden, T.W., R.W. Pamenter, P. Stanforth, T. Rotkis, and J.H. Wilmore. Sex steroids and endurance running in women. *Fertil. Steril.* 39:629–632, 1983.

20. Brook, C.G.D. Mechanisms of puberty. *Horm. Res.* 51(Suppl. 3):52–54, 1999.

21. Brooks-Gunn, J., M.P. Warren, J. Rosso, and J. Gargiulo. Validity of self-report measures of girls' pubertal status. *Child Develop.* 58:829–841, 1987.

22. Broulik, P.D., C.D. Kochakian, and J. Dubovsky. Influence of castration and testosterone propionate on cardiac output, renal blood flow, and blood volume in mice. *Proc. Soc. Exp. Med.* 144:671–673, 1973.

23. Buckley-Bleiler, R., R.J. Maughan, and P.M. Clarkson. Serum creatine kinase activity after isometric exercise in pre-menopausal and postmenopausal women. *Exp. Aging Res.* 15:195–198, 1989.

24. Caprio, S. Insulin: The other anabolic hormone of puberty. *Acta Paediatr. Suppl.* 433:84–87, 1999.

25. Charkoudian, N., and J.M. Johnson. Female reproductive hormones and thermoregulatory control of skin blood flow. *Exerc. Sport Sci. Rev.* 28:108–112, 2000.

26. Chehab F.F., K. Mounzih, R. Lu, and M.E. Lim. Early onset of reproductive function in normal female mice treated with leptin. *Science* 275:88–90, 1997.

27. Cheung, C.C., J.E. Thornton, J.L. Kuipjer, D. Weigle, D.K. Clifton, and R.A. Steiner. Leptin is a metabolic gate for the onset of puberty in the female rate. *Endocrinology* 138: 855–888, 1997.

28. Christiansen, K. Behavioral effects of androgens in men and women. *J. Endocrinol.* 170:39–48, 2001.

29. Chumlea, W.C., C.M. Schubert, A.F. Roche, H.E. Kulin, P.A. Lee, J.M. Himes, and S.S. Sun. Age at menarche and racial comparisons in US girls. *Pediatrics* 111:110–113, 2003.

30. Clark, P.A., and A.D. Rogol. Growth hormone and sex steroid interactions at puberty. *Endocrinol. Metab. Clin. North Am.* 25:665–681, 1996.

31. Clayton, P.E., and J.A. Trueman. Leptin and puberty. *Arch. Dis. Child.* 83:1–9, 2000.

32. Coleman, L., and J. Coleman. The measurement of puberty: A review. *J. Adolesc.* 25:535–550, 2002.

33. Creatsas, G., N. Salakos, M. Averkiou, K. Miras, and D. Aravantinos. Endocrinological profile of oligomenorrheic strenuously exercising adolescents. *Int. J. Gynecol. Obstet.* 38: 215–221, 1992.

34. Cumming, G.R., D. Everatt, and L. Hastman. Bruce treadmill test in children: Normal values in a clinic population. *Am. J. Cardiol.* 41:69–75, 1978.

35. Dallman, P.R., and M.A. Siimes. Percentile curves for hemoglobin and red cell volume in infancy and childhood. *J. Pediatr.* 94:26–31, 1966.

36. Daniels, S.R., T.R. Kimball, J.A. Morrison, P. Khoury, S. Witt, and R.A. Meyer. Effect of lean body mass, fat mass, blood pressure, and sexual maturation on left ventricular mass in children and adolescents: Statistical, biological, and clinical significance. *Circulation* 92:3249–3254, 1995.

37. De Ste Croix, M.B.A., N. Armstrong, J.R. Welsman, and P. Sharpe. Longitudinal changes in isokinetic leg strength in 10–14-year-olds. *Ann. Hum. Biol.* 29:50–62, 2002.

38. Dorn, L.D., E.J. Susman, E.D. Nottelmann, G. Inoff-Germain, and G.P. Chrousus. Perceptions of puberty: Adolescent, parent, and health professional. *Dev. Psychol.* 26: 322–329, 1990.

39. Fahey, T.D., A. Del Valle-Zuris, G. Dehlsen, M. Trieb, and J. Seymour. Pubertal stage differences in hormonal and haematological responses to maximal exercise in males. *J. Appl. Physiol.* 46:823–827, 1979.

40. Falgairette, G., M. Bedu, N. Fellman, E. Van Praagh, and J. Coudert. Bioenergetic profile in 144 boys aged from 6 to 15 years with special reference to sexual maturation. *Eur. J. Appl. Physiol.* 62:151–156, 1991.

41. Falk, B., and O. Bar-Or. Longitudinal changes in peak aerobic and anaerobic mechanical power in circumpubertal boys. *Pediatr. Exerc. Sci.* 5:318–331, 1993.

42. Frank, G.R. The role of estrogen in pubertal skeletal physiology: Epiphyseal maturation and mineralization of the skeleton. *Acta Paediatr.* 84:627–630, 1995.

43. Frisch, R.E. Body fat, menarche, fitness and fertility. *Hum. Reprod.* 2:521–533, 1987.

44. Gardner, F.H., D.G. Nathan, S. Piomeeli, and J.F. Cummins. The erythrocythaemic effects of androgen. *Br. J. Haem.* 14: 611–615, 1968.

45. Geithner, C.A., B. Woynarowska, and R.M. Malina. The adolescent spurt and sexual maturation in girls active and not active in sport. *Ann. Hum. Biol.* 25:415–423, 1998.

46. Genazzani, A.R., F. Bernardi, P. Monteleone, S. Luisi, and M. Luisi. Neuropeptides, neurotransmitters, neurosteroids, and the onset of puberty. *Ann. N.Y. Acad. Sci.* 90:1–9, 2000.

47. Georgopoulos, N., M. Markou, A. Theodoropoulou, P. Paraskevopoulou, L. Vakow, Z. Kazantzi, M. Leglise, et al. Growth and pubertal development in elite female rhythmic gymnasts. *J. Clin. Endocrinol. Metab.* 84:4525–4530, 1999.

48. Gill, M.S., C.M. Hall, V. Tilimann, and P.E. Clayton. Constitutional delay in growth and puberty (CDGP) is associated with hypoleptinemia. *J. Clin. Endocrinol. Metab.*, in press.

49. Goran, M.I., and B.A. Gower. Longitudinal study on pubertal insulin resistance. *Diabetes* 50:2444–2450, 2001.

50. Gruber, C.J., W. Tschugguel, C. Schneeberger, and J.C. Huber. Production and actions of estrogens. *New Engl. J. Med.* 346:340–352, 2002.

51. Gutin, B., L. Ramsey, P. Barbeau, W. Cannady, M. Ferguson, M. Litaker, and S. Owens. Plasma leptin concentrations in obese children: Changes during 4 mo periods with and without physical training. *Am. J. Clin. Nutr.* 69:388–394, 1999.

52. Hackney, A.C. The male reproductive system and endurance exercise. *Med. Sci. Sports Exerc.* 28:180–189, 1996.

53. Hayward, C.S., C.M. Webb, and P. Collins. Effect of sex hormones on cardiac mass. *Lancet* 357:1354–1356, 2001.

54. Hickey, M.S., and D.J. Calsbeek. Plasma leptin and exercise. *Sports Med.* 31:583–589, 2001.

55. Hilton, L.K., and A.B. Loucks. Low energy availability, not exercise stress, suppresses the diurnal rhythm of leptin in healthy young women. *Am. J. Physiol. Endocrinol. Metab.* 278: E43–E49, 2000.

56. Hindmarsh, P., P.J. Smith, C.G.D. Brook, and D.C. Mathews. The relation between height velocity and growth hormone secretion in short prepubertal children. *J. Clin. Endocrinol. Metab.* 27:581–591, 1987.

57. Horlick, M.B., M. Rosenbaum, M. Nicolson, L.S. Levine, B. Fedun, J. Wang, R.N. Pierson, et al. Effect of puberty on the relationship between circulating leptin and body composition. *J. Clin. Endocrinol. Metab.* 85:2509–2518, 2000.

58. Janz, K.F., T.L. Burns, J.D. Witt, and L.T. Mahoney. Longitudinal analysis of scaling $\dot{V}O_2$ for differences in body size during puberty: The Muscatine Study. *Med. Sci. Sports Exerc.* 30:1436–1444, 1998.

59. Janz, K.F., J.D. Dawson, and L.T. Mahoney. Predicting heart growth during puberty: The Muscatine Study. *Pediatrics* 105: E63, 2000.

60. Ji, C.-Y. Age at spermarche and comparison of growth and performance of pre- and postspermarcheal Chinese boys. *Am. J. Hum. Biol.* 13:35–43, 2001.

61. Katch, V.L. Physical conditioning of children. *J. Adolesc. Health Care* 3:241–246, 1983.

62. Kendall, B., and R. Eston. Exercise-induced muscle damage and the potential role of estrogen. *Sports Med.* 32:103–123, 2002.

63. Kendrick, Z.V., C.A. Steffen, and W.L. Ramsey. Effect of estradiol on tissue glycogen metabolism in exercised oophorectomized rats. *J. Appl. Physiol.* 63:492–496, 1987.

64. Kiess, W., A. Reich, K. Meyer, A. Glasgow, J. Deutscher, J. Klammt, Y. Yang, et al. A role for leptin in sexual maturation and puberty? *Horm. Res.* 51(Suppl. 3):55–63, 1999.

65. Klein, K.O., J. Baron, and M.J. Colli. Estrogen levels in childhood determined by an ultrasensitive recombinant cell bioassay. *J. Clin. Invest.* 94:2475–2480, 1994.

66. Kobayashi, K., K. Kitamura, M. Miura, H. Sodeyama, Y. Murase, M. Miyashita, and H. Matsui. Aerobic power as related to body growth and training in Japanese boys: A longitudinal study. *J. Appl. Physiol.* 44:666–672, 1978.
67. Koenig, H., A. Goldstone, and C.Y. Lu. Testosterone-mediated sexual dimorphism of the rodent heart. *Circ. Res.* 50: 782–787, 1982.
68. Koo, M.M., and T.E. Rohan. Accuracy of short-term recall of age at menarche. *Ann. Hum. Biol.* 24:61–64, 1997.
69. Kopp, W., W.F. Blum, S. von Prittwitz, A. Ziegler, H. Lubbert, G. Emmons, W. Herzog, et al. Low leptin levels predict amenorrhea in underweight and eating disordered females. *Mol. Psychiatry* 2:335–340, 1997.
70. Kraemer, R.R., E.O. Acevedo, L.B. Synovitz, E.P. Hebert, T. Gimpel, and V.D. Castracane. Leptin and steroid hormone responses to exercise in adolescent female runners over a 7-week season. *Eur. J. Appl. Physiol.* 86:85–91, 2001.
71. Krahenbuhl, G.S., J.S. Skinner, and W.M. Kohrt. Developmental aspects of maximal aerobic power in children. *Exerc. Sport Sci. Rev.* 13:503–538, 1985.
72. Kulin, H.E., and J. Muller. The biological aspects of puberty. *Pediatr. Rev.* 17:75–86, 1996.
73. Lamb, D.R. Androgens and exercise. *Med. Sci. Sports Exerc.* 7:105, 1975.
74. Litt, I.F. Amenorrhea in the adolescent athlete. *Postgrad. Med.* 80:245–253, 1986.
75. Loucks, A.B. Effects of exercise training on the menstrual cycle: Existence and mechanisms. *Med. Sci. Sports Exerc.* 22: 275–280, 1990.
76. Loucks, A.B. The reproductive system and physical activity in adolescents. In: *New horizons in pediatric exercise science.* C.J.R. Blimkie and O. Bar-Or (eds.). Champaign, IL: Human Kinetics, 1995, pp. 27–37.
77. Loucks, A.B., M. Verdun, and E.M. Heath. Low energy availability, not stress of exercise, alters LH pulsativity in exercising women. *J. Appl. Physiol.* 84:37–46, 1998.
78. Malina, R.M. Growth and maturation of young athletes—Is training for sport a factor? In: *Sports and children.* K.-M. Chan and L.J. Micheli (eds.). Hong Kong: Williams & Wilkins, 1998, pp. 133–161.
79. Malina, R.M., G. Beunen, J. Lefevre, and B. Woynarowska. Maturity-associated variation in peak oxygen uptake in active adolescent boys and girls. *Ann. Hum. Biol.* 24:19–31, 1997.
80. Malina, R.M., and C. Bouchard. *Growth, maturation, and physical activity.* Champaign, IL: Human Kinetics, 1991.
81. Mantzoros, C.S., J.S. Flier, and A.D. Rogol. A longitudinal assessment of hormonal and physical alterations during normal puberty in boys: V. Rising leptin levels may signal the onset of puberty. *J. Clin. Endocrinol. Metab.* 82:1066–1070, 1997.
82. Marin, G., H.M. Dormene, K.M. Barnes, B.J. Blackwell, F.G. Cassorla, and G.B. Cutler. The effect of estrogen priming and puberty on the growth hormone response to standardized treadmill exercise and arginine-insulin in normal boys and girls. *J. Clin. Endocrinol. Metab.* 79:537–541, 1994.
83. Martha, P.M., K.M. Gorman, R.M. Blizzard, A.D. Rogol, and J.D. Veldhuis. Endogenous growth hormone secretion and clearance rates in normal boys, as determined by deconvolution analysis: Relationship to age, pubertal status, and body mass. *J. Clin. Endocrinol. Metab.* 74:336–344, 1992.
84. Mauras, N., A.D. Rogol, and J.D. Veldhuis. Specific, time-dependent actions of low-dose ethinyl estradiol administration on the episodic release of growth hormone, follicle-stimulating hormone, and luteinizing hormone in prepubertal girls with Turner syndrome. *J. Clin. Endocrinol. Metab.* 69:1053–1058, 1989.
85. Meineke, H.A., and R.C. Crafts. Further observations on the mechanism by which androgens and growth hormone influence erythropoiesis. *Ann. N.Y. Acad. Sci.* 149:298–307, 1968.
86. Mero, A. Blood lactate production and recovery from anaerobic exercise in trained and untrained boys. *Eur. J. Appl. Physiol.* 57:660–666, 1988.
87. Moran, A., D.R. Jacobs, J. Steinberger, C.-P. Hong, R. Prineas, R.V. Luepken, and A.R. Sinaiko. Insulin resistance during puberty: Results from clamp studies in 357 children. *Diabetes* 48:2039–2044, 1999.
88. Neu, C.M., F. Rauch, J. Rittweger, F. Manz, and E. Schoenau. Influences of puberty on muscle development at the forearm. *Am. J. Physiol. Endocrinol. Metab.* 283:E103–E107, 2002.
89. Nottin, S., A. Vinet, F. Stecken, L.-D. Nguyen, F. Duniss, A.-M. Lecoq, and P. Obert. Central and peripheral cardiovascular adaptions during a maximal cycle exercise in boys and men. *Med. Sci. Sports Exerc.* 34:456–463, 2002.
90. Ong, K.K.L., A.L. Ahmed, and D.B. Dunger. The role of leptin in human growth and puberty. *Acta Paediatr. Suppl.* 433:95–98, 1999.
91. Palmert, M.R., S. Radovich, and P.A. Boepple. Leptin levels in children with central precocious puberty. *J. Clin. Endocrinol. Metab.* 83:2260–2265, 1998.
92. Parker, M.W., A.J. Johanson, A.D. Rogol, D.L. Kaiser, and R.M. Blizard. Effect of testosterone on somatomedin-C concentrations in prepubertal boys. *J. Clin. Endocrinol. Metab.* 58:87–90, 1984.
93. Pearce, G., S. Bass, N. Young, C. Formica, and E. Seeman. Does weight-bearing exercise protect against the effects of exercise induced oligomenorrhea on bone density? *Osteoporos. Int.* 6:448–452, 1996.
94. Reisner, E.H. Tissue culture of bone marrow: II. Effect of steroid hormones on hematopoiesis in vitro. *Blood* 27: 460–469, 1966.
95. Riley, J.L., M.E. Robinson, and E.A. Wise. A meta-analytic review of pain perception across the menstrual cycle. *Pain* 81:225–235, 1999.
96. Rinard, J., P.M. Clarkson, and L. Smith. Response of males and females to high-force eccentric exercise. *J. Sports Sci.* 18:229–236, 2000.
97. Roemmich, J.N., P.A. Clark, and S. Berr. Alterations in growth and body composition during puberty: II. Gender differences in leptin levels during puberty are related to the

subcutaneous fat depot and sex steroids. *Am. J. Physiol.* 275: E543, 1998.

98. Roemmich, J.N., and A.D. Rogol. Role of leptin during childhood growth and development. *Endocrinol. Metab. Clin. North Am.* 28:749–764, 1999.
99. Roemmich, J.N., and W.E. Sinning. Weight loss and wrestling training: II. Effects on growth-related hormones. *J. Appl. Physiol.* 82:1760–1764, 1997.
100. Rogers, M.A., G.A. Stull, and F.S. Apple. Creatine kinase isoenzyme activities in men and women following a marathon race. *Med. Sci. Sports Exerc.* 17:679–682, 1985.
101. Rogol, A.D. Growth at puberty: Interaction of androgens and growth hormone. *Med. Sci. Sports Exerc.* 26:767–770, 1994.
102. Rosenfield, R.L. Puberty in the female and its disorders. In: *Pediatric endocrinology.* 2nd edition. M.A. Sperling (ed.). Philadelphia: Saunders, 2002, pp. 455–518.
103. Round, J.M., D.A. Jones, J.W. Honour, and A.M. Nevill. Hormonal factors in the development of differences in strength between boys and girls during adolescence: A longitudinal study. *Ann. Hum. Biol.* 26:49–62, 1999.
104. Rowland, T.W. Adolescence: A "risk factor" for physical inactivity. *President's Council on Physical Fitness Sports Research Digest* June 1999, pp. 1–8.
105. Rowland, T., R. Bhargava, D. Parslow, and R. Heptulla. Cardiac responses to progressive cycle exercise in moderately obese adolescent females. *J. Adolesc. Health,* 32:422–427, 2003.
106. Rowland, T.W., G. Kline, D. Goff, L. Martel, and L. Ferrone. One mile run performance and cardiovascular fitness in children. *Arch. Pediatr. Adolesc. Med.* 153:845–849, 1999.
107. Rowland, T., K. Miller, P. Vanderburgh, D. Goff, L. Martel, and L. Ferrone. Cardiovascular fitness in premenarcheal girls and young women. *Int. J. Sports Med.* 20:117–121, 1999.
108. Rowland, T.W., A.H. Morris, J.F. Kelleher, B.L. Haag, and E.O. Reiter. Serum testosterone response to training adolescent runners. *Am. J. Dis. Child.* 142:165–169, 1988.
109. Rowland, T., B. Popowski, and L. Ferrone. Cardiac response to maximal upright cycle exercise in healthy boys and men. *Med. Sci. Sports Exerc.* 29:1146–1151, 1997.
110. Salbe, A.D., M. Nicolson, and E. Ravussin. Total energy expenditure and the level of physical activity correlate with plasma leptin concentrations in five-year-old children. *J. Clin. Invest.* 99:592–595, 1997.
111. Scaramella, T.J., and W.A. Brown. Serum testosterone and aggressiveness in hockey players. *Psychosom. Med.* 40: 262–265, 1978.
112. Scheuer, J., A. Malhotra, T.F. Schaible, and J. Capasso. Effects of gonadectomy and hormonal replacement on rat hearts. *Circ. Res.* 61:12–19, 1987.
113. Schoenau, E., C.M. Neu, E. Mokou, G. Wassmer, and F. Manz. Influence of puberty on muscle area and cortical bone of the forearm in boys and girls. *J. Clin. Endocrinol. Metab.* 85:1095–1098, 2000.
114. Scott, E.C., and F.E. Johnston. Critical fat, menarche, and the maintenance of menstrual cycles: A critical review. *J. Adolesc. Health Care* 2:249–260, 1982.
115. Shahidi, N.T. Androgens and erythropoiesis. *N. Engl. J. Med.* 289:72–80, 1973.
116. Strauss, R.H., R.R. Lanese, and W.B. Malarkey. Weight loss in amateur wrestlers and its effect on serum testosterone levels. *J. Am. Med. Assoc.* 254:3337–3338, 1985.
117. Tanner, J.M. Growth and endocrinology of the adolescent. In: *Endocrine and genetic disease of children and adolescence.* 2nd edition. L.I. Gardner (ed.). Philadelphia: Saunders, 1975, pp. 14–63.
118. Thong, F.S., C. McLean, and T.E. Graham. Plasma leptin in female athletes: Relation with body fat, reproductive, nutritional, and endocrine factors. *J. Appl. Physiol.* 88:2037–2044, 2000.
119. Ulloa-Aguirre, A., C.M. Christie, and E. Garcia-Rubi. Testosterone and oxandronlone (a non-aromatizable androgen) specifically amplify the mass and rate of growth hormone (GH) release secreted per burst without altering GH secretory burst duration or frequency of the GH half life. *J. Clin. Endocrinol. Metab.* 71:846–854, 1990.
120. Urban, R.J., Y.H. Bodengurg, C. Gilkison, J. Foxworth, A.R. Coggan, R.R. Wolfe, and A. Ferrnado. Testosterone administration to elderly men increases skeletal muscle strength and protein synthesis. *Am. J. Physiol.* 269:E820–E826, 1995.
121. Van der Meulen, J.H., H. Kuipers, and J. Drukker. Relationship between exercise-induced muscle damage and enzyme release in rats. *J. Appl. Physiol.* 71:999–1004.
122. Viru, A., L. Laaneots, K. Karelson, T. Smirnova, and M. Viru. Exercise-induced hormonal responses in girls at different stages of sexual maturation. *Eur. J. Appl. Physiol.* 77:401–408, 1998.
123. Vranic, M., and D. Wasserman. Exercise, fitness, and diabetes. In: *Exercise, fitness, and health: A consensus of current knowledge.* C. Bouchard, R.J. Shephard, T. Stephens, J.R. Sutton, and B.D. McPherson (eds.). Champaign, IL: Human Kinetics, 1990, pp. 467–490.
124. Warren, M.P. The effects of exercise on pubertal progression and reproductive function in girls. *J. Clin. Endocrinol. Metab.* 51:1150–1156, 1980.
125. Warren, M.P., and A. Stiehl. Exercise and female adolescents: Effects on the reproductive and skeletal systems. *J. Am. Med. Women's Assoc.* 54:115–120, 1999.
126. Weber, G., W. Kartodihardjo, and V. Klissouras. Growth and physical training with reference to heredity. *J. Appl. Physiol.* 40:211–220, 1976.
127. Weimann, E. Gender-related differences in elite gymnasts: The female athlete triad. *J. Appl. Physiol.* 92:2146–2152, 2002.
128. Weise, M., E. Graeme, and D.P. Merke. Pubertal and gender-related changes in the sympathoadrenal system in healthy children. *J. Clin. Endocrinol. Metab.* 87:5038–5043, 2002.
129. Welsman, J., N. Armstrong, and B. Kirby. Serum testosterone is not related to peak $\dot{V}O_2$ and submaximal blood lactate responses in 12–16-year-old males. *Pediatr. Exerc. Sci.* 6: 120–127, 1994.

130. Williams, J.R., and N. Armstrong. The influence of age and sexual maturation on child's blood lactate responses to exercise. *Pediatr. Exerc. Sci.* 3:111–120, 1991.

131. Williams, J.R., N. Armstrong, E.M. Winter, and N. Crichton. Changes in peak oxygen uptake with age and sexual maturation: Physiologic fact or statistical anomaly? In: *Children and exercise XVI*. J. Coudert and E. Van Praagh (eds.). Paris: Masson, 1992, pp. 35–37.

Chapter 4

1. Asano, K., and K. Hirakoba. Respiratory and circulatory adaptation during prolonged exercise in 10–12-year-old children and adults. In: *Children and sport*. J. Ilmarinen and I. Valimaki (eds.). Berlin: Springer-Verlag, 1984, pp. 119–128.
2. Åstrand, P.O. *Experimental studies of physical working capacity in relation to sex and age*. Copenhagen: Munksgaard, 1952.
3. Bar-Or, O. *Sports medicine for the practitioner.* New York: Springer-Verlag, 1983, pp. 311–314.
4. Bell, R.D., J.D. MacDougall, R. Billeter, and H. Howald. Muscle fiber types and morphometric analysis of skeletal muscle in six-year-old children. *Med. Sci. Sports Exerc.* 12: 28–31, 1980.
5. Berg, A., and J. Keul. Biochemical changes during exercise in children. In: *Young athletes: A biological, psychological, and educational perspective*. R.M. Malina (ed.). Champaign, IL: Human Kinetics, 1988, pp. 61–67.
6. Berg, A., S.S. Kim, and J. Keul. Skeletal muscle enzyme activities in healthy young subjects. *Int. J. Sports Med.* 7: 236–239, 1986.
7. Boisseau, N., and P. Delamarche. Metabolic and hormonal responses to exercise in children and adolescents. *Sports Med.* 30:405–422, 2000.
8. Booth, F.W. Physical activity as a stimulus to changes in gene expression in skeletal muscle. In: *Biological effects of physical activity.* Champaign, IL: Human Kinetics, 1989, pp. 91–104.
9. Brooks, G.A., and J. Mercier. Balance of carbohydrate and lipid utilization during exercise: The "crossover" concept. *J. Appl. Physiol.* 76:2253–2261, 1994.
10. Carlson, J.S., and G.A. Naughton. Assessing accumulated oxygen deficit in children. In: *Pediatric anaerobic performance.* E. Van Praagh (ed.). Champaign, IL: Human Kinetics, 1998, pp. 119–136.
11. Chia, M., N. Armstrong, and D. Childs. The assessment of children's anaerobic performance using modifications of the Wingate anaerobic test. *Pediatr. Exerc. Sci.* 9:80–90, 1997.
12. Cooper, D.M., and T.J. Barstow. Magnetic resonance imaging and spectroscopy in studying exercise in children. *Exerc. Sport Sci. Rev.* 24:475–499, 1996.
13. Couture, P., and A.J. Hulbert. On the relationship between body mass, tissue metabolic rate, and sodium pump activity in mammalian liver and kidney cortex. *Am. J. Physiol.* 268: R641–R650, 1995.
14. Coyle, E.F. Physical activity as a metabolic stressor. *Am. J. Clin. Nutr.* 72(Suppl.):512S–520S, 2000.
15. Coyle, E.F., A.E. Jeukendrup, A.J.M. Wagenmakers, and W.H.M. Saris. Fatty oxidation is directly regulated by carbohydrate metabolism during exercise. *Am. J. Physiol.* 273: E268–E275, 1997.
16. Cumming, G.R., L. Hastman, J. McCort, and S. McCullough. High serum lactates do occur in young children after maximal work. *Int. J. Sports Med.* 1:66–69, 1980.
17. Cunningham, L.N. Relationship of running economy, ventilatory threshold, and maximal O_2 consumption to running performance in high school females. *Res. Q. Exerc. Sport* 61: 369–374, 1990.
18. Cureton, K., R. Boileau, T. Lohman, and J. Misner. Determinants of distance running performance in children: Analysis of a path model. *Res. Q. Exerc. Sport* 48:270–279, 1971.
19. Davies, C.T.M., C. Barnes, and S. Godfrey. Body composition and maximal exercise performance in children. *Hum. Biol.* 44:195–214, 1972.
20. Delamarche, P., M. Monnier, and A. Gratas-Delamarche. Glucose and free fatty acid utilization during prolonged exercise in prepubertal boys in relation to catecholamine responses. *Eur. J. Appl. Physiol.* 65:66–72, 1992.
21. Docherty, D., and C.A. Gaul. Relationship of body size, physique, and composition to physical performance in young boys and girls. *Int. J. Sports Med.* 12:525–532, 1991.
22. Dunaway, G.A., T.P. Kasten, G.A. Nickols, and J.A. Chesky. Regulation of skeletal muscle 6-phosphofructo-1-kinase during aging and development. *Mech. Ageing Develop.* 36: 13–23, 1986.
23. Duncan, G.E., and E.T. Howley. Metabolic and perceptual responses to short-term cycle training in children. *Pediatr. Exerc. Sci.* 10:110–122, 1998.
24. Duncan, G.E., and E.T. Howley. Substrate metabolism during exercise in children and the "crossover concept." *Pediatr. Exerc. Sci.* 11:12–21, 1999.
25. Dyck, D.J., S.J. Peters, P.S. Wendly, A. Chesley, E. Hultman, and L.L. Spreit. Regulation of muscle glycogen phosphorylase activity during intense cycling with elevated FFA. *Am. J. Physiol.* 270:E116–E125, 1996.
26. Emmett, B., and P.W. Hochachka. Scaling of oxidative and glycolytic enzymes in mammals. *Resp. Physiol.* 45:261–272, 1981.
27. Eriksson, B.O. Physical training, oxygen supply, and muscle metabolism in 11–13-year-old boys. *Acta Physiol. Scand. Suppl.* 384:1–48, 1972.
28. Eriksson, B.O., B.D. Gollnick, and B. Saltin. Muscle metabolism and enzyme activities after training in boys 11–13 years old. *Acta Physiol. Scand.* 87:485–497, 1973.
29. Eriksson, B.O., J. Karlsson, and B. Saltin. Muscle metabolites during exercise in pubertal boys. *Acta Paediatr. Scand. Suppl.* 217:154–157, 1971.
30. Eriksson, B.O., and B. Saltin. Muscle metabolism in boys 11 to 16 years compared to adults. *Acta Paediatr. Belg.* 28(Suppl.): 257–265, 1974.

31. Falgairette, G., M. Bedu, N. Fellman, E. Van Praagh, and J. Coudert. Bio-energetic profile in 144 boys aged from 6 to 15 years. *Eur. J. Appl. Physiol.* 62:151–156, 1991.

32. Falgairette, G., P. Duche, M. Bedu, N. Fellman, and J. Coudert. Bioenergetic characteristics in prepubertal swimmers. *Int. J. Sports Med.* 14:444–448, 1993.

33. Falk, B., and O. Bar-Or. Longitudinal changes in peak aerobic and anaerobic mechanical power of circumpubertal boys. *Pediatr. Exerc. Sci.* 5:318–331, 1993.

34. Fellman, N., B. Beaum, and J. Coudert. Blood lactate after maximal and supramaximal exercise in 10- to 12-year-old Bolivian boys. *Int. J. Sports Med.* 15:S90–S95, 1994.

35. Gastin, P.B. Energy system interaction and relative contribution during maximal exercise. *Sports Med.* 31:725–741, 2001.

36. Gollnick, P.D., R.B. Armstrong, C.W. Saubert, K. Piehl, and B. Saltin. Enzyme activity and fiber composition in skeletal muscle of untrained and trained men. *J. Appl. Physiol.* 33: 312–319, 1972.

37. Haller, R.G., and S.F. Lewis. Human respiratory chain disorders: Implication for muscle oxidative metabolism. In: *Biochemistry of exercise VII.* A.L. Taylor, P.D. Gollnick, H.J. Green, C.D. Ianuzzo, E.G. Noble, G. Metivier, and J.R. Sutton. Champaign, IL: Human Kinetics, 1990, pp. 251–264.

38. Haralambie, G. Activites enzymatiques dans le muscle squelettique des enfants de divers ages [Enzymatic activity in the skeletal muscle of children of various ages]. In: *Le sports et l'énfant.* Montpellier: Euromed, 1980, pp. 243–258.

39. Haralambie, G. Enzyme activities in skeletal muscle of 13–15 year old adolescents. *Bull. Eur. Physiopathol. Respir.* 18:65–74, 1982.

40. Haralambie, G. Skeletal muscle enzyme activities in female subjects of various ages. *Bull. Eur. Physiopathol. Respir.* 15: 259–267, 1979.

41. Hargreaves, M. Interactions between muscle glycogen and blood glucose during exercise. *Exerc. Sport Sci. Rev.* 25:21–39, 1997.

42. Hebestreit, H., F. Meyer, H. Htay, G.J. Heigenhauser, and O. Bar-Or. Plasma metabolites, volume and electrolytes following 30-s high-intensity exercise in boys and men. *Eur. J. Appl. Physiol.* 72:563–569, 1996.

43. Heusner, A.A. Energy metabolism and body size: II. Dimensional analysis and energetic nonsimilarity. *Resp. Physiol.* 48: 13–25, 1982.

44. Hicks, A.L., J. Kent-Braun, and D.S. Ditor. Sex differences in human skeletal muscle fatigue. *Exerc. Sport Sci. Rev.* 29: 109–112, 2001.

45. Hoelzer, D.R., G.P. Dalsky, W.E. Clutter, S.D. Shah, J.H. Holloszy, and P.E. Cryer. Glucoregulation during exercise: Hypoglycemia is prevented by redundant glucoregulatory systems, sympathochromaffin activation, and changes in islet hormone secretion. *J. Clin. Invest.* 77:212–221, 1986.

46. Holliday, M.A., D. Potter, A. Jarrah, and S. Bearg. The relation of metabolic rate to body weight and organ size. *Pediatr. Res.* 1:185–195, 1967.

47. Holloszy, J.O. Utilization of fatty acids during exercise. In: *Biochemistry of exercise VII.* A.W. Taylor, P.D. Gollnick, H.J. Green, C.D. Ianuzzo, E.G. Noble, G. Metivier, and J.R. Sutton (eds.). Champaign, IL: Human Kinetics, 1990, pp. 314–323.

48. Holloszy, J.O., and E.F. Coyle. Adaptations of skeletal muscle to endurance exercise and their metabolic consequences. *J. Appl. Physiol.* 56:831–838, 1984.

49. Hoppeler, H., H. Howald, K. Conley, S.L. Lindstedt, H. Claassen, P. Vock, and E.R. Weibel. Endurance training in humans: Aerobic capacity and the structure of skeletal muscle. *J. Appl. Physiol.* 59:320–327, 1985.

50. Hoppeler, H., P. Luthi, H. Claasen, E.R. Weibel, and H. Howald. The ultrastructure of the normal human muscle: A morphometric analysis of untrained men, women, and well-trained orienteers. *Pflugers Arch.* 244:217–227, 1973.

51. Karlsson, J., L.-O. Nordesjö, L. Jorfeldt, and B. Saltin. Muscle lactate, ATP, and CP levels during exercise after physical training in man. *J. Appl. Physiol.* 33:199–203, 1972.

52. Knoebel, L.K. Energy metabolism. In: *Physiology.* E.E. Selkurt (ed.). Boston: Little Brown, 1963, pp. 564–579.

53. Komi, P.V., and J. Karlsson. Skeletal muscle fiber types, enzyme activities, and physical performance in young males and females. *Acta Physiol. Scand.* 103:210–218, 1978.

54. Krebs, H.A. Body mass and tissue respiration. *Biochim. Biophys. Acta* 4:249–269, 1950.

55. Kunkel, H.O., J.F. Spalding, G. de Franciscis, and M.F. Futrell. Cytochrome oxidase activity and body weight in rats and in three species of larger animals. *Am. J. Physiol.* 186:203–206, 1956.

56. Kuno, S., H. Takahashi, K. Fujimoto, H. Akikma, M. Miyamura, I. Nemoto, Y. Itai, et al. Muscle metabolism during exercise using phosphorus-31 nuclear magnetic resonance spectroscopy in adolescents. *Eur. J. Appl. Physiol.* 70:301–304, 1995.

57. Lehmann, M., J. Keul, and U. Korsten-Reck. The influence of graduated treadmill exercise on plasma catecholamines, aerobic and anaerobic capacity in boys and adults. *Eur. J. Appl. Physiol.* 47:301–311, 1981.

58. Lewis, S.F., and R.G. Haller. Disorders of muscle glycogenolysis/glycolysis: The consequences of substrate-limited oxidative metabolism in humans. In: *Biochemistry of exercise VII.* A.W. Taylor, P.D. Gollnick, H.J. Green, C.D. Ianuzzo, E.G. Noble, G. Metivier, and J.R. Sutton (eds.). Champaign, IL: Human Kinetics, 1990, pp. 211–226.

59. Lexell, J., M. Sjostrom, A. Nordlund, and C.C. Taylor. Growth and development of human muscle: A quantitative morphological study of whole vastus lateralis from childhood to adult age. *Muscle Nerve* 15:404–409, 1992.

60. Lundberg, A., B.O. Eriksson, and G. Mellgren. Metabolic substrates, muscle fiber composition, and fiber size in late walking and normal children. *Eur. J. Pediatr.* 130:79–92, 1979.

61. Macek, M., and J. Vavra. Prolonged exercise in prepubertal boys: I. Cardiovascular and metabolic adjustment. *Eur. J. Appl. Physiol.* 35:291–298, 1976.

62. Mahon, A.D., P. Del Corral, C.A. Howe, G.E. Duncan, and M. Ray. Physiological correlates of 3-kilometer running

performance in male children. *Int. J. Sports Med.* 17:580–584, 1996.

63. Malina, R.M., and C. Bouchard. *Growth, maturation, and physical activity.* Champaign, IL: Human Kinetics, 1991.
64. Martinez, L.R., and E.M. Haymes. Substrate utilization during treadmill running in prepubertal girls and women. *Med. Sci. Sports Exerc.* 24:975–983, 1991.
65. McArdle, W.D., F.I. Katch, and V.L. Katch. *Exercise physiology: Energy, nutrition, and human performance.* Philadelphia: Lea & Febiger, 1981.
66. Mero, A. Blood lactate production and recovery from anaerobic exercise in trained and untrained boys. *Eur. J. Appl. Physiol.* 57:660–666, 1988.
67. Murphy, S.E. *The relationship between aerobic and anaerobic power in untrained pre- and post-menarcheal females.* Unpublished master's thesis, University of Massachusetts, Amherst, 2000.
68. Naughton, G.A., and J.S. Carlson. The accumulated oxygen deficit measure and its application in pediatric exercise science. *Pediatr. Exerc. Sci.* 10:13–20, 1998.
69. Noakes, T.D. Physiological models to understand exercise fatigue and the adaptations that predict or enhance athletic performance. *Scand. J. Med. Sci. Sports* 10:123–145, 2000.
70. Oertel, G. Morphometric analysis of normal skeletal muscles in infancy, childhood, and adolescence: An autopsy study. *J. Neurol. Sci.* 88:303–313, 1988.
71. Paterson, D.H., and D.A. Cunningham. Development of anaerobic capacity in early and late maturing boys. In: *Children and exercise XI.* R.A. Binkhorst (ed.). Champaign, IL: Human Kinetics, 1985, pp. 119–128.
72. Peterson, S.R., C.A. Gaul, M.M. Stanton, and C.C. Hanstock. Skeletal muscle metabolism during short-term, high-intensity exercise in prepubertal and pubertal girls. *J. Appl. Physiol.* 87:2151–2156, 1999.
73. Pfitzinger, P., and P. Freedson. Blood lactate response to exercise in children: Part I. Peak lactate concentration. *Pediatr. Exerc. Sci.* 9:210–222, 1997.
74. Pianosi, P., L. Seargeant, and J.C. Haworth. Blood lactate and pyruvate concentrations, and their ratio during exercise in healthy children: Developmental perspectives. *Eur. J. Appl. Physiol.* 71:518–522, 1995.
75. Pluto, R., and P. Burger. Normal values of catecholamines in blood plasma determined by high-performance liquid chromatography with amperometric detection. *Int. J. Sports Med.* 9:75–78, 1988.
76. Porter, R.K., and M.D. Brand. Cellular oxygen consumption depends on body mass. *Am. J. Physiol.* 268:R641–R650, 1995.
77. Prasad, N., K.D. Coutts, D. Jespersen, L. Wolski, and T. Cooper. Relationship between aerobic and anaerobic exercise capacities in pre-pubertal children [abstract]. *Med. Sci. Sports Exerc.* 27:S115, 1995.
78. Randle, P.J., E.A. Newsholme, and P.B. Garland. Regulation of glucose uptake by muscle: Effects of fatty acids, ketone bodies, and pyruvate, of alloxan-diabetes and starvation, on the uptake and metabolic fate of glucose in rat heart and diaphragm muscles. *Biochem. J.* 93:652–664, 1964.
79. Ratel, S., P. Duché, A. Hennegrave, E. Van Praagh, and M. Bedu. Acid-base balance during repeated cycling sprints in boys and men. *J. Appl. Physiol.* 92:479–485, 2002.
80. Riddell, M.C., O. Bar-Or, H.P. Schwarcz, and G.J.F. Heigenhauser. Substrate utilization during exercise with [^{13}C]-glucose ingestion in boys. *Eur. J. Appl. Physiol. Occup. Physiol.,* in press. 83:441-448, 2000.
81. Riddell, M.C., O. Bar-Or, B. Wilk, M.L. Parolin, and G.J.F. Heigenhauser. Substrate utilization during exercise with glucose and glucose plus fructose ingestion in boys ages 10–14 yr. *J. Appl. Physiol.* 90:903–911, 2001.
82. Riner, W., and R. Boileau. Energy sources during prolonged physical activity in girls and boys [abstract]. *Pediatr. Exerc. Sci.* 13:286–287, 2001.
83. Rouslin, W. The mitochondrial ATPase in slow and fast heart rate hearts. *Am. J. Physiol.* 252:H622–H627, 1987.
84. Rowland, T.W., and L.N. Cunningham. Influence of aerobic and anaerobic fitness on ventilatory anaerobic threshold in children [abstract]. *Med. Sci. Sports Exerc.* 28(Suppl.):S147, 1996.
85. Rowland, T.W., G. Kline, D. Goff, L. Martel, and L. Ferrone. One-mile run performance and cardiovascular fitness in children. *Arch. Pediatr. Adolesc. Med.* 153:845–849, 1999.
86. Rowland, T.W., and T.A. Rimany. Physiological responses to prolonged exercise in premenarcheal and adult females. *Pediatr. Exerc. Sci.* 7:183–191, 1995.
87. Saltin, B., and P.D. Gollnick. Skeletal muscle adaptability: Significance for metabolism and performance. In: *Handbook of physiology.* L.D. Peachey, R. Adrian, and S.R. Geiger (eds.). Baltimore: Williams & Wilkins, 1983, pp. 555–631.
88. Saris, W.H.M., A.M. Noordeloos, B.E. Ringnalda, M.A. Van't Hof, and R.A. Binkhorst. In: *Children and exercise XI.* R.A. Binkhorst, H.C.G. Kemper, and W.H.M. Saris (eds.). Champaign, IL: Human Kinetics, 1985, pp. 151–160.
89. Schmidt-Nielsen, K. *Scaling: Why is animal size so important?* Cambridge: Cambridge University Press, 1984.
90. Shephard, R.J., C. Allen, O. Bar-Or, C.T.M. Davies, S. Degre, R. Hedman, K. Ishii, et al. The working capacity of Toronto schoolchildren. *Can. Med. Ass. J.* 100:560–566, 1969.
91. Shulman, R.G., and D.L. Rothman. The "glycogen shunt" in exercising muscle: A role for glycogen in muscle energetics and fatigue. *Proc. Nat. Acad. Sci.* 98:457–461, 2001.
92. Somero, G.N., and J.J. Childress. A violation of the metabolism–size scaling paradigm: Activities of glycolytic enzymes in muscle increase in larger-size fish. *Physiol. Zool.* 53: 322–337, 1980.
93. Spriet, L.L., R.A. Howlett, and G.J.F. Heigenhauser. An enzymatic approach to lactate production in human skeletal muscle during exercise. *Med. Sci. Sports Exerc.* 32:756–763, 2000.
94. Stear, K., J.S. Carlson, and G.A. Naughton. Developmental characteristics of anaerobic capacity in children. In: *Proceeding: International Conference of Science and Medicine in Sport.* Queensland: Australian Sports Medicine Federation, 1994, pp. 288–289.

95. Suarez, R.K. Upper limits to mass-specific metabolic rates. *Annu. Rev. Physiol.* 58:583–605, 1996.

96. Suei, K., R. McGillis, R. Calvert, and O. Bar-Or. Relationships among muscle endurance, explosiveness, and strength in circumpubertal boys. *Pediatr. Exerc. Sci.* 10:48–56, 1998.

97. Taylor, D.J., G.J. Kemp, C.H. Thompson, and G.K. Radda. Ageing effects on oxidative function of skeletal muscle in vivo. *Mol. Cell. Biochem.* 174:321–324, 1997.

98. Thorstensson, A., B. Sjodin, and J. Karlsson. Enzymatic activities and muscle strength after sprint training in man. *Acta Physiol. Scand.* 94:313–318, 1975.

99. Van Ekeren, G.J., E.A.M. Cornelissen, A.M. Stadhouders, and R.C.A. Sengers. Increased volume density of peripheral mitochondria in skeletal muscle of children with exercise intolerance. *Eur. J. Pediatr.* 150:744–750, 1991.

100. Van Praagh, E., M. Bedu, G. Falgairette, N. Fellman, and J. Coudert. *Oxygen uptake during a 30-s supramaximal exercise in 7- to 15-year-old boys.* Congress of Pediatric Work Physiology, September 11–15, 1989, Budapest, Hungary.

101. Van Praagh, E., N. Fellman, M. Bedu, G. Falgairette, and J. Coudert. Gender difference in the relationship of anaerobic power to body composition in children. *Pediatr. Exerc. Sci.* 2:336–348, 1990.

102. Vissing, J., H. Galbo, and R.G. Haller. Paradoxically enhanced glucose production during exercise in humans with blocked glycolysis caused by muscle phosphofructokinase deficiency. *Neurology* 47:766–771, 1996.

103. Weise, M., E. Graeme, and D.P. Merke. Pubertal and gender-related changes in the sympathoadrenal system in healthy children. *J. Clin. Endocrinol. Metab.* 87:5038–5043, 2002.

104. Welsman, J.R., and N. Armstrong. Assessing postexercise lactates in children and adolescents. In: *Pediatric anaerobic performance.* E. Van Praagh (ed.). Champaign, IL: Human Kinetics, 1998, pp. 137–190.

105. Williams, J.R., and N. Armstrong. The influence of age and sexual maturation on children's blood lactate responses to exercise. *Pediatr. Exerc. Sci.* 3:111–120, 1991.

106. Wilmore, J.H., and D.L. Costill. *Physiology of sport and exercise.* Champaign, IL: Human Kinetics, 1994.

107. Zanconato, S., S. Buchthal, T.J. Barstow, and D.M. Cooper. ^{31}P-magnetic resonance spectroscopy of leg muscle metabolism during exercise in children and adults. *J. Appl. Physiol.* 74:2214–2218, 1993.

Chapter 5

1. Alexander, R., A.S. Jayes, G.M.O. Maloiy, and E.M. Winter. Allometry of the leg muscles of mammals. *J. Zool. Soc. London* 194:539–552, 1981.

2. Allor, K.M., and J.M. Pivarnik. Relationship between aerobic fitness and physical activity in sixth grade girls [abstract]. *Pediatr. Exerc. Sci.* 13:91, 2001.

3. American Alliance for Health, Physical Education, Recreation and Dance. *Health related physical fitness: Test manual.* Reston, VA: Author, 1980.

4. Anderson, S.D., and S. Godfrey. Cardio-respiratory response to treadmill exercise in normal children. *Clin. Sci.* 40: 433–442, 1971.

5. Armon, Y., D.M. Cooper, R. Flores, S. Zanconato, and T.S. Barstow. Oxygen uptake dynamics during high-intensity exercise in children and adults. *J. Appl. Physiol.* 70:841–848, 1991.

6. Armstrong, N., and J. Welsman. Assessment and interpretation of aerobic fitness in children and adolescents. *Exerc. Sport Sci. Rev.* 22:435–475, 1994.

7. Armstrong, N., and J.R. Welsman. Cardiovascular responses to submaximal treadmill running in 11- to 13-year-olds. *Acta Paediatr.* 91:125–131, 2002.

8. Armstrong, N., and J. Welsman. Development of aerobic fitness. *Pediatr. Exerc. Sci.* 12:128–149, 2000.

9. Armstrong, N., and J.R. Welsman. Peak oxygen uptake in relation to growth and maturation in 11- to 17-year-old humans. *Eur. J. Appl. Physiol.* 85:546–551, 2001.

10. Armstrong, N., and J. Welsman. *Young people and physical activity.* Oxford: Oxford University Press, 1997.

11. Armstrong, N., J. Welsman, and B. Kirby. Physical activity, peak oxygen uptake, and performance on the Wingate anaerobic test in 12-year-olds. *Acta Kines. Univers. Tartu* 3: 7–21, 1998.

12. Armstrong, N., J. Welsman, and R. Winsley. Is peak $\dot{V}O_2$ a maximal index of children's aerobic fitness? *Int. J. Sports Med.* 17:356–359, 1996.

13. Asano, K., and K. Hirakoba. Respiratory and circulatory adaptation during prolonged exercise in 10–12-year-old children and in adults. In: *Children and sport.* J. Ilmarinen and I. Valimaki (eds.). Berlin: Springer-Verlag, 1984, pp. 119–128.

14. Benecke, R., H. Heck, V. Schwarz, and R. Leithauser. Maximal lactate steady state during the second decade of age. *Med. Sci. Sports Exerc.* 28:1474–1478, 1996.

15. Berthouze, S.E., P.M. Minaire, J. Castells, T. Busso, L.V. Bico, and J.-R. Lacour. Relationship between mean habitual daily energy expenditure and maximum oxygen uptake. *Med. Sci. Sports Exerc.* 27:1170–1179, 1995.

16. Billat, L.V. Use of blood lactate measurements for prediction of exercise performance and for control of training. *Sports Med.* 22:157–175, 1996.

17. Billat, L.V., V., A. Gratas-Delamarche, and M. Monmer. A test to approach maximal lactate steady state in 12-year-old boys and girls. *Arch. Physiol. Biochem.* 103:65–72, 1995.

18. Blimkie, C.J.R., P. Roche, and O. Bar-Or. The anaerobic-to-aerobic power ratio in adolescent boys and girls. In: *Children and exercise XII.* J. Rutenfranz, R. Mocellin, and F. Klimt (eds.). Champaign, IL: Human Kinetics, 1986, pp. 31–37.

19. Calder, W.A. *Size, function, and life history.* Cambridge, MA: Harvard University Press, 1984, pp. 21–22.

20. Campbell, M.K. *Biochemistry.* Philadelphia: Saunders College Publishing, 1991, pp. 216–227.

21. Cheatham, C.C., A.D. Mahon, J.D. Brown, and D.R. Bolster. Cardiovascular responses during prolonged exercise

at ventilatory thresholds in boys and men. *Med. Sci. Sports Exerc.* 32:1080–1087, 2000.

22. Conley, K.E., W.F. Kemper, and G.J. Crowther. Limits to sustainable muscle performance: Interaction between glycolysis and oxidative phosphorylation. *J. Exp. Biol.* 204: 3189–3194, 2001.
23. Cooper, D.M., C. Berry, L. Lamarra, and K. Wasserman. Kinetics of oxygen uptake at the onset of exercise as a function of growth in children. *J. Appl. Physiol.* 59:211–217, 1985.
24. Cooper, D.M., D. Weiler-Ravell, B.J. Whipp, and K. Wasserman. Aerobic parameters of exercise as a function of body size during growth of children. *J. Appl. Physiol.* 56:628–634, 1984.
25. Corbin, C.B., R.P. Pangrazi, and G.J. Welk. Toward an understanding of appropriate physical activity levels for youth. *President's Council on Physical Fitness and Sports Research Digest,* 1984.
26. Cumming, G.R., and L.M. Borysyk. Criteria for maximal oxygen uptake in men over 40 in a population survey. *Med. Sci. Sports Exerc.* 14:18–22, 1972.
27. Cureton, K., P. Bishop, P. Hutchinson, H. Newland, S. Vickery, and L. Zwiren. Sex differences in maximal oxygen uptake effect of equating hemoglobin concentration. *Eur. J. Appl. Physiol.* 54:656–660, 1986.
28. Cureton, K.J., M.A. Sloniger, D.M. Black, W.P. McCormack, and D.A. Rowe. Metabolic determinants of the age-related improvement in one-mile run/walk performance in youth. *Med. Sci. Sports Exerc.* 29:259–267, 1997.
29. Daniels, S.R., T.R. Kimball, J.A. Morrison, P. Khoury, S. Witt, and R.A. Meyer. Effect of lean body mass, fat mass, blood pressure, and sexual maturation on left ventricular mass in children and adolescents. *Circulation* 92:3249–3254, 1995.
30. Davies, K.J.A., L. Packer, and G.A. Brooks. Biochemical adaptation to mitochondria, muscle, and whole-animal respiration to endurance training. *Arch. Biochem. Biophys.* 209:539–554, 1981.
31. de Simone, G., R.B. Devereux, S.R. Daniels, G. Mureddu, M.J. Roman, T.R. Kimball, R. Greco, et al. Stroke volume and cardiac output in normotensive children and adults. *Circulation* 95:1837–1843, 1997.
32. Dishman, R.K., and M. Steinhardt. Reliability and concurrent validity for a 7-day recall of physical activity in college students. *Med. Sci. Sports Exerc.* 20:14–15, 1988.
33. Drinkwater, B.L. Women and exercise: Physiologic aspects. *Exerc. Sport Sci. Rev.* 12:21–51, 1984.
34. Eriksson, B.O., and G. Koch. Effect of physical training on hemodynamic response during submaximal and maximal exercise in 11–13-year-old boys. *Acta Physiol. Scand.* 87: 27–39, 1973.
35. Fawkner, S.G., and N. Armstrong. Assessment of critical power with children. *Pediatr. Exerc. Sci.* 14:259–268, 2002.
36. Fawkner, S.G., N. Armstrong, C.R. Potter, and J.R. Welsman. Oxygen uptake kinetics in children and adolescents after the onset of moderate-intensity exercise. *J. Sport Sci.* 20:319–326, 2002.
37. Gaesser, G.A., and D.C. Poole. The slow component of oxygen uptake kinetics. *Exerc. Sport Sci. Rev.* 24:35–70, 1996.
38. Gaesser, G.A., and L.A. Wilson. Effects of continuous and interval training on the parameters of the power–endurance time relationship for high-intensity exercise. *Int. J. Sports Med.* 9:417–421, 1988.
39. Godfrey, S., C.T.M. Davies, E. Wozniak, and C.A. Barnes. Cardio-respiratory responses to exercise in normal children. *Clin. Sci.* 40:419–431, 1971.
40. Hebestreit, H., S. Kriemler, R.L. Hughson, and O. Bar-Or. Kinetics of oxygen uptake at the onset of exercise in boys and men. *J. Appl. Physiol.* 85:1833–1841, 1998.
41. Hill, D.W. The critical power concept: A review. *Sports Med.* 16:237–254, 1993.
42. Hill, D.W., R.P. Steward, and C.J. Lane. Application of the critical power concept to young swimmers. *Pediatr. Exerc. Sci.* 7:281–293, 1995.
43. Hochachka, P.W. The biochemical limits of muscular work. In: *Biochemistry of exercise XII.* A.W. Taylor, P.D. Gollnick, H.J. Green, C.D. Ianuzzo, E.G. Noble, G. Metivier, and J.R. Sutton (eds.). Champaign, IL: Human Kinetics, 1990, pp. 1–9.
44. Holliday, M.A., D. Potter, A. Jarrah, and S. Bearg. The relationship of metabolic rate to body weight and organ size. *Pediatr. Res.* 1:185–195, 1967.
45. Honig, C.R., R.J. Connett, and T.E.J. Gayeski. O_2 transport and its interaction with metabolism: A systems view of aerobic capacity. *Med. Sci. Sports Exerc.* 24:47–53, 1992.
46. Jenkins, D.G., and B.M. Quigley. Endurance training enhances critical power. *Med. Sci. Sports Exerc.* 24:1283–1289, 1992.
47. Jenkins, D.G., and B.M. Quigley. The y-intercept of the critical power function as a measure of anaerobic work capacity. *Ergonomics* 34:13–22, 1991.
48. Kanaley, J.A., and R.A. Boileau. The onset of anaerobic threshold at three stages of physical maturity. *J. Sports Med.* 28:367–374, 1988.
49. Katsura, T. Influences of age and sex on cardiac output during submaximal exercise. *Ann. Physiol. Anthrop.* 5:39–57, 1986.
50. Kayser, K. *Height and weight in human beings: Autopsy report.* Munich: Oldenbourg, 1987.
51. Kemper, H.C.G., J.W.R. Twisk, L.L.J. Koppes, W. van Mechelen, and G.B. Post. A 15-year physical activity pattern is positively related to aerobic fitness in young males and females (13–27 years). *Eur. J. Appl. Physiol.* 84:395–402, 2001.
52. Kemper, H.C.G., R. Verschuur, and L. de Mey. Longitudinal changes of aerobic fitness in youth ages 12 to 23. *Pediatr. Exerc. Sci.* 1:257–270, 1989.
53. Krahenbuhl, G.S., D.W. Morgan, and R.P. Pangrazi. Longitudinal changes in distance running performance of young males. *Int. J. Sports Med.* 10:92–96, 1989.
54. Krahenbuhl, G.S., J.S. Skinner, and W.M. Kohrt. Developmental aspects of maximal aerobic power in children. *Exerc. Sport Sci. Rev.* 13:503–538, 1985.

55. Loftin, M., P. Strikmiller, B. Warren, L. Myers, L. Schroth, J. Pittman, D. Hatsha, et al. Comparison and relationship of $\dot{V}O_2$ peak and physical activity patterns in elementary and high school females. *Pediatr. Exerc. Sci.* 10:153–163, 1998.

56. Lohman, T.G. *Advances in body composition assessment*. Champaign, IL: Human Kinetics, 1992, p. 82.

57. Macek, M., and J. Vavra. Oxygen uptake and heart rate with transition from rest to maximal exercise in prepubertal boys. In: *Children and exercise IX*. K. Berg (ed.). Baltimore: University Park Press, 1980, pp. 64–68.

58. Macek, M., J. Vavra, and I. Novosadova. Prolonged exercise in prepubertal boys: I. Cardiovascular and metabolic adjustment. *Eur. J. Appl. Physiol.* 35:291–298, 1976.

59. Malcolm, D.D., T.L. Burns, L.T. Mahoney, and R.M. Lauer. Factors affecting left ventricular mass in childhood: The Muscatine Study. *Pediatrics* 92:703–709, 1993.

60. Malina, R.M., and C. Bouchard. *Growth, maturation, and physical activity.* Champaign, IL: Human Kinetics, 1991, pp. 115–132.

61. McLellan, T.M., and K.S.Y. Cheung. A comparative evaluation of the individual anaerobic threshold and the critical power. *Med. Sci. Sports Exerc.* 24:543–550, 1992.

62. Mocellin, R., M. Heusgen, and H.P. Gildein. Anaerobic threshold and maximal steady state blood lactate in prepubertal boys. *Eur. J. Appl. Physiol.* 62:56–60, 1991.

63. Morgan, D.W. Economy of locomotion. In: *Paediatric exercise science and medicine.* N. Armstrong and W. van Mechelen (eds.). Oxford: Oxford University Press, 2000, pp. 183–190.

64. Morrow, J.R., and P.S. Freedson. Relationship between habitual physical activity and aerobic fitness in adolescents. *Pediatr. Exerc. Sci.* 6:315–329, 1994.

65. Myers, J., D. Walsh, M. Sullivan, and V. Froelicher. Effect of sampling on variability and plateau in oxygen uptake. *J. Appl. Physiol.* 68:404–410, 1990.

66. Nagasawa, H., Y. Arakiki, and T. Nakajima. Longitudinal observations of left ventricular end diastolic dimension in children using echocardiography. *Pediatr. Cardiol.* 17: 169–174, 1996.

67. Nevill, A.M. Evidence of an increasing proportion of leg muscle mass to body mass in male adolescents and its implications on performance. *J. Sports Sci.* 12:163–164, 1994.

68. Noakes, T.D. Implications of exercise testing for prediction of athletic performance: A contemporary perspective. *Med. Sci. Sports Exerc.* 20:319–330, 1988.

69. Nottin, S., V. Agnes, F. Stecken, L.-D. Nguyen, F. Ounissi, A.-M. Lecoq, and P. Obert. Central and peripheral cardiovascular adaptations during maximal cycle exercise in boys and men. *Med. Sci. Sports Exerc.* 33:456–463, 2002.

70. Nottin, S., A. Vinet, F. Stecken, L.-D. N'Guyen, F. Ounissi, A.-M. Lecoq, and P. Obert. Central and peripheral cardiovascular adaptations to exercise in endurance-trained children. *Acta Physiol. Scand.* 175:85–92, 2002.

71. Obert, P., C. Cleziou, R. Candau, D. Courteix, A.-M. Lecoq, and P. Guenon. The slow component of $\dot{V}O_2$uptake kinetics during high-intensity exercise in trained and untrained prepubertal children. *Int. J. Sports Med.* 21:31–36, 2000.

72. Obert, P., S. Mandigout, S. Nottin, A. Vinet, L.-D. N'Guyen, and A.-M. Lecoq. Cardiovascular response to endurance training in children: Effect of gender. *Eur. J. Clin. Invest.*, 33: 199-208, 2003.

73. Paffenbarger, R.S., R.T. Hyde, and A.L. Wing. Physical activity and physical fitness as determinants of health and longevity. In: *Exercise, fitness, and health: A consensus of current knowledge.* C. Bouchard, R.J. Shephard, T. Stephens, J.R. Sutton, and B.D. McPherson (eds.). Champaign, IL: Human Kinetics, 1990, pp. 33–48.

74. Pettersen, S.A., P.M. Fredriksen, and F. Ingjer. The correlation between peak oxygen uptake ($\dot{V}O_2$peak) and running performance in children and adolescents: Aspects of different units. *Scand. J. Med. Sci. Sports* 11:223–228, 2001.

75. Poole, D.C., S.A. Ward, G.W. Gardner, and B.J. Whipp. Metabolic and respiratory profile of the upper limits for prolonged exercise in man. *Ergonomics* 31:1265–1279, 1988.

76. Reybrouck, T., M. Weymans, H. Stijns, J. Knops, and L. vander Hauwaert. Ventilatory anaerobic threshold for evaluating exercise performance in healthy children: Age and sex differences. *Eur. J. Appl. Physiol.* 54:278–284, 1985.

77. Riner, W.F., M. McCarthy, L.V. DeCillis, and D.S. Ward. Response of children and adolescents to onset of exercise. In: *Children and exercise XIX.* N. Armstrong, B. Kirby, and J. Welsman (eds.). London: Spon, 1997, pp. 248–252.

78. Riopel, D.A., A.B. Taylor, and A.R. Hohn. Blood pressure, heart rate, pressure–rate product, and electrocardiographic changes in healthy children during treadmill exercise. *Am. J. Cardiol.* 44:697–704, 1979.

79. Rowland, T.W. Does peak $\dot{V}O_2$ reflect $\dot{V}O_2$max in children? Evidence from supramaximal testing. *Med. Sci. Sports Exerc.* 25:689–693, 1993.

80. Rowland, T.W. Effect of prolonged inactivity on aerobic fitness of children. *J. Sports Med. Phys. Fitness* 34:147–155, 1994.

81. Rowland, T.W. Performance fitness in children as a model for fatigue, or, what good is allometry, anyway? *Pediatr. Exerc. Sci.* 7:1–4, 1995.

82. Rowland, T.W. Physical activity, fitness, and health in children: A close look. *Pediatrics* 93:669–671, 1994.

83. Rowland, T., and J.W. Blum. Cardiac dynamics during upright cycle exercise in boys. *Am. J. Hum. Biol.* 12:749–757, 2000.

84. Rowland, T.W., and L.N. Cunningham. Oxygen uptake plateau during maximal treadmill exercise in children. *Chest* 101:485–489, 1992.

85. Rowland, T., D. Goff, L. Martel, and L. Ferrone. Estimation of maximal stroke volume from submaximal values [abstract]. *Pediatr. Exerc. Sci.* 11:279, 1999.

86. Rowland, T., D. Goff, L. Martel, and L. Ferrone. Influence of cardiac functional capacity on gender differences in maximal oxygen uptake in children. *Chest* 117:629–635, 2000.

87. Rowland, T., D. Goff, L. Martel, L. Ferrone, and G. Kline. Normalization of maximal cardiovascular variables for body size in premenarcheal girls. *Pediatr. Cardiol.* 21:429–432, 2000.

88. Rowland, T., G. Kline, D. Goff, L. Martel, and L. Ferrone. One-mile run performance and cardiovascular fitness of children. *Arch. Pediatr. Adolesc. Med.* 153:845–849, 1999.

89. Rowland, T., G. Kline, D. Goff, L. Martel, and L. Ferrone. Physiologic determinants of maximal aerobic power in healthy 12-year-old boys. *Pediatr. Exerc. Sci.* 11:317–326, 1999.

90. Rowland, T., K. Miller, P. Vanderburgh, D. Goff, L. Martel, and L. Ferrone. Cardiovascular fitness in premenarcheal girls and young women. *Int. J. Sports Med.* 21:117–121, 1999.

91. Rowland, T., B. Popowski, and L. Ferrone. Cardiac responses to maximal upright cycle exercise in healthy boys and men. *Med. Sci. Sports Exerc.* 29:1146–1151, 1997.

92. Rowland, T.W., and T.A. Rimany. Physiological responses to prolonged exercise in premenarcheal and adult females. *Pediatr. Exerc. Sci.* 7:183–191, 1995.

93. Rowland, T., V. Unnithan, B. Fernhall, T. Baynard, and C. Lange. Left ventricular response to dynamic exercise in young cyclists. *Med. Sci. Sports Exerc.* 34:637–642, 2002.

94. Sady, S. Transient oxygen uptake and heart rate responses at the onset of relative endurance exercise in prepubertal boys and adult men. *Int. J. Sports Med.* 2:240–244, 1981.

95. Saltin, B. Cardiovascular and pulmonary adaptation to physical activity. In: *Exercise, fitness, and health: A consensus of current knowledge.* C. Bouchard, R.J. Shephard, T. Stephens, J.R. Sutton, and B.D. McPherson (eds.). Champaign, IL: Human Kinetics, 1990, pp. 187–203.

96. Saltin, B. Physiological adaptation to physical conditioning. *Acta Med. Scand. Suppl.* 711:11–24, 1986.

97. Schmidt-Nielsen, K. *Scaling: Why is animal size so important?* Cambridge: Cambridge University Press, 1984.

98. Shephard, R.J., C. Allen, and O. Bar-Or. The working capacity of Toronto school children. *Can. Med. Assoc. J.* 100:560–566, 1969.

99. Sparling, P.B. A meta-analysis of studies comparing maximal oxygen uptake in men and women. *Res. Q.* 51:542–552, 1980.

100. Sunnegardh, J., and L.-E. Bratteby. Maximal oxygen uptake, anthropometry, and physical activity in a randomly selected sample of 8- and 13-year-old children in Sweden. *Eur. J. Appl. Physiol.* 56:266–272, 1987.

101. Turley, K.R., and J.H. Wilmore. Cardiovascular responses to submaximal exercise in 7- to 9-yr-old boys and girls. *Med. Sci. Sports Exerc.* 29:824–832, 1997.

102. Vanden Eynde, B., D. Van Gerven, D. Vienne, M. Vuylsteke-Wauters, and J. Ghesquiere. Endurance fitness and peak height velocity in Belgian boys. In: *Children and exercise XIII.* S. Oseid and K.-H. Carlson (eds.). Champaign, IL: Human Kinetics, 1989, pp. 19–26.

103. Vinet, A., S. Mandigout, S. Nottin, L.-D. Nguyen, A.-M. Lecoq, D. Courteix, and P. Obert. Influence of body composition, hemoglobin concentration, cardiac size and function on gender differences in maximal oxygen uptake in prepubertal children. *Chest,* in press. 124: 1494-1499, 2003.

104. Washington, R.L., J.C. van Gundy, C. Cohen, H.M. Sondheimer, and R.R. Wolfe. Normal aerobic and anaerobic exercise data for North American school-age children. *J. Pediatr.* 112:223–233, 1988.

105. Welsman, J.R., N. Armstrong, M. Bell, P. Sharpe, and R.J. Winsley. Scaling the relationship between MRI determined leg volume, leg muscle volume, and peak $\dot{V}O_2$ in girls [abstract]. *Pediatr. Exerc. Sci.* 8:177–178, 1996.

106. Welsman, J.R., N. Armstrong, B.J. Kirby, A.M. Nevill, and E. Winter. Scaling peak $\dot{V}O_2$ for differences in body size. *Med. Sci. Sports Exerc.* 28:259–265, 1996.

107. Welsman, J.R., N. Armstrong, B.J. Kirby, R.J. Winsley, G. Parsons, and P. Sharpe. Exercise performance and magnetic resonance imaging–determined thigh muscle volume in children. *Eur. J. Appl. Physiol.* 76:92–97, 1997.

108. Weymans, M., T. Reybrouck, H. Stijns, and J. Knops. Influence of age and sex on the ventilatory anaerobic threshold in children. In: *Children and exercise XI.* R.B. Binkhorst, H.C.G. Kemper, and W.H.M. Saris (eds.). Champaign, IL: Human Kinetics, 1985, pp. 114–118.

109. Whipp, B.J. Developmental aspects of oxygen uptake kinetics in children. In: *Children and exercise XIX.* N. Armstrong, B. Kirby, and J. Welsman (eds.). London: Spon, 1997, pp. 233–247.

110. Williams, J.R., and N. Armstrong. Relationship of maximal lactate steady state to performance at fixed blood lactate reference values in children. *Pediatr. Exerc. Sci.* 3:333–341, 1991.

111. Zanconato, S., D.M. Cooper, and Y. Armon. Oxygen cost and oxygen uptake dynamics and recovery with 1 minute of exercise in children and adults. *J. Appl. Physiol.* 71:993–998, 1991.

112. Zanconato, S., G. Riedy, and D.M. Cooper. Calf muscle cross-sectional area and peak oxygen uptake and work rate in children and adults. *Am. J. Physiol.* 267:R720–R725, 1994.

Chapter 6

1. Adams, T.D., F.G. Yanowitz, A.G. Fisher, J.D. Ridges, A.G. Nelson, A.D. Hagan, R.R. Williams, et al. Hereditability of cardiac size: An echocardiographic and electrocardiographic study of monozygotic and dizygotic twins. *Circulation* 71: 39–44, 1985.

2. Alpert, B.S., N.L. Flood, W.B. Strong, E.V. Dover, R.H. DuRant, A.M. Martin, and D.L. Booker. Responses to ergometer exercise in a healthy biracial population of children. *J. Pediatr.* 101:538–545, 1982.

3. Alyono, D., W.S. Ring, and M.R. Anderson. Left ventricular adaptation to volume overload from large aortocaval fistula. *Surgery* 96:360–367, 1970.

4. Armstrong, N., and J.R. Welsman. Cardiovascular response to submaximal treadmill running in 11- to 13-year-olds. *Acta Paediatr.* 91:125–131, 2002.
5. Asano, K., and K. Hirakoba. Respiratory and circulatory adaptation during prolonged exercise in 10–12-year-old children and adults. In: *Children and sport*. J. Ilmarinen and I. Valimaki (eds.). Berlin: Springer-Verlag, 1984, pp. 119–128.
6. Baraldi, E., D.M. Cooper, S. Zanconato, and Y. Arman. Heart rate recovery from 1 minute of exercise in children and adults. *Pediatr. Res.* 29:575–579, 1991.
7. Barber, G. Cardiovascular function. In: *Paediatric exercise science and medicine.* N. Armstrong and W. van Mechelen (eds.). Oxford: Oxford University Press, 2000, pp. 57–64.
8. Bar-Or, O. *Pediatric sports medicine for the practitioner.* New York: Springer-Verlag, 1983.
9. Bar-Or, O., R.J. Shephard, and C.L. Allen. Cardiac output of 10- to 13-year-old boys and girls during submaximal exercise. *J. Appl. Physiol.* 30:219–223, 1971.
10. Batterham, A.M., K.P. George, G. Whyte, S. Sharma, and W. McKenna. Scaling cardiac structural data by body dimensions: A review of theory, practice, and problems. *Int. J. Sports Med.* 20:495–502, 1999.
11. Bevegard, B.S., and J.T. Shepherd. Regulation of the circulation during exercise in man. *Physiol. Rev.* 47:178–213, 1967.
12. Bielen, E.C., R.H. Fagard, and A.K. Amery. Inheritance of acute cardiac changes during bicycle exercise: An echocardiographic study in twins. *Med. Sci. Sports Exerc.* 23: 1254–1259, 1991.
13. Bielen, E., R. Fagard, and A. Amery. Inheritance of heart structure and physical exercise capacity: A study of left ventricular structure and exercise capacity in 7-year-old twins. *Eur. Heart J.* 11:7–16, 1990.
14. Binak, K., T.J. Rega, and R.C. Christensen. Arteriovenous fistula: Hemodynamic effects of occlusion and exercise. *Am. Heart J.* 60:495–502, 1960.
15. Blimkie, C.J.R., D.A. Cunningham, and P.M. Nichol. Gas transport capacity and echocardiographically determined cardiac size in children. *J. Appl. Physiol.* 49:994–998, 1980.
16. Braunwald, E., S.J. Sarnoff, and W.N. Stainsby. Determinants of duration and mean rate of ventricular ejection. *Circ. Res.* 6:319–325, 1958.
17. Burch, G.E., and T.D. Giles. A critique of the cardiac index. *Am. Heart J.* 32:425–426, 1971.
18. Cassone, R., G. Germano, S. Dalmaso, R. Corretti, C. Astarita, P. Chieco, and V. Corsi. Evaluation of cardiac dynamics during isometric exercise in young female athletes: An echocardiographic study. *J. Sports Med.* 21:359–364, 1981.
19. Cheatham, C.C., A.D. Mahon, J.D. Brown, and D.R. Bolster. Cardiovascular responses during prolonged exercise at ventilatory threshold in boys and men. *Med. Sci. Sports Exerc.* 32:1080–1087, 2000.
20. Clark, D.A., J.S. Schroder, and R.B. Griepp. Cardiac transplantation in man: Review of first three years' experience. *Am. J. Med.* 54:563–576, 1973.
21. Clausen, J.P. Circulatory adjustments to dynamic exercise and effect of physical training in normal subjects and in patients with coronary artery disease. *Progr. Cardiovasc. Dis.* 17:459–495, 1976.
22. Colan, S.D., I.A. Parness, P.J. Spevak, and S.P. Sanders. Developmental modulation of myocardial mechanics: Age- and growth-related alterations in afterload and contractility. *J. Am. Coll. Cardiol.* 19:619–629, 1992.
23. Cyran, S.E., F.W. James, S. Daniels, W. Mays, R. Shukla, and S. Kaplan. Comparison of the cardiac output and stroke volume response to upright exercise in children with valvular and subvalvular aortic stenosis. *J. Am. Coll. Cardiol.* 11: 651–658, 1988.
24. Daniels, S.R., T.R. Kimball, J.A. Morrison, P. Khoury, and R.A. Meyer. Indexing left ventricular mass to account for differences in body size in children and adolescents without cardiovascular disease. *Am. J. Cardiol.* 76:669–701, 1995.
25. Daniels, S.R., T.R. Kimball, J.A. Morrison, P. Khoury, S. Witt, and R.A. Meyer. Effect of lean body mass, fat mass, blood pressure, and sexual maturation on left ventricular mass in children and adolescents. *Circulation* 92:3249–3254, 1995.
26. Delp, M.D. Control of skeletal muscle perfusion at the onset of dynamic exercise. *Med. Sci. Sports Exerc.* 31:1011–1018, 1999.
27. de Simone, G., R.B. Devereux, S.R. Daniels, G.F. Mureddu, M.J. Roman, T.R. Kimball, R. Greco, et al. Stroke volume and cardiac output in normotensive children and adults. *Circulation* 95:1837–1843, 1997.
28. Donald, D.E., and J.T. Shepherd. Sustained capacity for exercise in dogs after complete denervation. *Am. J. Cardiol.* 14:853–859, 1964.
29. Ensing, G.J., C.T. Heise, and D.J. Driscoll. Cardiovascular response to exercise after the Mustard operation for simple and complex transposition of the great vessels. *Am. J. Cardiol.* 62:617–622, 1988.
30. Eriksson, B.O., and G. Koch. Effect of physical training on hemodynamic response during submaximal and maximal exercise in 11–13-year-old boys. *Acta Physiol. Scand.* 87: 27–39, 1973.
31. Falk, B. Temperature regulation. In: *Paediatric exercise science and medicine.* N. Armstrong and W. van Mechelen (eds.). Oxford: Oxford University Press, 2000, pp. 221–239.
32. Gotshall, R.W., T.A. Bauer, and S.L. Fahrner. Cycling cadence alters exercise hemodynamics. *Int. J. Sports Med.* 17:17–21, 1996.
33. Green, D.J., G. O'Driscoll, B.A. Blanksby, and R.R. Taylor. Control of skeletal muscle blood flow during dynamic exercise. *Sports Med.* 21:119–146, 1996.
34. Gumbiner, C.H., and H.P. Gutgesell. Response to isometric exercise in children and young adults with aortic regurgitation. *Am. Heart J.* 106:540–547, 1983.

35. Gutgesell, H.P., and C.M. Rembold. Growth of the human heart relative to body surface area. *Am. J. Cardiol.* 65:662–668, 1990.

36. Guyton, A.C. Regulation of cardiac output. *N. Engl. J. Med.* 277:805–812, 1967.

37. Guyton, A.C., B.H. Douglas, and J.B. Langston. Instantaneous increase in circulatory pressure and cardiac output at onset of muscular activity. *Circ. Res.* 11:431–441, 1962.

38. Hakumaki, M.O.K. Seventy years of the Bainbridge reflex. *Acta Physiol. Scand.* 130:177–185, 1987.

39. Hill, A.V., C.H.N. Long, and H. Lupton. Muscular exercise, lactic acid, and the supply and utilization of oxygen: Parts VII–VIII. *Proc. Royal Soc. Br.* 97:155–176, 1924.

40. Holloszy, J.O., and E.F. Coyle. Adaptations of skeletal muscle to endurance exercise and their metabolic consequences. *J. Appl. Physiol.* 56:831–838, 1984.

41. Holmgren, A., and C.O. Ovenfors. Heart volume at rest and during muscular work in the supine and sitting positions. *Acta Med. Scand.* 167:267–276, 1960.

42. Ianuzzo, C.D., S. Blank, N. Hamilton, P. O'Brien, V. Chen, S. Brotherton, and T.A. Salerno. The relationship of myocardial chronotropism to the biochemical capacities of mammalian hearts. In: *Biochemistry of exercise VII.* A.W. Taylor, P.D. Gollnick, H.J. Green, C.D. Ianuzzo, E.G. Noble, G. Metivier, and J.R. Sutton (eds.). Champaign, IL: Human Kinetics, 1990, pp. 145–163.

43. Karpovich, V. Textbook fallacies regarding the development of the child's heart. (Originally published in *Research Quarterly,* vol. 8, 1937.) Reprinted in *Pediatr. Exerc. Sci.* 3: 278–282, 1991.

44. Katori, R. Normal cardiac output in relation to age and body size. *Tokohu J. Exp. Med.* 128:377–387, 1979.

45. Keul, J., H.-H. Dickhuth, G. Simon, and M. Lehmann. Effect of static and dynamic exercise on heart volume, contractility, and left ventricular dimensions. *Circ. Res.* 48(Suppl. I): I162–I170, 1981.

46. Kimball, T.R., W.A. Mays, P.R. Khoury, R. Mallie, and R.P. Claytor. Echocardiographic determination of left ventricular preload, afterload, and contractility during and after exercise. *J. Pediatr.* 122:S89–S94, 1993.

47. Kirby, B.J., N. Armstrong, and J.R. Welsman. Cardiac output response to submaximal exercise in adolescents [abstract]. *Med. Sci. Sports Exerc.* 29(Suppl.):S2, 1997.

48. Koch, G., J. Mobert, and E.-M. Oyen. Cardiovascular adjustment to supine versus seated posture in prepubertal boys. In: *Children and exercise XIX.* N. Armstrong, B. Kirby, and J. Welsman (eds.). London: Spon, 1997, pp. 424–428.

49. Krovetz, L.J., T.G. McLoughlin, M.B. Mitchell, and G.L. Scheibler. Hemodynamic findings in normal children. *Pediatr. Res.* 1:122–130, 1967.

50. Laird, W.P., D.D. Fixler, and F.D. Huffines. Cardiovascular response to isometric exercise in normal adolescents. *Circulation* 59:651–654, 1979.

51. Laughlin, M.H., and W.G. Schrage. Effects of muscle contraction on skeletal muscle blood flow: When is there a muscle pump? *Med. Sci. Sports Exerc.* 31:1027–1035, 1999.

52. Lind, A.R., S.H. Taylor, P.W. Humphries, B.M. Kennelly, and K.W. Donald. The circulatory effects of sustained voluntary muscle contraction. *Clin. Sci.* 27:229–244, 1964.

53. Linden, R.J. The size of the heart. *Cardioscience* 5:225–233, 1994.

54. Locke, J.E., S. Einsig, and J.H. Moller. Hemodynamic response to exercise in normal children. *Am. J. Cardiol.* 41: 1278–1285, 1978.

55. Mahon, A.D., C.S. Anderson, K.A. Shores, and M.J. Brooker. Heart rate recovery from submaximal exercise in boys and girls [abstract]. *Pediatr. Exerc. Sci.* 12:319, 2001.

56. Malcolm, D.D., T.L. Burns, L.T. Mahoney, and R.M. Lauer. Factors affecting left ventricular mass in childhood: The Muscatine Study. *Pediatrics* 92:703–709, 1993.

57. Matthews, K.A., and C.M. Stoney. Influence of sex and age on cardiovascular responses during stress. *Psychosom. Med.* 50:46–56, 1988.

58. Mitchell, J.H. Neural control of the circulation during exercise. *Med. Sci. Sports Exerc.* 22:141–154, 1990.

59. Nau, K.L., V.L. Katch, R.H. Beekman, and M. Dick. Acute intra-arterial blood pressure response to bench press weight lifting in children. *Pediatr. Exerc. Sci.* 2:37–45, 1990.

60. Nelson, R.R., F.L. Gobel, C.R. Jorgensen, K. Wang, Y. Wang, and H.L. Taylor. Hemodynamic predictors of myocardial consumption during static and dynamic exercise. *Circulation* 50:1179–1189, 1974.

61. Nidorf, S.M., M.H. Picard, M.O. Triulz, J.D. Thomas, J. Newell, M.E. King, and A.E. Weyman. New perspectives in the assessment of cardiac chamber dimensions during development and adulthood. *J. Am. Coll. Cardiol.* 19:983–988, 1992.

62. Noakes, T.D. Physiological models to understand exercise fatigue and the adaptations that predict or enhance athletic performance. *Scand. J. Med. Sci. Sports* 10:123–145, 2000.

63. Notarius, C.F., and S. Magder. Central venous pressure during exercise: Role of the muscle pump. *Can. J. Physiol. Pharmacol.* 74:647–651, 1996.

64. Nottin, S., V. Agnes, F. Stecken, L.-D. N'Guyen, F. Ounissi, A.-M. Lecoq, and P. Obert. Central and peripheral cardiovascular adaptations during maximal cycle exercise in boys and men. *Med. Sci. Sports Exerc.* 33:456–463, 2002.

65. Ohuchi, H., H. Suzuki, K. Yasuda, Y. Arakaki, S. Echigo, and T. Kamija. Heart rate recovery after exercise and cardiac nervous activity in children. *Pediatr. Res.* 47:329–335, 2000.

66. O'Leary, D.S. Autonomic mechanisms of muscle metaboreflex control of heart rate. *J. Appl. Physiol.* 74:1748–1754, 1993.

67. Paavolainen, L.M., A.T. Nummela, and H.K. Rusko. Neuromuscular characteristics and muscle power as determinants of 5 km running performance. *Med. Sci. Sports Exerc.* 31: 124–130, 1999.

68. Palmer, G.J., M.G. Ziegler, and C.R. Lake. Responses of norepinephrine and blood pressure to stress increase with age. *J. Gerontol.* 33:482–487, 1978.

69. Patterson, S.W., and E.H. Starling. On the mechanical factors which determine the output of the ventricles. *J. Physiol.* 48: 357–359, 1914.

70. Petrofsky, J.S., and C.A. Phillips. The physiology of static exercise. *Exerc. Sport Sci. Rev.* 14:1–44, 1986.

71. Pokan, R., S.P. von Duvillard, P. Hofman, G. Smekal, F.M. Fruhwald, R. Gasser, H. Tschan, et al. Change in left atrial and ventricular dimensions during and immediately after exercise. *Med. Sci. Sports Exerc.* 32:1719–1728, 2000.

72. Raven, P.B., and G.H.J. Stevens. Cardiovascular function and prolonged exercise. In: *Perspectives in exercise science and sports medicine: Vol. 1. Prolonged exercise.* D.R. Lamb and R. Murray (eds.). Indianapolis: Benchmark Press, 1988, pp. 43–74.

73. Riopel, D.A., A.B. Taylor, and A.R. Hohn. Blood pressure, heart rate, pressure–rate product and electrocardiographic changes in healthy children during treadmill exercise. *Am. J. Cardiol.* 44:697–704, 1979.

74. Ross, J., J.W. Linhart, and E. Braunwald. Effects of changing heart rate in man by electrical stimulation of the right atrium. *Circulation* 32:549–558, 1965.

75. Rowell, L.B., and D.S. O'Leary. Reflex control of the circulation during exercise: Chemoreflexes and mechanoreflexes. *J. Appl. Physiol.* 69:407–418, 1990.

76. Rowell, L.B., D.S. O'Leary, and D.L. Kellogg. Integration of cardiovascular control systems in dynamic exercise. In: *Handbook of physiology: Regulation and integration of multiple systems.* L.B. Rowell and J.T. Shepherd (eds.). Bethesda, MD: American Physiological Society, 1996, pp. 771–781.

77. Rowland, T.W. Cardiovascular function. In: *Paediatric exercise science and medicine.* N. Armstrong and W. van Mechelen (eds.). Oxford: Oxford University Press, 2000, pp. 163–171.

78. Rowland, T.W. The circulatory response to exercise: Role of the peripheral pump. *Int. J. Sports Med.* 22:558–565, 2001.

79. Rowland, T.W. Post-exercise echocardiography in prepubertal boys. *Med. Sci. Sports Exerc.* 19:393–397, 1987.

80. Rowland, T., and J.W. Blum. Cardiac dynamics during upright cycle exercise in boys. *Am. J. Hum. Biol.* 12:749–757, 2000.

81. Rowland, T.W., and L.N. Cunningham. Heart rate deceleration during treadmill exercise in children [abstract]. *Pediatr. Exerc. Sci.* 5:463, 1993.

82. Rowland, T., A. Garrison, and A. DeIulio. Circulatory responses to progressive exercise: Insights from positional differences. *Int. J. Sports Med.,* 24:512-517, 2003.

83. Rowland, T., D. Goff, L. Martel, and L. Ferrone. Influence of cardiac functional capacity on gender differences in maximal oxygen uptake in children. *Chest* 117:629–635, 2000.

84. Rowland, T., D. Goff, L. Martel, L. Ferrone, and G. Kline. Normalization of maximal cardiovascular variables for body size in premenarcheal girls. *Pediatr. Cardiol.* 21:429–432, 2000.

85. Rowland, T., and R. Lisowski. Determinants of diastolic cardiac filling during exercise. *J. Sports Med. Phys. Fitness,* 43:380-385, 2003.

86. Rowland, T., and R. Lisowski. Hemodynamic responses to increasing cycling cadence in 11-year-old boys: Role of the skeletal muscle pump. *Int. J. Sports Med.* 22:405–409, 2001.

87. Rowland, T., E. Mannie, and L. Gawle. Dynamics of left ventricular diastolic filling during exercise: A Doppler echocardiographic study of 10–14-year-old boys. *Chest* 120:145–150, 2001.

88. Rowland, T., and P. Obert. Doppler echocardiography for the estimation of cardiac output with exercise. *Sports Med.* 32:973–986, 2002.

89. Rowland, T.W., and B. Popowski. Comparison of bioimpedance and Doppler cardiac output during exercise in children [abstract]. *Pediatr. Exerc. Sci.* 9:188–189, 1996.

90. Rowland, T., J. Potts, G. Sandor, D. Goff, and L. Ferrone. Cardiac responses to progressive exercise in normal children: A synthesis. *Med. Sci. Sports Exerc.* 31:253–259, 2000.

91. Rowland, T.W., and T.A. Rimany. Physiological responses to prolonged exercise in premenarcheal and adult females. *Pediatr. Exerc. Sci.* 7:183–191, 1995.

92. Rushmer, R.F., and O.A. Smith. Cardiac control. *Physiol. Rev.* 39:41–68, 1959.

93. Saltin, B. Cardiovascular and pulmonary adaptation to physical activity. In: *Exercise, fitness, and health: A consensus of current knowledge.* C. Bouchard, R.J. Shephard, T. Stephens, J.R. Sutton, and B.D. McPherson (eds.). Champaign, IL: Human Kinetics, 1990, pp. 187–203.

94. Schmidt-Nielsen, K. *Scaling: Why is animal size so important?* Cambridge: Cambridge University Press, 1984.

95. Senior, D.G., K.L. Waters, M. Cassidy, T. Crucitti, H. Shapiro, and A.L. Riba. Effect of aerobic training on left ventricular diastolic filling. *Conn. Med.* 53:67–70, 1989.

96. Smith, D.L., B.E. Kocher, A.L. Kolesnikoff, and T.W. Rowland. Cardiovascular responses to isometric contractions in girls and young women [abstract]. *Med. Sci. Sports Exerc.* 32: S95, 2000.

97. Smith, E.E., A.C. Guyton, and R.D. Manning. Integrated mechanisms of cardiovascular response and control during exercise in the normal human. *Progr. Cardiovasc. Dis.* 18: 421–443, 1976.

98. Sproul, A., and E. Simpson. Stroke volume and related hemodynamic data in normal children. *Pediatrics* 33: 912–918, 1964.

99. Starnes, J.W. Myocardial metabolism during exercise. In: *Cardiovascular response to exercise.* G.F. Fletcher (ed.). Mount Kisco, NY: Futura, 1994, pp. 3–13.

100. Stead, E.A., and J.V. Warren. Cardiac output in man. *Arch. Intern. Med.* 80:237–248, 1954.

101. Strong, W.B., M.D. Miller, M. Striplin, and M. Salehbhai. Blood pressure response to isometric and dynamic exercise in healthy black children. *Am. J. Dis. Child.* 132:587–591, 1978.

102. Takaishi, T., T. Sugiura, K. Katayama, Y. Sato, N. Shima, T. Yamamoto, and T. Moritani. Changes in blood volume and oxygenation level in a working muscle during a crank cycle. *Med. Sci. Sports Exerc.* 33:520–528, 2002.

103. Tschakovsky, M.E., J.K. Shoemaker, and R.L. Hughson. Vasodilation and muscle pump contribution to immediate exercise hyperemia. *Am. J. Physiol.* 271:H1697–H1701, 1996.

104. Turley, K.R. Cardiovascular responses to exercise in children. *Sports Med.* 24:241–257, 1997.

105. Turley, K.R., D.E. Martin, E.D. Marvin, and K.S. Cowley. Heart rate and blood pressure responses to static handgrip exercise of different intensities: Reliability and adult versus child differences. *Pediatr. Exerc. Sci.* 14:45–55, 2002.

106. Turley, K.R., and J.H. Wilmore. Cardiovascular responses to submaximal exercise in 7- to 9-year-old boys and girls. *Med. Sci. Sports Exerc.* 29:824–832, 1997.

107. Turley, K.R., and J.H. Wilmore. Ratio scaling of submaximal cardiovascular data: Is it appropriate? [abstract]. *Med. Sci. Sports Exerc.* 30(Suppl.):S242, 1998.

108. Udelson, J.F., S.L. Bacharach, P.O. Cannon, and R.O. Bonow. Minimum left ventricular pressure during beta-adrenergic stimulation in human subjects. *Circulation* 82:1174–1182, 1990.

109. Verhaaren, H.A., R.M. Schieken, P. Schwartz, M. Mosteller, D. Mathys, H. Maes, G. Beunen, et al. Cardiovascular reactivity in isometric exercise and mental arithmetic in children. *J. Appl. Physiol.* 76:146–150, 1994.

110. Vinet, A., S. Mandigout, S. Nottin, L.-D. Nguyen, A.-M. Lecoq, D. Courteix, and P. Obert. Influence of body composition, hemoglobin concentration, cardiac size and function on gender differences in maximal oxygen uptake in prepubertal children. *Chest,* 124:1494-1499, 2003.

111. Walloe, L., and J. Wesche. The course and magnitude of blood flow changes in the human quadriceps muscles during and following rhythmic exercise. *J. Physiol.* 405: 257–273, 1988.

112. Warburton, D.E.R., M.J.F. Haykowski, and H.A. Quinney. Reliability and validity of measures of cardiac output during incremental to maximal aerobic exercise: I. Conventional techniques. *Sports Med.* 27:23–41, 1999.

113. Warburton, D.E.R., M.J.F. Haykowski, and H.A. Quinney. Reliability and validity of measures of cardiac output during incremental to maximal exercise: II. Novel techniques and new advances. *Sports Med.* 27:241–260, 1999.

114. Washington, R.L., J.C. van Gundy, C. Cohen, H.M. Sondheimer, and R.R. Wolfe. Normal aerobic and anaerobic exercise data for North American school age children. *J. Pediatr.* 112:223–233, 1988.

115. Williamson, J.W., A.C.L. Nobrega, and P.K. Winchester. Instantaneous heart rate increase with dynamic exercise: Central command and muscle heart reflex contribution. *J. Appl. Physiol.* 78:1273–1279, 1995.

Chapter 7

1. Andersen, K.L., V. Seliger, J. Rutenfranz, and S. Messel. Physical performance capacity of children in Norway: Part III. Respiratory responses to graded exercise loadings—population parameters in a rural community. *Eur. J. Appl. Physiol.* 33:265–274, 1974.

2. Armon, Y., D.M. Cooper, and S. Zanconato. Maturation of ventilatory responses to 1-minute exercise. *Pediatr. Res.* 29: 362–368, 1991.

3. Armstrong, N., B.J. Kirby, A.M. McManus, and J.R. Welsman. Prepubescents' ventilatory responses to exercise with reference to sex and body size. *Chest* 112:1554–1560, 1997.

4. Asano, K., and K. Hirakoba. Respiratory and circulatory adaptation during prolonged exercise in 10–12-year-old children and in adults. In: *Children and sport.* J. Ilmarinen and I. Valimaki (eds.). Berlin: Springer-Verlag, 1984, pp. 119–128.

5. Asmussen, E. Control of ventilation in exercise. *Exerc. Sport Sci. Rev.* 11:24–54, 1983.

6. Asmussen, E., N.H. Secher, and E.A. Andersen. Heart rate and ventilation frequency in dimension-independent variables. *Eur. J. Appl. Physiol.* 46:379–386, 1981.

7. Åstrand, P.O. *Experimental studies of physical working capacity in relation to sex and age.* Copenhagen: Munksgaard, 1952.

8. Boileau, R.A., A. Bonen, V.H. Heyward, and B.H. Massey. Maximal aerobic capacity on the treadmill and bicycle ergometer of boys 11–14 years of age. *J. Sports Med.* 17: 153–162, 1977.

9. Boule, M., C. Gaultier, and F. Girard. Breathing pattern during exercise in untrained children. *Resp. Physiol.* 75: 225–234, 1989.

10. Byrne-Quinn, E., J.V. Weil, I.E. Sodal, G.F. Filley, and R.F. Grover. Ventilatory control in the athlete. *J. Appl. Physiol.* 30:91–98, 1971.

11. Cassels, D.E., and M. Morse. *Cardiopulmonary data for children and young adults.* Springfield, IL: Charles C Thomas, 1962, pp. 52–57.

12. Cooper, D.M., M.R. Kaplan, L. Baumgarten, D. Weiler-Ravell, B.J. Whipp, and K. Wasserman. Coupling of ventilation and CO_2 production during exercise in children. *Pediatr. Res.* 21:568–572, 1987.

13. Dempsey, J.A., E. Aaron, and B.J. Martin. Pulmonary function and prolonged exercise. In: *Perspectives in exercise science and sports medicine: Vol. 1. Prolonged exercise.* D.L. Lamb and R. Murray (eds.). Indianapolis: Benchmark Press, 1988, pp. 75–124.

14. Dempsey, J.A., S.K. Powers, and N. Gledhill. Discussion: Cardiovascular and pulmonary adaptation to physical activity. In: *Exercise, fitness, and health: A consensus of current knowledge.* C. Bouchard, R.J. Shephard, T. Stephens, J.R. Sutton, and B.D. McPherson (eds.). Champaign, IL: Human Kinetics, 1990, pp. 205–215.

15. Eriksson, B.O., G. Grimby, and B. Saltin. Cardiac output and arterial blood gases during exercise in pubertal boys. *J. Appl. Physiol.* 31:348–352, 1971.

16. Forster, H.V. Exercise hyperpnea: Where do we go from here? *Exerc. Sport Sci. Rev.* 28:133–137, 2000.

17. Gadhoke, S., and N.L. Jones. The responses to exercise in boys aged 9–15 years. *Clin. Sci.* 37:789–801, 1969.

18. Gaultier, C., L. Perret, M. Boule, A. Buvry, and F. Girard. Occlusion pressure and breathing pattern in healthy children. *Respir. Physiol.* 46:71–80, 1981.

19. Godfrey, S. *Exercise testing in children*. London: Saunders, 1974.

20. Gratas-Delamarche, A., J. Mercier, M. Ramonatxo, J. Dassonville, and C. Prefaut. Ventilatory response of prepubertal boys and adults to carbon dioxide at rest and during exercise. *Eur. J. Appl. Physiol.* 66:25–30, 1993.

21. Harms, C.A., S.R. McLaran, G.A. Nickele, D.F. Pegelow, W.B. Nelson, and J.A. Dempsey. Effect of exercise-induced arterial O_2 saturation on $\dot{V}O_2$max in women. *Med. Sci. Sports Exerc.* 32:1101–1108, 2000.

22. Harms, C.A., S.R. McLaran, G.A. Nickele, D.F. Pegelow, W.B. Nelson, and J.A. Dempsey. Exercise-induced arterial hypoxaemia in healthy young women. *J. Physiol.* 5–7: 619–628, 1998.

23. Hebestreit, H., F. Meyer, H. Htay, G.J. Heigenhauser, and O. Bar-Or. Plasma metabolites, volume and electrolytes following 30-s high intensity exercise in boys and men. *Eur. J. Appl. Physiol.* 72:563–569, 1996.

24. Johnson, R.L., H.F. Taylor, and W.H. Lawson. Maximal diffusing capacity of the lung for carbon monoxide. *J. Clin. Invest.* 44:349–355, 1965.

25. Kawakami, Y., T. Yoshikawa, A. Shida, Y. Asanuma, and M. Murao. Control of breathing in young twins. *J. Appl. Physiol.* 53:537–542, 1982.

26. Lanteri, C.J., and P.D. Sly. Changes in respiratory mechanics with age. *J. Appl. Physiol.* 74:369–378, 1993.

27. Laursen, P.B., G.C.K. Tsang, G.J. Smith, M.V. van Velzen, B.B. Ignatova, E.B. Sprules, K.S. Chu, et al. Incidence of exercise-induced arterial hypoxemia in prepubescent females. *Pediatr. Pulm.* 34:37–41, 2002.

28. Lyons, H.A., and R.W. Tanner. Total lung volume and its subdivisions in children: Normal standards. *J. Appl. Physiol.* 17:601–604, 1962.

29. Macek, M., and J. Vavra. Anaerobic threshold in children. In: *Children and exercise XI*. R.A. Binkhorst, H.C.G. Kemper, and W.H.M. Saris (eds.). Champaign, IL: Human Kinetics, 1985, pp. 110–113.

30. Mahler, D.A., E.D. Moritz, and J. Loke. Ventilatory responses at rest and during exercise in marathon runners. *J. Appl. Physiol.* 52:388–392, 1982.

31. McGurk, S.P., B.A. Blanksby, and M.J. Anderson. The relationship between carbon dioxide sensitivity and sprint or endurance performance in young swimmers. *Br. J. Sports Med.* 29:129–133, 1995.

32. Mercier, J., A. Varrag, M. Romaonatxo, B. Mecier, and C. Prefaut. Influence of anthropometric characteristics on changes in maximal exercise ventilation and breathing pattern during growth in boys. *Eur. J. Appl. Physiol.* 63: 235–241, 1991.

33. Morse, M., F.W. Schultz, and D.E. Cassells. Relation of age to physiological responses of the older boy (10–17 years) to exercise. *J. Appl. Physiol.* 1:683–709, 1949.

34. Nagano, Y., R. Baba, K. Kuraishi, T. Yasuda, M. Ikoma, K. Nishibata, M. Yokota, et al. Ventilatory control during exercise in normal children. *Pediatr. Res.* 43:704–707, 1998.

35. Ohkuwa, T., N. Fuisuka, T. Utuno, and M. Miyamura. Ventilatory response to hypercapnia in sprint and long-distance swimmers. *Eur. J. Appl. Physiol.* 43:235–241, 1980.

36. Ohuchi, H., K. Yoshihiro, H. Tasato, Y. Arakaki, and T. Kamiya. Ventilatory response and arterial blood gases during exercise in children. *Pediatr. Res.* 45:389–396, 1999.

37. Paterson, D.H., D.A. Cunningham, and A. Donnen. The effect of different treadmill speeds on the variability of $\dot{V}O_2$max in children. *Eur. J. Appl. Physiol.* 47:113–122, 1981.

38. Radford, E.P. Ventilation standards for use in artificial respiration. *J. Appl. Physiol.* 5:451–460, 1954.

39. Ratel, S., P. Duché, A. Hennegrave, E. van Praagh, and M. Bedu. Acid-base balance during repeated cycling sprints in boys and men. *J. Appl. Physiol.* 92:479–485, 2002.

40. Robinson, S. Experimental studies of physical fitness in relation to age. *Arbeitsphysiologie* 10:318–323, 1938.

41. Rowland, T.W., and L.N. Cunningham. Development of ventilatory responses to exercise in normal white children. *Chest* 11:327–332, 1997.

42. Rowland, T.W., and G.M. Green. The influence of biological maturation and aerobic fitness on ventilatory responses to treadmill exercise. In: *Exercise physiology: Current selected research*. C.O. Dotson and J.H. Humphrey (eds.). New York: AMS Press, 1990, pp. 51–59.

43. Rowland, T.W., and T.A. Rimany. Physiological responses to prolonged exercise in premenarcheal and adult females. *Pediatr. Exerc. Sci.* 7:183–191, 1995.

44. Schmidt-Nielsen, K. *Scaling: Why is animal size so important?* Cambridge: Cambridge University Press, 1984, pp. 99–103.

45. Shephard, R.J., C. Allen, O. Bar-Or, C.T.M. Davies, S. Degre, R. Hedman, and K. Ishi. The working capacity of Toronto schoolchildren, part II. *Can. Med. Assoc. J.* 100: 705–714, 1969.

46. Shephard, R.J., and O. Bar-Or. Alveolar ventilation in near maximum exercise: Data on pre-adolescent children and young adults. *Med. Sci. Sports* 2:83–92, 1970.

47. Taussig, L.M., K. Cota, and W. Kaltenborg. Different mechanical properties of the lung in boys and girls. *Am. Rev. Resp. Dis.* 123:640–643, 1981.

48. Welsman, J., S. Fawkner, and N. Armstrong. Respiratory response to non-steady-state exercise in children and adults [abstract]. *Pediatr. Exerc. Sci.* 13:263–264, 2001.

49. Whipp, B.J., and S.A. Ward. Respiratory responses of athletes to exercise. In: *Oxford textbook of sports medicine*. M. Harries, C. Williams, W.D. Stanish, and L.J. Micheli (eds.). Oxford: Oxford University Press, 1994, pp. 13–26.

50. Zanconato, S., D.M. Cooper, T.J. Barstow, and E. Landaw. $^{13}CO_2$ washout dynamics during intermittent exercise in children and adults. *J. Appl. Physiol.* 73:2476–2482, 1992.

Chapter 8

1. Ariens, G.A.M., W. van Mechelen, H.C.G. Kemper, and J.W.R. Twisk. The longitudinal development of running

economy in males and females aged between 13 and 27 years: The Amsterdam Growth and Health Study. *Eur. J. Appl. Physiol.* 76:214–220, 1997.

2. Armstrong, N., J.R. Welsman, and B.J. Kirby. Submaximal exercise and maturation in 12-year-olds. *J. Sport Sci.* 17: 107–114, 1999.
3. Åstrand, P.O. *Experimental studies of physical working capacity in relation to sex and age.* Copenhagen: Munksgaard, 1952.
4. Ayub, B.V., and O. Bar-Or. Relative contribution of body mass and adiposity in energy cost of walking in children [abstract]. *Pediatr. Exerc. Sci.* 11:79–80, 1999.
5. Basmajian, J. Motor learning and control: A working hypothesis. *Arch. Phys. Med. Rehabil.* 58:38–41, 1977.
6. Bosco, C., and P.V. Komi. Influence of aging on the mechanical behavior of leg extensor muscles. *Eur. J. Appl. Physiol.* 45:200–219, 1980.
7. Bowen, T.R., S.R. Cooley, P.W. Castagno, F. Miller, and J. Richards. A method of normalization of oxygen cost and consumption in normal children while walking. *J. Pediatr. Orthop.* 18:589–593, 1998.
8. Calder, W.A. *Size, function, and life history.* Cambridge, MA: Harvard University Press, 1984.
9. Cooper, D.M., D. Weiler-Ravell, B.J. Whipp, and K. Wasserman. Aerobic parameters of exercise as a function of growth in children. *J. Appl. Physiol.* 56:628–634, 1984.
10. Cunningham, L.N. Relationship of running economy, ventilatory threshold, and maximal oxygen consumption to running performance in high school females. *Res. Q. Exerc. Sport* 61:369–374, 1990.
11. Cureton, K.J., M.A. Sloniger, D.M. Black, W.P. McCormack, and D.A. Rowe. Metabolic determinants of the age related improvement in one-mile run/walk performance in youth. *Med. Sci. Sports Exerc.* 29:259–267, 1997.
12. Davies, C.T.M. Metabolic cost of exercise and physical performance in children with some observations on external loading. *Eur. J. Appl. Physiol.* 45:95–102, 1980.
13. Davies, M.J., M.T. Mahar, and L.N. Cunningham. Running economy: Comparison of body mass adjustment methods. *Res. Q. Exerc. Sport* 68:177–181, 1997.
14. Ebbeling, C.J., J. Hamill, P.S. Freedson, and T.W. Rowland. An examination of efficiency during walking in children and adults. *Pediatr. Exerc. Sci.* 4:36–49, 1992.
15. Frost, G., J. Dowling, K. Dyson, and O. Bar-Or. Cocontraction in three age groups of children during treadmill locomotion. *J. Electromyogr. Kinesiol.* 7:179–186, 1997.
16. Grossner, C.M., E.M. Johnson, and M.E. Cabrera. Economy and efficiency in female adolescents and young adults matched for height and weight [abstract]. *Pediatr. Exerc. Sci.* 15:103–104, 2003.
17. Hamar, D., L. Komadel, and O. Kuthanova. Mechanical efficiency of muscular work and economy of walking and running. In: *Children and exercise XIII.* S. Oseid and K.-H. Carlson (eds.). Champaign, IL: Human Kinetics, 1989, pp. 39–45.
18. Holliday, M.A., D. Potter, A. Jarrah, and S. Bearg. The relation of metabolic rate to body weight and organ size. *Pediatr. Res.* 1:185–195, 1967.
19. Knutzen, K.M., and L. Martin. Using biomechanics to explore children's movement. *Pediatr. Exerc. Sci.* 14:222–247, 2002.
20. Krahenbuhl, G.S., D.W. Morgan, and R.P. Pangrazi. Longitudinal changes in distance running performance of young males. *Int. J. Sports Med.* 10:92–96, 1989.
21. Krahenbuhl, G.S., and R. Pangrazi. Characteristics associated with running performance in young boys. *Med. Sci. Sports Exerc.* 15:486–490, 1983.
22. Krahenbuhl, G.S., R.P. Pangrazi, and E.A. Chomokas. Aerobic responses of young boys to submaximal running. *Res. Q. Exerc. Sport* 50:413–421, 1979.
23. Kram, R. Muscular force or work: What determines the metabolic energy cost of running? *Exerc. Sport Sci. Rev.* 28: 138–143, 2000.
24. Kram, R., and C.R. Taylor. Energetics of running: A new perspective. *Nature* 346:265–267, 1990.
25. Maffeis, C., Y. Schutz, F. Schena, M. Zafanello, and L. Pinelli. Energy expenditure during walking and running in obese and nonobese prepubertal children. *J. Pediatr.* 123:193–199, 1993.
26. Maliszewski, A.F., and P.S. Freedson. Is running economy different between adults and children? *Pediatr. Exerc. Sci.* 8: 351–360, 1996.
27. McDougall, J.D., P.D. Roche, O. Bar-Or, and J.R. Moroz. Maximal aerobic capacity of Canadian school children: Prediction based on age-related oxygen cost of running. *Int. J. Sports Med.* 4:194–198, 1983.
28. McMurray, R.G., J.S. Harrell, S.I. Bangdiwala, S. Deng, and C. Baggett. Factors contributing to the energy expenditure of youth during cycling and running. *Pediatr. Exerc. Sci.* 11: 122–128, 1999.
29. Minetti, A.E., L.P. Ardigo, and F. Saibene. Mechanical determinants of the minimum cost of gradient running in humans. *J. Exp. Biol.* 195:211–225, 1994.
30. Morgan, D.W. Economy of locomotion. In: *Paediatric exercise science and medicine.* N. Armstrong and W. van Mechelen (eds.). Oxford: Oxford University Press, 2000, pp. 183–190.
31. Morgan, D.W., W. Tseh, W. Caputo, J.L. Craig, D.J. Keefer, and P.E. Martin. Effect of step length manipulation on the aerobic demand of walking in young children [abstract]. *Pediatr. Exerc. Sci.* 11:271–272, 1999.
32. Morgan, D.W., W. Tseh, W. Caputo, J.L. Craig, D.J. Keefer, and P.E. Martin. Sex differences in running economy of younger children. *Pediatr. Exerc. Sci.* 11:122–128, 1999.
33. Moritani, T., L. Oddson, A. Thorstensson, and P.O. Åstrand. Neural and biomechanical differences between men and young boys during a variety of motor tasks. *Acta Physiol. Scand.* 137:347–355, 1989.
34. Perkins, C.D., K.M. Allor, L.J. Sam, and J.M. Pivarnik. Treadmill economy in girls and women matched for height and weight [abstract]. *Med. Sci. Sports Exerc.* 32:S71, 2000.

35. Petray, C.K., and G.S. Krahenbuhl. Running training, instruction on running technique, and running economy in 10-year-old males. *Res. Q. Exerc. Sport* 56:251–255, 1985.

36. Pivarnik, J.M., and N.W. Sherman. Responses of aerobically fit men and women to uphill/downhill walking and slow jogging. *Med. Sci. Sports Exerc.* 22:127–130, 1990.

37. Rall, J.A. Energetic aspects of skeletal muscle contraction: Implications of fiber types. *Exerc. Sport Sci. Rev.* 13:313–374, 1985.

38. Roberts, T.J., M.S. Chen, and C.R. Taylor. Energetics of bipedal running: I. Metabolic cost of generating force. *J. Exp. Biol.* 201:2745–2751, 1998.

39. Robinson, S. Experimental studies of physical fitness in relation to age. *Arbeitsphysiologie* 10:251–253, 1938.

40. Rogers, D.M., K.R. Turley, K.I. Kujawa, K.M. Harper, and J.H. Wilmore. Allometric scaling factors for oxygen uptake during exercise in children. *Pediatr. Exerc. Sci.* 7: 12–25, 1995.

41. Rowland, T.W., J.A. Auchinachie, T.J. Keenan, and G.M. Green. Physiological responses to treadmill running in adult and prepubertal males. *Int. J. Sports Med.* 8: 292–297, 1987.

42. Rowland, T.W., J.A. Auchinachie, T.J. Keenan, and G.M. Green. Submaximal running economy and treadmill performance in prepubertal boys. *Int. J. Sports Med.* 9:187–194, 1988.

43. Rowland, T.W., and A. Boyajian. Aerobic response to endurance exercise training in children. *Pediatrics* 96:654–658, 1995.

44. Rowland, T.W., L.N. Cunningham, L. Martel, P. Vanderburgh, T. Manos, and N. Charkoudian. Gender effects on submaximal energy expenditure in children. *Int. J. Sports Med.* 18:420–425, 1997.

45. Rowland, T.W., J.S. Staab, V.B. Unnithan, J.M. Rambusch, and S.F. Siconolfi. Mechanical efficiency during cycling in prepubertal and adult males. *Int. J. Sports Med.* 11:452–455, 1990.

46. Schepens, B., P.A. Willems, and G.A. Cavagna. The mechanics of running in children. *J. Physiol.* 509:927–940, 1998.

47. Schmidt-Nielsen, K. *Scaling: Why is animal size so important?* Cambridge: Cambridge University Press, 1984, pp. 165–181.

48. Sjodin, B., and J. Svedenhag. Oxygen uptake during running as related to body mass in circumpubertal boys: A longitudinal study. *Eur. J. Appl. Physiol.* 65:150–157, 1992.

49. Taylor, C.R., N.C. Heglund, and G.M.O. Maloiy. Energetics and mechanics of terrestrial locomotion: I. Metabolic energy consumption as a function of speed and body size in birds and mammals. *J. Exp. Biol.* 97:1–21, 1982.

50. Thompson, E.M. *A study of the energy expenditure and mechanical efficiency of young girls and adult women.* Doctoral dissertation, Columbia University Press, New York, 1940.

51. Thorstensson, A. Effects of moderate external loading on the aerobic demand of submaximal running in men and 10-year-old boys. *Eur. J. Appl. Physiol.* 55:569–574, 1986.

52. Unnithan, V.B., and R.G. Eston. Stride frequency and submaximal treadmill running economy in adults and children. *Pediatr. Exerc. Sci.* 2:149–155, 1990.

53. Unnithan, V.B., J.A. Timmons, R.T. Brogan, J.Y. Paton, and T.W. Rowland. Submaximal running economy in run-trained prepubertal boys. *J. Sports Med. Phys. Fitness* 36: 16–23, 1996.

54. van Mechelen, W., H.C.G. Kemper, and J. Twisk. The development of running economy from 13–27 years of age [abstract]. *Med. Sci. Sports Exerc.* 26:S205, 1994.

55. Walker, J.L., T.D. Murray, A.S. Jackson, J.R. Morrow, and T.J. Michaud. The energy cost of horizontal walking and running in adolescents. *Med. Sci. Sports Exerc.* 31:311–322, 1999.

56. Waters, R.L., H.J. Hislop, L. Thomas, and J. Campbell. Energy cost of walking in normal children and teenagers. *Dev. Med. Child Neurol.* 25:184–188, 1983.

57. Webber, L.M., W.C. Byrnes, T.W. Rowland, and V.L. Foster. Serum creatine kinase activity and delayed onset muscle soreness in prepubescent children: A preliminary study. *Pediatr. Exerc. Sci.* 1:351–359, 1989.

58. Welsman, J.R., and N. Armstrong. Longitudinal changes in submaximal oxygen uptake in 11- to 13-year-olds. *J. Sport Sci.* 18:183–189, 2000.

59. Wickstrom, R.L. *Fundamental motor patterns.* 3rd edition. Philadelphia: Lea & Febiger, 1983.

60. Workman, J.M., and B.W. Armstrong. Oxygen cost of treadmill walking. *J. Appl. Physiol.* 18:798–803, 1963.

61. Zanconato, S., D.M. Cooper, and Y. Armon. Oxygen cost and oxygen uptake dynamics and recovery with one minute of exercise in children and adults. *J. Appl. Physiol.* 71:993–998, 1991.

Chapter 9

1. Almuzaini, K.S. Optimal peak and mean power on the Wingate test: Relationship with sprint ability, vertical jump, and standing long jump in boys. *Pediatr. Exerc. Sci.* 12:349–359, 2000.

2. American Alliance for Health, Physical Education, Recreation and Dance. *Youth fitness testing manual.* Washington, DC: Author, 1980.

3. Armstrong, N., J.R. Welsman, and M.Y.H. Chia. Short term power output in relation to growth and maturation. *Br. J. Sports Med.* 35:118–124, 2001.

4. Asmussen, E. Growth in muscular strength and power. In: *Physical activity, human growth and development.* G.L. Rarick (ed.). New York: Academic Press, 1973, pp. 60–79.

5. Asmussen, E., and K.R. Heeboll-Nielsen. A dimensional analysis of physical performance and growth in boys. *J. Appl. Physiol.* 7:593–603, 1955.

6. Bailey, R.C., J. Olson, S.L. Pepper, J. Porszasz, T.J. Barstow, and D.M. Cooper. The level and tempo of children's physical activities: An observational study. *Med. Sci. Sports Exerc.* 27: 1033–1041, 1995.

7. Baranowski, T., P. Hooks, Y. Tsong, C. Cieslik, and P.R. Nader. Aerobic physical activity among third- to sixth-grade children. *J. Dev. Behav. Pediatr.* 8:203–206, 1987.

8. Bar-Or, O. Anaerobic performance. In: *Measurement in pediatric exercise science.* D. Docherty (ed.). Champaign, IL: Human Kinetics, 1996, pp. 161–181.

9. Bar-Or, O. *Pediatric sports medicine for the practitioner.* New York: Springer-Verlag, 1983, pp. 311–314.

10. Bar-Or, O. The Wingate anaerobic test: An update on methodology, reliability, and validity. *Sports Med.* 4:381–394, 1987.

11. Bar-Or, O., V. Unnithan, and C. Illescas. Physiologic considerations in age group swimming. In: *Medicine and science in aquatic sports.* M. Miyashita, Y. Mutoh, and A.B. Richardson (eds.). Basel: Karger, 1994, pp. 199–205.

12. Berg, A., S.S. Kim, and J. Keul. Skeletal muscle enzyme activities in healthy young subjects. *Int. J. Sports Med.* 7: 236–239, 1986.

13. Beunen, G., M. Thomis, H. Maes, R. Loos, M. Peeters, and R. Vlietinck. Genetics of isometric strength and power [abstract]. *Pediatr. Exerc. Sci.* 13:263, 2001.

14. Blimkie, C.J.R. Age- and sex-associated variation in strength during childhood: Anthropometric, morphological, neurological, biomechanical, endocrinologic, genetic, and physical activity correlates. In: *Perspectives in exercise science and sports medicine: Vol. 2. Youth, exercise, and sport.* C.V. Gisolfi and D.R. Lamb (eds.). Indianapolis: Benchmark Press, 1989, pp. 99–164.

15. Blimkie, C.J.R., P. Roache, J.T. Hay, and O. Bar-Or. Anaerobic power of arms in teenage boys and girls: Relationship to lean tissue. *Eur. J. Appl. Physiol.* 57:677–683, 1988.

16. Bobbert, M.F., and A.J. van Soest. Why do people jump the way they do? *Exerc. Sport Sci. Rev.* 29:95–102, 2001.

17. Chelly, S.M., and C. Denis. Leg power and hopping stiffness: Relationship with sprint running performance. *Med. Sci. Sports Exerc.* 33:326–333, 2001.

18. Chia, M., N. Armstrong, and D. Childs. The assessment of children's anaerobic performance using modifications of the Wingate anaerobic test. *Pediatr. Exerc. Sci.* 9:80–90, 1997.

19. Costill, D.L., J. Daniels, W. Evans, W. Fink, and G. Krahenbuhl. Skeletal muscle enzymes and fiber composition in male and female track athletes. *J. Appl. Physiol.* 40:149–154, 1976.

20. Counil, F.-P., A. Varray, C. Karila, M. Hayot, M. Voisin, and C. Prefaut. Wingate test performance in children with asthma: Aerobic or anaerobic limitation? *Med. Sci. Sports Exerc.* 29: 430–435, 1997.

21. Cumming, G.R. Correlation of athletic performance and aerobic power in 12–17-year-old children with bone age, calf muscle, total body potassium, heart volume, and two indices of anaerobic power. In: *Pediatric work physiology.* O. Bar-Or (ed.). Natanya, Israel: Wingate Institute, 1973, pp. 109–134.

22. De Ste Croix, M.B.A., N. Armstrong, M.Y.H. Chia, J.R. Welsman, G. Parsons, and P. Sharpe. Changes in short-term power output in 10- to 12-year-olds. *J. Sport Sci.* 19: 141–148, 2001.

23. Doré, E., M. Bedu, N.M. Franca, and E. Van Praagh. Anaerobic cycling performance characteristics in prepubescent, adolescent, and young adult females. *Eur. J. Appl. Physiol.* 84:476–481, 2001.

24. Doré, E., O. Diallo, N.M. Franca, M. Bedu, and E. Van Praagh. Dimensional changes cannot account for all differences in short-term cycling power during growth. *Int. J. Sports Med.* 21:360–365, 2000.

25. Dowson, M.N., M.E. Nevill, H.K. Lakomy, A.M. Nevill, and R.J. Hazeldine. Modelling the relationship between isokinetic muscle strength and sprint running performance. *J. Sport Sci.* 16:257–265, 1998.

26. Emmett, B., and P.W. Hochachka. Scaling of oxidative and glycolytic enzymes in mammals. *Resp. Physiol.* 45:261–272, 1981.

27. Eriksson, B.O., and B. Saltin. Muscle metabolism in boys 11 to 16 years compared to adults. *Acta Paediatr. Belg.* 28(Suppl.): 257–265, 1974.

28. Fawkner, S.G., N. Armstrong, D.J. Childs, and J.R. Welsman. Reliability of the visually identified ventilatory threshold and v-slope in children. *Pediatr. Exerc. Sci.* 14:181–192, 2002.

29. Ferretti, G., M.V. Narici, T. Binzoni, L. Gariod, J.F. Le Bas, H. Reutenauer, and P. Cerretelli. Determinants of peak muscle power: Effects of age and physical conditioning. *Eur. J. Appl. Physiol.* 68:111–115, 1994.

30. Gamstorp, J. *Paediatric neurology.* 2nd edition. London: Butterworths, 1985, pp. 49–50.

31. Garcia, A., J. Calleja, F.M. Antolin, and J. Berciano. Peripheral motor and sensory nerve conduction studies in normal infants and children. *Clin. Neurophysiol.* 111:513–520, 2000.

32. Gastin, P.B. Energy system interaction and relative contribution during maximal exercise. *Sports Med.* 31:725–741, 2001.

33. Gaul, C.A., D. Docherty, and R. Cicchini. Differences in anaerobic performance between boys and men. *Int. J. Sports Med.* 16:451–455, 1995.

34. Gilliam, T.B., P.S. Freedson, D.L. Geenan, and B. Shahraray. Physical activity patterns determined by heart rate monitoring in 6- to 7-year-old children. *Med. Sci. Sports Exerc.* 13:65–67, 1981.

35. Haralambie, G. Activites enzymatiques dans le muscle squelettique des enfants de divers ages [Enzymatic activity in the skeletal muscle of children of various ages]. In: *Le sport et l'enfant.* Montpelier: Euromed, 1980, pp. 243–258.

36. Hebestreit, H., K.-I. Mimura, and O. Bar-Or. Recovery of muscle power after high intensity short-term exercise: Comparing boys and men. *J. Appl. Physiol.* 74:2875–2880, 1993.

37. Hebestreit, H., B. Staschen, and A. Hebestreit. Ventilatory threshold: A useful method to determine aerobic fitness in children? *Med. Sci. Sports Exerc.* 32:1964–1969, 2000.

38. Klausen, K., B. Schibye, and B. Rasmussen. A longitudinal study of changes in physical performance of 10- to 15-year-old girls and boys. In: *Children and exercise XIII.* S. Oseid

and K.-H. Carlsen (eds.). Champaign, IL: Human Kinetics, 1989, pp. 113–122.

39. Kukolj, M., R. Ropret, D. Ugarkovic, and S. Jaric. Anthropometric, strength, and power predictors of sprinting performance. *J. Sports Med. Phys. Fitness* 39:120–122, 1990.

40. Kurokawa, K., E. Tanaka, H. Yamashita, T. Nakayama, H. Maruyama, M. Yukawa, T. Kohriyama, et al. Age-related changes in amplitude ratio, duration ratio, and area ratio in nerve conduction studies. *Jpn. J. Geriat.* 32:547–552, 1995.

41. Lexell, J., M. Sjostrom, A. Nordlund, and C.C. Taylor. Growth and development of human muscle: A quantitative morphological study of whole vastus lateralis from childhood to adult age. *Muscle Nerve* 15:404–409, 1992.

42. Mahon, A.D., and C.C. Cheatham. Ventilatory threshold in children: A review. *Pediatr. Exerc. Sci.* 14:16–29, 2002.

43. Malina, R.M., and C. Bouchard. *Growth, maturation, and physical activity.* Champaign, IL: Human Kinetics, 1991.

44. Martin, J.C., and R.M. Malina. Developmental variation in anaerobic performance associated with age and sex. In: *Pediatric anaerobic performance.* E. Van Praagh (ed.). Champaign, IL: Human Kinetics, 1998, pp. 45–64.

45. Mercier, B., J. Mercier, P. Granier, D. LeGallais, and C. Prefaut. Maximal anaerobic power: Relation to anthropometric characteristics during growth. *Int. J. Sports Med.* 13:21–26, 1992.

46. Mero, A. Power and speed training in childhood. In: *Pediatric anaerobic performance.* E. Van Praagh (ed.). Champaign, IL: Human Kinetics, 1998, pp. 241–267.

47. Mero, A., L. Jaakola, and P.V. Komi. Relationships between muscle fibre characteristics and physical performance capacity in trained athletic boys. *J. Sports Sci.* 9:161–171, 1991.

48. Mero, A., H. Kauhanen, E. Peltola, T. Vuorimaa, and P.V. Komi. Physiological performance capacity in different prepubescent athletic groups. *J. Sports Med. Phys. Fitness* 30:57–66, 1990.

49. Mero, A., P.V. Komi, and R.J. Gregor. Biomechanics of sprint running. *Sports Med.* 13:376–392, 1992.

50. Mero, A., P. Luhtanen, J.T. Viitasalo, and P.V. Komi. Relationships between the maximal running velocity, muscle fibre characteristics, force production, and force relaxation of sprinters. *Scand. J. Sport Sci.* 3:16–22, 1981.

51. Nummela, A., A. Mero, J. Stray-Gunderson, and H. Rusko. Important determinants of anaerobic running performance in male athletes and nonathletes. *Int. J. Sports Med.* 17 (Suppl. 2):S91–S96, 1996.

52. Oertel, G. Morphometric analysis of normal skeletal muscles in infancy, childhood and adolescence: An autopsy study. *J. Neurol. Sci.* 88:303–313, 1988.

53. Ohuchi, H., T. Nakajima, M. Kawade, M. Matsuda, and T. Kamiya. Measurement and validity of the ventilatory threshold in patients with congenital heart disease. *Pediatr. Cardiol.* 17:7–14, 1996.

54. Ross, A., and M. Leveritt. Long-term metabolic and skeletal muscle adaptations to short-sprint training. *Sports Med.* 31: 1063–1082, 2001.

55. Rowland, T.W. *Exercise and children's health.* Champaign, IL: Human Kinetics, 1990.

56. Santos, A.M.C., J.R. Welsman, M.B.A. De Ste Croix, and N. Armstrong. Age- and sex-related differences in optimal peak power. *Pediatr. Exerc. Sci.* 14:202–212, 2002.

57. Sargeant, A.J. Anaerobic performance. In: *Paediatric exercise science and medicine.* N. Armstrong and W. van Mechelen (eds.). Oxford: Oxford University Press, 2000, pp. 143–152.

58. Sargent, L.W. The physical test of a man. *Am. Phys. Educ. Rev.* 26:188–194, 1921.

59. Schmidt-Nielsen, K. *Scaling: Why is animal size so important?* Cambridge: Cambridge University Press, 1984, pp. 176–180.

60. Shephard, R.J. *Physical activity and growth.* Chicago: Year Book Medical, 1982, chapter 6:107-125.

61. Shephard, R.J., H. Lavallee, and R. LaBarre. On the basis of data standardization in prepubescent children. In: *Kinanthropometry II.* M.Ostyn (ed.) Basel: Karger, 1980, pp. 306–316.

62. Simoneau, J.-A., and C. Bouchard. The effects of genetic variation on anaerobic performance. In: *Pediatric anaerobic performance.* E. Van Praagh (ed.). Champaign, IL: Human Kinetics, 1998, pp. 5–22.

63. Singh, T.P., V. Joshi, N. Sullivan, and B. Perry. A comparison of methodologies to detect ventilatory anaerobic threshold in children [abstract]. *Pediatr. Exerc. Sci.* 9:93–94, 1997.

64. Thorland, W.G., G.O. Johnson, C.J. Cisar, T.J. Housh, and G.D. Tharp. Strength and anaerobic responses of elite young female sprint and distance runners. *Med. Sci. Sports Exerc.* 19:56–61, 1987.

65. Van Ingen Schenau, G.J., J.J. de Koning, and G. de Groot. Optimisation of sprinting performance in running, cycling and speed skating. *Sports Med.* 17:259–275, 1994.

66. Van Praagh, E. Development of anaerobic function during childhood and adolescence. *Pediatr. Exerc. Sci.* 12:150–173, 2000.

67. Van Praagh, E., M. Bedu, G. Falgairette, N. Fellman, and J. Coudert. *Oxygen uptake during a 30-s supramaximal test in 7- to 15-year-old boys.* Congress of Pediatric Work Physiology, September 11–15, 1989, Budapest, Hungary.

68. Van Praagh, E., and E. Doré. Short-term muscle power during growth and maturation. *Sports Med.* 32:701–728, 2002.

69. Van Praagh, E., G. Falgairette, M. Bedu, N. Fellman, and J. Coudert. Laboratory and field tests in 7-year-old boys. In: *Children and exercise XIII.* S. Oseid and K.-H. Carlsen (eds.). Champaign, IL: Human Kinetics, 1989, pp. 11–17.

70. Van Praagh, E., N. Fellman, M. Bedu, M. Delaitre, B. Beaune, and J. Coudert. Analysis of "anaerobic fitness" in 7- and 12-year-old boys [abstract]. *Pediatr. Exerc. Sci.* 8:92, 1996.

71. Van Praagh, E., N. Fellman, M. Bedu, G. Falgairette, and J. Coudert. Gender difference in the relationship of anaerobic power output to body composition in children. *Pediatr. Exerc. Sci.* 2:336–348, 1990.

72. Van Praagh, E., and N.M. França. Measuring maximal short-term power output during growth. In: *Pediatric anaerobic performance.* E. Van Praagh (ed.). Champaign, IL: Human Kinetics, 1998, pp. 155–189.

73. Welsman, J.R., N. Armstrong, B.J. Kirby, R.J. Winsley, G. Parsons, and P. Sharpe. Exercise performance and magnetic resonance imaging–determined thigh muscle volume in children. *Eur. J. Appl. Physiol.* 76:92–97, 1997.

74. Weyand, P.G., C.S. Lee, R. Martinez-Ruiz, M.W. Bundle, M.J. Bellizzi, and S. Wright. High-speed running performance is largely unaffected by hypoxic reductions in aerobic power. *J. App. Physiol.* 86:2059–2064, 1999.

75. Weyand, P.G., D.B. Sternlight, M.J. Bellizzi, and S. Wright. Faster top running speeds are achieved with greater ground forces not more rapid leg movements. *J. Appl. Physiol.* 89: 1991–1999, 2000.

76. Wilkie, D.R. Man as a source of mechanical power. *Ergonomics* 3:1–8, 1960.

77. Young, W., B. McLean, and J. Ardagna. Relationship between strength qualities and sprinting performance. *J. Sports Med. Phys. Fitness* 35:13–19, 1995.

Chapter 10

1. Apple, F.S., M.A. Rogers, D.C. Casal, W.M. Sherman, and J.L. Ivy. Creatine kinase-MB isoenzyme adaptations in stressed human skeletal muscle of marathon runners. *J. Appl. Physiol.* 59:149–153, 1985.

2. Asmussen, E. Growth in muscular strength and power. In: *Physical activity: Human growth and development.* G. Rarick (ed.). New York: Academic Press, 1973, pp. 60–79.

3. Asmussen, E., and K.R. Heeboll-Nielsen. A dimensional analysis of physical performance and growth in boys. *J. Appl. Physiol.* 7:593–603, 1955.

4. Backman, E., and K.G. Henriksson. Skeletal muscle characteristics in children 9–15 years old: Force, relaxation rate, and contraction time. *Clin. Physiol.* 8:521–527, 1988.

5. Baltzopoulos, V., and E. Kellis. Isokinetic strength during childhood and adolescence. In: *Pediatric anaerobic performance.* E. Van Praagh (ed.). Champaign, IL: Human Kinetics, 1998, pp. 225–240.

6. Belanger, A.Y., and A.J. McComas. Contractile properties of human skeletal muscle in childhood and adolescence. *Eur. J. Appl. Physiol.* 58:563–567, 1989.

7. Beunen, G., M. Thomis, M. Peeters, H. Maes, A.L. Claessens, and R. Vlietinck. Genetics of strength and power characteristics in children and adolescents. *Pediatr. Exerc. Sci.* 15: 128-138, 2003.

8. Blimkie, C.J.R. Age- and sex-associated variation in strength during childhood: Anthropometric, morphological, neurologic, biochemical, endocrinologic, genetic, and physical activity correlates. In: *Perspectives in exercise science and sports medicine: Vol. 2. Youth, exercise, and sport.* C.V. Gisolfi and D.R. Lamb (eds.). Indianapolis: Benchmark Press, 1989, pp. 99–164.

9. Blimkie, C.J.R., and D.G. Sale. Strength development and trainability during childhood. In: *Pediatric anaerobic performance.* E. Van Praagh (ed.). Champaign, IL: Human Kinetics, 1998, pp. 193–224.

10. Brodie, D.A., J. Burnie, R.G. Eston, and J.A. Royce. Isokinetic strength and flexibility characteristics in preadolescent boys. In: *Children and exercise XII.* J. Rutenfranz, R. Mocellin, and F. Klimt (eds.). Champaign, IL: Human Kinetics, 1986, pp. 309–319.

11. Byrnes, W.C., and P.C. Clarkson. Delayed onset muscle soreness and training. *Clin. Sports Med.* 5:605–613, 1986.

12. Davies, C.T.M. Strength and mechanical properties of muscle in children and young adults. *Scand. J. Sports Sci.* 7: 11–15, 1985.

13. Davies, C.T.M., M.J. White, and K. Young. Muscle function in children. *Eur. J. Appl. Physiol.* 52:111–114, 1983.

14. De Ste Croix, M.B.A., N. Armstrong, and J.R. Welsman. Concentric isokinetic leg strength in pre-teen, teenage, and adult males and females. *Biol. Sport* 16:75–86, 1999.

15. De Ste Croix, M.B.A., N. Armstrong, J.R. Welsman, and P. Sharpe. Longitudinal changes in isokinetic leg strength in 10–14-year-olds. *Ann. Hum. Biol.* 29:50–62, 2002.

16. Duarte, J.A., J.F. Magalhaes, L. Monteiro, A. Almeida-Dias, J.M.C. Soares, and H.J. Appell. Exercise-induced signs of muscle overuse in children. *Int. J. Sports Med.* 20:103–108, 1999.

17. Eyre, J.A., S. Miller, and V. Rambusch. Constancy of central conduction delays during development in man: Investigation of motor and somatosensory pathways. *J. Physiol.* 434: 441–452, 1991.

18. Fietzek, U.M., F. Heinen, S. Berweck, S. Maute, A. Hufschmidt, J. Schulte-Motning, C.H. Lucking, et al. Development of the corticospinal system and hand motor function: Central conduction times and motor performance tests. *Dev. Med. Child Neurol.* 42:220–227, 2000.

19. Froberg, K., and O. Lammert. Development of muscle strength during childhood. In: *The child and adolescent athlete.* O. Bar-Or (ed.). Oxford: Blackwell Science, 1996, pp. 42–53.

20. Gandevia, S.C. Spinal and supraspinal factors in human muscle fatigue. *Physiol. Rev.* 81:1725–1789, 2001.

21. Garcia, A., J. Calleja, F.M. Antolin, and J. Berciana. Peripheral motor and sensory nerve conduction studies in normal infants and children. *Clin. Neurophys.* 111:513–520, 2000.

22. Gaul, C. Muscular strength and endurance. In: *Measurement in pediatric exercise science.* D. Docherty (ed.). Champaign, IL: Human Kinetics, 1996, pp. 225–258.

23. Going, S.B., B.H. Massey, T.B. Hoshizaki, and T.G. Lohman. Maximal voluntary static strength production characteristics of skeletal muscle in children 8–11 years of age. *Res. Q. Exerc. Sport* 58:115–123, 1987.

24. Green, H.J. Manifestations and sites of neuromuscular fatigue. In: *Biochemistry of exercise VII.* A.W. Taylor, P.D. Gollnick, H.J. Green, C.O. Ianuzzo, E.G. Noble, G. Metivier, and J.R. Sutton (eds.). Champaign, IL: Human Kinetics, 1990, pp. 13–36.

25. Hebestreit, H., F. Meyer, G.J.F. Heigenhauser, and O. Bar-Or. Plasma metabolites, volume and electrolytes following

30-s high intensity exercise in boys and men. *Eur. J. Appl. Physiol.* 72:563–569, 1996.

26. Heinen, F., U.M. Fietzek, S. Berweck, A. Hufschmidt, G. Deuschl, and R. Korinthenberg. Fast corticospinal system and motor performance in children: Conduction proceeds skill. *Pediatr. Neurol.* 19:217–221, 1998.
27. Jansson, E., and C. Sylven. Creatine kinase MB and citrate synthetase in type I and type II muscle fibers in trained and untrained men. *Eur. J. Appl. Physiol.* 54:207–209, 1985.
28. Jaric, S. Role of body size in the relation between muscle strength and movement performance. *Exerc. Sport Sci. Rev.* 31:8–12, 2003.
29. Jaric, S., D. Ugarkovic, and M. Kukolj. Evaluation of methods for normalizing muscle strength in elite and young athletes. *J. Sports Med. Phys. Fitness* 42:141–151, 2002.
30. Jones, D.A., and J.M. Round. Strength and muscle growth. In: *Paediatric exercise science and medicine.* N. Armstrong and W. van Mechelen (eds.). Oxford: Oxford University Press, 2000, pp. 133–142.
31. Kanehisa, H., S. Ikegawa, N. Isunoda, and T. Fukunaga. Strength and cross-sectional area of knee extensor muscles in children. *Eur. J. Appl. Physiol.* 68:402–405, 1994.
32. Kanehisa, H., H. Okuyama, S. Ikegawa, and T. Fukunaga. Fatigability during repetitive maximal knee extensions in 14-year-old boys. *Eur. J. Appl. Physiol.* 72:170–174, 1995.
33. Lang, H.A., A. Puusa, P. Hynninen, V. Kuusela, V. Jantti, and M. Sillanpaa. Evolution of nerve conduction velocity in later childhood and adolescence. *Muscle Nerve* 8:38–43, 1985.
34. Malina, R.M., and C. Bouchard. *Growth, maturation and physical activity.* Champaign, IL: Human Kinetics, 1991.
35. Malmstrom, J.-E., and L. Lindstrom. Propagation velocity of muscle action potentials in the growing normal child. *Muscle Nerve* 20:403–410, 1997.
36. Marginson, V.F., R.G. Eston, and C.G. Parfitt. A comparison of soreness and strength loss in children and adults following high-impact eccentric exercise [abstract]. *Pediatr. Exerc. Sci.* 13:87, 2001.
37. McComas, A.J., V. Galea, and H. de Bruin. Motor unit populations in healthy and diseased muscles. *Phys. Ther.* 73: 868–877, 1993.
38. McComas, A.J., R.E.P. Sica, and F. Petito. Muscle strength in boys of different ages. *J. Neurol. Neurosurg. Psychiatr.* 36: 171–173, 1973.
39. Neu, C.M., F. Rauch, J. Rittweger, F. Manz, and E. Schoenau. Influence of puberty on muscle development at the forearm. *Am. J. Physiol. Endocrinol. Metab.* 283:E103–E107, 2002.
40. Nevill, A.M., R.L. Holder, A. Baxter-Jones, J.M. Round, and D.A. Jones. Modeling developmental changes in strength and aerobic power in children. *J. Appl. Physiol.* 84:963–970, 1998.
41. Noakes, T.D. Effect of exercise on serum enzyme activities in humans. *Sports Med.* 4:245–267, 1987.
42. Oertel, G. Morphometric analysis of normal skeletal muscles in infancy, childhood and adolescence. *J. Neurol. Sci.* 88: 303–313, 1988.
43. Patten, C., G. Kamen, and D.M. Rowland. Adaptations in maximal motor unit discharge rate to strength training in young and older adults. *Muscle Nerve* 24:542–550, 2001.
44. Pette, D. Plasticity in skeletal cardiac and smooth muscle. Historical perspectives: Plasticity of mammalian skeletal muscle. *J. Appl. Physiol.* 90:1119–1124, 2001.
45. Ramos, E., W.R. Frontera, A. Llopart, and D. Feliciano. Muscle strength and hormonal levels in adolescence: Gender related differences. *Int. J. Sports Med.* 19:526–531, 1998.
46. Rauch, F., C.M. Neu, G. Wassmer, B. Beck, G. Rieger-Wettengl, E. Rietschel, F. Manz, et al. Muscle analysis by measurement of maximal isometric grip force: New reference data and clinical applications in pediatrics. *Pediatr. Res.* 51: 505–510, 2002.
47. Rowe, S.A., K.G. Zahka, N. Hu, E.B. Clark, and W.E. Jacobus. Cardiac function and creatine kinase in the developing chick heart [abstract]. *Am. J. Cardiol.* 60:635, 1987.
48. Seger, J.Y., and A. Thorstensson. Muscle strength and electromyogram in boys and girls followed through puberty. *Eur. J. Appl. Physiol.* 81:54–61, 2000.
49. Seger, J.Y., and A. Thorstensson. Muscle strength and myoelectric activity in prepubertal and adult males and females. *Eur. J. Appl. Physiol.* 69:81–87, 1994.
50. Shephard, R.J., H. Lavallee, and R. LaBarre. On the basis of data standardization in prepubescent children. In: *Kinanthropometry II.* M. Ostyn (ed.). Basel: Karger, 1980, pp. 306-316.
51. Sinaki, M., P.J. Limburg, P.C. Wollan, J.W. Rogers, and P.A. Murtaugh. Correlation of trunk muscle strength with age in children 5 to 18 years old. *Mayo Clin. Proc.* 71:1047–1054, 1996.
52. Soares, J.M.C., P. Mota, J.A. Duarte, and H.J. Appell. Children are less susceptible to exercise-induced muscle damage than adults: A preliminary investigation. *Pediatr. Exerc. Sci.* 8:361–367, 1996.
53. Webber, L.M., W.C. Byrnes, T.W. Rowland, and V.L. Foster. Serum creatine kinase activity and delayed onset muscle soreness in prepubescent children: A preliminary study. *Pediatr. Exerc. Sci.* 1:351–359, 1989.
54. Wickiewicz, T.L., R.R. Roy, P.L. Powell, and V.R. Edgerton. Muscle metabolism of the human lower limb. *Clin. Orthop. Rel. Res.* 179:275–283, 1983.
55. Winter, D.A. *Biomechanics of human movement.* New York: Wiley, 1983, chapter 6, pp. 108-126.
56. Woods, J.A., R.R. Pate, and M.L. Burgess. Correlates to performance on field tests of muscular strength. *Pediatr. Exerc. Sci.* 4:302–311, 1992.

Chapter 11

1. Aagaard, P. Training-induced changes in neural function. *Exerc. Sport Sci. Rev.* 31:61–67, 2003.
2. Aagaard, P., E.B. Simonsen, J.L. Andersen, P. Magnusson, and P. Dyhre-Poulson. Increased rate of force development and neural drive of human skeletal muscle following resistance training. *J. Appl. Physiol.* 93:1318–1326, 2002.

3. Bar-Or, O. Trainability of the prepubescent child. *Phys. Sportsmed.* 17:65–81, 1989.

4. Bawa, P. Neural control of motor output: Can training change it? *Exerc. Sport Sci. Rev.* 30:59–63, 2002.

5. Bhasin, S., T.W. Storer, N. Berman, C. Callegari, B. Clevenger, J. Phillips, T.J. Brunnell, et al. The effects of supraphysiologic doses of testosterone on muscle size and strength in normal men. *N. Engl. J. Med.* 335:107, 1996.

6. Blimkie, C.J.R., and O. Bar-Or. Trainability of muscle strength, power and endurance in childhood. In: *The child and adolescent athlete.* O. Bar-Or (ed.). Oxford: Blackwell Science, 1996, pp. 113–129.

7. Blimkie, C.J.R., and D.G. Sale. Strength development and trainability in children. In: *Pediatric anaerobic performance.* E. Van Praagh (ed.). Champaign, IL: Human Kinetics, 1998, pp. 193–224.

8. Booth, F.W. Perspectives on molecular and cellular exercise physiology. *J. Appl. Physiol.* 65:1461–1471, 1988.

9. Brand, T., and M.D. Schneider. Peptide growth factors as determinants of myocardial development and hypertrophy. In: *Cardiovascular response to exercise.* G.F. Fletcher (ed.). Mount Kisco, NY: Futura, 1994, pp. 59–99.

10. Broulik, P.D., C.D. Kochakian, and J. Dubovsky. Influence of castration and testosterone propionate on cardiac output, renal blood flow, and blood volume in mice. *Proc. Soc. Exp. Biol. Med.* 144:671–673, 1973.

11. Buttrick, P.M. Role of hemodynamic load in the genesis of cardiac hypertrophy. In: *Cardiovascular response to exercise.* G.F. Fletcher (ed.). Mount Kisco, NY: Futura, 1994, pp. 101–110.

12. Cahill, B.R., J.E. Misner, and R.A. Boileau. The clinical importance of the anaerobic energy system and its assessment in human performance. *Am. J. Sports Med.* 25:863–872, 1997.

13. Cameron, J.D., and A.M. Dart. Exercise training increases total systemic arterial compliance in humans. *Am. J. Physiol.* 266:H693–H701, 1994.

14. Colan, S.D. Mechanics of left ventricular systolic and diastolic function in physiologic hypertrophy of the athlete's heart. *Cardiol. Clin.* 15:355–372, 1997.

15. Convertino, V.A. Blood volume responses to training. In: *Cardiovascular response to exercise.* G.F. Fletcher (ed.). Mount Kisco, NY: Futura, 1994, pp. 207–221.

16. Convertino, V.A., P.J. Brock, L.C. Keil, E.M. Bernauer, and J.E. Greenleaf. Exercise training–induced hypervolemia: Role of plasma albumen, rennin, and vasopressin. *J. Appl. Physiol.* 48:665–669, 1990.

17. Cooper, D.M., D. Moromisato, S. Zanconato, M. Moromisato, S. Jensen, and J.A. Brasel. Effect of growth hormone suppression on exercise training and growth responses in young rats. *Pediatr. Res.* 35:223–227, 1994.

18. Cronin, J.B., P.J. McNair, and R.N. Marshall. Is velocity-specific strength training important in improving functional performance? *J. Sports Med. Phys. Fitness* 42:267–273, 2002.

19. Delecluse, C. Influence of strength training on sprint running performance. *Sports Med.* 24:147–156, 1997.

20. DeMaria, A.N., A. Neumann, P.J. Schubart, G. Lee, and D.T. Mason. Systematic correlation of cardiac chamber size and ventricular performance determined with echocardiography and alterations in heart rate in normal persons. *Am. J. Cardiol.* 43:1–9, 1979.

21. Deschenes, M.R., C.M. Maresh, J.F. Crivello, L.E. Armstrong, W.J. Kraemer, and J. Covault. The effect of exercise training of different intensities on neuromuscular junction morphology. *J. Neurocytol.* 22:603–615, 1993.

22. Diallo, O., E. Doré, C. Hautier, P. Duché, and E. Van Praagh. Effects of jump and sprint training on athletic performance in prepubertal boys [abstract]. *Med. Sci. Sports Exerc.* 31(Suppl.):S317, 1999.

23. Diallo, O., E. Doré, C. Hautier, P. Duché, and E. Van Praagh. Effects of 10-week training and 8-week detraining on athletic performance in prepubertal boys [abstract]. *Pediatr. Exerc. Sci.* 11:287–288, 1999.

24. Eliakim, A., T. Scheet, N. Allmendinger, J.A. Brasel, and D.M. Cooper. Training, muscle volume, and energy expenditure in nonobese American girls. *J. Appl. Physiol.* 90:35–44, 2001.

25. Eriksson, B.O., P.D. Gollnick, and B. Saltin. Muscle metabolism and enzyme activities after training in boys 11–13 years old. *Acta Physiol. Scand.* 87:485–497, 1973.

26. Eriksson, B.O., and G. Koch. Effect of physical training on hemodynamic response during submaximal and maximal exercise in 11–13-year-old boys. *Acta Physiol. Scand.* 87:27–39, 1973.

27. Fagard, F.H. Impact of different sports and training on cardiac structure and function. *Cardiol. Clin.* 15:397–412, 1997.

28. Fahey, T.D., A.D. Valle-Zuris, G. Oehlsen, M. Trieb, and J. Seymour. Pubertal stage differences in hormonal and hematological responses to maximal exercise in males. *J. Appl. Physiol.* 46:823–827, 1979.

29. Faigenbaum, A.D. Strength training for children and adolescents. *Clin. Sports Med.* 19:593–619, 2000.

30. Falk, B., O. Bar-Or, and J.D. McDougall. Aldosterone and prolactin response to exercise in the heat in circumpubertal boys. *J. Appl. Physiol.* 71:1741–1745, 1991.

31. Falk, B., and G. Tenenbaum. The effectiveness of resistance training in children: A meta-analysis. *Sports Med.* 22:176–186, 1996.

32. Ferguson, S., N. Gledhill, V.K. Jamnik, C. Wiebe, and N. Payne. Cardiac performance in endurance-trained and moderately active young women. *Med. Sci. Sports Exerc.* 33:1114–1119, 2001.

33. Fleck, S.J. Cardiovascular adaptations to resistance training. *Med. Sci. Sports Exerc.* 20(Suppl.):S146–S151, 1988.

34. Fleck, S.J., and W.J. Kraemer. *Designing resistance training programs.* Champaign, IL: Human Kinetics, 1987, chapter 7, pp. 149-175.

35. Follard, L., B. Leach, T. Little, K. Hawkes, S. Myerson, H. Montgomery, and D. Jones. Angiotensin-converting enzyme genotype affects the response of human skeletal muscle to functional overload. *Exp. Physiol.* 85:575–579, 2000.

36. Fournier, M., J. Ricci, A.W. Taylor, R.J. Ferguson, R.R. Montpetit, and B.R. Chaitman. Skeletal muscle adaptation in adolescent boys: Sprint and endurance training and detraining. *Med. Sci. Sports Exerc.* 14:453–456, 1982.

37. Fukunaga, T., K. Funato, and S. Ikegawa. The effects of resistance training on muscle area and strength in prepubertal age. *Ann. Physiol. Anthropol.* 11:357–364, 1992.

38. Geenen, D.L., A. Malhotra, P.M. Buttrick, and J. Scheuer. Increased heart rate prevents the isomyosin shift after cardiac transplantation in the rat. *Circ. Res.* 70:554–558, 1992.

39. Goldspink, G., and S.Y. Yang. Effects of activity on growth factor expression. *Int. J. Sports Nutr. Metab.* 11:S21–S27, 2001.

40. Gollnick, P.D., R.B. Armstrong, C.W. Saubert, K. Piehl, and B. Saltin. Enzyme activity and fiber composition in skeletal muscle of untrained and trained men. *J. Appl. Physiol.* 33: 312–319, 1972.

41. Grodjinovsky, A., O. Inbar, R. Dotan, and O. Bar-Or. Training effect on the anaerobic performance of children as measured by the Wingate anaerobic test. In: *Children and exercise IX.* K. Berg and B.O. Eriksson (eds.). Baltimore: University Park Press, 1980, pp. 139–145.

42. Gustafsson, T., and W.E. Kraus. Exercise-induced angiogenesis-related growth and transcription factors in skeletal muscle and their modification in muscle pathology. *Front. Biosci.* 6:75–89, 2001.

43. Gutin, B., N. Mayers, J.A. Levy, and M.V. Herman. Physiologic and echocardiographic studies in age-group runners. In: *Competitive sports for children and youth.* E.W. Brown and C.F. Branta (eds.). Champaign, IL: Human Kinetics, 1988, pp. 117–128.

44. Hakkinen, K., and A. Pakarinen. Acute hormonal responses to heavy resistance exercise in men and women at different ages. *Int. J. Sports Med.* 16:507–513, 1995.

45. Hefti, M.A., B.A. Harder, H.M. Eppenberger, and M.C. Schaub. Signaling pathways in cardiac myocyte hypertrophy. *J. Mol. Cell Cardiol.* 29:2873–2892, 1997.

46. Holloszy, J.O., and E.F. Coyle. Adaptations of skeletal muscle to endurance exercise and their metabolic consequences. *J. Appl. Physiol.* 56:831–838, 1984.

47. Hoppeler, H., H. Howald, K. Conley, S.L. Lindstedt, H. Claasen, P. Vock, and E.R. Weibel. Endurance training in humans: Aerobic capacity and structure of skeletal muscle. *J. Appl. Physiol.* 59:320–327, 1985.

48. Hopper, M.K., A.R. Coggin, and E.F. Coyle. Exercise stroke volume relative to plasma volume expansion. *J. Appl. Physiol.* 64:404–408, 1988.

49. Housh, T.J., R.J. Hughes, G.O. Johnson, D.J. Housh, L.L. Wagner, J.P. Weir, and S.A. Evans. Age-related increases in the shoulder strength of high school wrestlers. *Pediatr. Exerc. Sci.* 2:240–243, 1988.

50. Housh, T.J., J.R. Stout, D.J. Housh, and G.O. Johnson. The covariate influence of muscle mass on isokinetic peak torque in high school wrestlers. *Pediatr. Exerc. Sci.* 7: 176–182, 1995.

51. Ianuzzo, C.D., P.J. O'Brien, T.A. Salerno, and M.H. Laughlin. Effects of chronic tachycardia on the myocardium. In: *Cardiovascular response to exercise.* G.F. Fletcher (ed.). Mount Kisco, NY: Futura, 1994, pp. 111–140.

52. Ignico, A.A., and A.D. Mahon. The effects of a physical fitness program on low-fit children. *Res. Q. Exerc. Sport* 66: 85–90, 1995.

53. Ikai, M., and T. Fukunaga. A study on training effect on strength per unit cross-sectional area of muscle by means of ultrasonic measurement. *Eur. J. Appl. Physiol.* 28:173–180, 1970.

54. Jacobs, I., M. Esbjornsson, C. Sylven, I. Holm, and E. Jansson. Sprint training effects on muscle myoglobin, enzymes, fiber types, and blood lactate. *Med. Sci. Sports Exerc.* 19:368–374, 1987.

55. Jansson, E., M. Esbjornsson, I. Holm, and E. Jacobs. Increase in the proportion of fast-twitch muscle fibers by sprint training in males. *Acta Physiol. Scand.* 140:359–363, 1990.

56. Ji, L.L., D.L.F. Lennon, R.G. Kochan, F.J. Nagle, and H.A. Lardy. Enzymatic adaptation to physical training under beta-blockade in the rat. Evidence of a beta-adrenergic mechanism in skeletal muscle. *J. Clin. Invest.* 78:771–778, 1986.

57. Jungersten, L., A. Ambring, B. Wall, and A. Wennmalm. Both physical fitness and acute exercise regulate nitric oxide formation in healthy humans. *J. Appl. Physiol.* 82: 760–764, 1997.

58. Kamen, G., C.A. Knight, D.P. Laroche, and D.G. Asermley. Resistance training increases vastus lateralis motor unit firing rates in young and old adults [abstract]. *Med. Sci. Sports Exerc.* 30(Suppl.):S337, 1998.

59. Katch, V.L. Physical conditioning of children. *J. Adolesc. Health Care* 3:241–246, 1983.

60. Kingwell, B.A. Nitric oxide as a metabolic regulator during exercise: Effects of training in health and disease. *Clin. Exp. Pharmacol. Physiol.* 27:239–250, 2000.

61. Kobayashi, K., K. Kitamura, M. Miura, H. Sodeyama, Y. Murase, M. Moyashita, and H. Matsui. Aerobic power as related to body growth and training in Japanese boys: A longitudinal study. *J. Appl. Physiol.* 44:666–672, 1978.

62. Koch, G. Aerobic power, lung dimensions, ventilatory capacity, and muscle blood flow in 12–16-year-old boys with high physical activity. In: *Children and exercise IX.* K. Berg and B.O. Eriksson (eds.). Baltimore: University Park Press, 1980, pp. 99–108.

63. Koch, G., and L. Rocker. Plasma volume and intravascular protein masses in trained boys and fit young men. *J. Appl. Physiol.* 43:1085–1088, 1977.

64. Kraemer, W.J. Endocrine responses to resistance exercise. *Med. Sci. Sports Exerc.* 20(Suppl.):S152–S157, 1988.

65. Kraemer, W., A. Fry, P. Frykman, B. Conroy, and J. Hoffman. Resistance training and youth. *Pediatr. Exerc. Sci.* 1:336–350, 1989.

66. Krip, B., N. Gledhill, V. Jamnik, and D. Warburton. Effect of alterations in blood volume on cardiac function during maximal exercise. *Med. Sci. Sports Exerc.* 29:1469–1476, 1997.

67. Leveritt, M., P.J. Abernethy, B.K. Barry, and P.A. Logan. Concurrent strength and endurance training: A review. *Sports Med.* 28:413–427, 1999.

68. Linossier, M.T., D. Denis, D. Dormois, A. Geyssant, and J.R. Lacour. Ergometric and metabolic adaptation to a 5 s sprint training programme. *Eur. J. Appl. Physiol.* 68:408–414, 1993.

69. Lortie, G., J.A. Simoneau, P. Hamel, M.R. Boulay, F. Landry, and C. Bouchard. Responses of maximal aerobic power and capacity to aerobic training. *Int. J. Sports Med.* 5:232–236, 1984.

70. Malina, R.M. Darwinian fitness, physical fitness and physical activity. In: *Application of biological anthropology to human affairs.* C.G.N. Mascie-Taylor and G.W. Lasker (eds.). Cambridge: Cambridge University Press, 1991, pp. 143–184.

71. Mandigout, S., A. Melin, A.M. Lecoq, D. Courteix, and P. Obert. Effect of gender in response to an aerobic training programme in prepubertal children. *Acta Paediatr.* 90:9–15, 2001.

72. Marcus, C., P.C. Gillette, and A. Garson. Intrinsic heart rate in children and young adults: An index of sinus node function isolated from autonomic control. *Am. Heart J.* 112: 911–916, 1990.

73. Maroun, M.J., S. Mehta, R. Turcotte, M.G. Cosio, and S.N. Hussain. Effects of physical conditioning on endogenous nitric oxide output during exercise. *J. Appl. Physiol.* 79: 1219–1225, 1995.

74. McArdle, W.D., K.I. Katch, and V.L. Katch. *Exercise physiology: Energy, nutrition, and human performance.* Philadelphia: Lea & Febiger, 1981, chapter 20, pp. 266-285.

75. McManus, A.M., N. Armstrong, and C.A. Williams. Effect of training on the anaerobic power and anaerobic performance of prepubertal girls. *Acta Paediatr.* 86:456–459, 1997.

76. Mero, A. Blood lactate production and recovery from anaerobic exercise in trained and untrained boys. *Eur. J. Appl. Physiol.* 57:660–666, 1988.

77. Mero, A. Power and speed training during childhood. In: *Pediatric anaerobic performance.* E. Van Praagh (ed.). Champaign, IL: Human Kinetics, 1998, pp. 241–267.

78. Mero, A., H. Kauhanen, E. Peltola, T. Vuorimaa, and P.V. Komi. Physiological performance capacity in different prepubescent athletic groups. *J. Sports Med. Phys. Fitness* 30:57–66, 1990.

79. Mersch, F., and H. Stoboy. Strength training and muscle hypertrophy in children. In: *Children and exercise XIII.* S. Oseid and K.-H. Carlsen (eds.). Champaign, IL: Human Kinetics, 1989, pp. 165–182.

80. Mobert, J., G. Koch, O. Humplik, and E.-M. Oyen. Cardiovascular adjustment to supine and seated postures: Effect of physical training. In: *Children and exercise XIX.* N. Armstrong, B.J. Kirby, and J.R. Welsman (eds.). London: Spon, 1997, pp. 429–433.

81. Mosher, R.E., E.C. Rhodes, H.A. Wenger, and B. Filsinger. Interval training: The effects of a 12-week programme on elite prepubertal male soccer players. *J. Sports Med.* 25:5–9, 1985.

82. Nottin, S., A. Vinet, F. Stecken, L.-D. N'guyen, F. Ounissi, A.-M. Lecoq, and P. Obert. Central and peripheral cardiovascular adaptations to exercise in endurance-trained children. *Acta Physiol. Scand.* 175:85–92, 2002.

83. Obert, P., S. Mandigout, S. Nottin, A. Vinet, L.-D. N'Guyen, and A.-M. Lecoq. Cardiovascular response to endurance training in children: Effect of gender. *Eur. J. Clin. Invest.,* 33:199-208, 2003.

84. Obert, P., S. Mandigout, A. Vinet, and D. Courteix. Effect of a 13-week aerobic training programme on the maximal power developed during a force-velocity test in prepubertal boys and girls. *Int. J. Sports Med.* 22:442–446, 2001.

85. Ogawa, T., R.J. Spina, W.H. Martin, W.M. Kohrt, K.B. Schechtman, J.O. Holloszy, and A.A. Ehsani. Effects of aging, sex, and physical training on cardiovascular responses to exercise. *Circulation* 86:494–503, 1992.

86. Ozmun, J.C., A.E. Mikesky, and P.R. Surburg. Neuromuscular adaptations following prepubescent strength training. *Med. Sci. Sports Exerc.* 26:510–514, 1994.

87. Pate, R.R., and D.S. Ward. Endurance exercise trainability in children and youth. In: *Advances in sports medicine and fitness, vol. 3.* W.A. Grana, J.A. Lombardo, B.J. Sharkey, and J.A. Stone (eds.). Chicago: Year Book Medical, 1990, pp. 37–55.

88. Patten, C., G. Kamen, and D.M. Rowland. Adaptations in maximal motor unit discharge rate to strength training in young and older adults. *Muscle Nerve* 24:542–550, 2001.

89. Payne, V.G., and J.R. Morrow. The effect of physical training on prepubescent $\dot{V}O_2$max: A meta-analysis. *Res. Q.* 64: 305–313, 1993.

90. Payne, V.G., J.R. Morrow, L. Johnson, and S.N. Dalton. Resistance training in children and youth: A meta-analysis. *Res. Q. Exerc. Sport* 68:80–88, 1997.

91. Perrault, H.M., and R.A. Turcotte. Do athletes have the "athlete heart"? *Prog. Pediatr. Cardiol.* 2:40–50, 1993.

92. Prado, L.S. Lactate, ammonia and catecholamine metabolism after anaerobic training. In: *Children and exercise XIX.* N. Armstrong, B. Kirby, and J. Welsman (eds.). London: Spon, 1997, pp. 306–312.

93. Pullinen, T., A. Mero, P. Huttunen, A. Pakarinen, and P.V. Komi. Resistance exercise–induced hormonal responses in men, women, and pubescent boys. *Med. Sci. Sports Exerc.* 34: 806–813, 2002.

94. Ramsay, J.A., C.J.R. Blimkie, K. Smith, S. Garner, J.D. MacDougall, and D.G. Sale. Strength training effects in prepubescent boys. *Med. Sci. Sports Exerc.* 22:605–614, 1970.

95. Rich, C., and E. Cafarelli. Submaximal motor unit firing rates after 8 weeks of isometric resistance training. *Med. Sci. Sports Exerc.* 32:190–196, 2000.

96. Rotstein, A., R. Dotan, O. Bar-Or, and G. Tenenbaum. Effects of training on anaerobic threshold, maximal aerobic power and anaerobic performance of preadolescent boys. *Int. J. Sports Med.* 7:281–286, 1986.

97. Rowland, T.W. Aerobic response to endurance training in prepubescent children: A critical analysis. *Med. Sci. Sports Exerc.* 17:493–497, 1985.

98. Rowland, T.W. Cardiac characteristics of the child endurance athlete. In: *Youth sports—Perspectives for a new century.* R.M. Malina and M.A. Clark (eds.). Monterey, CA: Coaches Choice, pp. 53-68, 2003.

99. Rowland, T.W. The "trigger hypothesis" for aerobic trainability: A 14-year follow-up. *Pediatr. Exerc. Sci.* 9:1–9, 1997.

100. Rowland, T.W., and A. Boyajian. Aerobic response to endurance training in children. *Pediatrics* 96:654–658, 1995.

101. Rowland, T.W., D. Goff, and B. Popowski. Cardiac responses to exercise in child distance runners. *Int. J. Sports Med.* 19: 385–390, 1998.

102. Rowland, T.W., L. Martel, P. Vanderburgh, T. Manos, and N. Charkoudian. The influence of short-term aerobic training on blood lipids in healthy 10–12-year-old children. *Int. J. Sports Med.* 17:487–492, 1996.

103. Rowland, T.W., and M.W. Roti. Cardiac responses to progressive upright exercise in adult male cyclists. *J. Sports Med. Phys. Fitness,* in press.

104. Rowland, T.W., M. Wehnert, and K. Miller. Cardiac responses to exercise in competitive child cyclists. *Med. Sci. Sports Exerc.* 32:747–752, 2000.

105. Sadoshima, J.I., Y. Xu, H.S. Slater, and S. Izumo. Autocrine release of angiotensin II mediates stretch-induced hypertrophy of cardiac myocytes in vitro. *Cell* 75:977–984, 1993.

106. Sale, D.G. Neural adaptation to resistance training. *Med. Sci. Sports Exerc.* 20(Suppl.):S135–S145, 1988.

107. Sale, D. Strength training in children. In: *Perspectives in exercise science and sports medicine: Vol. 2. Youth, exercise, and sport.* G. Gisolfi and D. Lamb (eds.). Indianapolis: Benchmark Press, 1989, pp. 165–216.

108. Sale, D.G., and L.L. Spreit. Skeletal muscle function and energy production. In: *Perspectives in exercise science and sports medicine: Vol. 9. Exercise and the female—A life span approach.* O. Bar-Or, D.R. Lamb, and P.M. Clarkson (eds.). Carmel, IN: Cooper, 1996, pp. 289–363.

109. Saltin, B., L.H. Hartely, A. Kilbom, and I. Åstrand. Physical training in sedentary middle-aged and older men: II. Oxygen uptake, heart rate and blood lactate concentrations at submaximal and maximal exercise. *Scand. J. Clin. Lab. Invest.* 24:323–334, 1969.

110. Sargeant, A.J., P. Dolan, and A. Thorne. Effects of supplementary physical activity on body composition, aerobic, and anaerobic power in 13-year-old boys. In: *Children and exercise XI.* R.A. Binkhorst, H.C.G. Kemper, and W.H. Saris (eds.). Champaign, IL: Human Kinetics, 1985, pp. 140–150.

111. Schaible, T.F., G. Malhotra, G. Ciambrone, and J. Scheuer. The effects of gonadectomy on left ventricular function and cardiac contractile proteins in male and females rats. *Circ. Res.* 54:38–49, 1984.

112. Scheuer, J. Factors contributing to the myocardial adaptations of long-term physical exercise. In: *Cardiovascular response to exercise.* G.F. Fletcher (ed.). Mount Kisco, NY: Futura, 1994, pp. 141–151.

113. Selye, H. *The physiology and pathology of exposure to stress.* Montreal: Acta Medica, 1950.

114. Servidio, F.J., R.L. Bartels, R.L. Hamlin, D. Teske, T. Shaffer, and A. Servidio. The effects of weight training using Olympic style lifts on various physiological variables in pre-pubertal boys [abstract]. *Med. Sci. Sports Exerc.* 17:288, 1985.

115. Shapiro, L.M. The morphologic consequences of systemic training. *Cardiovasc. Clin.* 15:373–379, 1997.

116. Shore, S., and R.J. Shephard. Immune responses to exercise and training: A comparison of children and young adults. *Pediatr. Exerc. Sci.* 10:210–226, 1998.

117. Simoneau, J.-A., D.A. Hood, and D. Pette. Species-specific responses in enzyme activities of anaerobic and aerobic energy metabolism to increased contractile activity. In: *Biochemistry of exercise VII.* A.W. Taylor, P.D. Gollnick, H.J. Green, C.D. Ianuzzo, E.G. Noble, G. Metivier, and J.R. Sutton (eds.). Champaign, IL: Human Kinetics, 1990, pp. 95–104.

118. Sinha-Hikim, I., J. Artza, L. Woodhouse, N. Gonzalez-Cadavid, A.B. Singh, M.I. Lee, T.W. Storer, et al. Testosterone-induced increase in muscle size in healthy young men is associated with muscle fiber hypertrophy. *Am. J. Physiol. Endocrinol. Metab.* 283:E154–E164, 2002.

119. Sjodin, B., and J. Svedenhag. O_2 uptake during running as related to body mass in circumpubertal boys: A longitudinal study. *Eur. J. Appl. Physiol.* 65:150–157, 1992.

120. Stachenfeld, N.S., H.S. Taylor, and D.L. Keefe. Mechanisms for estrogen and progesterone effects on plasma volume [abstract]. *Med. Sci. Sports Exerc.* 35:S198, 2003.

121. Tesch, P.A. Skeletal muscle adaptations consequent to long-term heavy resistance exercise. *Med. Sci. Sports Exerc.* 20(Suppl.):S132–S134, 1988.

122. Thomis, M.A., G.P. Beunen, H.H. Maes, C.J. Blimkie, M. Van Leemputte, A.L. Claessens, G. Marchal, et al. Strength training: Importance of genetic factors. *Med. Sci. Sports Exerc.* 30:724–731, 1998.

123. Tolfrey, K., I.G. Campbell, and A.M. Batterham. Aerobic trainability of prepubertal boys and girls. *Pediatr. Exerc. Sci.* 10:248–263, 1998.

124. Turcotte, L.P. Mitochondria: Biogenesis, structure, and function—Symposium introduction. *Med. Sci. Sports Exerc.* 35: 82–85, 2003.

125. Van Praagh, E. Development of anaerobic function during childhood and adolescence. *Pediatr. Exerc. Sci.* 12:150–173, 2000.

126. Varnauskas, E., P. Bjorntorp, M. Fahlen, J. Prerovsky, and J. Stenberg. Effects of physical training on exercise blood flow and enzymatic activity in skeletal muscle. *Cardiovasc. Res.* 4:418–422, 1970.

127. Vrijens, J. Muscle strength development in the pre- and post-pubertal age. *Med. Sport* 11:152–158, 1978.

128. Wagner, P.D. Skeletal muscle angiogenesis: A possible role for hypoxia. *Adv. Exp. Med. Biol.* 502:21–38, 2001.

129. Watt, P.W., F.J. Kelly, D.F. Goldspink, and G. Goldspink. Exercise-induced morphological and biochemical changes in skeletal muscles of the rat. *J. Appl. Physiol.* 53:1144–1151, 1982.

130. Weber, G., W. Kartodihardjo, and V. Klissouras. Growth and physical training with reference to heredity. *J. Appl. Physiol.* 40:211–215, 1976.

131. Welsman, J.R., N. Armstrong, and S. Withers. Responses of young girls to two modes of aerobic training. *Br. J. Sports Med.* 31:139–142, 1997.

132. Williams, C.A., N. Armstrong, and J. Powell. Aerobic responses of prepubertal boys to two modes of training. *Br. J. Sports Med.* 34:168–173, 2000.

133. Williford, H.N., D.L. Blessing, and W.J. Duey. Exercise training in black adolescents: Changes in blood lipids and $\dot{V}O_2$max. *Ethnicity Dis.* 6:279–285, 1996.

134. Wilmore, J.H., and D.L. Costill. *Physiology of sport and exercise.* Champaign, IL: Human Kinetics, 1994, chapter 10, pp. 215-238.

135. Woods, D.R., S.E. Humphries, and H.E. Montgomery. The ACE I/D polymorphism and human physical performance. *Trends Endocrinol. Metab.* 11:416–420, 2000.

136. Yoshizawa, S., H. Honda, N. Nakamura, K. Itoh, and N. Watanabe. Effects of an 18-month endurance run training program on maximal aerobic power in 4- to 6-year-old girls. *Pediatr. Exerc. Sci.* 9:33–43, 1997.

Chapter 12

1. Armstrong, L.E., D.L. Costill, and W.J. Fink. Influence of diuretic-induced dehydration on competitive running performance. *Med. Sci. Sports Exerc.* 17:456–461, 1985.

2. Armstrong, L.E., and C.M. Maresh. Exercise-heat tolerance of children and adolescents. *Pediatr. Exerc. Sci.* 7:239–252, 1995.

3. Asano, K., and K. Hirakoba. Respiratory and circulatory adaptation during prolonged exercise in 10–12-year-old children and adults. In: *Children and sport.* J. Ilmarinen and I. Valimaki (eds.). Berlin: Springer-Verlag, 1984, pp. 119–128.

4. Bar-Or, O. Temperature regulation during exercise in children and adolescents. In: *Perspectives in exercise science and sports medicine: Vol. 2. Youth, exercise, and sport.* C.V. Gisolfi and D.R. Lamb (eds.). Indianapolis: Benchmark Press, 1989, pp. 335–368.

5. Bar-Or, O. Thermoregulation in females from a life span perspective. In: *Perspectives in exercise science and sports medicine: Vol. 9. Exercise and the female—A lifespan approach.* O. Bar-Or, D.R. Lamb, and P.M. Clarkson (eds.). Carmel, IN: Cooper, 1996, pp. 249–288.

6. Bar-Or, O., R. Dotan, O. Inbar, A. Rotshtein, and H. Zonder. Voluntary hypohydration in 10- to 12-year-old boys. *J. Appl. Physiol.* 48:104–108, 1980.

7. Brenner, I.K.M., S. Thomas, and R.J. Shephard. Spectral analysis of heart rate variability during heat exposure and repeated exercise. *Eur. J. Appl. Physiol.* 76:145–156, 1997.

8. Caputa, M., G. Feistkorn, and C. Jessen. Effects of brain and trunk temperatures on exercise performance in goats. *Pflugers Arch.* 406:184–189, 1986.

9. Cheatham, C.C., A.D. Mahon, J.D. Brown, and D.R. Bolster. Cardiovascular responses during prolonged exercise at ventilatory thresholds in boys and men. *Med. Sci. Sports Exerc.* 32:1080–1087, 2000.

10. Cooper, T., V.L. Williams, and C.R. Hanlon. Cardiac and peripheral vascular responses to hyperthermia induced by blood stream heating. *J. Thorac. Cardiovasc. Surg.* 44:667–673, 1962.

11. Coyle, E.F., and J. Gonzalez-Alonso. Cardiovascular drift during prolonged exercise: New perspectives. *Exerc. Sport Sci. Rev.* 29:88–92, 2001.

12. Craig, F.V., and E.G. Cummings. Dehydration and muscular work. *J. Appl. Physiol.* 21:670–674, 1966.

13. Davies, C.T.M. Thermal responses to exercise in children. *Ergonomics* 24:55–61, 1981.

14. Drinkwater, B.L., I.C. Kepprat, J.E. Denton, J.L. Crist, and S.M. Horvath. Response of prepubertal girls and college women to work in the heat. *J. Appl. Physiol.* 43:1046–1053, 1977.

15. Falk, B. Temperature regulation. In: *Paediatric exercise science and medicine.* N. Armstrong and W. van Mechelen (eds.). Oxford: Oxford University Press, 2000, pp. 223–242.

16. Falk, B., O. Bar-Or, R. Calvert, and J.D. MacDougall. Sweat gland response to exercise in the heat among pre-, mid-, and late pubertal boys. *Med. Sci. Sports Exerc.* 24:313–319, 1992.

17. Faulkner, J.A. Heat and contractile properties of skeletal muscle. In: *Environmental physiology: Ageing, heat and altitude.* S.M. Horvath and M.K. Yosek (eds.). New York: Elsevier/North Holland, 1981, pp. 191–203.

18. Febbraio, M.A. Does muscle function and metabolism affect exercise performance in the heat? *Exerc. Sport Sci. Rev.* 28: 171–176, 2000.

19. Febbraio, M.A., M.F. Carey, R.J. Snow, C.G. Stathis, and M. Hargreaves. Influence of elevated muscle temperature on metabolism during intense dynamic exercise. *Am. J. Physiol.* 271:R1251–R1255, 1996.

20. Fink, W.J., D.L. Costill, and P.J. Handel. Leg muscle metabolism during exercise in the heat and cold. *Eur. J. Appl. Physiol. Occup. Physiol.* 34:183–190, 1975.

21. Fritzche, R.G., T.W. Switzer, B.J. Hogkinson, and E.F. Coyle. Stroke volume decline during prolonged exercise is influenced by the increase in heart rate. *J. Appl. Physiol.* 86:799–805, 1999.

22. Galloway, S.D.R., and R.J. Maughan. Effects of ambient temperature on the capacity to perform prolonged exercise in man. *J. Physiol.* 489:35–36, 1995.

23. Gonzalez-Alonso, J., R. Mora-Rodriguez, P.R. Below, and E.F. Coyle. Dehydration markedly impairs cardiovascular function in hyperthermic endurance athletes during exercise. *J. Appl. Physiol.* 82:1229–1236, 1997.

24. Gonzalez-Alonso, J., C. Teller, S.L. Anderson, F.B. Jensen, T. Hyldig, and B. Nielsen. Influence of body temperature on

the development of fatigue during prolonged exercise in the heat. *J. Appl. Physiol.* 86:1032–1039, 1999.

25. Greenleaf, J.E. Problem: Thirst, drinking behavior, and involuntary dehydration. *Med. Sci. Sports Exerc.* 24:645–656, 1992.
26. Greenleaf, J.E., P.J. Brock, L.C. Keil, and J.T. Morse. Drinking and water balance during exercise and heat acclimation. *J. Appl. Physiol.* 54:414–419, 1983.
27. Hargreaves, M., and M. Febbraio. Limits to exercise in the heat. *Int. J. Sports Med.* 19:S115–S116, 1998.
28. Haymes, E.M., E.R. Buskirk, J.L. Hodgson, H.M. Lundegren, and W.C. Nicholas. Heat tolerance of exercising lean and heavy prepubertal girls. *J. Appl. Physiol.* 36:566–571, 1974.
29. Jokinen, E., I. Valimaki, K. Antila, A. Seppanen, and J. Tuominen. Children in sauna: Cardiovascular adjustment. *Pediatrics* 86:282–288, 1990.
30. Kenney, W.L., and P. Chiu. Influence of age on thirst and fluid intake. *Med. Sci. Sports Exerc.* 33:1524–1532, 2001.
31. Meyer, F., and O. Bar-Or. Fluid and electrolyte loss during exercise: The paediatric angle. *Sports Med.* 18:4–9, 1994.
32. Meyer, F., O. Bar-Or, D. MacDougall, and G.J.F. Heigenhauser. Sweat electrolyte loss during exercise in the heat: Effects of gender and maturation. *Med. Sci. Sports Exerc.* 24: 776–781, 1992.
33. Moore, F.T., S.A. Marable, and E. Ogden. Contractility of the heart in abnormal temperatures. *Ann. Thorac. Surg.* 2: 446–450, 1966.
34. Nadel, E.R. Temperature regulation and prolonged exercise. In: *Perspectives in exercise science and sports medicine: Vol. 1. Prolonged exercise.* D.R. Lamb and R. Murray (eds.). Indianapolis: Benchmark Press, 1988, pp. 125–151.
35. Nadel, E.R., E. Cafarelli, M.F. Roberts, and C.B. Wenger. Circulatory regulation during exercise in different ambient temperatures. *J. Appl. Physiol.* 46:430–437, 1979.
36. Nielsen, B., J.R.S. Hales, S. Strange, K.J. Christensen, J. Warberg, and B. Saltin. Human circulatory and thermoregulatory adaptations with acclimation and exercise in a hot, dry environment. *J. Physiol.* 460:467–485, 1993.
37. Nielsen, B., T. Hyldig, F. Bidstrup, J. Gonzalez-Alonso, and G.R.J. Christoffersen. Brain activity and fatigue during prolonged exercise in the heat. *Pflugers Arch.* 442:41–48, 2001.
38. Nielsen, B., G. Savard, E.A. Richter, M. Hargreaves, and B. Saltin. Muscle blood flow and muscle metabolism during exercise and heat stress. *J. Appl. Physiol.* 69:1040–1046, 1990.
39. Noakes, T.D. Physiological models to understand exercise fatigue and the adaptations that predict or enhance athletic performance. *Scand. J. Med. Sci. Sports* 10:123–145, 2000.
40. Nose, H., and T. Akira. Integrative regulation of body temperature and body fluid in humans exercising in a hot environment. *Int. J. Biometeorol.* 40:42–49, 1997.
41. Parkin, J.M., M.F. Carey, S. Zhao, and M.A. Febbraio. Effect of ambient temperature on human skeletal muscle metabolism during fatiguing submaximal exercise. *J. Appl. Physiol.* 86:902–908, 1999.
42. Rees, J., and S. Shuster. Pubertal induction of sweat gland activity. *Clin. Sci.* 60:689–692, 1981.
43. Rivera Brown, A.M., R. Gutierrez, J.C. Gutierrez, W.R. Frontera, and O. Bar-Or. Drink composition, voluntary drinking, and fluid balance in exercising, trained, heat acclimatized boys. *J. Appl. Physiol.* 86:78–84, 1999.
44. Rivera Brown, A.M., T.W. Rowland, F. Ramirez-Marrero, G. Santacama, and A. Vann. Exercise tolerance in hot and humid climates: Comparison between active, heat-acclimated girls and women [abstract]. *Med. Sci. Sports Exerc.* 35:S198, 2003.
45. Rivera Brown, A.M., M. Torres, F. Ramirez-Marrero, and O. Bar-Or. Drink composition, voluntary drinking, and fluid balance in exercising trained heat-acclimatized girls [abstract]. *Med. Sci. Sports Exerc.* 31:S92, 1999.
46. Rowell, L.B. Human cardiovascular adjustments to exercise and thermal stress. *Physiol. Rev.* 54:75–159, 1974.
47. Rowell, L.B., H.J. Marx, R.A. Bruce, R.D. Conn, and F. Kusumi. Reductions in cardiac output, central blood volume, and stroke volume with thermal stress in normal men during exercise. *J. Clin. Invest.* 45:1801–1816, 1966.
48. Rowland, T.W., J.A. Auchinachie, T.J. Keenan, and G.M. Green. Physiologic responses to treadmill running in adult and prepubertal males. *Int. J. Sports Med.* 8:292–297, 1987.
49. Rowland, T.W., and T.A. Rimany. Physiological responses to prolonged exercise in premenarcheal and adult females. *Pediatr. Exerc. Sci.* 7:183–191, 1995.
50. Saltin, B., and J. Stenberg. Circulatory response to prolonged severe exercise. *J. Appl. Physiol.* 19:833–838, 1964.
51. Savard, G.K., B. Nielsen, I. Laszczynska, B.E. Larsen, and B. Saltin. Muscle blood flow is not reduced in humans during moderate exercise and heat stress. *J. Appl. Physiol.* 64:649–657, 1988.
52. Sawka, M.N., S.J. Mountain, and W.A. Latzka. Hydration effects on thermoregulation and performance in the heat. *Comp. Biochem. Physiol.* 128:679–690, 2001.
53. Sawka, M.N., and A.J. Young. Physical exercise in hot and cold climates. In: *Exercise and sport science.* W.E. Garrett and D.J. Kirkendall (eds.). Philadelphia: Lippincott Williams & Wilkins, 2000, pp. 385–400.
54. Shibasaki, M., Y. Inoue, N. Kondo, and A. Iwata. Thermoregulatory responses of prepubertal boys and young men to moderate exercise. *Eur. J. Appl. Physiol.* 75:212–218, 1997.
55. Suzuki, Y. Human physical performance and cardiocirculatory response to hot environments during upright cycling. *Ergonomics* 23:527–542, 1980.
56. Wagner, J.A., S. Robinson, S .P. Tzankoff, and R.P. Marino. Heat tolerance and acclimatization to work in the heat in relation to age. *J. Appl. Physiol.* 33:616–622, 1972.
57. Yang, K., C. Mei, Q. Zhou, Y. Jia, and C. Yu. Investigation on the hemodynamic alterations and their mechanisms during

heat stroke under hot environment. *J. Yongji Med. Univ.* 6: 48–52, 1986.

Chapter 13

1. Arai, Y., J.P. Saul, and P. Albrecht. Modulation of cardiac autonomic activity during and immediately after exercise. *Am. J. Physiol.* 256:H132–H141, 1989.
2. Arnold, R.W., J.A. Dwyer, A.B. Gould, G.G. Hohberger, and P.A. Low. Sensitivity to vasovagal maneuvers in normal children and adults. *Mayo Clin. Proc.* 66:797–804, 1991.
3. Asmussen, E. Growth in muscular strength and power. In: *Physical activity: Human growth and development.* G. Rarick (ed.). New York: Academic Press, 1973, pp. 60–79.
4. Bar-Or, O. Age-related changes in exercise perception. In: *Physical work and effort.* G. Borg (ed.). Oxford: Pergamon Press, 1977, pp. 255–266.
5. Bar-Or, O., and D.S. Ward. Rating of perceived exertion in children. In: *Advances in pediatric sport sciences: Vol. 3. Biological issues.* O. Bar-Or (ed.). Champaign, IL: Human Kinetics, 1989, pp. 151–168.
6. Beunen, G., and M. Thomis. Genetic determinants of sports participation and daily physical activity. *Int. J. Obes. Relat. Metab. Disord.* 23:S55–S63, 1999.
7. Borg, G. Perceived exertion as an indicator of somatic stress. *J. Rehabil. Med.* 2:92–98, 1970.
8. Cheatham, C.C., A.D. Mahon, J.D. Brown, and D.R. Bolster. Cardiovascular responses during prolonged exercise at ventilatory threshold in boys and men. *Med. Sci. Sports Exerc.* 32:1080–1087, 2000.
9. Cooper, D.M., R.C. Bailey, T.J. Barstow, and N. Berman. Spectral analysis of spontaneous patterns of physical activity in children [abstract]. *Med. Sci. Sports Exerc.* 27(Suppl.):S165, 1995.
10. Darr, K.C., D.R. Bassett, B.J. Morgan, and D.P. Thomas. Effect of age and training status on heart rate recovery after peak exercise. *Am. J. Physiol.* 254:H340–H343, 1988.
11. Epstein, L.H., and R.R. Wing. Aerobic exercise and weight. *Addict. Behav.* 5:371–388, 1980.
12. Eston, R., and K.L. Lamb. Effort perception. In: *Paediatric exercise science and medicine.* N. Armstrong and W. van Mechelen (eds.). Oxford: Oxford University Press, 2000, pp. 85–94.
13. Febbraio, M.A. Does muscle function and metabolism affect exercise performance in the heat? *Exerc. Sport Sci. Rev.* 28: 171–176, 2000.
14. Finley, J.P., S.T. Nugent, and W. Hellenbrand. Heart rate variability in children: Spectral analysis of developmental changes between 5 and 24 years. *Can. J. Physiol. Pharmacol.* 65:2048–2052, 1987.
15. Hunt, J.R., C.A. Zito, J. Erjavec, and L.K. Johnson. Severe or marginal iron deficiency affects spontaneous physical activity in rats. *Am. J. Clin. Nutr.* 59:413–418, 1994.
16. Lamb, K.L. Children's ratings of effort during cycle ergometry: An examination of the validity of two effort rating scales. *Pediatr. Exerc. Sci.* 7:407–421, 1995.
17. Lehmann, M., J. Keul, and U. Korsten-Reck. The influence of graduated treadmill exercise on plasma catecholamines, aerobic and anaerobic capacity in boys and adults. *Eur. J. Appl. Physiol.* 47:301–311, 1981.
18. Mahon, A.D. Assessment of perceived exertion during exercise in children. In: *Children and exercise XIX, vol. 2.* J. Welsman, N. Armstrong, and B. Kirby (eds.). Exeter, UK: Washington Singer Press, 1997, pp. 25–32.
19. Mahon, A.D., G.E. Duncan, C.A. Howse, and P. Del Corrall. Blood lactate and perceived exertion relative to ventilatory threshold: Boys versus men. *Med. Sci. Sports Exerc.* 29: 1332–1337, 1997.
20. Mahon, A.D., J.A. Gay, and K.Q. Stolen. Differentiated ratings of perceived exertion at ventilatory threshold in children and adults. *Eur. J. Appl. Physiol.* 18:115–120, 1998.
21. Mandigout, S., A. Melin, L.D. Nguyen, L. Fauchier, and P. Obert. Effect of an endurance training program on heart rate variability in prepubertal boys and girls [abstract]. *Pediatr. Exerc. Sci.* 13:281–287, 2001.
22. McManus, A.M., N. Armstrong, B.J. Kirby, and J.R. Welsman. Ratings of perceived exertion in prepubescent girls and boys. In: *Children and exercise XIX.* N. Armstrong, B. Kirby, and J. Welsman (eds.). London: Spon, 1997, pp. 253–257.
23. Melanson, E.L. *Heart rate variability: Relationship to physical activity level, response to training, and effect of maturation.* Doctoral thesis, University of Massachusetts, Amherst, 1998.
24. Myashita, M., K. Onodera, and I. Tabata. How Borg's scale has been applied to Japanese. In: *The perception of exertion in physical work.* G. Borg and D. Ottoson (eds.). Basingstoke, UK: Macmillan, 1986, pp. 27–34.
25. Nielsen, B., T. Hyldig, F. Bidstrup, J. Gonzalez-Alonso, and G.R.J. Christoffersen. Brain activity and fatigue during prolonged exercise in the heat. *Pflugers Arch.* 442:41–48, 2001.
26. Noakes, T.D. Physiological models to understand exercise fatigue and the adaptations that predict or enhance athletic performance. *Scand. J. Med. Sci. Sports* 10:123–145, 2000.
27. Ohuchi, H., H. Suzuki, K. Yasuda, Y. Arakaki, S. Echigo, and T. Kamiya. Heart rate recovery after exercise and cardiac autonomic nervous activity in children. *Pediatr. Res.* 47: 329–335, 2000.
28. Palmer, G.J., M.G. Ziegler, and C.R. Lake. Responses of norepinephrine and blood pressure to stress increase with age. *J. Gerontol.* 33:482–487, 1978.
29. Panksepp, J., S. Siviy, and L. Normansell. The psychobiology of play: Theoretical and methodological perspectives. *Neurosci. Biobehav. Rev.* 8:465–492, 1984.
30. Rowland, T.W. The biological basis of physical activity. *Med. Sci. Sports Exerc.* 30:392–399, 1998.
31. Rowland, T.W. *Exercise and children's health.* Champaign, IL: Human Kinetics, 1990.

32. Rowland, T.W., C.M. Maresh, N. Charkoudian, P.M. Vanderburgh, J.W. Castellani, and L.E. Armstrong. Plasma norepinephrine responses to cycle exercise in boys and men. *Int. J. Sports Med.* 17:22–26, 1996.

33. Shephard, R.J., J.-C. Jequier, H. Lavallee, R. Labarre, and M. Rajic. Habitual physical activity: Effects of sex, milieu, season and required activity. *J. Sports Med.* 20:55–66, 1980.

34. Thor, D.H., and W.R. Holloway. Social play in juvenile rats: A decade of methodological and experimental research. *Neurosci. Biobehav. Rev.* 8:455–464, 1984.

35. Thorburn, A.W., and J. Proietto. Biological determinants of spontaneous physical activity. *Obes. Rev.* 1:87–94, 2000.

36. Tolfrey, K., and J. Mitchell. Rating of perceived exertion at standard and relative exercise intensities in prepubertal, teenage and young adult males. *J. Sports Sci.* 14:101–102, 1996.

37. Wade, M.G., M.J. Ellis, and R.E. Bohrer. Biorhythms in the activity of children during free play. *J. Exp. Anal. Behav.* 20: 155–162, 1973.

38. Weise, M., G. Eisenhofer, and D.P. Merke. Pubertal and gender-related changes in the sympathoadrenal system in healthy children. *J. Clin. Endocrinol. Metab.* 87:5038–5043, 2002.

39. Winsley, R. Acute and chronic effects of exercise on heart rate variability in adults and children: A review. *Pediatr. Exerc. Sci.* 14:328–344, 2002.

40. Yamamoto, K., M. Miyachi, T. Saitoh, A. Yoshoka, and S. Onodera. Effects of endurance training on resting and post-exercise cardiac autonomic control. *Med. Sci. Sports Exerc.* 33:1496–1502, 2001.

41. Yeragani, V.K., R. Pohl, R. Berger, R. Balon, and K. Srinivasan. Relationship between age and heart rate variability in supine and standing postures: A study of spectral analysis of heart rate. *Pediatr. Cardiol.* 15:14–20, 1994.

INDEX

Note: The italicized *f* and *t* following page numbers refer to figures and tables, respectively.

T

V

ABOUT THE AUTHOR

Thomas W. Rowland, MD, is director of pediatric cardiology at the Baystate Medical Center in Springfield, Massachusetts, where he established an exercise testing laboratory. The author of *Exercise and Children's Health* and editor of the journal *Pediatric Exercise Science* for the past 15 years, he has extensive research experience in exercise physiology of children.

Dr. Rowland has served as president of the North American Society for Pediatric Exercise Medicine (NASPEM) and was on the board of trustees of the American College of Sports Medicine (ACSM). He is a past president of the New England chapter of the ACSM and received the ACSM Honor Award in 1993.

Since receiving BS and MD degrees from the University of Michigan in 1965 and 1969, Dr. Rowland has been an assistant and associate professor of pediatrics at the University of Massachusetts Medical School in Worchester (1977 to 1990) and an assistant and associate clinical professor of pediatrics at Tufts University School of Medicine in Boston (1975 to the present). He is professor of pediatrics at Tufts University School of Medicine and adjunct professor of exercise science at the University of Massachusetts.

In addition to conducting extensive research, Dr. Rowland has written and spoken about developmental exercise physiology, the effects of lifestyle on cardiovascular function in children, iron deficiency in adolescent athletes, and the determinants of exercise performance in children.

DATE DUE

SEP 2 6 2007			
SEP 0 9 2007			

GAYLORD PRINTED IN U.S.A.